CIRCUIT ANALYSIS: AN INTEGRATED APPROACH

CIRCUIT ANALYSIS: AN INTEGRATED APPROACH

Jerome Zornesky
Technical Career Institute
New York, NY

Stephen H. Maybar
Technical Career Institute
New York, NY

Prentice Hall
Upper Saddle River, New Jersey *Columbus, Ohio*

Editor: Scott Sambucci
Production Editor: Rex Davidson
Design Coordinator: Karrie Converse-Jones
Cover Photo: © Westlight
Cover Designer: Jason Moore
Production Manager: Patricia A. Tonneman
Production Coordination: York Production Services
Marketing Manager: Ben Leonard
Illustrations: York Production Services

Library of Congress Cataloging-in-Publication Data
Zornesky, Jerome.
Circuit analysis: an integrated approach / Zornesky & Maybar.
p. cm.
ISBN 0-13-727587-0
1. Electric circuit analysis. I. Maybar, Steve. II. Title.
TK454.Z67 2000
621.319′2—dc21 99-32118
CIP

This book was set in Times Roman by York Graphic Services, Inc., and was printed and bound by R. R. Donnelley & Sons Company. The cover was printed by Phoenix Color Corp.

Printed in the United States of America
10 9 8 7 6 5 4 3 2 1

ISBN: 0-13-727587-0

Prentice-Hall International (UK) Limited, *London*
Prentice-Hall of Australia Pty. Limited, *Sydney*
Prentice-Hall of Canada, Inc., *Toronto*
Prentice Hall Hispanoamericana, S. A., *Mexico*
Prentice-Hall of India Private Limited, *New Delhi*
Prentice-Hall of Japan, Inc., *Tokyo*
Prentice-Hall (Singapore) Pte. Ltd., *Singapore*
Editora Prentice-Hall do Brasil, Ltda., *Rio de Janeiro*

PREFACE

The subject of *Circuit Analysis: An Integrated Approach* has been presented by community colleges in basically the same manner for at least the last 50 years. One semester is used to teach what may be called "DC Circuit Analysis" and a second semester to teach "AC Circuit Analysis." Many fine textbooks have been written over the years in a format corresponding to this manner of presentation.

I have always felt that this method of presentation, while adequate, had several drawbacks. First of all, it presented the student with the artificial implication that there were two different subjects, ac and dc circuit analysis, when in fact there is only one subject, circuit analysis, with different sources. Further, in view of the fact that *most* circuits include *both ac and dc sources,* the student was getting a somewhat distorted view of the majority of industrial circuitry.

As time progressed, another important factor became apparent to me. The continuing and rapid changes in technology were, and still are, creating a problem for all technical schools. New courses must be introduced to give our students the background required to work in the new fields emerging as a result of all these changes in technology.

The continuing problem is how to incorporate these new courses into existing curriculums. It is impossible to simply keep adding new courses and still maintain a two-year program.

Many schools, therefore, have had to re-evaluate existing courses in order to determine which ones could be shortened or even eliminated. This provides a realistic way in which new courses can be introduced without sacrificing overall program quality.

I soon realized that by teaching circuit analysis using the integrated approach, in which dc is presented as a special case of ac, the traditional two-semester course in circuit analysis could be effectively shortened to one semester. Topics such as Thevenin's and Norton's Theorems, Mesh and Node Analysis, Series, Parallel and Series Parallel Circuit Laws, among others, can be presented once instead of twice. Students can learn how to apply these laws and techniques for circuits with both ac and dc sources from the very beginning.

I decided to try this technique at Technical Career Institute, where I have been teaching since 1991. My co-author, Stephen Maybar, and I began teaching the subject of circuit analysis using the integrated approach in the spring of 1995. Our traditional two-semester course was reduced to one semester, enabling us to free up several hours for new courses.

Stephen and I are extremely gratified to be able to report that our integrated approach is very successful. Not only have our students received this new approach very favorably, but we also have seen a significant increase in student retention, which I am convinced is at least in part due to this new approach.

With the success of our integrated approach, Steve and I decided to try the same technique in other courses. We constructed a similar course outline in the subjects of semiconductor and integrated circuit analysis. In this manner, a three-semester electronics sequence was reduced to two, again enabling us to introduce new courses. Once again, the new approach was very favorably received by our students.

As of the writing of this book, we have been successfully using the integrated approach for over 4 years in both circuit analysis and electronics. This has served to greatly increase our students' capabilities in these subjects. It has also given us the opportunity to modify our curriculum by introducing important new courses without sacrificing quality or burdening our students with unnecessary additional hours of classwork.

Organization of This Text

This text starts in the traditional manner of introducing the student to the basic principles underlying the concepts of current and voltage in Chapter 1. The common prefixes are reviewed as well as scientific notation.

Chapter 2 then covers the topics of voltage and current sources. Both dc and ac sources are discussed. Periodic waveforms are introduced with emphasis on the sinusoidal waveform.

The mathematics of complex numbers is covered in Chapter 3, enabling the student to become acquainted with this topic early in the course. The "rotating vector" concept is used to introduce phasors.

Chapter 4 familiarizes the student with the concept of impedance. The resistor, capacitor, and inductor are then introduced and their impedances are discussed in both ac and dc, with their dc "impedances" derived from the ac using zero frequency.

Series and parallel circuits are analyzed in Chapter 5. Both ac and dc circuits are analyzed using the concepts discussed in Chapter 4.

Chapter 6 continues the integrated approach by analyzing series-parallel circuits using the same techniques as in the previous two chapters. The behavior of circuit elements in dc circuits is always advanced by the "zero frequency special case" concept.

Chapter 7 next introduces the student to mesh and nodal analysis techniques. Several important theorems, including Thevenin's Theorem and Superposition, are included and dependent sources are introduced.

The concept of frequency response is discussed in Chapter 8. Logarithms are reviewed before introducing the topics of transfer functions and filters.

Chapter 9 covers the subject of resonance. The series and parallel R–L–C circuits are used to demonstrate this subject.

Magnetic induction and transformers are introduced in Chapter 10. Ideal transformers are used to develop the basic transformer equations, after which practical transformers are discussed. Equivalent circuits and reflection are covered with a discussion of the standard industry "short circuit" and "open circuit" tests.

The subject of "Power" is covered in Chapter 11. A graphical approach is used to introduce the concepts of real and reactive power. The maximum power transfer theorem is introduced here as well as power factor and power factor correction.

Circuit transients are discussed in Chapter 12. The Laplace transform is used as an analytical tool to analyze the behavior of circuits in the transient stage,

The physical properties of components are described in Chapter 13. The properties of linearity and temperature effects are reviewed.

The last chapter discusses the important topic of circuit measurements. The use of standard laboratory equipment such as the oscilloscope and other meters is described. The use of circuit analysis software, specifically Electronics Workbench, is demonstrated using specific examples.

Numerous examples are included in each chapter. There are a great many problems at the end of each chapter with answers to odd-numbered problems included at the end.

Acknowledgments

We wish to thank the following reviewers of the manuscript:

Mauro Caputi, Hofstra University
J. W. Roberts, West Georgia Technical Institute
Kenneth Reid, Purdue University—IUPUI
Mike O'Rear, Chattahoochee Technical Institute

Shamala Chickamenahalli, Wayne State University
Julio Garcia, San Jose State University
Kenneth Markowitz, New York City Technical College
Hirak Patangia, University of Arkansas at Little Rock
Michael Chier, Milwaukee School of Engineering
William Mack, Harrisburg Area Community College
Nazar Karzay, Ivy Tech State College
Lee Rosenthal, Fairleigh Dickinson University
Frederick Berry, Rose–Hulman Institute of Technology
Phillip Anderson, Muskegon Community College

Stephen and I want to thank the people at Prentice Hall for their patience with new authors as well as their contributions to the text format.

Jerry Zornesky

CONTENTS

CHAPTER **1**

PHYSICAL ELECTRONICS

INTRODUCTION

Electrical circuits come in all sizes and shapes. From the simplest circuit to the most complex system using thousands of circuits one thing remains true: All of these circuits obey a small number of fundamental principles. Once you understand these principles you will be able to analyze any electrical circuit completely. The purpose of this text is to explain these principles one at a time.

1.1 SCIENTIFIC NOTATION AND POWERS OF 10

In the field of circuit analysis, measurement is king. We must be able to measure the properties of the circuits we discuss, and we must also be able to do calculations using the answers we obtain. Measured electrical quantities span many orders of magnitude. For example, the number of atoms in a piece of copper is very large, while the size of an individual atom is very small. When you are working with many of *these different magnitude numbers* at the same time it is very easy to make mistakes in calculations. In seeking to simplify the calculation process, mathematicians created a system of representing numbers that separated the number value from the size of the number. This system is called the system of ***scientific notation*** and is based upon expressing all numbers as being of magnitude between 1 and 10 multiplied by factors of 10 to some power.

Consider the multiplication of factors of 10. For example, $10 \times 10 \times 10 \times 10$ represents four factors of 10 multiplied together. A shorthand way of indicating this product is to write:

$$10 \times 10 \times 10 \times 10 = 10^4$$

The number 4 is called the ***exponent of the number 10*** and represents the fact that the number 10 is repeated four times in the multiplication. Another way of expressing the same fact is to say that the number 10 has been raised to the ***fourth power.*** Using this ***exponential notation*** we are able to write products of the factor 10 in compact form. Thus:

$$10 \times 10 \times 10 \times 10 \times 10 \times 10 \times 10 \times 10 \times 10$$

can be written more simply as 10^9. This compact notation scheme makes it easy to write repeated factors of 10 in a multiplication problem in a simple way. It also eliminates errors in counting the number of times that the factor of 10 is repeated because the exponent tells us how many times the factor is repeated.

Exponents force rules upon mathematicians based on the way that they are written. Consider the product:

$$(10 \times 10 \times 10 \times 10) \times (10 \times 10 \times 10 \times 10 \times 10 \times 10)$$

Obviously, there are 10 repetitions of the factor 10. This means that the product of ten factors of 10 can be written as 10^{10}. On the other hand we can also write this as:

$$10^4 \times 10^6 = 10^{10}$$

because there are four factors of 10 multiplied by six factors of 10. By the associative law of mathematics we know that the way we group these factors should not have any effect on the answer to the multiplication problem. Thus we see that in dealing with factors of 10 in a multiplication problem, we can simply add exponents to find the resulting exponent.

Consider a problem in which factors of 10 are divided as well as multiplied. How do we handle this problem? For example, consider:

$$\frac{10 \times 10 \times 10}{10 \times 10}$$

This can be written as:

$$\frac{10^3}{10^2} = 10^1$$

We see from this example that in the division problem we can get the correct answer by subtracting the exponent of the denominator from the exponent of the numerator. This leads us to the following rules for exponents:

1.1.1 Rules for Exponents

1. To multiply factors of 10 simply add the corresponding exponents of each of the factors.
2. To divide factors of 10 subtract the exponent of the denominator from the exponent of the numerator.

These two rules imply other rules when working with exponents. If we apply rules 1 and 2 to the following example:

$$\frac{10 \times 10 \times 10}{10 \times 10 \times 10 \times 10}$$

we find

$$\frac{10 \times 10 \times 10}{10 \times 10 \times 10 \times 10} = \frac{10^3}{10^4} = 10^{-1}$$

This result leads to rule 3.

3. A negative exponent means that the factor in the denominator is larger than the factor in the numerator.

This rule means that when we see a factor such as 10^{-3}, we can also write this same factor as:

$$10^{-3} = \frac{1}{10^3}$$

Consider a factor such as $1/10^{-3}$. If we multiply the top and bottom of the fraction by 10^3 and apply rules 1 and 2 we find:

$$\frac{1}{10^{-3}} \times \frac{10^3}{10^3} = \frac{10^3}{10^0} = 10^3 \quad (\text{since } 10^0 = 1)$$

The effect of this mathematical procedure is to demonstrate rule 4 of our set of rules.

4. To move an exponent from the numerator to the denominator of a fraction (or vice versa), simply switch the sign of its exponent.

Using the properties of exponents and factors of 10 shown here, it is straightforward to write any number as the product of a number whose magnitude is greater than 1 and less than 10 times a factor of 10 to some power. Thus:

$$135 = 1.35 \times 10^2$$
$$0.0167 = 1.67 \times 10^{-2}$$
$$3.99 = 3.99 \times 10^0$$

Numbers expressed in this way are said to be written in ***scientific notation.*** This method of writing numbers makes the process of calculation easier because we are working with numbers between 1 and 10 and exponents of powers of 10.

EXAMPLE 1.1

Multiply 0.00000035 × 13,400 × 0.00000234/0.000126.

Solution Changing this to scientific notation we find:

$$3.5 \times 10^{-7} \times 1.34 \times 10^4 \times 2.34 \times 10^{-6}/1.26 \times 10^{-4}$$
$$= 3.5 \times 1.34 \times 2.34 \times 10^{-7+4-6}/1.26 \times 10^{-4}$$
$$= 8.71 \times 10^{-5}$$

Notice how the use of scientific notation made the problem easier. One can almost guess at the size of the answer with this type of number representation.

1.1.2 Measurement of Physical Properties

All physical measurements involve questions of size. We must be able to measure things that we see so that they can be described to others. Early measurement systems involved units of length, such as cubits, that were only approximately the same size. The cubit, which was a measure of length, was defined as the distance between the elbow and the end of the middle finger. Every person's cubit was a different size because no two people's physical dimensions are equal. When countries decided to standardize the cubit within the country, the problem got no better. The ancient Egyptians defined the cubit as 20.54 inches long, the Roman cubit was 17.4 inches long, and the English cubit was defined as 18 inches long. With these differences in measurement between countries, it was impossible to agree on measurements made. Thus it was impossible to standardize measurement throughout the world.

In modern times, nations have joined together to establish measurement standards that all can agree on. Two competing systems emerged from the early days of measurement standards. One was the English system and the other was the (French) metric system. Both systems relied on a small number of fundamental standards as the basis for all measurements and then used physical laws to define secondary standards. The major difference between the two systems is that the metric system uses units related by powers of 10 whereas the English system uses units that are not generally related by factors of 10. Thus, there are 36 inches in a yard in the English system but 100 centimeters in a meter in the metric system. The decade relationship between corresponding metric units makes conversions between related units easier.

The modern version of the metric system is called the ***International System of Units.*** In this system, measurement names use Greek prefixes to adjust the size of the basic unit. For example the prefix ***kilo*** is placed in front of the unit meter to produce the new unit kilometer, which is simply 1000 meters in length. Each new unit is some power of ten times the basic unit. This makes it easy to convert from one unit to another because all common units share the same name stem. It is easy to compare milligrams, grams, and kilograms, for example, as units of mass because they all have the same stem, namely *gram.* Contrast this with the English unit of ounces, pounds, and tons. Table 1.1 lists the common Greek prefixes used in the International System of Units.

TABLE 1.1
Greek prefixes of the international system of measurement

Prefix		Multiplier	Prefix		Multiplier
deci	=	10^{-1}	deca	=	10^{1}
centi	=	10^{-2}	hecto	=	10^{2}
milli	=	10^{-3}	kilo	=	10^{3}
micro	=	10^{-6}	mega	=	10^{6}
nano	=	10^{-9}	giga	=	10^{9}
pico	=	10^{-12}	tera	=	10^{12}

As you work with these units you will get to know their values and names almost automatically. In addition, the factor of 10 relationship means that any unit can be expressed in terms of the fundamental one. Thus, 1 kilometer can also be thought of as 1000 meters. All units large and small can be expressed as the fundamental unit times some power of 10 as a multiplier. We therefore have the freedom to express length, for example, as powers of 10 times the meter or by using a prefix. We can write 1000 meters as 10^3 meters, we can elect to call it 1 kilometer, understanding that they represent the same quantity.

1.2 THE ATOMIC NATURE OF MATTER

We live in a world made up of physical matter such as rocks, water, and air. Physical matter is something that occupies space. In our everyday world, physical matter may be a solid, a liquid, or a gas. Early scientists discovered that all physical matter could be made from a small group of substances they called ***elements.*** This was based upon the observation that these elements, unlike other substances they examined, could not be broken down into simpler substances through any means that they tried.

All elements have the unique property that they can't be changed into simpler elements by any physical processes such as mixing, heating, pressure, and so on. In addition, they differ from each other in their chemical properties. Some are gases, some are liquids, and some are solids at room temperature. Each element interacts with all other elements in a unique way. These were the building blocks of matter to early scientists. As they continued their studies of elements they found new elements, but always, the number of elements was small (less than 100) compared to the thousands of substances they tested.

1.2.1 Properties of Elements

Elements always combine in simple, fixed ratios by weight to form more complex combinations of elements that scientists call ***compounds.*** For example, two parts of hydrogen (a gas) combine with one part of oxygen (a gas) to form water (a liquid). Every compound, like every element, has a set of unique properties. Thus the same compounds, with the same properties, result whenever we combine the same elements in the same way. In addition, ***each of these compounds differs from every other compound that we can form from any other combination of elements.*** Thus the compounds formed, like the elements themselves, are all unique in their properties. This uniqueness must be related to the way in which the elements combine.

One way to explain the simple combining ratios of elements and the uniqueness of their properties is to assume that each element is made up of a very large number of extremely small particles with identical properties. These particles each have the same properties as the original element. When the elements interact, it is these small particles that are individually interacting. Even the smallest amount of an element one can see is made up of many, much smaller particles. If we were to reduce the number of particles in a sample, we would arrive at a point where there is only one particle left.

The name ***atom*** is given to the smallest particle that still has the same properties as the original element. This comes from the Greek word ***atomos*** meaning "indivisible." Atoms are very small. They are so small that they can not be directly seen by even the most powerful optical microscope. If we could place atoms next to each other it would take about ***10 million*** of them to form a line ***only one millimeter*** long.

Every atom of an element is the same size as any other atom of the element and interacts in the same way as other atoms of that element. Thus, interactions between elements take place on an atomic level with an atom of one element interacting with an atom of another. Compounds formed when elements combine are also very small. The number of atoms involved in an interaction between even the smallest visible amounts of elements is astronomical. The resulting compounds are therefore made up of a huge number of identical clusters of atoms. This accounts for the uniformity of properties of any compound regardless of the physical amount of the compound available.

1.2.2 The Periodic Table

If the properties of different elements are not the same, then obviously their atoms must somehow differ. Atoms of a heavier element should weigh more than atoms of a lighter element. How can we explain the properties of various atoms through the studies of their weight? This difficult problem was tackled by many scientists. In 1869, a Russian chemist named Mendeleev presented a proposed solution to this problem in which he described his ***Periodic Table*** (Fig. 1.1).

FIGURE 1.1
Mendeleev's Periodic Table

I	II	III	IV	V	VI	VII	VIII
R_2O	RO	R_2O_3	RO_2	R_2O_5	RO_3	R_2O_7	RO_4
H							
Li	Be	B	C	N	O	F	
Na	Mg	Al	Si	P	S	Cl	
K	Ca		Ti	V	Cr	Mn	Fe, Co, Ni
Cu	Zn			As	Se	Br	Ru, Rh, Pd
Ag	Cd	In	Sn	Sb	Te	I	
Cs	Ba						

Mendeleev listed the known elements according to their relative atomic weights and properties, listing lighter elements first and heavier elements later. He placed all elements with similar properties into one column on his table. His finished "periodic" table had eight different columns. Every known element found a place in the table. Strangely, however, there appeared to be blank spaces in the table with no known element that fit into the space available. Mendeleev recognized these blank spaces as belonging to elements that had not yet been discovered. By comparing the properties of the missing elements to the properties of the known elements in his table he predicted the properties of the missing elements. With his Periodic Table as a guide, these missing elements were subsequently found. The properties of the missing elements matched closely those predicted by Mendeleev. This outstanding success cemented the acceptance of the Periodic Table and the picture of how atoms combine.

1.2.3 Static Electricity

How and why do atoms interact the way that they do? Clearly there must be some sort of underlying structure to the atom. Scientists postulated that atoms must be made up of more fundamental components. Since 400 B.C. it has been known that static electricity existed. It was thought that static electricity was some sort of fluid that somehow passed from one substance to another. This assumption arose from the observation that if a glass rod is stroked with a piece of silk, both the glass rod and the silk become ***charged.*** Two different materials that had both been previously uncharged interchange something called charge so that their charge properties are both modified. Experimenters established that the charges on the silk and on the glass rod were opposite in behavior. Two pieces of charged silk or two pieces of charged glass rod would repel each other, but the glass rod attracted the silk and vice versa. To distinguish between the two different types of charge, scientists said that the glass become positively charged and the silk became negatively charged. This choice was arbitrary—only the difference between the types of charges was important. We can sum up the behavior of static electricity with the ***Law of Static Electricity.***

Like charges repel and unlike charges attract.

Static electricity clearly is a property of the atomic structure of atoms. Something about the way atoms are structured controls the behavior of static electricity. Pieces of the puzzle started to fall into place in 1874, when electric charge was discovered to be made up of small, individual charges called electronic charges. These electronic charges were shown to all have exactly the same value of charge. As in the case of elements, it turned out that total charge could only be divided until one reached an indivisible fundamental charge. This means that ***no fractional charges exist and every value of electric charge must be an integer multiple of the fundamental electronic charge.***

The value of the fundamental charge depends on the set of units chosen. Just as the inch and the meter are both units of length but not equal to each other, so the unit of charge must somehow be related to the other units in the measurement set. In the International System of Units the unit of charge defined is much larger than the elementary electronic charge. The unit of charge is called the ***coulomb*** (C). In terms of the coulomb it is experimentally found that the fundamental unit of charge equals approximately 1.602×10^{-19} C. Since this is the value of the fundamental charge in the International System of Units, no charge value smaller than this one can exist in the International System of Units. This is a different concept than that of the meter for example. The meter in principle can be divided down to any size we desire. Not so with the fundamental electronic charge!

1.2.4 The Bohr Model of the Atom

Based on the discovery of a fundamental electronic charge, many different models were proposed for the structure of the atom. Each model explained some of the properties of the atom, but not all of them. One of these models, however, proposed by Niels Bohr in 1913, was so simple in concept and so successful in prediction that it is still used today to explain many simple concepts about the structure of atoms. The Bohr model (Fig. 1.2) proposed a positively charged heavy central part of the atom called a ***nucleus*** surrounded by lighter negatively charged particles circling the nucleus called ***electrons.*** The magnitude of the charge on these electrons is the fundamental electronic charge (1.602×10^{-19} C).

FIGURE 1.2
Bohr atom

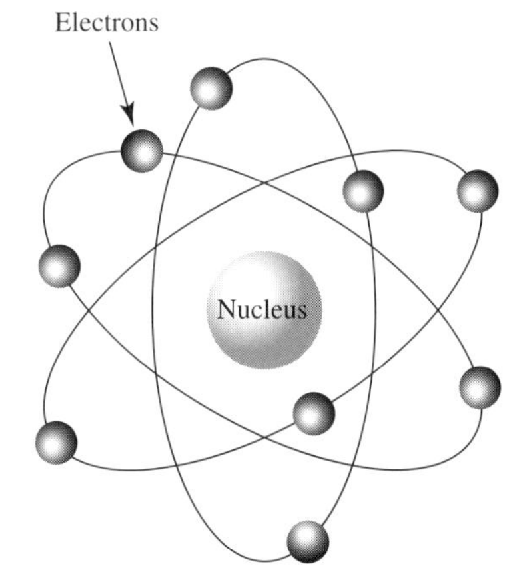

The components of an atom are held in place by the forces between the nucleus and the electrons. The Bohr atom thus looks like a mini solar system in which the nucleus, like the sun, is at the center of the system and the electrons, like the planets, surround the center of the system.

1.2.5 Modern Atomic Theory

Today it is known that the nucleus is made up of ***positively*** charged particles called ***protons*** and uncharged particles called ***neutrons.*** A proton has the same magnitude of charge as the electron, the fundamental electronic charge, but it is approximately 1840 times heavier than an electron. Neutrons have approximately the same weight as protons and can be thought of as protons with no charge. The nucleus of an atom occupies a very small part of the space of an atom. Less than 1% of the volume of an atom is occupied by the nucleus. The electrons that surround the nucleus are also extremely small. This means that most of the volume of an atom is empty space, just like our solar system. The size of an atom is determined by the orbit of the outermost electrons.

In comparison to protons, electrons are practically weightless and are held to the nucleus by electrostatic forces. Since all atoms are electrically neutral, atoms can have no net electronic charge. The neutrality of the atom along with the indivisibility of the fundamental charge value then means that ***there must be as many protons in the nucleus as there are electrons circling the nucleus.*** Mendeleev's Periodic Table was adjusted for this new view of the atom. Today we also know that atoms that have the same number of protons can have different numbers of neutrons in the nucleus. These atoms with additional neutrons behave virtually identically to the atom with the same number of protons but fewer neutrons. They are considered variations of the same element and are said to be ***isotopes*** of the element. Instead of dealing with the weight of atoms to decide their location in the Periodic Table, today we use the number of protons in the nucleus. The number of protons is called the ***atomic number*** of the element. The modern version of the Periodic Table lists the elements by their atomic number rather than their atomic weight (Fig. 1.3).

The fact that all atoms are made up of the same subcomponents implies that the weight of an atom can be found by multiplying the weight of each of the subcomponents by the number of subcomponents that there are. Thus if there are eight protons, eight electrons, and eight neutrons in an atom of oxygen, for example, then the weight of the oxygen atom can be found by multiplying the weight of a proton by eight, the weight of an electron by eight, and the weight of a neutron by eight, and adding them all together to get the total weight of an atom of oxygen. If each atom of an element weighs the same, it follows that the number of atoms in a gram atomic weight of any element must be the same. For example, oxygen has an atomic weight of 16 while sulfur has an atomic weight of 32. The number of atoms in 16 g of oxygen will be the same as the number of atoms in 32 g of sulfur. The number of atoms in this gram atomic weight has been measured very accurately and is called Avogadro's number. The numerical value of this number is 6.023×10^{23} atoms. Thus we can determine the atomic weight of any element by measuring the gram weight of 6.023×10^{23} atoms of the element. This number plays a very valuable role in chemical calculations of all kinds.

Using modern atomic theory, the concept of static electricity can also be easily explained in terms of electrons and protons. When silk is used to rub a glass rod, the electrons on the glass rod are transferred to the silk. The electrons have a negative charge, so the silk becomes negatively charged due to the extra electrons. At the same time, when electrons are removed from the glass rod their removal causes the rod to have a net positive charge due to the unbalanced charge of the excess protons. The magnitude of the negative charge must equal the magnitude of the positive charge.

Electrons located around a nucleus are restricted in the positions that they can occupy because electrons repel each other but are attracted to the protons in the nucleus. Amazingly, the electrons form regions or shells where they can exist and regions where they can't exist. Modern atomic theory demonstrates that the number of electrons that can fit in any one shell is limited. For example, the shell closest to the nucleus can hold only two electrons while shells more distant from the nucleus can hold more. Electrons in the shell closest to the nucleus are held most tightly to the nucleus while electrons in a shell farther from the nucleus are less tightly held. This is caused by the closer electrons partially screening out the effect of the proton in the nucleus.

The outermost shell of electrons surrounding the nucleus is called the ***valence shell,*** and the electrons in this shell are called ***valence electrons.*** It is these valence electrons that interact with valence electrons of other elements to form compounds and therefore determine the interacting characteristics of their atom. The number of electrons in this valence shell is limited to eight. This limitation on the number of valence electrons accounts

I	II a	III b	IV b	V b	VI b	VII b	VIII b			I b	II b	III a	IV a	V a	VI a	VII a	VIII a
1 H Hydrogen																	2 He Helium
3 Li Lithium	4 Be Beryllium											5 B Boron	6 C Carbon	7 N Nitrogen	8 O Oxygen	9 F Fluorine	10 Ne Neon
11 Na Sodium	12 Mg Magnesium											13 Al Aluminum	14 Si Silicon	15 P Phosphorous	16 S Sulfur	17 Cl Chlorine	18 Ar Argon
19 K Potassium	20 Ca Calcium	21 Sc Scandium	22 Ti Titanium	23 V Vanadium	24 Cr Chromium	25 Mn Manganese	26 Fe Iron	27 Co Cobalt	28 Ni Nickel	29 Cu Copper	30 Zn Zinc	31 Ga Gallium	32 Ge Germanium	33 As Arsenic	34 Se Selenium	35 Br Bromine	36 Kr Krypton
37 Rb Rubidium	38 Sr Strontium	39 Y Yttrium	40 Zr Zirconium	41 Nb Niobium	42 Mo Molybdenum	43 Tc Technetium	44 Ru Ruthenium	45 Rh Rhodium	46 Pd Palladium	47 Ag Silver	48 Cd Cadmium	49 In Indium	50 Sn Tin	51 Sb Antimony	52 Te Tellurium	53 I Iodine	54 Xe Xenon
55 Cs Cesium	56 Ba Barium	57–71 Rare Earth Elements	72 Hf Hafnium	73 Ta Tantalum	74 W Tungsten	75 Re Rhenium	76 Os Osmium	77 Ir Iridium	78 Pt Platinum	79 Au Gold	80 Hg Mercury	81 Tl Thallium	82 Pb Lead	83 Bi Bismuth	84 Po Polonium	85 At Astatine	86 Rn Radon
87 Fr Francium	88 Ra Radium	89–103 Rare Earth Elements															

FIGURE 1.3
Modern Periodic Table

for the eight columns of the periodic table. Although the modern periodic table is not the same as the original one that Mendeleev proposed, the similarities are striking. The eight columns indicate that only eight different categories of atoms can exist. These are defined by reference to the number of valence electrons that they have. Thus the column number is related to how many electrons there are in the valence shell of the named element.

One more consideration must be added in the understanding of modern atomic theory. Temperature is a measure of the heat content of any material. As atoms are heated they gain kinetic energy. The more they are heated, the more kinetic energy they gain. Kinetic energy is energy of motion and represents movement and vibration of atoms and the protons, neutrons, and electrons that they are made up from. Everything above the temperature of absolute zero (the lowest theoretical temperature possible) has some kinetic energy (energy of motion). As a material is heated, the atoms that make up the material vibrate harder and harder and bang against each other harder. This explains why substances expand as they are heated. Properties of matter change as they are heated. In general, materials go from solid to liquid to gas as they are heated. Thus, the state of a particular substance is related to the temperature that it is at.

1.2.6 Conductors

Let us consider an element that has less than four valence electrons. Elements of this type are called ***metals, and are found in columns I, II, or III of the Periodic Table.*** The valence electrons in metals are only loosely bound to the atoms that they circle. Because of the kinetic energy that they possess, the electrons vibrate. If an electron from one atom comes close to an electron from another atom, their interaction may force one of the electrons loose from its valence shell. When this happens, the dislodged electron is no longer held by its original atom, but is free to drift in the empty space between atoms. This electron will drift at random, banging into nuclei and electrons, until it strikes another electron and knocks that electron away from its parent atom or recombines with an atom that is missing an electron. At any instant of time there are many electrons that are not attached to any particular atom in the metal and so are free to drift throughout the empty space between atoms in the metal. A material in which free charge moves easily is called a ***conductor.*** Therefore, all metals are conductors because they all contain free electrons roaming between the atoms of the metal.

TABLE 1.2

Metal	Relative Resistance (compared to silver)
Aluminum	1.7263
Copper	1.0526
Gold	1.4905
Iron	5.8947
Nickel	5.3158
Silver	1.0000 (best conductor)
Tin	7.0526
Tungsten	3.4211

While all metals are conductors, they do not allow electrons to drift through them equally well. An ideal conductor would allow electrons to drift through it with no opposition to the flow of the electrons. In a typical metal the electrons have to drift around atoms and other electrons to be able to move at all. This opposition to electron flow is called ***resistance.*** Metals are typically low-resistance conductors of electricity. The best conductor among the common metals is silver. When other metals are compared to silver the results are as shown in Table 1.2.

Since silver has the lowest resistance its ratio of resistance to the resistance of silver is 1. Other metals have more resistance, so their resistance ratio is greater than 1. The unit of resistance in the International System of Units is called the ***ohm.*** Wires have very few ohms of resistance for a given length and size.

1.2.7 Insulators

From the Periodic Table we know that the maximum number of valence electrons that an atom can have is limited to eight. An atom with eight valence electrons positioned around it is said to be inert. That is, it does not react with other elements to form compounds. The eighth column of the modern periodic table contains these elements. These elements hold onto their electrons very tightly and so there is virtually no chance for an electron to break free of its atom. Such an element would have no free electrons that could wander through the space between atoms and therefore would not be a conductor. An element of this type is called an insulator because it does not allow electrons to flow through easily. All atoms with more than four valence electrons are found to be insulators. These atoms, like the inert atoms of column VIII, hold on to their valence electrons and therefore are not able to allow electrons to flow easily through them. From this we see that the value of resistance associated with insulators is very large and can be in the millions of millions of ohms. It is not unusual for good insulators to have almost infinite resistance.

1.2.8 Semiconductors

One important class of elements is found in the fourth column of the Periodic Table. These elements have only four valence electrons and are neither insulators nor conductors. They belong to the class of elements called semiconductors. We associate the name with electronic devices called transistors, diodes, and integrated circuits. The study of this special kind of element is covered in other texts and is not the subject of the present book.

1.3 INTRODUCTION TO BASIC ELECTRICAL CONCEPTS

1.3.1 Electrical Charge

The unit of charge in the international system of measurement is the coulomb. The relationship between this unit of charge and the charge on a single electron or proton has been determined by measurements to be:

$$\text{charge per electron} = 1.602 \times 10^{-19}\ \text{C}$$

This equation implies that there are a huge number of electrons in one coulomb. If we divide both sides of the equation by 1.602×10^{-19} we find that

$$6.242 \times 10^{18}\ \text{electrons} = 1\ \text{C}$$

It takes over 6 billion billion electrons to make up one coulomb of charge. Each electron weighs 9.1×10^{-21} kg. That means that ***the weight of one coulomb is only:***

$$9.1 \times 10^{-31}\ \text{kg/electron} \times 6.242 \times 10^{18}\ \text{electrons}$$
$$= 5.68 \times 10^{-12}\ \text{kg}$$

From this calculation we can see that the mass of a coulomb of electrons is negligible.

1.3.2 Interactions Between Charged Particles: Coulomb's Law

The interaction between charged particles was studied by Charles Coulomb. This English scientist, for whom the unit of charge was named, discovered that two charged bodies separated by a distance r exert a force on each other that is proportional to the product of the charge magnitudes and inversely proportional to their separation squared. This relationship, called ***Coulomb's Law,*** can be written formally as:

$$\text{Force (newtons)} = k\,Q_1 \times Q_2/r^2$$

where k is a constant that depends on the system of units chosen. For the International System of Units the value of k equals $9 \times 10^9\ \text{Nm}^2/\text{C}^2$.

A negative sign for the force represents an attractive force between the charges while a positive force means that the two charges repel each other. This law quantifies the experience of the ancients in their studies of static electricity. With it we can determine the electrostatic force that exists between charges.

EXAMPLE 1.2

What is the force that exists between an electron and a proton when they are separated by one angstrom ($=10^{-10}$ meters)? (This represents the electrostatic force that exists in a hydrogen atom.) How does this force compare to the force of gravity?

Solution The charge on an electron or a proton is 1.602×10^{-19} C. This means that the force is equal to:

$$F = \frac{(9 \times 10^{9})(-1.602 \times 10^{-19})(1.602 \times 10^{-19})}{(10^{-10})^2} = -2.31 \times 10^{-8}\ \text{N}$$

Note that the force between an electron and a proton is attractive because the sign of the force is negative. While this force does not seem large, it is enormous when compared to the effect of gravity on the electron.

The weight of an electron is:

$$\begin{aligned} \text{Weight} &= \text{Mass} \times \text{Acceleration of gravity} \\ &= 9.1 \times 10^{-81} \times 9.8 \\ &= 8.9 \times 10^{-30}\ \text{N} \end{aligned}$$

This means that the electrostatic force between an electron and a proton compared to the gravity effect on the electron is:

Electrostatic force/Gravity force $= 2.31 \times 10^{-8}/8.9 \times 10^{-30} = 2.6 \times 10^{21}$.

The electrostatic force is 2.6×10^{21} times larger than the effect of gravity!

The large difference in effect between the force of gravity and the electrostatic force means that the forces on electrons tend to be electrostatic in nature rather than gravity related. Thus, in studying electrons we can normally neglect the effect of gravity with little effect on the accuracy of the answer obtained.

1.3.3 Electrical Current

When charge moves it creates an electrical current. This current is expressed in a unit called amperes. By definition, if one coulomb of charge passes through a circuit in one second of time the resultant current is defined as one ampere. Expressed as an equation this becomes:

1 ampere = 1 coulomb/second

Huge numbers of charged particles, electrons or protons, must move for even the smallest current to flow. A one-microampere (μA) current represents one millionth the amount of a one-ampere current yet still represents the motion of 6,242,000,000,000 electrons moving past a point in a circuit in a second of time. When we work with electric circuits we are always dealing with the motion of a huge number of electrons, even when the magnitude of the current flowing seems small.

The direction of current flow tells us which way the charges are moving. Since both positive and negative charges can move, we must be careful how current is defined. To uniquely define the direction of current flow we say that the direction of current flow is the direction that positive charges would flow if they could. Because electrons have a negative charge, this means that the direction of current flow is opposite to the direction that electrons flow. Where both electrons and positive charges move, the current is the sum of the two currents because they move in opposite directions when both move.

We can express the relationship between total charge passing a point in a circuit and current through it by using the relationship:

$$I = Q/T$$

where I is the current in amperes, Q is the charge in coulombs passing a given point, and T is the time the electrons take to pass the given point in the circuit.

EXAMPLE 1.3

A current of 5 A flows for 3 min. How many coulombs pass through a wire carrying the current? How many electrons does that represent? What is the total weight of the electrons?

Solution 5 A means that 5 C pass through a point in the conductor per second. Thus the total number of coulombs is:

$$Q = 5 \text{ C/s} \times 180 \text{ s} \quad (3 \text{ min})$$
$$Q = 900 \text{ C}$$

Each coulomb contains 6.242×10^{18} electrons, so there are:

$900 \times 6.242 \times 10^{18}$ electrons $= 5.618 \times 10^{21}$ electrons

The weight of each electron is 9.1×10^{-31} kg. Therefore the total weight of the electrons is:

$$\begin{aligned} \text{Weight} &= 9.1 \times 10^{-31} \times 5.618 \times 10^{21} \\ &= 5.1 \times 10^{-9} \text{ kg} \\ &= 5.1\ \mu\text{g} \end{aligned}$$

The electronic charge moving through a volume can also be expressed in terms of the number of free electrons per unit volume that the substance contains times the charge times the volume swept out by the moving electrons. This representation will allow us to find out how fast electrons move in a given wire. Expressed mathematically this expression becomes:

$$Q = Nevat$$

where Q = total charge passing through the volume
N = the number of free electrons/cm^3
$e = 1.602 \times 10^{-19}$ C (charge per electron)
v = velocity of free electrons in centimeters/second
a = cross-sectional area of conductor in centimeters^2
t = time in seconds that the charge moves

TABLE 1.3
Value of N for various metals

Metal	Free Electrons/cm^3
Lithium	4.6×10^{22}
Sodium	2.5×10^{22}
Potassium	1.3×10^{22}
Copper	8.5×10^{22}
Silver	5.8×10^{22}
Gold	5.9×10^{22}

If we divide Q by t we get the current passing through the volume as

$$I = Neva$$

This equation gives us the current flowing through a conductor in terms of the metal properties and the dimensions of the wire carrying the electrons of the circuit. The value of N depends on the material being used for the conductor. Table 1.3 gives values of N for several common metals.

Note that the number of free electrons per cubic centimeter is in the order of a number $\times 10^{22}$, so the speed obtained for any of these metals is approximately the same independent of the metal being used. The following example demonstrates this fact.

EXAMPLE 1.4

A copper wire with a cross-sectional area of 1 mm^2 has 10 amperes flowing through it. What is the average velocity, v, of the electrons traveling through the wire?

Solution

$$I = 10 \text{ A} = 8.5 \times 10^{22} \times 1.602 \times 10^{-19} \times v \times 1 \times 10^{-2} = 1.36 \times 10^{2}\, v$$

$$v = \frac{10}{1.36 \times 10^{2}} = 0.735 \text{ cm/s}$$

From this calculation we see that the average velocity of the electrons traveling in any wire is very slow even when the current is substantial. This is because of the huge numbers of free electrons that move. It is important to understand that the number of electrons moving in a circuit is dependent on the material and its dimensions and is independent of the current value. Current only affects the speed with which the electrons move and therefore the current that flows in the circuit.

When the electrons are forced to move more rapidly, they strike nuclei more often and harder, and therefore transfer more energy to the material they are moving through. This results in an increase of the temperature of the conductor due to the energy being shed by the moving electrons. This heat uses energy, which comes from the electron moving force pushing the electrons along.

1.3.4 Voltage

In the International System of Units, the unit of electron moving force (emf) is the volt. Thus we say that we have so many volts available to move the electrons through the circuit. This force is electrical in nature rather than mechanical and operates because of the electron's charge rather than its mass.

Anything that provides an electrical force causing electrons to move is called a voltage source. We are familiar with certain of these sources that we see every day. Batteries are examples of voltage sources. In our studies of circuit analysis we shall see there are many different types of sources, but they all share common properties, making them easy to analyze.

When a voltage source is applied across a circuit, the voltage causes electrons to flow through the various conductors that make up the circuit. ***All*** of the free electrons in each of the conductors move under the influence of the voltage source. This does not mean that the number of electrons moving is the same in each part of the circuit but rather that the product of their number and their velocity is a constant.

Voltage is the force that drives the electrons. As the electrons speed up, they strike the bound electrons and the nuclei more often. An increase in current causes the electrons to speed up which in turn causes the conductor to heat up due to the greater number of impacts on the atoms making up the conductor. This mechanism accounts for the heating of conductors by current passing through them. We shall discuss this matter further when we talk of power in electric circuits.

■ SUMMARY

- In working with electrical circuits we use scientific notation. In this notation all numbers are expressed as a number between 1 and 10 times 10 to some integer power.
- All physical measurements require a system of consistent measurements. Modern science uses the International System of Units. In this set of units the measurement of length is the meter, the unit of mass is the kilogram, the unit of time is the second, and the unit of charge is the coulomb.
- All physical matter is made from a group of approximately 100 different types of matter called elements. These elements are uniform in their properties and cannot be broken down into simpler elements. All materials made from these elements are called compounds and are a combination of the elements in fixed proportions.
- The smallest amount of an element is called an atom. All atoms are made from positively charged particles called protons, uncharged particles called neutrons, and, circling the center or nucleus of the atom, negatively charged very light particles called electrons. All protons, neutrons, and electrons have exactly the same properties.
- The Law of Static Electricity states that like particles repel each other and unlike particles attract each other.
- A conductor is a material that contains electrons that are not tightly held to the individual atoms but are free to roam in response to an applied electric force field.
- An insulator is a material that has very few electrons that can move in the material.
- A semiconductor is a material that has properties between those of a conductor and an insulator. Semiconductors have some but not many electrons that are free to move in the material.
- An electron has a charge of 1.6×10^{-19} C. It takes 6.242×10^{18} electrons to make up a coulomb of charge.
- Coulomb's Law expresses the interaction relationship between charges. The force between charges is proportional to the product of the individual charges and inversely proportional to the distance between them.

- Electrical current is considered to be a flow of charge. Generally the current is measured in amperes. One ampere represents a charge of one coulomb moving through a plane in one second of time. The direction of current is the direction in which positive charge would move if the charge were positive. Electrons, having negative charge, move in the opposite direction to conventional current.
- The quantity pushing the electrons in a conductor is called voltage. The higher the voltage, the faster the electrons move. If we know what material the electrons are moving in, we can describe how fast they move. Their motion in a conductor is very slow due to the large number of electrons contained in a conductor.

PROBLEMS

1.1 Convert the following numbers to scientific notation.
(a) 3579
(b) 0.0000456
(c) 457,000,000
(d) 0.93700000
(e) 3,985,678,000,123
(f) 135.89723

1.2 Evaluate the following expressions by first converting them to scientific notation and then carrying out the operations.
(a) $34.9(567)(234)^2/(568)^3$
(b) $(0.000764)^2(89.7)$
(c) $(389.2)(78.43)$
(d) $(894)(456)/(0.0009)^3$
(e) $(6.54)(678.000)(.00093)$
(f) $(789.7)(56)^{0.73}/(123)$

1.3 Convert the following sets of values as described.
(a) 34,795 g to kilograms
(b) 78.6 kg to micrograms
(c) 14.7 m to millimeters
(d) 89.9 cm to kilometers

1.4 Convert the following sets of values as described.
(a) 4.7 MΩ to kilohms
(b) 475 Ω to megohms
(c) 678 kV to megavolts
(d) 547.9 μA to milliamperes
(e) 43 mV to volts
(f) 798 mA to amperes

1.5 The charge on a proton is 1.602×10^{-19} C.
(a) How many protons are there in 75 C?
(b) How many electrons are there in 75 C?
(c) How many protons are there in 96,368 C?
(d) How many neutrons does it take to make a coulomb? Explain.

1.6 Resistance is the name given to the opposition to flow of electrons. How would the following happenings affect the value of resistance? Explain.
(a) Doubling the length that the electrons must travel
(b) Multiplying the length that electrons must flow by a constant
(c) Lowering the opposition to the flow of electrons
(d) Increasing the speed of the electrons

1.7 A copper wire is measured and found to have 10 Ω of resistance. It is then replaced by a wire of identical size made from another metal. Use the information in Table 1.2 to find how much resistance would the new wire have if it were made of:
(a) Nickel **(b)** Silver **(c)** Gold **(d)** Tungsten

1.8 An electron weighs approximately 9.1×10^{-31} kg. How many electrons are there in the following quantities?
(a) A gram
(b) A pound (1 pound = 454 grams)
(c) 5 C

1.9 A spaceship that runs on electrons lands on Earth to refuel. The spaceship holds 5 kg of electrons for fuel,
(a) How many electrons will they need to fill their ship?
(b) How many coulombs is this?
(c) How long will it take to refuel the ship if they use a 100-A source?

1.10 Two metal plates are each loaded with 1 C of electrons and placed 1 mm apart.
(a) Using Coulomb's Law, calculate the force between the two plates.
(b) Is this force attractive or repulsive? Explain.
(c) One of the charges is removed and replaced with a coulomb of protons. Explain what difference this makes in the problem.
(d) How far apart must the plates be moved to reduce the force by a factor of 100?

1.11 Assume that the weight of the electrons in a human body is 1/4000 the total weight of the body. If a man weighs 75 kg:
(a) What is the weight of the electrons in this man's body?
(b) How many electrons does it take to make this weight?

1.12 What is the speed of an electron traveling through a 0.1 mm^2 wire made of silver that has 20 A of current and is 10 m long?
(a) How long would it take the electron to make the trip from one end of the wire to the other?

1.13 A conductor is found to have a value of N (free electrons/cm^3) of 3×10^{14}. Calculate the speed of the electrons through this conductor if the current flowing is 10 mA and the cross-sectional area is 0.01 mm^2.

1.14 How fast must a billion electrons in a square box of side 1 mm travel to represent a current of
(a) 1 mA? **(b)** 1 A?

1.15 What happens to the resistance of a wire as the cross-sectional area of the wire increases? Explain using the concept of flowing electrons and equation $I = Neva$.

CHAPTER 2

VOLTAGE AND CURRENT SOURCES: THE SINUSOIDAL WAVEFORM

INTRODUCTION

Voltage and current sources provide the energy that is required for electrical and electronic circuits to function in the various modes that are the subject of this text. An understanding of these sources is fundamental to the understanding of circuit operation in general.

2.1 THE INDEPENDENT VOLTAGE SOURCE

An ***independent voltage source*** is a source of electron moving force whose output voltage is independent of the value of any other circuit parameter. (***Note:*** The output of a ***dependent*** voltage source does depend on other circuit parameters. These sources will be discussed in Chapter 7. Unless otherwise specified, all the voltage sources discussed in this text will be independent).

The symbol for an independent voltage source is shown in Fig. 2.1. Figure 2.1(a) represents an *ac* voltage source and Figure 2.1(b) represents a *dc* voltage source. An ac source is one whose output voltage magnitude changes with time. A *dc* source is one whose output voltage is constant and therefore does not change with time.

FIGURE 2.1
Independent voltage sources

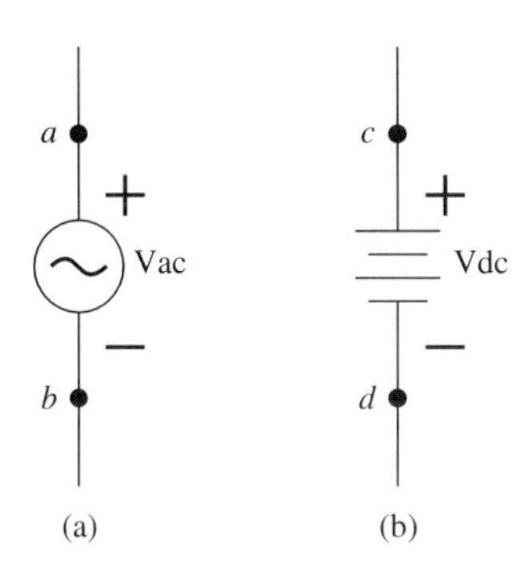

The + sign on a dc source shows which of the two terminals is always at a higher potential than the other. Since the dc source is constant in magnitude, this value is placed next to the dc source symbol. In Fig. 2.1(b), the value is labeled Vdc. (Vdc could be 20 V, 10 mV, 5 μV, etc.).

The + sign for an ac source shows which terminal is higher in potential when the voltage is designated as having a positive value. This same terminal would be lower in potential when the voltage is designated as having a negative value.

Note the label Vac next to the ac voltage source. Since the ac voltage source magnitude varies with time by definition, the value Vac represents a number that is obtained from the voltage source waveform either by an equation or by referencing specific points on the waveform. This will be discussed later on in this chapter.

The independent voltage source, whether ac or dc, *fixes* the voltage between two "points," or terminals, in a circuit. If the source is *dc,* this means the voltage between the two points is a fixed magnitude. If the source is *ac,* the voltage is a fixed waveform.

All the circuits discussed in this text consist of voltage and/or current sources and circuit elements (introduced in Chapter 4). These sources and elements are connected by wires that are considered as ***ideal conductors.*** Recall from Chapter 1 that ideal conductors offer no resistance to the flow of electric current.

Each individual ideal conductor is considered a point or "terminal." In Fig. 2.1(a), the ac voltage source Vac fixes the voltage waveform between two ideal conductors represented by the letters *a* and *b.* In Fig. 2.1(b), the dc voltage source Vdc fixes the voltage magnitude, and polarity, between two ideal conductors represented by the letters *c* and *d.*

2.2 PERIODIC WAVEFORMS

A ***periodic waveform*** is one in which a portion of the waveform is continuously repeated in a consecutive manner. Some examples of periodic waveforms are shown in Fig. 2.2. Figure 2.2(a) is called a ***triangular*** wave. Figure 2.2(b) is known as a ***rectangular*** wave. Figures 2.2(c) and (d) are known, respectively, as a ***ramp*** wave and a ***sinusoidal*** wave.

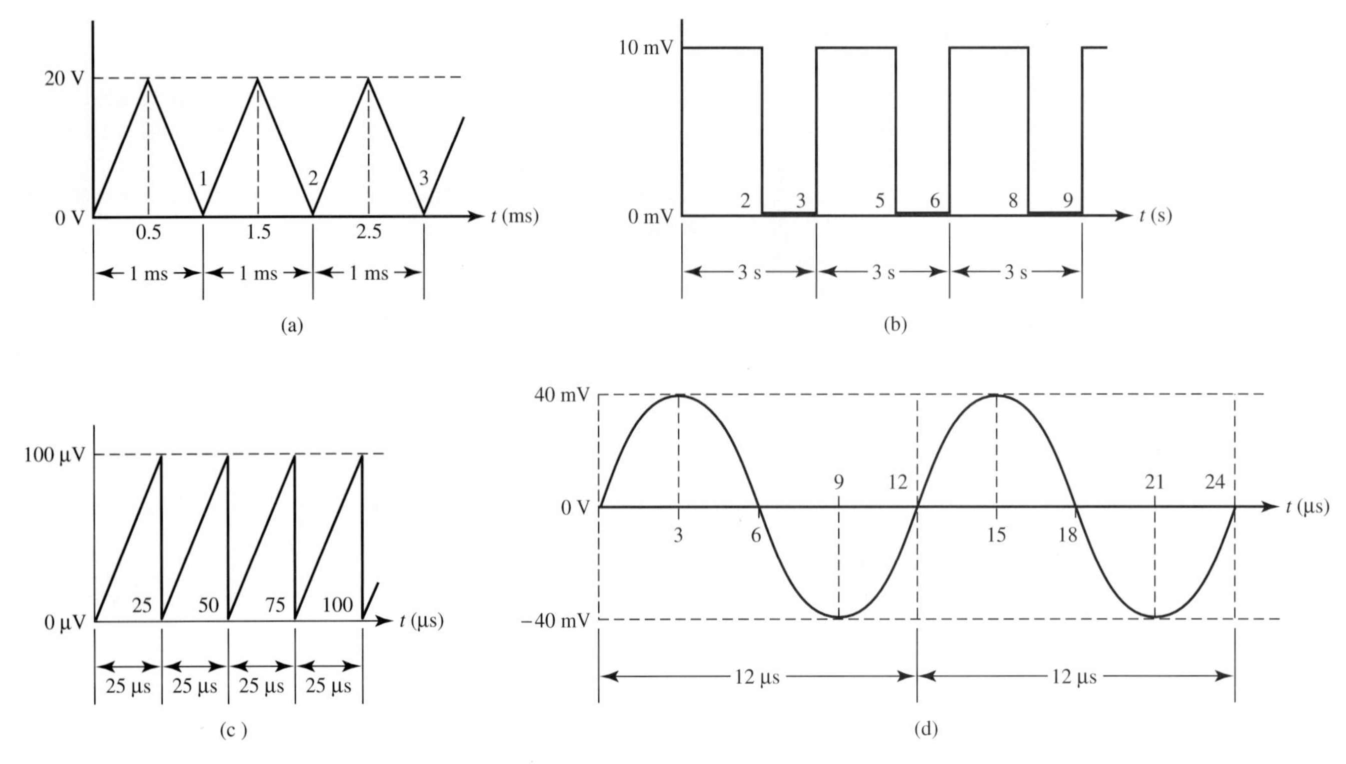

FIGURE 2.2
Periodic waveforms

Although these waveforms all have different shapes, they are all considered periodic. This is because they all have the "repeating" characteristic just noted in which a portion of the wave is continuously, and consecutively, repeated.

Periodic waveforms can all be mathematically described using specific terminology. In the next section we will discuss some of the terminology used in describing such waveforms.

2.2.1 Periodic Waveform Definitions

In this section we will define and explain some common terms associated with periodic waveforms. An understanding of these terms is fundamental to the understanding of circuit behavior with periodic input voltage (or current) sources.

> Cycle: A ***cycle*** is the smallest, continuous portion of a periodic waveform that is continuously, and consecutively, repeated.

> Period: The ***period*** of a periodic waveform is the duration of one cycle. The period is generally denoted by the letter T.

Each of the periodic waveforms of Fig. 2.2 is divided into cycles by vertical lines. To identify a cycle in each case consider the following:

> In the ***triangular*** wave of Fig. 2.1(a), the voltage is 0 V at $t = 0$ ms, increases linearly to 20 V in 0.5 ms, then decreases linearly to 0 V in another 0.5 ms. The voltage varies in exactly the same manner from $t = 1$ ms to $t = 2$ ms and again from $t = 2$ ms to $t = 3$ ms. This variation keeps repeating over and over again. Each cycle is therefore shown as a portion of the waveform between two vertical lines.

> In the ***rectangular*** wave of Fig. 2.2(b), consider the voltage from $t = 0$ to $t = 3$ s. This voltage is 10 mV for 2 s and 0 mV for 1 s. This same variation is repeated from $t = 3$ s to $t = 6$ s, again from $t = 6$ s to $t = 9$ s, and so on.

The ***ramp*** wave of Fig. 2.2(c) shows the voltage varying from 0 μV at $t = 0$ μs to 100 μV at $t = 25$ μs, then back to 0 μV in zero time. This variation is repeated exactly from $t = 25$ μs to $t = 50$ μs, from $t = 50$ μs to $t = 75$ μs, and from $t = 75$ μs to $t = 100$ μs. This process then continues in exactly the same manner.

The waveform of Fig. 2.2(d) is a sinusoid. A single cycle of the sinusoid cannot be as easily identified as the previous three waveforms because the variation is not linear. A comparison of values at different times would show where a cycle begins and ends. Two cycles are shown in this figure as indicated by the vertical lines.

Recall that the ***period*** of the wave is the duration of one cycle. Accordingly, the period of each of these waveforms is as follows:

$T = 1$ ms for the triangular waveform of Fig. 2.2(a)

$T = 3$ s for the rectangular waveform of Fig. 2.2(b)

$T = 25$ μs for the ramp waveform of Fig. 2.2(c)

$T = 12$ μs for the sinusoidal waveform of Fig. 2.2(d)

Note that because the period, T, represents the time for one cycle, its units may be expressed as seconds per cycle. The reciprocal, or inverse, of the period has a special meaning and is noted as the frequency of the waveform.

Frequency: The reciprocal of the period of a periodic waveform is known as the frequency and is denoted by the letter f. The frequency is given by the following equation:

$$f = 1/T \qquad \textbf{(Eq. 2.1)}$$

Because the frequency is the reciprocal of the period, its unit may be expressed as the reciprocal of the unit for period, or as cycles per second (cps). In fact, cps was used as the unit for frequency for many years.

To honor the German physicist Heinrich Rudolph Hertz, who performed groundbreaking research on electromagnetic waves in the nineteenth century, the unit for frequency was subsequently changed to hertz (Hz). The unit of Hz is used worldwide, and we will use it in this text. It is important to remember, however, that frequency represents the number of cycles that occur in one second.

$$1 \text{ hertz (Hz)} = 1 \text{ cps} \qquad \textbf{(Eq. 2.2)}$$

For the waveforms of Fig. 2.2:

$f = 1/(1 \times 10^{-3}) = 100$ Hz for the triangular waveform of Fig. 2.2(a)

$f = 1/3 = 0.33$ Hz for the rectangular waveform of Fig. 2.2(b)

$f = 1/(25 \times 10^{-6}) = 40$ kHz for the ramp waveform of Fig. 2.2(c)

$f = 1/(12 \times 10^{-6}) = 83.3$ kHz for the sinusoid of Fig. 2.2(d)

Note that the frequency of a periodic waveform ***may be less than one*** as in the case of Fig. 2.2(b). This simply means that one cycle of the wave has a duration (period, T) ***greater than one second.***

The frequency of a voltage, or current, source waveform is one of the primary factors in determining how a circuit will behave in response to the source. Beginning with Chapter 4 we will see how circuits are analyzed by considering the effect of frequency on the behavior of circuit elements.

Amplitude: The ***amplitude*** of a periodic waveform is the maximum value of the waveform.

The amplitude is usually denoted by the subscript p or m. Periodic voltage waveforms, for example, will have amplitudes denoted by V_p or V_m while current waveforms will have amplitudes denoted by I_p or I_m.

Since some periodic waveforms have both positive and negative values, this definition may be divided into two parts as follows:

Positive Amplitude: The maximum ***positive*** value of the waveform.

Negative Amplitude: The maximum ***negative*** value of the waveform.

Peak Value: The ***peak value*** of a periodic waveform is the same as the amplitude.

Both terms, ***amplitude*** and ***peak,*** are used interchangeably. We may therefore use the terms ***positive peak*** and ***negative peak*** interchangeably with the terms ***positive amplitude*** and ***negative amplitude,*** respectively.

For the waveforms of Fig. 2.2:

$V_p = 20$ V = amplitude of the triangular waveform of Fig. 2.2(a)

$V_p = 10$ mV = amplitude of the rectangular wave of Fig. 2.2(b)

$V_p = 100\ \mu$V = amplitude of the ramp waveform of Fig. 2.2(c)

$(V_p)_+ = 40$ mV = ***positive*** amplitude of the sinusoid of Fig. 2.2(d)

$(V_p)_- = -40$ mV = ***negative*** amplitude of the sinusoid of Fig. 2.2(d)

When a waveform has both positive and negative values, it is sometimes desirable to refer to its peak-to-peak value.

Peak-to-Peak Value: The ***peak-to-peak value*** of a waveform, denoted by the subscript pp, is the difference in value between its positive and negative amplitudes.

The peak to peak value of a voltage waveform may be denoted by V_{pp}, while that for a current waveform may be denoted by I_{pp}. The peak to peak value is the difference between the positive and negative values and may be given by the following equation:

$$V_{pp} = (V_p)_+ - (V_p)_- \qquad \textbf{(Eq. 2.3)}$$

Referring once again to Fig. 2.2, note that only the sinusoidal waveform of Fig. 2.2(d) has both positive and negative values. Be aware, however, that ***any*** waveshape may have both positive and negative values.

Therefore, $(V_p) - = 0$ for the waveforms of Figs. 2.2(a), (b), and (c) and:

$V_{pp} = V_p = 20$ V for Fig. 2.2(a)

$V_{pp} = V_p = 10$ mV for Fig. 2.2(b)

$V_{pp} = V_p = 100\ \mu$V for Fig. 2.2(c)

$V_{pp} = (V_p)_+ - (V_p)_- = 40 \text{ mV} - (-40 \text{ mV}) = 80$ mV for Fig. 2.2(d)

EXAMPLE 2.1

For each of the waveforms of Fig. 2.3 find the following:

(a) Number of cycles shown
(b) Period
(c) Frequency
(d) Positive amplitude
(e) Negative amplitude
(f) Peak-to-peak value

(a) Voltage in Millivolts

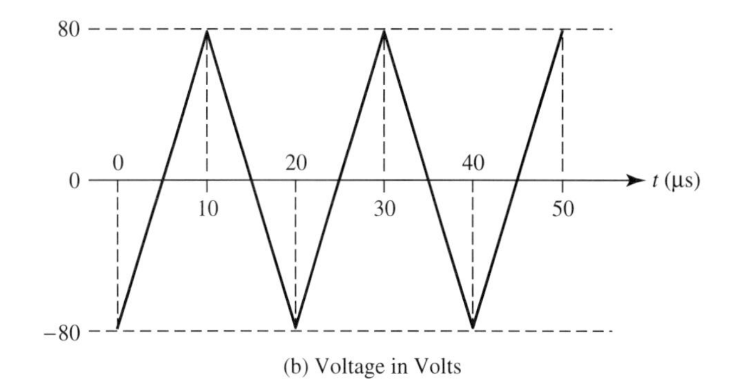

(b) Voltage in Volts

FIGURE 2.3
Periodic voltage waveforms

Solution

(a) ***Answer*** Fig. 2.3(a), 3 cycles; Fig. 2.3(b), 2.5 cycles.

(b) Since the period is the time for one cycle:

Answer Fig. 2.3(a), $T = 10$ ms; Fig. 2.3(b), $T = 20\ \mu$s.

(c) Using Equation 2.1:

$$\text{Fig. 2.3(a)} \quad f = 1/T = 1/(10 \times 10^{-3}) = 100$$
$$\text{Fig. 2.3(b)} \quad f = 1/T = 1/(20 \times 10^{-6}) = 50 \times 10^{3}$$

Answer Fig. 2.3(a), $f = 100$ Hz; Fig. 2.3(b), $f = 50$ kHz.

(d) Recall that the positive amplitude of a periodic waveform is its maximum positive value. Therefore:

Answer Fig. 2.3(a), $(V_p)_+ = 240$ mV; Fig. 2.3(b), $(V_p)_+ = 80$ V.

(e) Recall that the negative amplitude of a periodic waveform is its maximum negative value. Accordingly:

Answer Fig. 2.3(a), $(V_p)- = 0$; Fig. 2.3(b), $(V_p)- = -80$ V.

(f) Using Equation 2.3:

$$\text{Fig. 2.3(a),} \quad V_{pp} = 240 \times 10^{-3} - 0 = 240 \times 10^{-3}$$
$$\text{Fig. 2.3(b),} \quad V_{pp} = 80 - (-80) = 160$$

Answer Fig. 2.3(a), $V_{pp} = 240$ mV; Fig. 2.3(b), $V_{pp} = 160$ V.

There are two more periodic waveform parameters that must be introduced. These two parameters, ***radian frequency*** and ***phase angle,*** will be reviewed in Section 2.6, where the sinusoidal waveform is discussed in detail.

2.2.2 The DC Waveform

As previously mentioned, the output voltage of a dc waveform does not vary with time. Waveforms of dc voltages or currents will therefore simply be horizontal lines because such lines represent constant values.

We will see in later chapters that analysis of dc circuits can be accomplished by considering dc voltages and currents as special cases of ac voltages and currents in which the "period" T, is infinite. If T_{dc} represents the period of a dc waveform, then

$$T_{dc} = \infty \qquad \textbf{(Eq. 2.4)}$$

Using Equation 2.4 in Equation 2.1 yields the following important result:

$$f_{dc} = 1/T_{dc} = 1/\infty \equiv 0 \qquad \textbf{(Eq. 2.5)}$$

f_{dc} is the frequency of any dc waveform and is identically equal to zero.

Equation 2.5 is the basis for the analysis of dc circuits that we will perform in this text. Many of the equations that we develop for circuit responses to dc sources will be obtained by substituting zero for frequency wherever it appears in the equations of responses to ac sources. This concept not only simplifies the analysis of dc circuits but also aids in our understanding of the results obtained.

2.3 THE INDEPENDENT CURRENT SOURCE

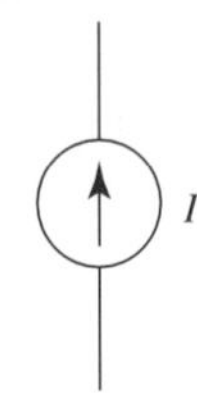

FIGURE 2.4
Independent current source symbol

The independent current source is one whose output current is independent of the value of any other circuit parameter. The symbol for an independent current source is shown in Fig. 2.4. This same symbol is used for both ac and dc sources.

For *dc* current sources, the I next to the symbol represents the magnitude of the dc current source output. The arrow points in the direction of the current supplied by the source.

For *ac* sources the I represents a complex number (discussed in the next chapter) that has a specific relationship to the current waveform. The arrow in the symbol shows the direction in which the source will supply current when it is connected to an external circuit ***and*** the current is positive. (Conversely, when the current is negative, the source will supply current to an external circuit in the reverse direction.)

All the comments made previously for the definitions of ***cycle, amplitude, frequency,*** and ***period*** of periodic voltage waveforms apply equally to current waveforms. The only difference is, of course, that the unit for voltage is the volt while the unit for current is the ampere. Positive, negative, and peak-to-peak amplitudes of current will therefore be expressed in amperes.

It should be emphasized that the current source is primarily useful as an analytical tool that simplifies the analysis of certain types of circuits. Most laboratories will include commercial voltage sources as part of their standard equipment but will seldom carry commercial current sources.

2.4 VOLTAGE SOURCE CURRENTS

As discussed in the first chapter, current in electric circuits consists generally of electrons, which are ***negatively charged*** particles. The flow of electrons therefore constitutes negative current.

In engineering fields, it is common practice to work with ***positive*** current rather than negative current. This means that we consider current as flowing in the opposite direction from "electron current." This would be in the direction that positive charge, having the same magnitude as the negative electrons, would flow. The previous discussion of current directions in current sources is in keeping with this definition of positive current. Macroscopically, it makes no difference in the final results obtained when analyzing electric circuits.

> All references to current throughout this text will mean positive current unless specifically stated otherwise.

Voltage sources will also generate current when connected into circuits. The following rule describes the direction of the current that voltage sources generate:

VOLTAGE SOURCE CURRENT DIRECTION RULE

The direction of positive current generated by a voltage source is through the source from the negative terminal to the positive terminal.

Very often, a voltage source is connected within a circuit that contains other voltage and/or current sources. In such cases, the actual magnitude and direction of current flowing through the source may be different from that specified by the source. This is because of the effect that the other sources have on the circuit and is the case for both ac and dc sources.

Fortunately, however, we need not concern ourselves with the actual direction of current through a voltage source. As we will see in later chapters, the analysis techniques used will automatically give us the actual magnitude and direction of ***all*** the currents in the circuit. We can assume any direction when starting the analysis, and the analysis itself will give us the correct values.

Since the polarity of a *dc* voltage source never changes, the direction of the current generated by the dc source is ***always*** in the same direction and follows the preceding rule. In an *ac* voltage source, however, the direction of the current generated will change each time the voltage polarity changes in accordance with the preceding rule.

2.5 COMMERCIAL VOLTAGE SOURCES

2.5.1 Commercial AC Voltage Sources

Low-level commercial ac voltage sources are available for use in circuit testing. These are generally known as function generators.

Function generators provide periodic voltage waveforms at their output terminals. Generally, three different waveforms are made available: the sinusoidal waveform, the triangular waveform, and the rectangular waveform.

The user generally has the ability to vary the voltage amplitude from about 20 mV to about 10 V. The frequency can also be varied over a wide range, generally from a few hertz to more than 20 MHz.

Commercial function generators are a basic tool of most electronic laboratories. Most modern units have digital displays, which enable the user to easily determine the frequency of the waveform being generated. The amplitude is set with a knob but generally can only be checked using either a voltmeter or an oscilloscope. In any case, if accuracy is critical both the magnitude and frequency of a voltage waveform should be checked with an oscilloscope.

2.5.2 Commercial DC Voltage Sources

Commercial dc voltage sources are generally known as power supplies. These voltage sources provide dc voltages ranging from tens of millivolts to about 30 V.

DC voltage sources are called power supplies because their main application is to provide the power necessary to allow a circuit to perform its function. You will gain more insight into this when you study electronics.

2.6 THE SINUSOIDAL WAVEFORM

The sinusoidal waveform is a particularly significant waveform because, as discussed in the next chapter, the sinusoidal function is extremely useful in the analysis of linear electric circuits. It is therefore important to become familiar with the characteristics of this type of waveform.

2.6.1 The Basic Sine Wave Equation

Consider Equation 2.6:

$$v = V_p \sin\theta = V_p \sin\omega t \qquad \textbf{(Eq. 2.6)}$$

Equation 2.6 describes the sine wave drawn in Fig. 2.5. As discussed in Section 2.2.1, V_p is the amplitude of this function. The angle $\theta = \omega t$ is plotted on the abscissa. The symbol ω is known as the angular velocity or radian frequency associated with the sinusoidal waveform. This parameter arises from the cyclical nature of the periodic waveform.

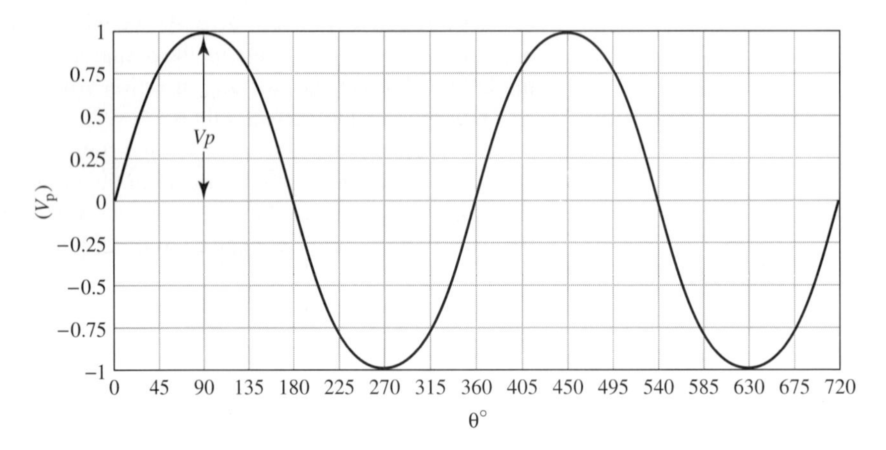

FIGURE 2.5
$v = V_p \sin(\omega t)$

The repeating cycle of the sine wave is analogous to moving around in a circle. Each time a point moves around a complete circle it has traveled an angle of 360° or 2π radians. If it moves around the circle a second time it has traveled a total of $2 \times 360° = 720°$ ***or*** $2 \times 2\pi = 4\pi$ radians. Figure 2.5 shows two complete cycles of the sine wave of Equation 2.6.

The parameter ω represents an angular velocity, which indicates how fast a point moves around the circle in units of radians per second (rad/s). In a similar manner, this same parameter is used to indicate how "fast" the sine wave changes. One cycle of a sine wave may be considered as having a "period" of 360°, or 2π radians. Therefore, ω, like f, is an indication of how many cycles occur in one second but using different units (rad/s). Accordingly, ω is also known as ***radian frequency.*** The radian frequency ω is related to the frequency f by recognizing that because one cycle represents 2π radians

$$\omega = 2\pi f \text{ radians/second (r/s)} \qquad \textbf{(Eq. 2.7)}$$

Note the following characteristics of the sine wave in Fig. 2.5.

For a single cycle:

1. The voltage goes from 0 to V_p in 1/4 cycle, or 90° or $\pi/4$ radians.
2. The voltage goes from V_p to 0 in 1/4 cycle, or 90° or $\pi/4$ radians.
3. The voltage goes from 0 to $-V_p$ in 1/4 cycle, or 90° or $\pi/4$ radians.
4. The voltage goes from $-V_p$ to 0 in 1/4 cycle, or 90° or $\pi/4$ radians.

This "division" of the cycle into quarters makes the sine wave relatively easy to sketch. It is convenient to identify each of the four points noted and connect them in the general manner of a sine wave.

EXAMPLE 2.2

Given the equation $v = 12 \sin 2000t$ V, find each of the following:
(a) ω = radian frequency in radians per second
(b) f = frequency in hertz
(c) V_p = amplitude in volts
(d) T = period in seconds

Sketch the waveform of v with the abscissa in:
(e) Degrees **(f)** radians **(g)** seconds

Solution

(a) Comparing this equation to the general form of Equation 2.6, we note that ω is the coefficient of time.

Answer $\omega = 2000$ rad/s.

(b) From Equation 2.7: $f = \omega/2\pi = 2000/2\pi = 3.18 \times 10^2$.

Answer $f = 318$ Hz.

(c) Noting that V_p is the coefficient of the sine in Equation 2.6:

Answer $V_p = 12$ V.

(d) Using Equation 2.1 in Equation 2.7 and solving for T:

$$T = 2\pi/\omega = 2\pi/2000 = 3.14 \times 10^{-3}$$

Answer $T = 3.14$ ms.

(e), **(f)**, and **(g)** The waveform of Fig. 2.5 serves as the basis for the sketches asked for in each of these parts. Only one waveform need be drawn. The abscissa of this waveform must be calibrated in each of the three units requested—degrees, radians, and seconds.

Answer See Fig. 2.6.

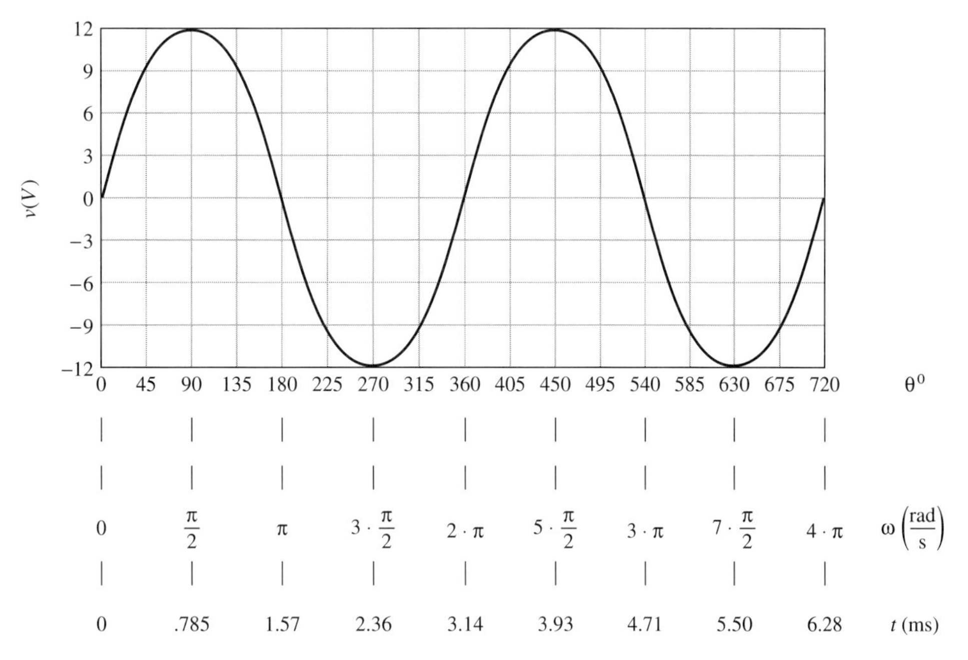

FIGURE 2.6
$v = 12 \sin(2000t)$ V

The sine wave of Equation 2.6, shown in Fig. 2.5, is valid as long as the phase angle of the wave is zero. The phase angle is the angular distance between the origin of the axes and the point where the "first" zero value of the sine wave (the zero value from which the sine wave increases) occurs. The general form of the sine wave equation would be:

$$v = V_p \sin(\omega t + \alpha) \qquad \textbf{(Eq. 2.8)}$$

where α is the phase angle of this wave.

The phase angle, α, may be positive or negative. Fig. 2.7(a) shows the waveform representing Equation 2.8 with positive α, and Fig. 2.7(b) shows the waveform representing Equation 2.8 with negative α. Note that if $\alpha = 0$, Equation 2.8 reduces to Equation 2.6.

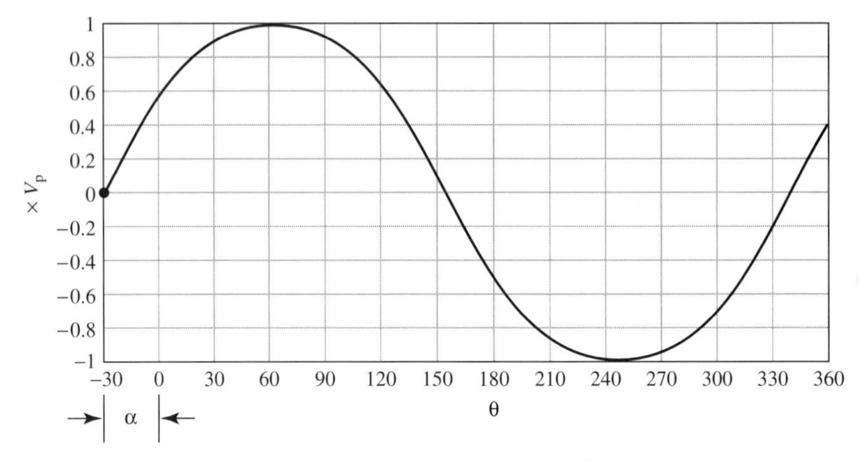

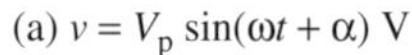
(a) $v = V_p \sin(\omega t + \alpha)$ V

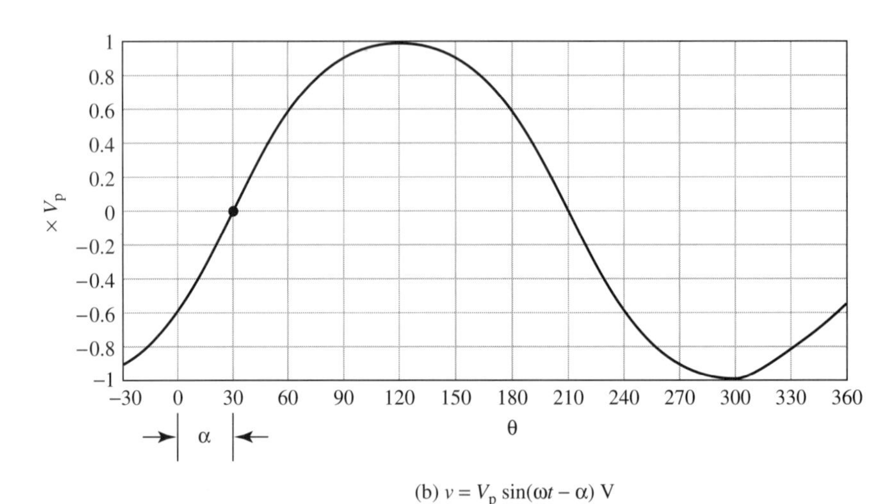

(b) $v = V_p \sin(\omega t - \alpha)$ V

FIGURE 2.7
Positive and negative phase angles

Sine waves with nonzero phase angles may be sketched simply by identifying the "first" zero and continually adding 1/4 period to find the points identified for a single cycle.

EXAMPLE 2.3

Sketch the waveform form $i = 200 \sin(12{,}000t - 45°)$ mA.

Solution The waveform is plotted by noting that it is of the form of Fig. 2.7(b) with a negative phase angle. Accordingly, the first zero is identified at $\theta = 45°$ and the four additional points are identified as follows:

$i = 200$ mA at $\theta = 45 + 90 = 135°$
$i = 0$ at $\theta = 135 + 90 = 225°$
$i = -200$ mA at $\theta = 225 + 90 = 315°$
$i = -200$ mA at $45 - 90 = -45°$

Answer The waveform is plotted in Fig. 2.8.

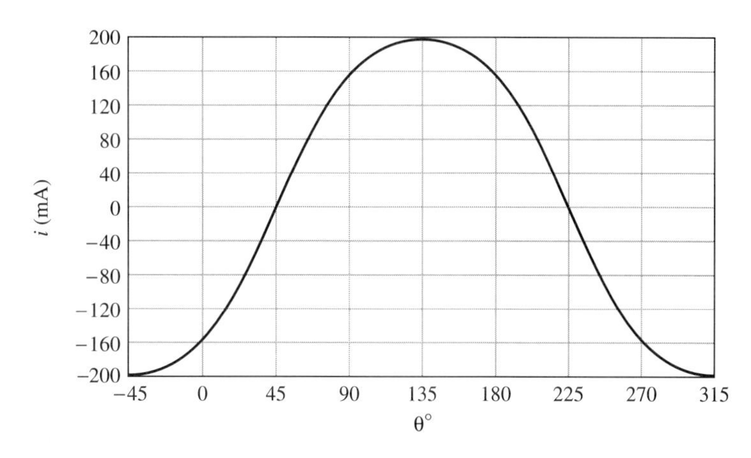

FIGURE 2.8
$i = 200 \sin(12{,}000t - 45°)$ **mA**

2.6.2 The RMS Value of the Sine Wave

Another parameter that is useful in sinusoidal circuit analysis is the ***rms*** or ***effective value*** of the sine wave. The rms value of a periodic waveform is defined by an equation involving an integration process, and determining this value therefore requires some knowledge of integral calculus. Fortunately, however, finding the rms value of a sinusoidal waveform is easy because the integral involved reduces to a simple equation.

The rms value of a sinusoidal waveform is equal to the peak value of the wave divided by the square root of 2.

For the sinusoidal waveform defined by Equation 2.8:

$$V_{\text{rms}} = V_p/\sqrt{2} = 0.707\ V_p \qquad \textbf{(Eq. 2.9)}$$

Note from the definition above and from Equation 2.9 that the rms value of any sinusoid is independent of the frequency or phase angle of the wave. It is a function ***only*** of the amplitude of the wave. We will see later on that the rms value is significant in power calculations.

EXAMPLE 2.4

Find the rms value of the sinusoidal current of Example 2.3.

Solution $I_{\text{rms}} = 0.707\ I_p = 0.707(200 \times 10^{-3}) = 141.4 \times 10^{-3}$ A

Answer $I_{\text{rms}} = 141.4$ mA.

2.6.3 Phase Relationship Between Waveforms

Another important factor in periodic waveforms is the phase relationship betwen two periodic waves. The phase relationship is a measurement of the angle between two corresponding points on both waves.

> If a point on one cycle of waveform 1 occurs $\alpha°$ before its corresponding point on the same cycle of waveform 2 and both waveforms have the same frequency, then ***waveform 1 is said to lead waveform 2 by $\alpha°$***. Alternatively, ***waveform 2 is said to lag waveform 1 by $\alpha°$***.

A key point in this definition of phase relationship is that both waves ***must*** have the same frequency. Most of the circuits considered in this text have a single frequency.

Consider the two waveforms plotted in Fig. 2.9. The equations for these waveforms are:

$$v = 10 \sin(\omega t - 30°) \text{ mV}$$
$$i = 20 \sin(\omega t + 45°) \text{ mA}$$

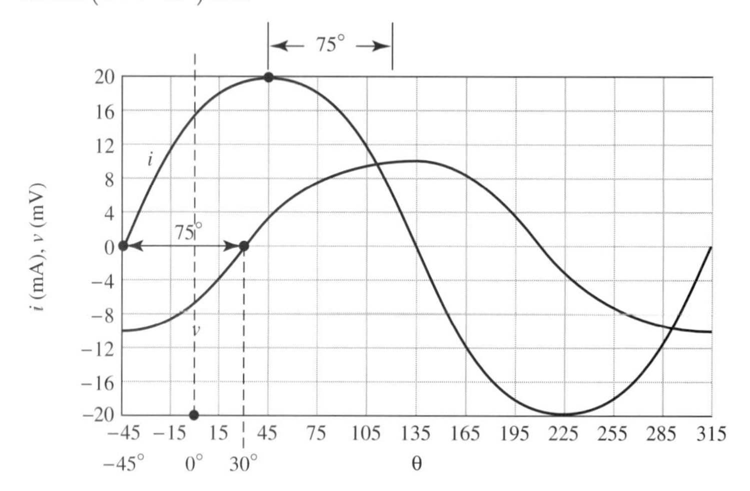

FIGURE 2.9
$v = 10 \sin(\omega t - 30°)$ **mV**
$i = 20 \sin(\omega t + 45°)$ **mA**

Consider the relation between these waves and the origin. The "first" zero on the positive slope portion of the voltage occurs 30° to the right, or ***after,*** the origin. The corresponding point on the current waveform occurs 45° to the left, or ***before*** the origin.

The "first" zero on the current waveform therefore occurs 75° "before" the corresponding point on the voltage waveform. Therefore:

i leads v by 75° or *v lags i* by 75°

Recall that the definition of phase relationship given here states that two corresponding points on the two waveforms can be used to determine phase relationship. For the two waveforms of Fig. 2.9 we used the first zero to determine the phase relationship.

Let us repeat the measurement using the "first" positive peaks of both waves. The first positive peak of the current waveform occurs 45° to the right, or after, the origin. The first positive peak of the voltage waveform occurs 120° to the right of the origin. Again, however, the peak of the current waveform occurs before the peak of the voltage waveform, leading to the same conclusion as earlier (i leads v by 120° − 45°, or 75°).

Note that both the voltage and current waveforms were written as ***sine*** equations. The preceding comparison, made by subtracting the two phase angles $45° - (-30°) = 75°$ or $(120° - 45° = 75°)$ can only be used if both waveforms are expressed as sines or both are expressed as cosines. If one waveform is expressed as a sine and the other as a cosine, one equation must be converted to the other form before subtracting phase angles.

EXAMPLE 2.5

Find the phase relationship between the following sets of waveforms.

(a) $v = 350 \sin(2000t - 60°)$ V, $i = 12 \sin(2000t - 40°)$ A
(b) $v = 4 \times 10^{-3} \cos(\omega t + 80°)$ mV, $i = 80 \cos(\omega t + 40°)$ A
(c) $v_1 = 1000 \sin(300t - 45°)$ μV, $v_2 = 40 \sin(300t + 80°)$ V
(d) $i_1 = 12 \cos(8000t + 70°)$ mA, $i_2 = 100 \cos(8000t - 80°)$ A
(e) $v = 12 \sin(500t + 20°)$ V, $i = 10 \cos(500t - 40°)$ A

Solution

(a) Since both waveforms are sines, we need only subtract the two phase angles to determine the magnitude of the phase relationship. The "first" zero of the current occurs at 40° to the right of the origin and the "first" zero of the voltage occurs at 60° to the right of the origin. The current zero therefore occurs "first" and leads. The value of the angle that it leads by is the absolute value of $-60° - (-20°)$, or 20°.

Answer i leads v by 20° or v lags i by 20°.

(b) Since both waveforms are cosines, again we need only subtract the two phase angles to determine the magnitude of the phase relationship. The "first" positive peak of the voltage occurs 80° to the left of the origin, and the first positive peak of the current occurs 40° to the left of the origin. The voltage peak therefore occurs first and leads by $(80° - 40°)$, or 40°.

Answer v leads i by 40° or i lags v by 40°.

(c) Again, both waveforms are sines. v_1 has its "first" zero occurring 45° to the right of the origin and v_2 has its "first" zero occurring 80° to the left of the origin. The phase difference has a magnitude of $80° - (-45°)$, or 125°.

Answer v_2 leads v_1 by 125° or v_1 lags v_2 by 125°.

(d) Both waveforms are cosines. i_1 has its first positive peak occurring 70° to the left of the origin. i_2 has its "first" positive peak occurring 80° to the right of the origin. The phase angle between the two has a magnitude of $70° - (-80°)$, or 150°.

Answer i_1 leads i_2 by 150° or i_2 lags i_1 by 150°.

(e) Here, one waveform is a sine and the other is a cosine. We must therefore convert one form to the other using one of the following trigonometric identities:

$$\cos\theta = \sin(\theta + 90°) \quad \text{or} \quad \sin\theta = \cos(\theta - 90°)$$

From these identities we may state the following rules:

To convert a sine wave to a cosine wave, ***subtract*** 90°.
To convert a cosine wave to a sine wave, ***add*** 90°.

Accordingly, let us convert the cosine wave to a sine wave. Using the preceding trigonometric identity:

$$i = 10\cos(500t - 40°) = 10\sin(500t - 40° + 90°)$$

or

$$i = 10\sin(500t + 50°)$$

Now that both waves are sines, we may use the same technique of the previous parts (a) through (d).

Answer i leads v by 30° or v lags i by 30°.

Notice that it was not necessary to sketch the waveforms to determine the phase relationship. You may want to sketch them on your own to clarify in your own mind what the different phase relationships represent.

■ SUMMARY

- An independent voltage (current) source is one whose output voltage (current) is independent of any other circuit parameter.
- A periodic waveform is one in which a portion of the wave, called a cycle, is repeated continuously and consecutively.
- A period is the duration of one cycle of a periodic wave.
- Frequency (f) is the reciprocal of the period in seconds and represents the number of cycles that occur in one second ($f = 1/T$).
- The unit of frequency is the hertz (Hz).
- Positive and negative amplitude represent, respectively, the maximum positive and negative values of a periodic wave.
- Radian frequency (ω), or angular velocity, is the frequency in radians per second.
- $\omega = 2\pi f$ rad/s.
- Current in a circuit is always considered ***positive*** current.
- A voltage source fixes the voltage between two points in a circuit.
- Each individual ideal conductor is considered a "point" or a "terminal."
- A current source fixes the current through a path in a circuit.
- The phase relationship between two waveforms is described using two quantities:
  1. The magnitude of the angle between two corresponding points on each waveform
  2. Which of these points occurs first (leads) or later (lags).

■ PROBLEMS

2.1 What is a periodic waveform?

2.2 Define each of the following terms as they pertain to a periodic waveform.
- **(a)** Period
- **(b)** Cycle
- **(c)** Frequency (f)
- **(d)** Radian frequency
- **(e)** Angular velocity
- **(f)** Phase angular

2.3 A sinusoidal current is described by the following equation:

$$i = 6400 \sin(4000t + 60°) \text{ mA}$$

Find each of the following quantities for this current.
- **(a)** Amplitude
- **(b)** Phase angle in degrees
- **(c)** Phase angle in radians
- **(d)** Radian frequency (rad/s)
- **(e)** Frequency (Hz)
- **(f)** RMS value
- **(g)** Draw this current waveform with the horizontal axis calibrated in units of degrees.
- **(h)** Repeat part (g) with the horizontal axis calibrated in units of radians.
- **(i)** Repeat part (g) with the horizontal axis calibrated in units of time in ms.

2.4 A sinusoidal voltage is described by the following equation:

$$v = 20 \cos(4000t - 20°)\ \mu\text{V}$$

Repeat parts (a) through (i) of problem 2.3 for this voltage.
- **(j)** What is the phase relationship between this voltage and the current of problem 2.3?

2.5 **(a)** What is the period of any sinusoidal waveform in degrees?
- **(b)** What is the period of any sinusoidal waveform in radians?

2.6 Write the equation for a sinusoidal current with the following characteristics assuming it is a ***sine*** wave: $V_{rms} = 80$ mV, $f = 1800$ Hz, and α = phase angle = 50°.

2.7 What is the rule for the direction of current supplied by a dc voltage source?

2.8 What is the rule for the direction of current supplied by an ac voltage source?

2.9 What determines the ***actual magnitude and direction*** of current through a voltage source?

2.10 Is the voltage across an independent voltage source affected by the circuit to which it is connected?

2.11 Is the current through an independent current source affected by the circuit to which it is connected?

2.12 What waveforms are generally available from a commercial ac function generator?

2.13 On the same set of axes, sketch the waveforms for each of the following two electrical parameters with the horizontal axis in degrees: $i = 48 \cos(\omega t + 45°)$ A, and $v = 600 \sin(\omega t - 60°)$ mV.

2.14 Find each of the following parameters for the voltage $v = 10 \cos(\omega t + 80°)$ V.
- **(a)** Positive amplitude
- **(b)** Negative amplitude
- **(c)** Peak-to-peak value
- **(d)** RMS value
- **(e)** Phase angle

2.15 For the rectangular current shown in Fig. P2.15, find each of the following parameters:

(a) T = period
(b) f = frequency
(c) Positive amplitude
(d) Negative amplitude
(e) Peak-to-peak value

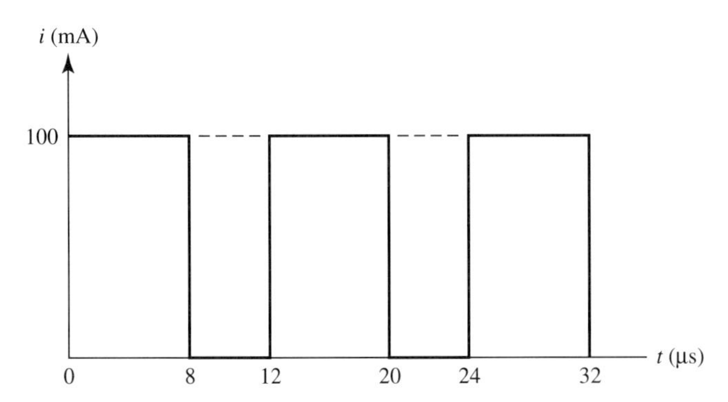

FIGURE P2.15

2.16 Repeat problem 2.15 for the triangular voltage shown in Fig. P2.16.

2.17 Find the phase relationship between each of the following sets of sinusoidal electrical parameters.

(a) $i = 350 \sin(\omega t)$ mA, $v = 24 \sin(\omega t - 65°)$ V
(b) $v = 38 \sin(200t + 80°)$ mV,
$i = 880 \cos(200t - 40°)$ A
(c) $v_1 = 20 \sin(4000t - 30°)\ \mu$V,
$v_2 = 400 \sin(4000t - 40°)$ mV
(d) $i_1 = 86 \cos(377t + 45°)$ A,
$i_2 = 12 \cos(377t + 50°)$ mA

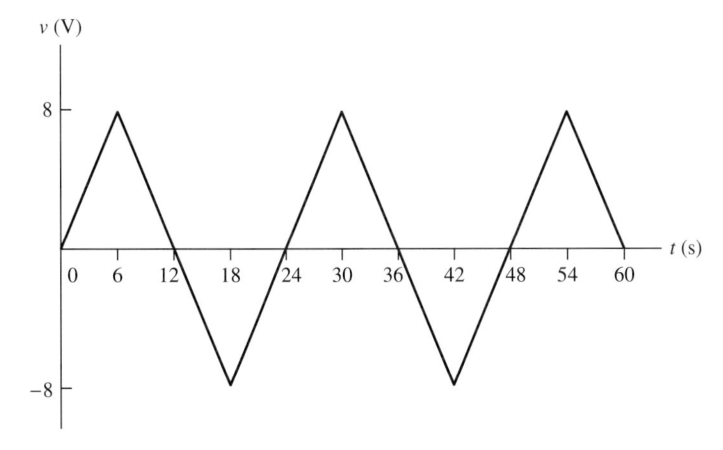

FIGURE P2.16

CHAPTER 3

MATHEMATICAL BACKGROUND

INTRODUCTION

Early studies of mathematics generally begin with a study of numbers and counting. This is a one-dimensional world. We learn to add, subtract, multiply, and divide these numbers in our early studies of arithmetic. This chapter expands those concepts to a two-dimensional world, which is required for us to understand more advanced topics in mathematics such as relationships and equivalences. In all that follows we will be discussing concepts related to a two-dimensional world known popularly as plane geometry. In this world all objects exist on a plane, which is an infinite flat surface. Straight lines, circles, and triangles are just some of the objects that exist in this world. Everything we discuss exists on this plane.

3.1 ANGLES

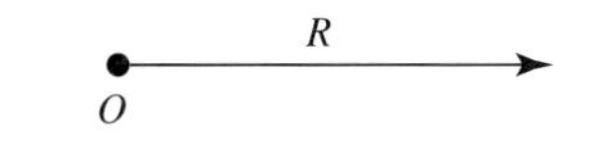

FIGURE 3.1

Let us consider a straight line of length R as shown in Fig. 3.1.

If we consider the line to be fixed at point O, then we can rotate the line counterclockwise about point O. The end point of the line (R) traces out a geometric figure called a circle. By its construction, every point on the circle is the same distance from the point O as every other point on the circle because the line is fixed in length. When the line has returned to its starting point, we say that the line has made one revolution about the point O and traced out a circle in the process. This line, of length R, is called the ***radius*** of the circle (see Fig. 3.2).

Figure 3.3 shows the line after it has rotated around the point O but not yet returned to its starting position. Point O is known as the ***vertex*** of the angle O. The starting position of the radius is called the initial radius position; the line in the other position is called the terminal radius position. In moving from its original position to its terminal position, the radius tip traces out a segment of a circle. This segment is made up of the two radii (plural of ***radius***) and the circular arc (piece of the circle). The two radii define a geometrical quantity called an angle. ***An angle is a measure of the amount of rotation.*** As the terminal radius rotates it finally returns to the original position that it started from. At this point it is said to have completed one revolution and traced out a complete circle. Rotation of more than one revolution causes the terminal radius to retrace points on the circle already passed through the first time. We measure angles by counting the number of revolutions and fractional revolutions that the terminal radius makes.

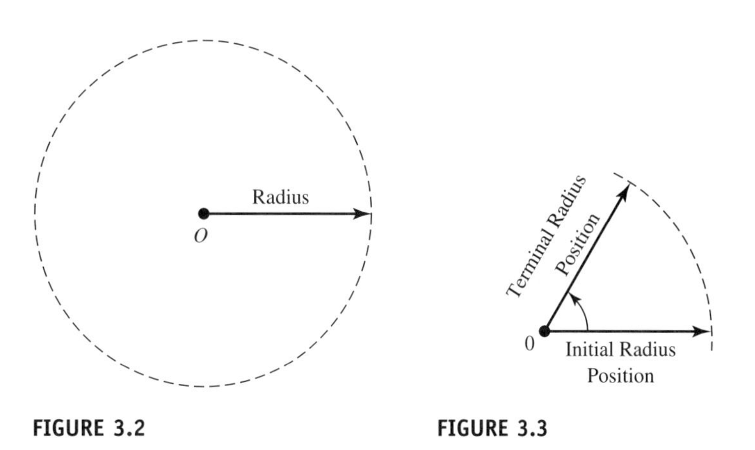

FIGURE 3.2 FIGURE 3.3

The ***principal value of an angle is defined as the fractional part of the total revolution.*** With this definition, angles become single-valued functions of the position of the terminal radius. Any angle is thus made up of the total number of revolutions added to the principal value of the angle. The position of the terminal radius is thus cyclic and repeats itself as the number of total revolutions adds up.

3.1.1 Degree Measurement of Angles

If we divide a circle into equal segments, then each of these segments will contain identical angles and identical arc lengths. In short, these segments will be identical to each other. We can cut a circle into however many pieces we wish. One standard measurement technique divides the circle into 360 equal parts. Each of these parts is said to contain a 1-degree angle (1°). We use the superscript ° as a shorthand notation for degrees. With angles defined in terms of degrees we can express any point on a circle in terms of the radius of the circle and the number of degrees between the terminal radius position and the initial radius position.

RULE FOR CONVERSION OF REVOLUTIONS INTO DEGREES

To convert revolutions into degrees multiply the number of revolutions by 360. Conversely, to convert degrees into revolutions simply divide the number of degrees by 360.

EXAMPLE 3.1

How many degrees are there in 1/4 revolution? In 1/5 revolution?

Solution

In 1/4 revolution there are $1/4 \times 360 = 90$ degrees $= 90°$.

In 1/5 revolution there are $1/5 \times 360 = 72$ degrees $= 72°$.

EXAMPLE 3.2

What fraction of a revolution is there in 22.5 degrees? In 600 degrees?

Solution

In 22.5 degrees there is $22.5/360 = 1/16$ revolution.

In 600 degrees there are $600/360 = 1.6667$ revolutions $= .6667$ revolutions.

3.1.2 Radian Measurement of Angles

In our discussion of segments it was pointed out that for a given radius R, equal segments of a circle have equal angles and equal arc lengths. A complete circle has an arc length called a circumference equal to:

$$\text{Circumference} = 2\pi R \qquad \textbf{(Eq. 3.1)}$$

Where π is a number approximately equal to 3.14159 and R is the radius of the circle. Note that when we cut a circle into equal segments we also cut the circumference into equal segments. Thus the arc length of any segment is related to the angle of the segment. There is therefore a direct relationship between the angle associated with a segment and the arc length associated with the segment.

EXAMPLE 3.3

A circle has a radius 10 cm. What is the length of its circumference? What is the length of a 1° segment of the circle? What is the length of a 30° segment of the circle?

Solution

Circumference $= 2\pi(10) = 20\pi$ cm $= 62.83$ cm in length.

A 1° segment of the circle has a 62.83/360-cm arc $= 0.1745$-cm arc length.

A 30° segment has a $30(0.1745 \text{ cm}) = 5.235$-cm arc length.

Taking a cue from Example 3.3, the arc length of a segment can be expressed as:

$$\begin{aligned}\text{Arc length} &= 2\pi(\theta°/360°)R \qquad \textbf{(Eq. 3.2)}\\ &= (2\pi/360)\theta° R\end{aligned}$$

where θ equals the angle of the segment in degrees and R is the radius of the circle. This equation demonstrates that arc length is proportional to the product of the angle and the radius of the segment. The proportionality constant equals $2\pi/360 = \pi/180$.

The angle of a segment whose arc length equals the radius of the circle has a special name. ***We call this angle one radian.*** By setting arc length equal to R in Equation 3.2 we find that this angle is $\theta = 180/\pi = 57.295°$. Thus by definition:

1 radian $= 57.295°$

If we divide the circumference, which is an arc length, by the radius we can find the number of radians in a complete circle. Thus:

Circumference/Radius $= 2\pi$ radians

This means that a circle, (one complete revolution), contains 2π radians.

EXAMPLE 3.4

How many radians are there in 1/4 revolution?

Solution

In 1/4 revolution there are $1/4 \times 2\pi$ radians $= \pi/2$ radians $= 1.57$ radians approximately.

By comparing this example to the results of Example 3.1 we see that $\pi/2$ radians equals 90 degrees. Correspondingly, π radians equals 180 degrees. This means that there is a fixed relationship between an angle measured in degrees and the same angle measured in radians. Instead of using degrees to measure angles we can express all angles in radians. If we do this, then Equation 3.2 becomes simply:

Arc length $= \theta R$ **(Eq. 3.3)**

where θ is the segment angle expressed in radians and R is the radius of the segment.

CONVERTING ANGLES EXPRESSED IN DEGREES TO ANGLES EXPRESSED IN RADIANS

1. To convert an angle expressed in degrees to an equivalent angle expressed in radians multiply the number of degrees by the factor $\pi/180$.
2. To convert an angle expressed in radians to an equivalent angle expressed in degrees multiply the number of radians by the factor $180/\pi$.

EXAMPLE 3.5

(a) Convert an angle of 60 degrees to radian measurement.

Solution 60 degrees $= 60 \times \pi/180$ radians $= 1.047$ radians.

(b) Convert an angle of one radian to degree measurement.

Solution 1 radian $= 1 \times 180/\pi$ degrees $= 57.295°$.

3.2 COORDINATE SYSTEMS

If we are to use planes in our work we must be able to measure the properties of the geometric figures that exist in planes. We have already introduced the ideas of angle and distance to our concept of planes. This next section introduces the concept of coordinate systems. A coordinate system is an aid in the measurement of angles and distances. It provides reference points for measurements and comparisons as well as making calculations easier. Many different coordinate systems exist in a two-dimensional plane; however, we will study only two of them, namely, polar coordinates and rectangular coordinates.

3.2.1 Polar Coordinates

If we choose an origin O on a plane and an initial radius position, then any point on that plane can be expressed as being a distance R from the origin at an angle θ from the initial radius position.

> The angle θ is said to be positive when the terminal radius position is measured counterclockwise from the initial radius position.
>
> The angle θ is said to be negative when the terminal radius position is measured clockwise from the initial radius position.

Figure 3.4 shows the convention used to measure angles. This convention holds whether we use degrees or radians to measure the angle in question.

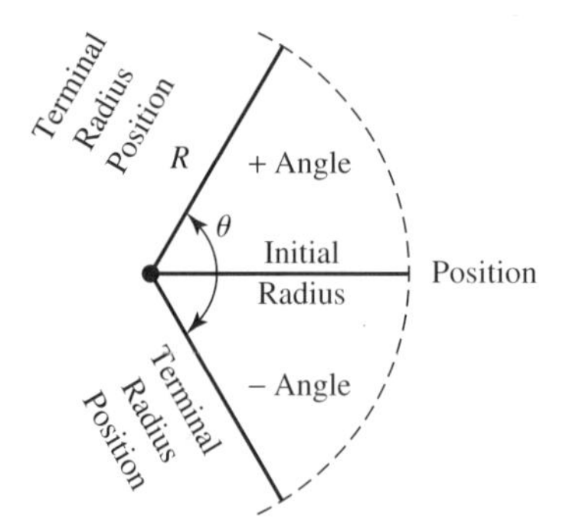

FIGURE 3.4

Note that in order to identify any point in a plane two parameters are needed. These two parameters are called coordinates. The measurement system that we have just been describing uses the coordinates R and θ to describe the location of points in the plane. These coordinates are known as ***polar coordinates.*** The coordinate R is the radius or distance to a point, and the coordinate θ is the angle that R makes with the initial radius position. We thus can find any point in the plane by drawing a straight line to the point from the origin and measuring the length of the line and the angle that the line makes with the original radius position.

3.2.2 Rectangular Coordinates

Another coordinate system commonly employed for the location of points in a plane uses two axes at 90° angles to each other as shown in Fig. 3.5. The reason for the choice of 90° lies in the observation that the shortest distance between a point and a given straight line is found by drawing the line that makes a 90° angle with the given line and measuring the length of the line drawn. The line thus drawn is said to be perpendicular to the given line (see Fig. 3.6).

By convention the horizontal axis is called the x axis and is said to be positive to the right of the intersection of axes and negative to the left of the intersection of axes. The vertical axis is called the y axis and is perpendicular to the x axis. The y axis is considered positive above the intersection of the two axes and is considered negative below. The intersection of the two axes is called the origin of the coordinate axes and is said to have the value $x = 0$ and $y = 0$. Any point in the plane can be specified by giving the value of the x and y coordinates of the point.

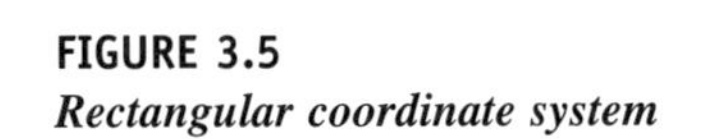

FIGURE 3.5
Rectangular coordinate system

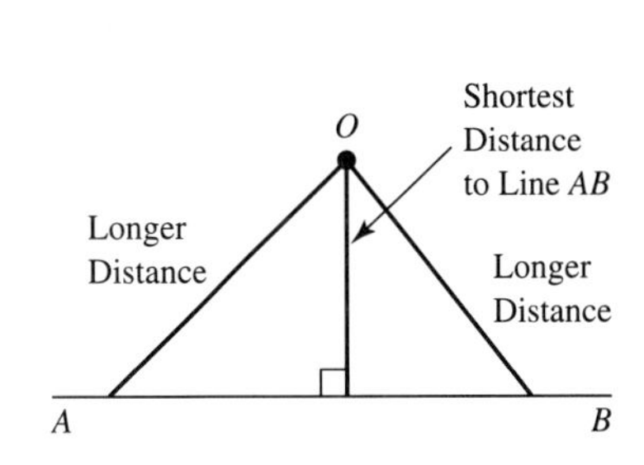

FIGURE 3.6
Shortest distance point O to line AB

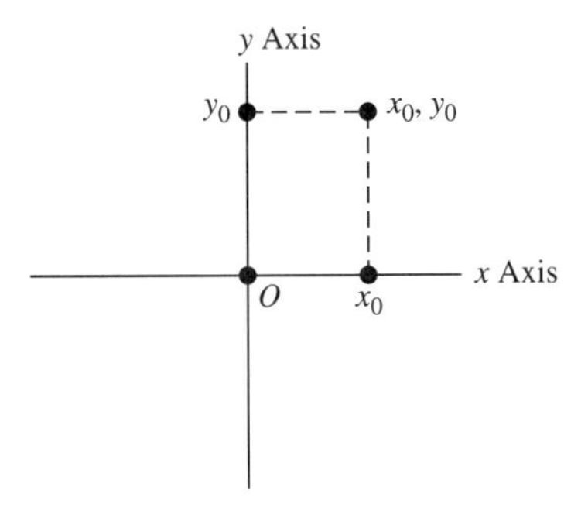

FIGURE 3.7

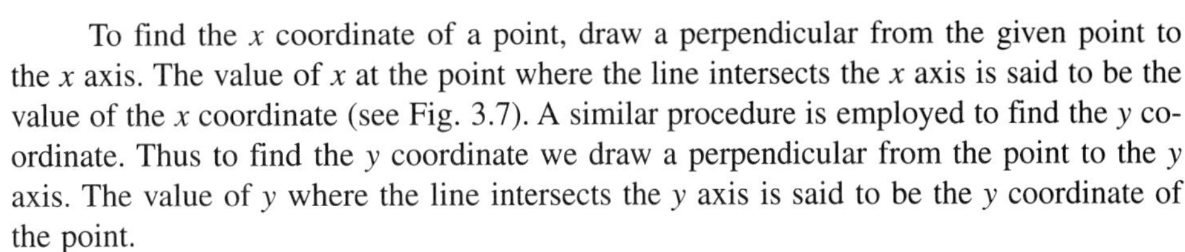

To find the x coordinate of a point, draw a perpendicular from the given point to the x axis. The value of x at the point where the line intersects the x axis is said to be the value of the x coordinate (see Fig. 3.7). A similar procedure is employed to find the y coordinate. Thus to find the y coordinate we draw a perpendicular from the point to the y axis. The value of y where the line intersects the y axis is said to be the y coordinate of the point.

Points in a rectangular coordinate system are specified by giving the value of the x coordinate and then the value of the y coordinate separated by a comma and contained in a pair of parentheses. Thus the coordinate value (5, 7) is understood to represent a point whose x value is 5 units and whose y value is 7 units. This standard notation applies whenever a rectangular coordinate system is used.

3.2.3 Conversion Between Polar and Rectangular Coordinates

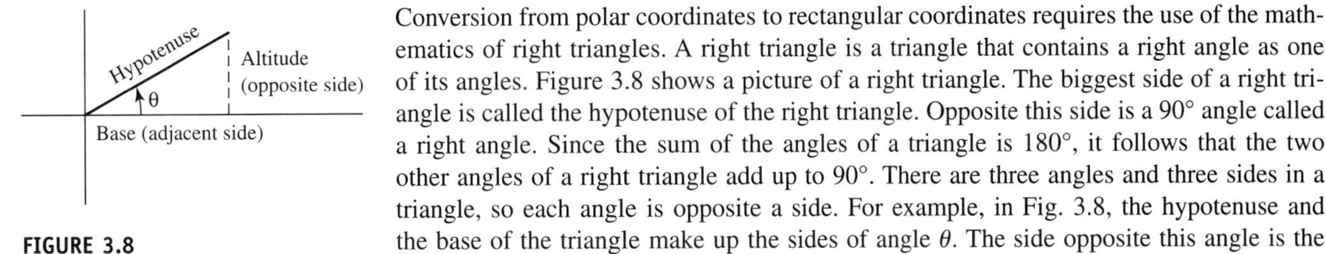

FIGURE 3.8

Conversion from polar coordinates to rectangular coordinates requires the use of the mathematics of right triangles. A right triangle is a triangle that contains a right angle as one of its angles. Figure 3.8 shows a picture of a right triangle. The biggest side of a right triangle is called the hypotenuse of the right triangle. Opposite this side is a 90° angle called a right angle. Since the sum of the angles of a triangle is 180°, it follows that the two other angles of a right triangle add up to 90°. There are three angles and three sides in a triangle, so each angle is opposite a side. For example, in Fig. 3.8, the hypotenuse and the base of the triangle make up the sides of angle θ. The side opposite this angle is the side labeled the altitude. The base of the triangle, which is one of the sides of angle θ, is said to be adjacent to the angle θ.

For every triangle there are relationships that exist between the sides of the triangle and the angles that the triangle contains. For example, we note that the largest side is always opposite the largest angle and the smallest side is opposite the smallest angle. For a right triangle the relationships that exist between the sides and the angles simplify. Pythagoras, an early Greek mathematician, discovered that for a right triangle the relationship between the two sides and the hypotenuse of a right triangle simplified to:

$$(\text{Hypotenuse})^2 = (\text{Opposite side})^2 + (\text{Adjacent side})^2 \qquad \textbf{(Eq. 3.5)}$$

This means that the hypotenuse can be expressed as the square root of the sum of the squares of the other two sides:

$$\text{Hypotenuse} = \sqrt{(\text{Opposite side})^2 + (\text{Adjacent side})^2} \qquad \textbf{(Eq. 3.6)}$$

Using this formula we can convert rectangular coordinates into the polar coordinate, R.

EXAMPLE 3.6

How far from the origin is the point $x = 6$ units, $y = 8$ units?

Solution Distance R = Hypotenuse = $\sqrt{6^2 + 8^2} = \sqrt{36 + 64} = \sqrt{100} = 10$ units

The conversion from rectangular coordinates to the polar angle θ cannot be done with a simple formula. Consider two right triangles with exactly the same angles but with one triangle twice as big as the other. Since the angles are the same, it follows that the size of any one side must be related to the size of any other side. If the hypotenuse is twice the size, then both the opposite and adjacent sides must likewise be twice the size. Note that the ratio of one side to another stays constant independent of the actual size of any side. These ratios are related to the size of the angle under consideration. Change the angle and you change the ratio. Every angle has a different ratio of sides. This means that you must know the angle to find the ratio or the ratio to find the angle.

3.2.3.1 Functions of Angles: Trigonometry

The ratio of any two sides of a triangle is called a ***trigonometric function.*** There are three sides and so there are three functions, each function being defined in terms of two of the three sides. We can also take the reciprocal of each of our defined ratios as being a different trigonometric function, but fundamentally, there are only three trigonometric functions. These are called sine, cosine, and tangent functions. Each of these functions is dependent on the angle associated with the function, and so each of these functions depends on the value of the angle of the triangle.

The trigonometric sine function is defined as the ratio of the side opposite the angle divided by the hypotenuse. We write this in mathematical shorthand as:

$$\sin(\theta) = \text{Opposite side/Hypotenuse} \qquad \textbf{(Eq. 3.7)}$$

For every value of θ there is a corresponding value of $\sin(\theta)$. We write the sine function in this way to remind us that it is a function of the angle θ. ***It must be understood that the sine function is not sin times θ, but rather that sine is a function of θ.*** Note that if we have a table of sine function values and know the angle and either the opposite side or the hypotenuse of the triangle, then this formula allows us to find the missing element in the equation. On the other hand, if we know the opposite side and the hypotenuse, then we know the sine of the angle and using a table of sine values we can find the angle.

Modern scientific calculators have removed the need for sine tables because they contain programs that enable them to compute values of sine functions given the angle expressed in either degrees or radians. We simply select the sine function and then enter the angle. The calculator then displays the value of the sine of the given angle.

EXAMPLE 3.7

Find the $\sin(40°)$. If the hypotenuse is 5 units long, how long is the opposite side?

Solution From the calculator we find that $\sin(40°) = 0.6428$, but from the formula we know that $\sin(40°) = 0.6428$ = opposite side/hypotenuse = opposite side/5. Therefore,

$$\text{Opposite side} = 0.6428 \times 5 = 3.214 \text{ units long}$$

If we form the ratio of the adjacent side divided by the hypotenuse we call this the cosine function. This is abbreviated as:

$$\cos(\theta) = \text{Ad acent side/Hypotenuse} \quad \textbf{(Eq. 3.8)}$$

Where once again the function is written to stress the dependence of the function on the angle of the triangle. *Without a cosine table or a calculator we cannot solve for an angle even if we know the ratio of the adjacent side divided by the hypotenuse.*

EXAMPLE 3.8

Find the cos(40°). If the hypotenuse is 5 units long how long is the adjacent side?

Solution From the calculator we find that cos(40°) = 0.7660 = adjacent side/hypotenuse. Therefore, 0.7660 = adjacent side/5, from which the adjacent side is:

$$0.7660 \times 5 = 3.830 \text{ units long}$$

The third function commonly employed, namely the tangent function, is written as $\tan(\theta)$ and is defined as the ratio of the opposite side divided by the adjacent side. Thus:

$$\tan(\theta) = \text{Opposite side/Ad acent side} \quad \textbf{(Eq. 3.9)}$$

We can see from the definitions of $\sin(\theta)$ and $\cos(\theta)$ that $\tan(\theta)$ can also be written as

$$\tan(\theta) = \sin(\theta)/\cos(\theta) \quad \textbf{(Eq. 3.10)}$$

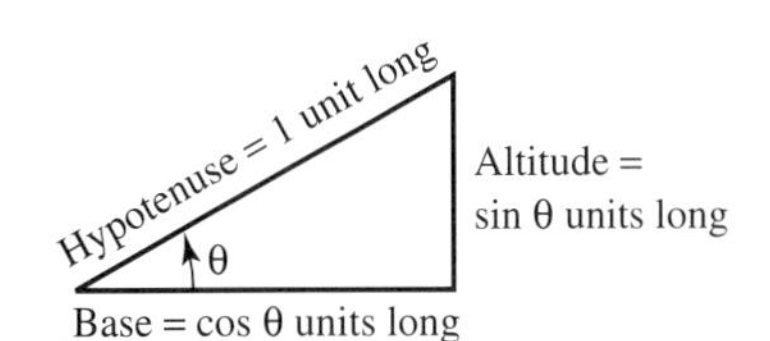

FIGURE 3.9

From this relationship it follows that the tangent value can be derived from the value of the sine and cosine functions. Consider a triangle which has a hypotenuse of 1 unit long. Then from the preceding definitions it follows that the size of the opposite side is $\sin(\theta)$ units long and that the adjacent side is $\cos(\theta)$ units long. This is shown in Fig. 3.9. If we apply the Pythagorean relationship to this triangle we find that:

$$1 = (\sin\theta)^2 + (\cos\theta)^2 \quad \textbf{(Eq. 3.11)}$$

This trigonometric identity shows that even $\cos(\theta)$ can be expressed in terms of $\sin(\theta)$. Thus while we use the three trigonometric formulas for convenience it is not necessary to use more than the sine table to derive all of the other trigonometric values.

Note that $(\sin\theta)^2 = (\sin\theta)(\sin\theta) = \sin^2(\theta)$

Thus we can rewrite Equation 3.11 to get:

$$\cos(\theta) = \sqrt{1 - \sin^2(\theta)} \quad \textbf{(Eq. 3.12)}$$

3.2.3.2 Inverse Trigonometric Functions

Consider the equation $\tan(\theta) = N$. We read this equation as: "The tangent of the angle theta is the value *N*." Suppose that we know the value of *N* but we do not know the value of θ. In this case the mathematical statement should read: "Theta is the angle whose tangent is *N*." The second mathematical statement is the inverse of the first one. In the first case we are looking for *N* given theta, while in the second case we are looking for theta given *N*. We write the expression: "Theta is the angle whose tangent is *N*," by introducing the inverse trigonometric function arctan. Thus the expression we seek is written as:

$$\theta = \arctan N \quad \textbf{(Eq. 3.13)}$$

This mathematical equation is read as "Theta equals the angle whose tangent is N." In a similar way we can write the inverse sine and cosine functions as arcsin and arccos. Thus:

If $\sin(\theta) = A$, then $\theta = \arcsin A$. If $\cos(\theta) = B$, then $\theta = \arccos B$.

Some mathematics books use the inverse sign (-1) to indicate the inverse function. This means that the following expressions are identical ways of writing the same thing:

$$\arcsin(\theta) = \sin^{-1}(\theta) \qquad \arccos(\theta) = \cos^{-1}(\theta) \qquad \arctan(\theta) = \tan^{-1}(\theta)$$

Be very careful when using the (-1) notation. It means ***inverse,*** not ***reciprocal.***

3.2.3.3 Coordinate Transformations

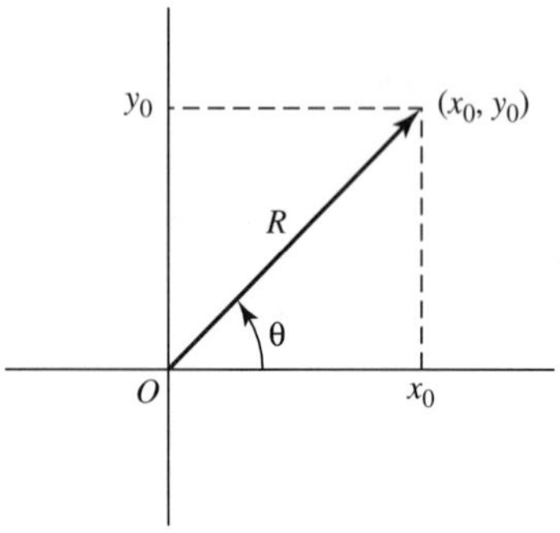

A comparison of the polar and rectangular coordinate systems, as shown in Fig. 3.10, along with the trigonometric identities and Equations 3.5 and 3.6, leads to the following rules.

y_0 (x_0, y_0) R θ O x_0

FIGURE 3.10

CONVERSION RULES

Polar to Rectangular Coordinates

$$x_0 = R\cos(\theta) \qquad \textbf{(Eq. 3.13)}$$

$$y_0 = R\sin(\theta) \qquad \textbf{(Eq. 3.14)}$$

Rectangular to Polar Coordinates

$$R = \sqrt{x_0^2 + y_0^2} \qquad \textbf{(Eq. 3.15)}$$

$$\tan(\theta) = y_0/x_0 \qquad \textbf{(Eq. 3.16)}$$

EXAMPLE 3.9

Find the polar coordinates for a point expressed as (5,12) in rectangular coordinates. Express the answer for the angle in degrees as well as radians.

Solution We can use Equation 3.15 to find R. Thus:

$$R = \sqrt{5^2 + 12^2} = \sqrt{25 + 144}$$

$$R = \sqrt{169} = 13$$

$$\tan(\theta) = 12/5 = 2.4$$

$$\theta = 67.38° = 1.176 \text{ radians}$$

EXAMPLE 3.10

Find the rectangular coordinates for a point in polar coordinates given as $R = 7$ and $\theta = 0.6$ radians.

Solution $x_0 = R\cos(0.6) = 7(0.8253) = 5.773$

$y_0 = R\sin(0.6) = 7(0.5646) = 3.952$

3.3 COMPLEX NUMBERS

Mathematicians define the square root of a number as a number that when multiplied by itself produces the original number. For example, if we consider the number 4 we can say that the number 2 is a square root of 4 because 2 times 2 equals 4. If, instead of 2, we use -2, then we find that (-2) times (-2) equals 4. Thus -2 is also a square root of 4. Every ***positive*** number including 0 has two square roots, one positive and one negative. We can express this mathematically as:

$$x = \pm\sqrt{y} \quad y \geq 0 \qquad \textbf{(Eq. 3.17)}$$

where we understand that x may be a positive or negative number but that y must be a positive number.

If we consider the case where y is a negative number we find that no *real* number exists that when multiplied by itself will be a negative number. The square root of a negative number, although not ***real,*** has some valuable properties that will prove useful in our study of electrical circuits.

Let us study some of the properties of these numbers. Consider the number -4. We may write the square root of -4 as follows:

$$\sqrt{-4} = \sqrt{4}\sqrt{-1} = \pm 2\sqrt{-1}$$

We can do this because any negative number can be considered to be the product of a positive number and (-1). Further, the square root of a product of numbers is the product of the square roots of the numbers. The square roots of negative numbers are called ***imaginary*** numbers. When working with imaginary numbers the $\sqrt{-1}$ always appears as a factor of the imaginary number. We shall adopt the shorthand notation that:

$$j = \sqrt{-1}$$

With this notation we can write:

$$\sqrt{-4} = \pm 2\sqrt{-1} = \pm 2j$$

Real numbers and imaginary numbers are not the same, but they are both numbers. We know from arithmetic that we can add, subtract, multiply, and divide real numbers. Everything that is true for real numbers is also true for imaginary numbers. This means that we can add, subtract, multiply, and divide imaginary numbers. Let us observe the results of these operations.

Consider the sum of two imaginary numbers $2j$ and $3j$. We can write this in the form:

$$x = 2j + 3j = (2 + 3)j = 5j$$

This indicates that the sum of two imaginary numbers is an imaginary number that is found by adding the real numbers (coefficients) of j together. This operation is an extension of the same operation we do in algebra.

If the two numbers are of opposite sign then we find that

$$y = 2j - 3j = (2 - 3)j = -j$$

This is again an extension of the operation of subtraction that we perform in algebra. If we consider j as an unknown then this operation is exactly the same as in algebra operation.

The product of the two imaginary numbers yields an interesting result. Thus, if we multiply the two imaginary numbers, we find:

$$Z = 2j \text{ times } 3j = 6j^2$$

But j^2 is equal to -1 by definition so we find that:

$$Z = 6j^2 = -6$$

This indicates that ***the product of two imaginary numbers is a real number.*** In a like manner we can show that ***the division of an imaginary number by an imaginary number is a real number.***

We have seen that imaginary numbers behave in much the same way as real numbers. In our study of circuit analysis it will be necessary to add real numbers to imaginary numbers. The result of this operation produces a number that is neither real nor imaginary. We call such a number a ***complex number.*** We see that a complex number has both real and imaginary parts. We can write any complex number as of the form $x + jy$, where x is the real part and y is the magnitude of the imaginary part.

3.3.1 Mathematics of Complex Numbers (Addition and Subtraction)

We can perform all of the same mathematical operations on complex numbers that we do on real and imaginary numbers. One such operation is addition. ***To add two complex numbers we simply add the real parts to get the new real part and we add the imaginary parts to get the new imaginary part.***

EXAMPLE 3.11

Find the sum of the two complex numbers $A = 3 + j4$ and $B = 5 + j2$.

Solution

$$C = A + B = (3 + 5) + j(4 + 2)$$
$$C = 8 + j6$$

We can subtract complex numbers by subtracting the real part of one from the real part of the other and the corresponding imaginary part of one from the imaginary part of the other. The answer is given as the sum of the real part and the imaginary part. Note that if we subtract the real part of B from the real part of A we must subtract the imaginary part of B from the imaginary part of A. Example 3.12 demonstrates this point.

EXAMPLE 3.12

Find the difference $A - B$ for the two complex numbers $A = 5 + j8$ and $B = 3 + j6$.

Solution

$$C = A - B = (5 - 3) + j(8 - 6)$$
$$C = 2 + j2$$

Note that we can generalize the addition and subtraction of complex numbers in the same way as we have for real numbers. Thus, we can view a subtraction of two numbers as an addition of a positive number and a negative number. We can repeat Example 3.12 and view B as a complex number with negative real and imaginary parts. Thus:

$$\text{if} \quad B = 3 + j6, \quad \text{then} \quad -B = -3 - j6$$

and $A - B$ can be viewed as $A + (-B)$. Viewed in this way addition and subtraction are seen to be the same operation.

3.3.2 Complex Numbers: Multiplication

Multiplication is performed in the same way algebraic multiplication is performed. It has the following rules:

1. A real number times a real number is a real number.
2. A real number times an imaginary number is an imaginary number.
3. An imaginary number times an imaginary number is a real number.
4. j^2 is equal to -1 because j is equal to $\sqrt{-1}$.

EXAMPLE 3.13

Find the product of the two complex numbers $A = 3 + j5$ and $B = 2 + j4$.

$$\begin{aligned} C = A \times B &= (3 + j5) \times (2 + j4) \\ &= (3 \times 2) + (3 \times j4) + (j5 \times 2) + (j5 \times j4) \\ &= 6 + j12 + j10 + j^2 \times 20 \end{aligned}$$

but $j^2 = -1$, so:

$$C = (6 - 20) + j(12 + 10) = -14 + j22$$

3.3.3 Complex Numbers: Complex Conjugates

Two numbers whose real parts are the same but whose imaginary parts are of opposite sign are called ***complex conjugates.*** The complex conjugate of any complex number can be formed by changing the sign of ***only*** the imaginary part of the number. ***If we multiply a number by its complex conjugate the answer is real and positive.*** It is equal to the sum of the real part squared plus the imaginary part squared.

EXAMPLE 3.14

Find the product of the number $(3 + j4)$ and its complex conjugate.

Solution The complex conjugate of $(3 + j4)$ is $(3 - j4)$. Therefore:

$$\begin{aligned} (3 + j4)(3 - j4) &= 3^2 - j12 - j12 + 4^2 \\ &= 3^2 + 4^2 = 25 \end{aligned}$$

3.3.4 Complex Numbers: Division Using Complex Conjugates

Division can be carried out with the aid of the complex conjugate. If we multiply the top and bottom of a complex fraction by the complex conjugate of the denominator, the denominator becomes real (see Example 3.14) and the numerator becomes the product of the original numerator and the complex conjugate of the original denominator.

EXAMPLE 3.15

Given that $A = 5 + j5$ and $B = 3 + j4$, find A/B using complex conjugates.

Solution

$$\begin{aligned} A/B &= \frac{(5 + j5)}{(3 + j4)} \\ &= \left[\frac{(5 + j5)}{(3 + j4)}\right] \cdot \left[\frac{(3 - j4)}{(3 - j4)}\right] \\ &= \frac{[(5 + j5)(3 - j4)]}{[(3 + j4)(3 - j4)]} \\ &= \frac{[(5 + j5)(3 - j4)]}{(3^2 + 4^2)} \\ &= \frac{(15 + j15 - j20 + 20)}{25} \\ &= \frac{(35 - j5)}{25} \\ &= \frac{(7 - j)}{5} = 1.4 + j0.2 \end{aligned}$$

3.4 GRAPHICAL REPRESENTATIONS OF COMPLEX NUMBERS

If we consider the set of all real numbers—positive, negative, and zero we can represent this set of numbers as a straight line. We plot the positive numbers to the right side of zero and the negative numbers to the left side. Figure 3.11 shows this representation. When numbers are plotted in this way we see that the more we move to the right the larger the number is. A representation of this form represents all real numbers. This is a one-dimensional world and does not require a two-dimensional space.

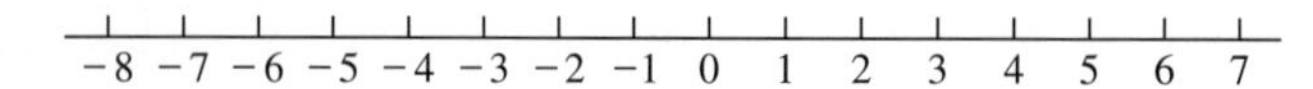

FIGURE 3.11
Real number line

Complex numbers are two-dimensional in structure. The real part of the complex number represents one dimension, and the imaginary component represents the other dimension. If we plot the real part of the complex number along the x axis and the imaginary part of the complex number along the y axis, then any complex number can be represented as a point in this plane. Because it is a plane of complex numbers we call this plane ***the complex number plane or complex plane*** for short. This plane will form a vital part of our understanding of circuit analysis. Figure 3.12 shows this complex plane with the x axis labeled real and the y axis labeled imaginary.

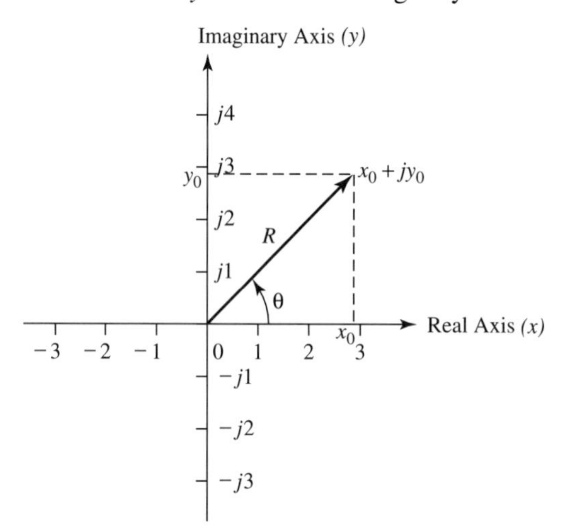

FIGURE 3.12
Complex plane system

3.4.1 Rectangular Coordinate Representation of Complex Numbers

Let us choose any point in the complex plane $(x_0 + jy_0)$. We can identify its location by observing the value of its x coordinate (real part) and its y coordinate (imaginary part). To find the x coordinate we can draw a line through the chosen point that is parallel to the y axis. This line crosses the x axis at the value of the x coordinate of the point. Similarly, a line drawn through the chosen point parallel to the x axis cuts the y axis at the value of the y coordinate of the point. The values of the coordinates allow us to reconstruct the point by reversing the process. This is shown in Fig. 3.12.

3.4.2 Polar Coordinates (Phasors)

If we choose any point in the complex plane we can identify its location by giving the value of its x and y coordinates. That is, the point is identified by its real and imaginary components as shown earlier. If we draw a line from the origin of coordinates to the chosen point, then we notice that the point can also be identified by its distance R from the origin and the angle θ that it makes with the positive x axis. A point specified in this way is expressed in ***polar coordinates.***

From the Pythagorean Theorem we see that the distance R is equal to the magnitude of the complex number. Observe also that the tangent of the angle between the line and the x axis is given by the ratio of the imaginary component of the point to the real component of the point (y_0/x_0). If the magnitude of the complex number is R, then the x axis component (the real coordinate) of the point A is given as $x_0 = R\cos\theta$ and the y axis coordinate (imaginary part) is given as $y_0 = R\sin\theta$, where R represents the distance from the coordinate origin to the point A and θ is the angle between the line drawn to the point A from the origin and the x axis.

We call the line drawn from the origin to the complex number a ***radius vector*** because it represents a complex number in terms of the number's ***magnitude*** and ***angle.*** If we know the magnitude R and angle θ of the radius vector, we can find the point in the complex plane that represents it.

3.5 THE EULER FORMULA

Leonard Euler, a famous Swiss mathematician, demonstrated a mathematical identity many years ago that is very helpful in working with complex numbers. This formula states:

$$e^{j\theta} = 1\underline{/\theta} = \cos\theta + j\sin\theta \qquad \textbf{(Eq. 3.17)}$$

provided that θ is expressed in radians. If we multiply $e^{j\theta}$ by $e^{-j\theta}$ we can find the magnitude of $e^{j\theta}$. Thus:

$$\begin{aligned}(e^{j\theta})(e^{-j\theta}) &= (\cos\theta - j\sin\theta)(\cos\theta - j\sin\theta)\\ &= \cos^2\theta + \sin^2\theta = 1\end{aligned}$$

From this it follows that a complex number of magnitude A and angle θ can be written as $Ae^{j\theta}$. This notation is usually shortened to $A\underline{/\theta}$ because only the magnitude and angle are important and it is quicker to write in this way.

EXAMPLE 3.16

Convert $3 + j4$ to polar form.

Solution

$$\begin{aligned}\text{Magnitude of the complex number} &= \sqrt{3^2 + 4^2}\\ &= \sqrt{25}\\ \text{Angle} &= \arctan 4/3 = 0.927 \text{ rad}\\ 3 + j4 \quad \text{(rectangular notation)} &= 5e^{j.927} \quad \text{(phasor notation)}\\ &= 5\underline{/0.927} \text{ rad}\end{aligned}$$

The rectangular representation of a complex number (real and imaginary parts) is very useful for addition and subtraction of complex numbers because we can add the real parts and the imaginary parts of the numbers separately. We can also multiply and divide in rectangular coordinates, but it is not as simple as performing the operation using polar coordinate notation. For multiplication we can use the polar notation more easily. If we have two complex numbers A and B and wish to find the product of A and B, then in the polar coordinate notation we see that the answer is as follows:

1. The magnitude of the resultant is the product of the magnitudes of the complex numbers A and B.
2. The angle of the resultant is the algebraic sum of the angles of A and B.

EXAMPLE 3.17

Given $A = 3\underline{/30^\circ}$ and $B = 5\underline{/45^\circ}$. Find $(A)\cdot(B)$.

Solution $(A)\cdot(B) = 3\underline{/30^\circ} \times 5\underline{/45^\circ} = 15\underline{/30^\circ + 45^\circ} = 15\underline{/75}$

The quotient of two complex numbers is found in a similar way except that the magnitude is the quotient of the two magnitudes and the angle is the difference of the two angles. The following example demonstrates this point.

EXAMPLE 3.18

Given $A = 3\angle 30°$ and $B = 5\angle 45°$. Find A/B.

$$\frac{A}{B} = A \cdot \left(\frac{1}{B}\right) = \frac{(3\angle 30°)}{(5\angle 45°)}$$

but $1/(1\angle 45°) = 1\angle 45°$;therefore, it follows that

$$A/B = (3/5)\angle 30° - 45° = .6\angle -15°$$

3.6 TIME-VARYING COMPLEX NUMBERS: PHASORS

Consider the complex number $A = Me^{j\theta}(M\angle\theta)$.If θ increases with time in a linear fashion, then we can write that $\theta = \omega t + \theta_0$, where ω is a positive number, t is time, and θ_0 represents the value of θ at $t = 0$. In the complex plane this would represent a complex number of magnitude M at an angle θ. Since θ is changing with time, the point representing the complex number moves counterclockwise around the complex plane, always at a distance M from the origin. If we know the value of ω and the value of t, we can find the position of the point at each instant of time.

The point keeps rotating on the complex plane and returns to its starting point every time that ωt is some integer multiple of 2π. ***We call a complex number that varies in time a phasor.*** If the phasor takes T seconds to go through 2π radians, then we can rewrite the equation for θ as:

$$\theta = \frac{2\pi t}{T} \quad \textbf{(Eq. 3.18)}$$

If time starts at $t = 0$, then the corresponding value for θ is θ_0 rad. After T s the value for θ is:

$$\theta = 2\pi + \theta_0 \text{ rad} \quad \textbf{(Eq. 3.19)}$$

Note that the phasor is back at its original position and in fact is back at its original position every T s. T is called the period of the phasor because it is the value that tells how long it takes to make one complete cycle of its path. The units for T are seconds per cycle (revolution). If we take the reciprocal of T we get a unit with dimensions of cycles per second. This unit, called frequency (abbreviated f), tells how many times the circular path is completed in one second. Thus:

$$f = 1/T \quad \textbf{(Eq. 3.20)}$$

EXAMPLE 3.19

A phasor passes through π radians in 5 ms. Find T, the time per cycle, and f, the frequency of the phasor.

Solution Note that π radians is the same as half a circle because a circle contains 2π radians. Thus: $T = 10$ ms per cycle. The frequency of the phasor is the reciprocal, hence $f = 1/(10 \times 10^{-3}) = 100$ cycles per second.

By definition, the unit cycles per second is called hertz (abbreviated *Hz*) after the famous scientist by the same name. Thus in Example 3.19 the answer for the frequency is:

100 cycles/second = 100 Hz

If we use the definition of frequency given earlier we may write the equation for θ as:

$$\theta = 2\pi ft + \theta_0 \qquad \textbf{(Eq. 3.21)}$$

where f is the frequency and t is the time. The quantity $2\pi f$ appears very often in circuit analysis. It represents the number of radians that a phasor passes through in one second. The similarity to the definition of frequency is so strong that $2\pi f$ is given the name ***radian frequency*** and is denoted by the symbol ω (omega). Substituting back into the equation for θ we find:

$$\theta = \omega t + \theta_0 \qquad \textbf{(Eq. 3.22)}$$

where $\omega = 2\pi f$, t is time expressed in seconds, and θ_0 is the initial value of θ. If we place this expression for θ back into the original polar coordinates complex number expression, then:

$$A = Me^{j\theta} = M\underline{/\ \omega t + \theta_0} \qquad \textbf{(Eq. 3.23)}$$

which represents a phasor that rotates in time counterclockwise in the complex plane. If Euler's Equation is applied to this phasor we find:

$$A = M\underline{/\ \omega t + \theta_0} = M\cos(\omega t + \theta_0) + jM\sin(\omega t + \theta_0) \qquad \textbf{(Eq. 3.24)}$$

The real coordinate traces out $x = M\cos(\omega t + \theta_0)$ while the imaginary coordinate traces out $y = M\sin(\omega t + \theta_0)$.

The equation $y = M\sin(\omega t + \theta_0)$ is called a sine wave. A sine wave repeats its basic waveform every 2π radians. This means that the value of x (or y) is the same every time that ωt is an integer multiple of 2π radians. A waveform that repeats every 2π radians is called a ***repeating*** or ***periodic waveform.*** As the value of t increases, the waveform goes on, forever repeating its shape every 2π radians. The shape of this waveform is shown in Fig. 3.13.

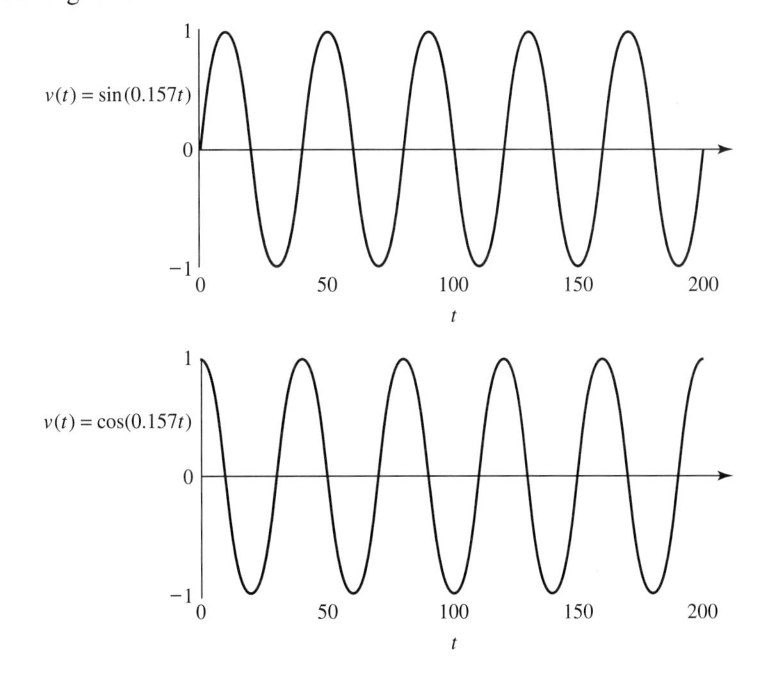

FIGURE 3.13
Sinusoidal waveforms

3.7 PHASORS AS SOURCES

Both sine wave and cosine wave sources are used in circuit analysis because of their unique properties, which make analysis simple. Sine and cosine waveforms are identical in shape but differ in angular dependence. The relationship that exists between sine and cosine curves is

$$\sin(\theta) = \cos(\theta - 90°) \quad \textbf{(Eq. 3.25)}$$

$$\sin(\theta) = \cos(\theta - \tfrac{\pi}{2}) \quad \textbf{(Eq. 3.25a)}$$

This means that the cosine waveform is always 90° ($\frac{\pi}{2}$ rad) ahead of the sine waveform.

Since sine and cosine waveforms are exactly repetitive in time, it follows that each cycle of the waveform repeats exactly the same set of values as the previous cycle. If the circuit has been operating for a long period of time, the circuit's response to the signal will likewise exactly repeat for each cycle of the waveform. ***The response waveform need not have the same shape as the driving waveform,*** but it will be repetitive. There is a correspondence between the source and the response. If the source repeats its shape, then the response waveform must likewise repeat its shape.

Linear circuit analysis deals with circuits that respond linearly to an input signal. If we double the input signal level to a linear circuit, the response will likewise double. If we place two ***different*** input signals into a linear circuit at the same time, the response to the two signals will be the same as the sum of the two responses to each signal separately applied. Only circuits that obey these properties are said to be linear. The other restriction on the circuits we evaluate is that they are time invariant.

Sine waves have many properties in addition to their being repetitive. The most important of these properties is the fact that if you apply a sine wave to a linear circuit, the response to that sine wave is a sine wave of exactly the same frequency. Circuit analysis consists of finding responses to inputs. ***If the input is a sinusoid (generic term for a sine or cosine wave) then the response is likewise a sinusoid. Sinusoids are the only repetitive waveform for which this is true.***

Since the response to the input is at the same frequency as the input, the two phasors representing the input and the output both rotate at the same speed in the complex plane. This means that they are not moving relative to each other. We define a complex number plane in which these two sinusoids rotate as a *phase space.* Since the two sinusoids rotate at the same rate, we can choose any instant of time to *freeze* the picture. It is always the same with regard to the positions of the two phasors. When we suppress the picture of the two sinusoids rotating, we are left with two phasors in phase space. The magnitudes of the two phasors are the magnitudes of the input source and the output response to the source. The angles associated with each phasor are the initial phase angles of each sinusoid.

To sum up: When the source applied to the circuit (be it a current or a voltage source) is a sinusoid, the response is also a sinusoid of exactly the same frequency. Therefore, we need only concern ourselves with the magnitude and phase angle of the input signal source. The solution to the problem will be an identical waveform with some magnitude and phase angle. The problem of circuit analysis is to solve the circuit for the values of the magnitude of the response and its phase angle.

■ SUMMARY

- When two straight lines intersect they produce a vertex, which is their point of intersection. The two lines can be made to coincide by rotating one line with respect to the other. The amount of rotation is called an angle.
- Angles can be measured in quantities called degrees or radians. In a circle there are 360 degrees or 2π radians. Degrees can be converted into radians by multiplying the number of degrees by $\pi/180$. Conversely radians can be changed to degrees by multiplying the number of radians by $180/\pi$.

- Angles are measured starting with a horizontal line called the initial side of the angle and rotating counterclockwise until the other side of the angle is reached. This side is called the terminal side. Angles measured in this way are said to be positive angles. Angles measured in the clockwise direction are called negative angles.
- A coordinate system is a method used to identify points in a two-dimensional flat surface called a plane. Two different coordinate systems are in general use. They are called polar coordinates and rectangular coordinates.
- In polar coordinates any point in a plane can be specified by giving its distance from the origin of the axes and the angle that the line drawn from the origin makes with the horizontal reference direction, called the initial side of the angle.
- In rectangular coordinates any point in a plane can be specified by giving its distances along two axes arranged to be at 90 degrees to each other. One axis is horizontal and is called the x axis. The other is called the y axis and is vertical.
- Simple conversion formulas link the two different coordinate systems. The text describes them in great detail.
- Trigonometry is a mathematical way that people have created to express the relationships that exist between the sides of a two-dimensional figure called a triangle. A triangle has three sides and three angles. Relationships between the sides are expressed in terms of ratios called sine, cosine, and tangent. The ratios depend on the angles that the triangle contains. Every angle has a unique set of ratios. These rates depend on the size of the angle and are interrelated.
- Complex numbers are numbers that contain real and imaginary parts. The real and imaginary parts of complex numbers are independent of each other. Complex numbers can be added, subtracted, multiplied, and divided. They obey the rules of complex number arithmetic.
- Complex numbers can be plotted on a rectangular coordinate plane by plotting the real parts along the x axis and the imaginary part along the y axis. This representation of the complex numbers is called a complex plane. Every point in such a plane has both real and imaginary parts.
- Any point in a complex plane can be located using polar coordinates by giving the distance to the point and the angle that the point makes with the real axis of the rectangular coordinate system of the complex plane.
- The path formed by all points that are the same distance from the origin but at different angles is called a circle. The distance from the origin to any point is called the radius of the circle. A radius that rotates counterclockwise at a constant angular rate (so many degrees per second) is called a phasor.
- The phasor (rotating radius) can be used to represent a time-varying waveform called a sine wave. The voltages and currents we use to analyze our electrical circuits vary in time in this sine wave pattern. The response of an electrical circuit to an applied sine wave is a sine wave that behaves in the same way as the original applied sine wave.

■ PROBLEMS

3.1 Find the number of degrees in the following rotations and draw a picture of the resulting rotations expressed in radians.

(a) 1/3 revolution
(b) $-1/2$ revolution
(c) 0.35 revolution
(d) 0.732 revolution
(e) -0.25 revolution

3.2 Convert the following rotations expressed in degrees to revolutions.

(a) 315 degrees
(b) 45 degrees
(c) 75 degrees
(d) -43.5 degrees
(e) -735 degrees

3.3 For the following circles find the arc length of:

(a) A circle of radius 10 cm
(b) A semicircle of radius 12 cm
(c) A circle of radius 5 cm
(d) A quarter circle of radius 15 m
(e) A 10-degree piece of a 5-cm circle

3.4 Change the following angles expressed in degrees to radians.

(a) 45 degrees
(b) 75 degrees
(c) -90 degrees
(d) 146 degrees
(e) -270 degrees

3.5 Change the following angles expressed in radians into degrees.

(a) 35 radians
(b) 0.275 radians
(c) 4.8 radians
(d) $\pi/6$ radians
(e) $\pi/15$ radians

3.6 Find the magnitude of the radius vector corresponding to the following rectangular coordinate values.

(a) $(12, -5)$ **(b)** $(9, 6)$ **(c)** $(12, 8)$
(d) $(-4, -6)$ **(e)** $(6, 6)$

3.7 Find the value of sine, cosine, and tangent functions for the following angles.

(a) $40°$ **(b)** $-70°$ **(c)** 1.1 rad **(d)** -0.8 rad
(e) $153°$

3.8 For the given values of rectangular coordinates find the corresponding polar coordinates.

(a) $(3, 6)$ **(b)** $(4, 10)$ **(c)** $(-3, 4)$ **(d)** $(5, -5)$
(e) $(-7, -8)$

3.9 For the given values of polar coordinates find the corresponding rectangular coordinates.

(a) $5\angle 40°$ **(b)** $7\angle 120°$ **(c)** $4\angle 180°$
(d) $12\angle 1.2$ rad **(e)** $6\angle 0.85$ rad

3.10 For each set of complex numbers find $A + B$ and $A - B$.

(a) $A = 3 + j5$, $B = 9 - j4$
(b) $A = -4 + j6$, $B = 7 + j7$
(c) $A = 2 + j - 7$, $B = 6 - j6$
(d) $A = 6 + j8$, $B = 8 + j3$
(e) $A = 3 - j9$, $B = -6 - j14$

3.11 For each set of complex numbers find AB and A/B.

(a) $A = 3 + j5$, $B = 9 - j4$
(b) $A = -4 + j6$, $B = 7 + j7$
(c) $A = 2 + j - 7$, $B = 6 - j6$
(d) $A = 6 + j8$, $B = 8 + j3$
(e) $A = 3 - j9$, $B = -6 - j14$

3.12 For each set of complex numbers find the complex conjugate.
(a) $A = 3 + j5$
(b) $A = -4 + j6$
(c) $A = 2 + j - 7$
(d) $A = 6 + j8$
(e) $A = 3 - j9$

3.13 Convert the following sets of complex numbers in rectangular form into polar form.
(a) $A = 3 + j5, \quad B = 9 - j4$
(b) $A = -4 + j6, \quad B = 7 + j7$
(c) $A = 2 + j - 7, \quad B = 6 - j6$
(d) $A = 6 + j8, \quad B = 8 + j3$
(e) $A = 3 - j9, \quad B = -6 - j14$

3.14 Find the product AB and the quotient A/B for the following sets of phasors.
(a) $A = 5\angle 23°, \quad B = 17\angle 47°$
(b) $A = 6.5\angle 42°, \quad B = 24\angle 23°$
(c) $A = 4.35\angle 30°, \quad B = 1.3\angle 16°$
(d) $A = 27\angle 65°, \quad B = 4.25\angle -53°$
(e) $A = 3.42\angle -28°, \quad B = 6.7\angle 33°$

3.15 Find the product AB and the quotient A/B for the following set of phasors.
(a) $A = 5\angle 0.23$ rad, $\quad B = 17\angle 0.47$ rad
(b) $A = 6.5\angle 4.2$ rad, $\quad B = 24\angle 2.6$ rad
(c) $A = 4.35\angle 3$ rad, $\quad B = 1.3\angle 1.6$ rad
(d) $A = 27\angle 6$ rad, $\quad B = 4.25\angle -5.3$ rad
(e) $A = 3.42\angle -2.8$ rad, $\quad B = 6.7\angle 3.3$ rad

3.16 For the given values of radian frequency find the frequency f in Hz and period T in seconds and milliseconds.
(a) $\omega = 200$ rad/s
(b) $\omega = 120$ rad/s
(c) $\omega = 1000$ rad/s
(d) $\omega = 2500$ rad/s
(e) $\omega = 10{,}000$ rad/s

3.17 For the given values of frequency f, find ω (rad/s) and T (period).
(a) $f = 200$ Hz
(b) $f = 1200$ Hz
(c) $f = 2200$ Hz
(d) $f = 200{,}000$ Hz
(e) $f = 20$ Hz

CHAPTER 4

STEADY STATE BEHAVIOR OF CIRCUIT ELEMENTS

INTRODUCTION

In this chapter, the steady state behavior of the three basic electrical circuit elements—the ***resistor, capacitor,*** and ***inductor***—is discussed. By behavior, we mean how each of these elements responds to voltage and current sources.

The term ***response*** in an electrical circuit refers to a ***voltage*** or a ***current*** or both. The response of a specific element therefore refers to the voltage across the element or the current through the element or both. The term ***steady state*** may require some explanation at this point. Every circuit must be either energized or de-energized by opening or closing a switch, whether the switch is mechanical or electronic. Immediately after the closing of a switch, the circuit begins to change from its "open circuit" response to its "closed circuit" response.

The time it takes for the circuit to completely change from its open circuit response to its closed circuit response is known as the ***transient*** period. In general, the transient period is a relatively small period of time during which the circuit response is constantly changing. Once the transient period is complete, the circuit response stops changing and remains fixed. This is known as the ***steady state*** period, and all circuit responses remain unchanged.

The subject of ***transient circuit analysis*** requires some additional mathematical background. This is covered to a degree in later chapters. It is important to understand that we are now discussing ***steady state*** responses only.

4.1 PHASORS

To simplify the calculation of voltages and currents in electric circuits, the concept of ***phasors*** is introduced. A phasor is a complex number used to represent a sinusoidal quantity. We will see that phasors allow us to perform mathematical calculations involving sine and cosine waves with relative ease. Such calculations are critical to the analysis and understanding of the behavior of circuit elements.

The reader should, at this point, be familiar with the mathematics of complex numbers as discussed in Chapter 3. A thorough understanding of complex numbers will greatly simplify the analyses performed in this and all of the following chapters.

4.1.1 Sine and Cosine Waves as Phasors

Consider the sinusoidal Equation 4.1:

$$v = V_p \sin(\omega t + \theta) \qquad \textbf{(Eq. 4.1)}$$

Recall that this equation is a voltage of ***peak*** amplitude V_p and phase angle θ. In phasor form, this sine wave is represented by Equation 4.2:

$$V = V_p \underline{/\theta} \qquad \textbf{(Eq. 4.2)}$$

This phasor may be plotted on the complex plane with the understanding that the real axis represents a sine wave with a 0° phase angle and the imaginary axis represents a sine wave with a 90° phase angle, or cosine wave. (Recall that a cosine wave leads a sine wave by 90°.)

EXAMPLE 4.1

Consider the sinusoidal voltage Equation 4.3:

$$v = 20 \sin(100t + 30°) \text{ V} \qquad \textbf{(Eq. 4.3)}$$

This equation may be represented by the phasor given in Equation 4.4.

$$V = 20\underline{/30°} \qquad \textbf{(Eq. 4.4)}$$

In Equation 4.4, the numbers 20 and 30° represent, respectively, the 20-V amplitude V_p and the 30° phase angle θ corresponding to the same quantities in the sine wave of Equation 4.1.

Equation 4.4 may be plotted on the complex plane as shown in Fig. 4.1. Note that the ***phase angles*** are plotted using the same rules as discussed previously in Chapter 3 for any complex number. That is, positive angles are plotted in a counterclockwise (CCW) direction from the real axis and negative angles are plotted in a clockwise (CW) direction from the real axis.

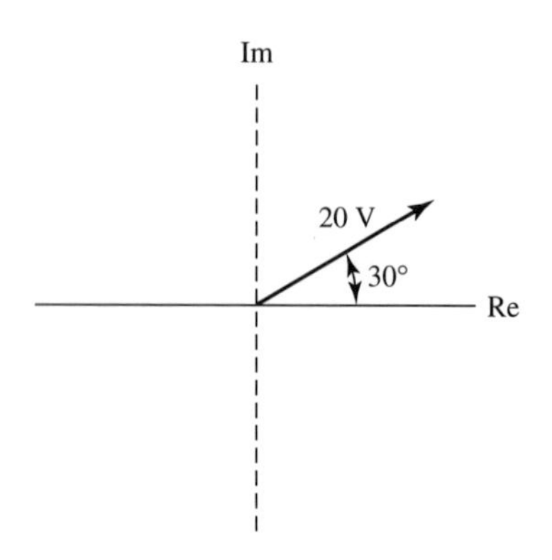

FIGURE 4.1
Phasor of Equation 4.4

Similarly, the ***relative length*** of the phasor represents its ***magnitude.*** A 10-V amplitude would be half as long as the 20-V phasor shown in Fig. 4.1.

The sinusoidal voltage of Equation 4.3 is plotted in Fig. 4.2. Note the relationship between the phase angle of Fig. 4.2 and the same phase angle as shown in Fig. 4.1.

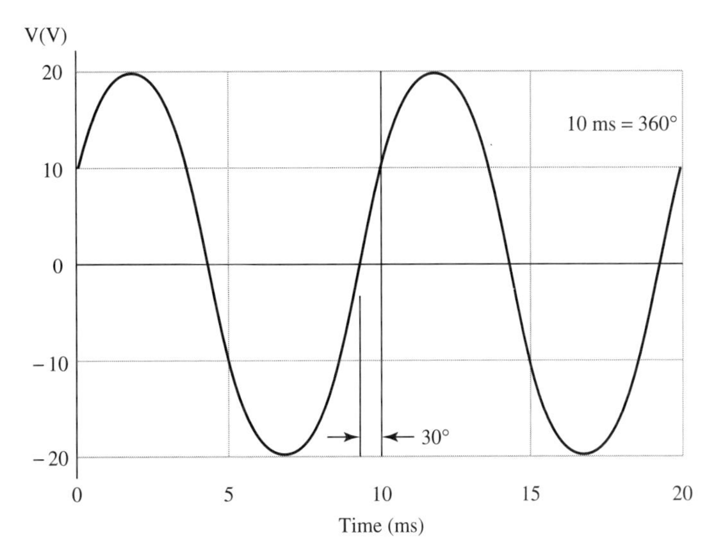

FIGURE 4.2
Voltage of Equation 4.3

EXAMPLE 4.2

Consider the sinusoidal Equation 4.5:

$$i = 40 \sin(\omega t - 45°) \text{ A} \qquad \textbf{(Eq. 4.5)}$$

This sinusoidal current has an amplitude of 40 A and a phase angle of $-45°$. It is represented by the phasor of Equation 4.6:

$$I = 40\angle{-45°} \text{ A} \qquad \textbf{(Eq. 4.6)}$$

In Equation 4.6, the 40 represents the 40-A amplitude of the sinusoidal current of Equation 4.5. The $-45°$ represents the phase angle of the same current.

This phasor is plotted on the complex plane of Fig. 4.3. Note that because this phase angle is negative, it is plotted moving CW from the real axis, opposite to the direction of the positive angle of Equation 4.4.

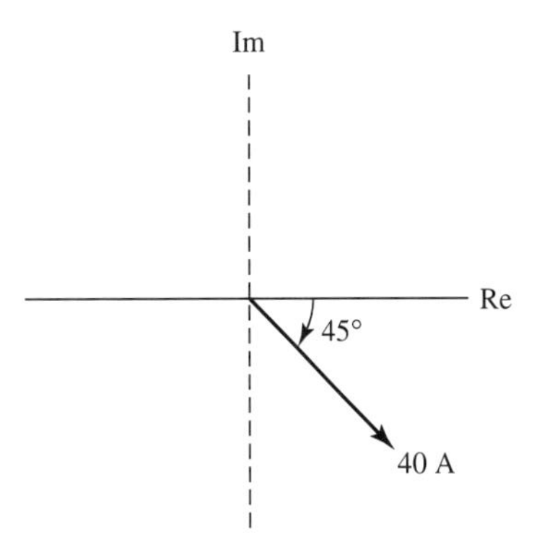

FIGURE 4.3
Phasor of Equation 4.6

The sine wave of Equation 4.5 is plotted in Fig. 4.4. Note how the negative phase angle of Equation 4.5, as shown in Fig. 4.4, compares to the same phase angle as shown in the phasor diagram of Fig. 4.3.

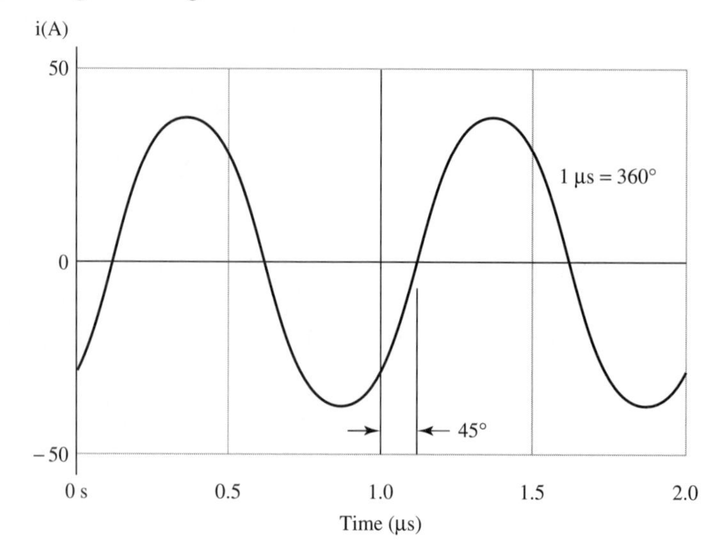

FIGURE 4.4
Current of Equation 4.5

A cosine equation may similarly be represented by a phasor and plotted on the complex plane. It must be remembered, however, that the angle of a phasor is always with respect to the positive real axis, which represents a ***sine*** wave with a 0° phase angle. When expressing a ***cosine*** wave as a phasor, therefore, it is necessary to add 90° to the phase angle of a cosine wave equation to find the angle of its phasor because a cosine wave leads its corresponding sine wave expression by 90°. Once the phasor is properly written, it may then be plotted on the complex plane in the normal manner.

EXAMPLE 4.3

Consider Equation 4.7:

$$v = 30\cos(\omega t + 60°)\text{ V} \qquad \textbf{(Eq. 4.7)}$$

This cosine voltage equation is represented by the phasor of Equation 4.8:

$$V = 30\underline{/150°}\text{ V} \qquad \textbf{(Eq. 4.8)}$$

Notice that the angle of the phasor is $60° + 90° = 150°$ because the cosine wave leads the sine wave by 90°. The 30-V amplitude of the cosine voltage in Equation 4.7 is represented by the magnitude 30 of the phasor. Accordingly, this phasor is plotted on the complex plane as shown in Fig. 4.5.

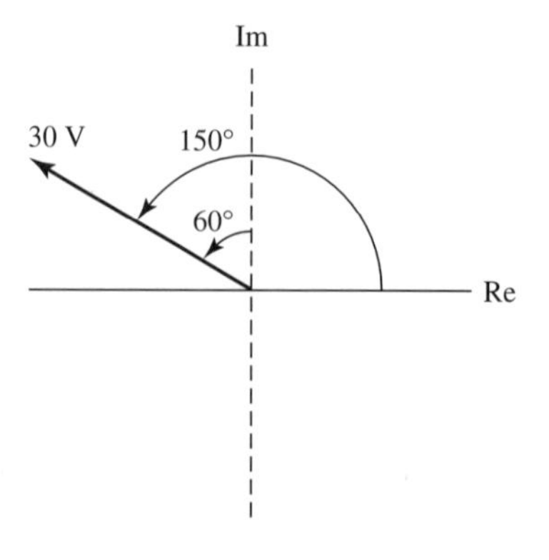

FIGURE 4.5
Phasor of Equation 4.8

Note that in ***plotting*** the angle of a cosine wave, it is equally correct to ***measure*** the angle of the cosine wave phasor directly from the Im axis (60° in this example) because the Im axis actually represents a ***cosine*** wave with a 0° phase angle. The ***phasor equation,*** however, must be ***written*** with the proper angle measured from the Re axis in order to be consistent in our calculations. For this reason, it is recommended that the phasor be written first and then plotted on the complex plane.

The cosine voltage of Equation 4.7 is plotted in Fig. 4.6. Note the positive 60° phase angle of this plot. Compare it with the 60° angle between the phasor and the Im axis of Fig. 4.5.

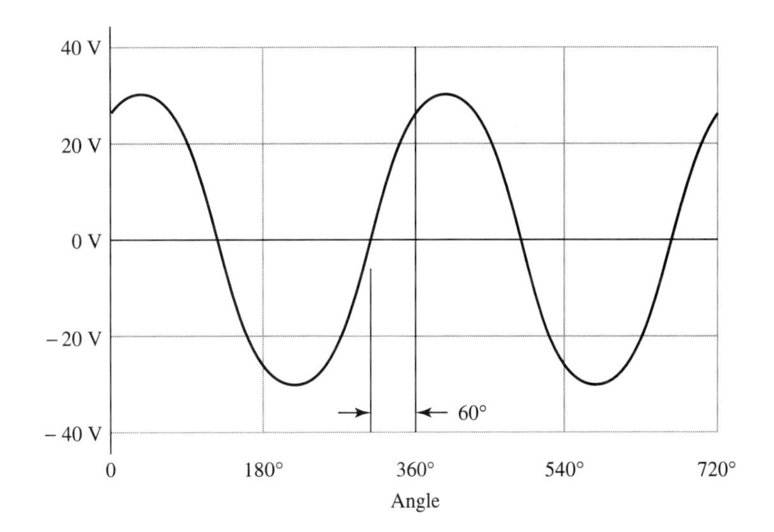

FIGURE 4.6
Voltage of Equation 4.7

4.1.2 Mathematical Computations Using Phasors

Sinusoidal circuit analysis includes the mathematical manipulation of sinusoidal equations. This manipulation is made much easier by the use of phasors to represent the sinusoidal quantities.

Now that we can represent sinusoidal waveforms by phasors, we will examine how to use these phasors in performing calculations involving sinusoids. In all cases, we will use the rules and procedures discussed earlier and remember that sinusoids manipulated using phasors are ***all*** of the ***same*** frequency.

The ***addition*** of sinusoidal equations is often required in circuit analysis. We will now look at an example of how to add two sinusoidal equations.

EXAMPLE 4.4

Add the two sinusoids of Equation 4.9 to find the voltage, v.

$$v = [12 \sin(\omega t + 45°) + 15 \cos(\omega t - 30°)] \text{ V} \qquad \textbf{(Eq. 4.9)}$$

Solution Note that Equation 4.9 is the ***sum*** of two sinusoidal waveforms. Without the use of phasors, this addition could only be performed graphically. This would obviously be a time-consuming exercise.

To add these waveforms using phasors we follow this procedure:

1. Represent each individual sinusoidal expression by a phasor.
2. Add the two phasors as you would any two complex numbers. This means converting each phasor to rectangular form and adding the real parts together and the imaginary parts together to obtain the result in rectangular form.

3. Convert the answer from rectangular form to ***polar*** form.
4. Obtain the final answer by converting the phasor in polar form back into its sinusoidal counterpart.

From step 1, the individual sinusoidal waves are represented by phasors as shown in Equation 4.10:

$$\begin{aligned} 12\sin(\omega t + 45°) &\rightarrow 12\angle 45° \\ 15\cos(\omega t - 30°) &\rightarrow 15\angle 60° \end{aligned} \qquad \textbf{(Eq. 4.10)}$$

Note that the phasor representing the 15-V cosine wave has a phase angle of +60°. This was obtained by adding 90° to the −30° angle of the cosine wave as discussed earlier.

From step 2, the two phasors are added by first breaking them into real and imaginary parts as shown:

$$\begin{aligned} 12\angle 45° &= 12\cos 45° + j12\sin 45° = 8.46 + j8.46 \\ 15\angle 60° &= 15\cos 60° + j15\sin 60° = 7.5 + j13.0 \end{aligned} \qquad \textbf{(Eq. 4.11)}$$

In accordance with complex number addition, we now add the real parts together to obtain the real part of the answer and the imaginary parts together to obtain the imaginary part of the answer. This is shown in Equation 4.12:

$$12\angle 45° + 15\angle 60° = (8.46 + 7.5) + j(8.46 + 13.0) = 15.96 + j21.46 \qquad \textbf{(Eq. 4.12)}$$

Next, as per step 3, we convert this phasor from its rectangular form to its polar form as shown in Equation 4.13:

$$15.96 + j21.46 = 26.74\angle 53.4° \qquad \textbf{(Eq. 4.13)}$$

Finally, as per step 4, we now write the phasor of Equation 4.13 as a sinusoid and arrive at the answer of Equation 4.14:

$$12\sin(\omega t + 45°) + 15\cos(\omega t - 30°) = 26.74\sin(\omega t + 53.4°) \qquad \textbf{(Eq. 4.14)}$$

Answer $v = 26.74\sin(\omega t + 53.4°)$ V

Subtraction, multiplication, and division of sinusoids are all performed in a similar manner. There are basically four steps involved in these mathematical computations:

1. Convert each of the sinusoids into a phasor.
2. Perform the indicated mathematical operation on the phasors, treating them as you would any complex number.
3. Convert the result into a phasor of polar form if it is not already so.
4. Convert the phasor result back into its sinusoidal counterpart.

Note that in all the computations performed here using phasors, we did not use the radian frequency, ω, because the frequency does not affect any of these computations. In all the examples done using phasor notation, the frequency of all sinusoids in a given problem ***must*** be the ***same.*** When dealing with linear electrical circuits in which the voltage and/or currents involved have ***different*** frequencies, it is necessary to invoke a theorem, which we will discuss in a later chapter, known as the Superposition Theorem.

Until we reach the discussion of the Superposition Theorem, we will only consider circuits in which all voltage and current sources have the ***same frequency.*** In this case, all voltages and currents in the same circuit will have ***exactly*** the same frequency as the sources. The differences will lie in the ***magnitude and phase*** of the voltages and currents associated with each component.

EXAMPLE 4.5

Perform the subtraction of two sinusoidal currents indicated in Equation 4.15.

$$i = [240\cos(\omega t - 20°) - 100\cos(\omega t + 40°)]\text{ mA} \quad \textbf{(Eq. 4.15)}$$

Step 1. Represent each cosine expression by its phasor.

$$240\cos(\omega t - 20°) \rightarrow 240\underline{/70°}$$
$$100\cos(\omega t + 40°) \rightarrow 100\underline{/130°} \quad \textbf{(Eq. 4.16)}$$

Step 2. Subtract the phasors. To do so, we will first convert them from their phasor form to rectangular form in Equation 4.17:

$$240\underline{/70°} = 240\cos 70° + j240\sin 70° = 82.08 + j225.53$$
$$\textbf{(Eq. 4.17)}$$
$$100\underline{/130°} = 100\cos 130° + j100\ \sin 130° = -64.28 + j76.60$$

Now we subtract the real parts and the imaginary parts.

$$240\underline{/70°} - 100j\underline{/130°} = [82.08 - (-64.28) + j(225.53 - 76.60)]$$
$$= 146.36 + j148.93$$

Step 3. Convert the resulting complex number into a phasor.

$$146.36 + j148.93 = 208.81\underline{/45.50°} \quad \textbf{(Eq. 4.18)}$$

Step 4. Convert the phasor into its sinusoidal form. This gives us our final answer of Equation 4.19.

$$240\cos(\omega t + 70°) - 100\cos(\omega t + 40°) = 208.81\sin(\omega t + 45.50)° \quad \textbf{(Eq. 4.19)}$$

Answer $208.81\sin(\omega t + 45.50°)$ mA.

Note that the answer is in sine wave form even though both terms in the original equation were cosine expressions. This will always be the case due to the need to express phasor angles in terms of the sine wave, or Re, axis. If we wanted to express this answer as a cosine wave, we would merely subtract 90° from the phase angle in the sine wave equation. This would result in the expression:

$$208.81\cos(\omega t - 44.50°)\text{ mA}$$

4.2 OHM'S LAW AND IMPEDANCE

The term ***impedance*** comes from the word ***impede,*** which means "to slow down the progress of something." In the case of electric circuits, impedance refers to the element, or groups of elements, in a current path that determine the current in that path. We define the impedance between two points in a circuit, a and b, by Equation 4.20:

$$Z_{ab} = V_{ab}/I_{ab} \quad \textbf{(Eq. 4.20)}$$

where I_{ab} = phasor current through the path ab (amps)
V_{ab} = phasor voltage of a with respect to b (volts)
Z_{ab} = impedance between points a and b (ohms)

Equation 4.20 is known as ***Ohm's Law.*** This is probably the most basic law of electric circuits and must be thoroughly understood. Note that when the current is expressed in ***amps*** and the voltage in ***volts,*** the unit for impedance is the ***ohm.*** The symbol Ω is used to represent ohms.

Note also that since the voltage and current phasors are complex numbers, the impedance is also a complex number. The ***magnitude*** of the impedance is the ratio of voltage magnitude to current magnitude. The ***angle*** of the impedance, θ_Z, is the angle by which the voltage phasor ***leads*** the current phasor. We may therefore express the angle θ_Z by Equation 4.21.

$$\theta_Z = \theta_V - \theta_I \qquad \textbf{(Eq. 4.21)}$$

where θ_V is the phase angle of the V_{ab} phasor
θ_I is the phase angle of the I_{ab} phasor

EXAMPLE 4.6

In the circuit shown in Fig. 4.7, find the impedance across terminals 1 and 2 if:

$$V_{12} = 100\text{ V}\underline{/60^\circ} \quad \text{and} \quad I = 5\text{ A}\underline{/45^\circ}$$

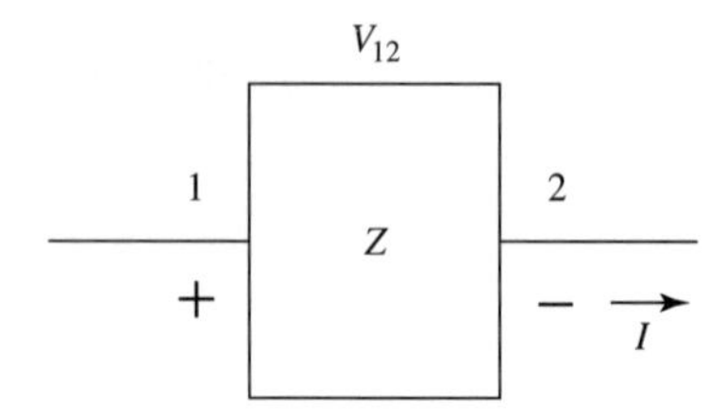

FIGURE 4.7
Impedance symbol

Solution From Equation 4.20:

$$Z_{12} = V_{12}/I_{12} = 100\underline{/60^\circ}/5\ \underline{/45^\circ} = 20\underline{/15^\circ}$$

Answer $Z_{12} = 20\ \Omega\underline{/15^\circ}$.

The answer to Example 4.5 tells us that the ***magnitude*** of the impedance, $|Z|$, is 20 Ω and the ***angle*** of the impedance, θ_Z, is 15°. This means that the voltage phasor leads the current phasor by 15°.

Note that while the voltage and current phasors represent sinusoidal functions, the impedance is a simple complex number. No function of time is implied by a complex impedance. The meaning of the magnitude and angle of a complex impedance was mentioned in the previous example and will become clearer in the following discussion.

Since there are three variables in Ohm's Law—Z, V, and I—we can solve for any one of these variables as long as we know the other two. In Example 4.6, we were given the voltage and current and asked to solve for the impedance. In Example 4.7, we are given the impedance and the current and asked to solve for the voltage.

EXAMPLE 4.7

Find the voltage, v, across terminals a and b, in the impedance shown in Fig. 4.8 if:

$$i = 40 \sin(\omega t + 70^\circ)\text{ mA} \quad \text{and} \quad Z = 20\underline{/60^\circ}\ \Omega$$

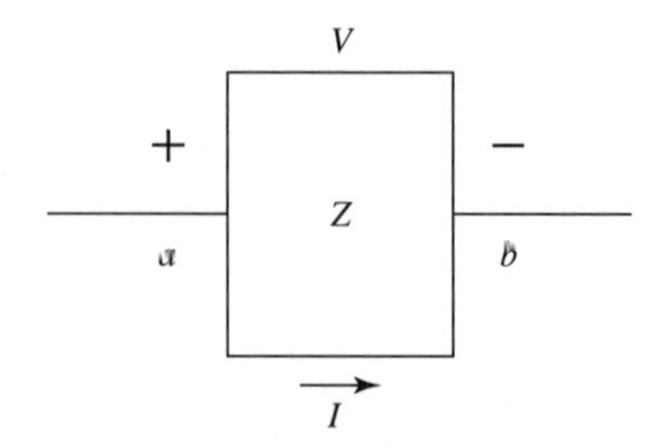

FIGURE 4.8
Using Ohm's Law

Solution Before we can apply Equation 4.20 to find the voltage, we must convert the sinusoidal form of the current to a phasor quantity. From the given current:

$$I = 40\underline{/70^\circ}$$

Using Ohm's Law, Equation 4.20:

$$V = IZ = 40 \times 10^{-3}\underline{/70^\circ} \times 20\underline{/60^\circ} = 800 \times 10^{-3}\underline{/130^\circ}$$
$$V = 800\underline{/130^\circ}$$

Answer $v = 800 \sin(\omega t + 130^\circ)$ mV.

The voltage and current phasors can be drawn on the complex plane to demonstrate the angular relationship between the two. The impedance expression informed us that the voltage would lead the current by θ_Z. From Equation 4.21:

$$\theta_V = \theta_Z + \theta_I = 60^\circ + 70^\circ = 130^\circ$$

This is confirmed by the answer, which shows that the phase angle of the voltage is 130°. Fig. 4.9 shows the voltage and current phasors drawn on the complex plane.

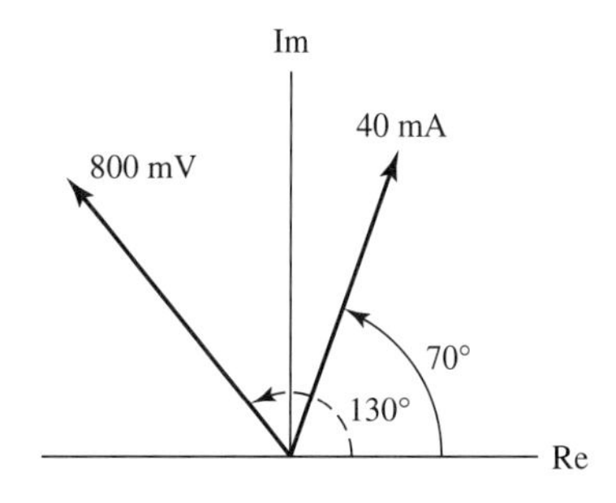

FIGURE 4.9
Voltage and current phasors for Example 4.7

4.3 PASSIVE CIRCUIT ELEMENTS

Throughout this course we will be analyzing circuits that consist of voltage and/or current sources and impedances formed by passive circuit elements. There are three basic types of passive circuit elements: ***resistors, capacitors,*** and ***inductors.*** (A passive element is one that always has a net energy draw of zero or more from a given circuit.)

Each of these elements behaves differently in a circuit. The behavior of an element is described by its response, which, as discussed in the introduction to this chapter, refers to the voltage across the element and/or the current through the element. Therefore, the response of a resistor to a given voltage or current source will differ from the response of an inductor to that same source. The response of a capacitor to the same source will differ from the responses of both the resistor and the inductor.

Since we have introduced the concept of impedance, each of these elements is discussed by specifying its impedance. Once the impedance of an element or, as we will see later, the impedance of a combination of elements is known, then the response of that element can be found by Ohm's Law.

4.3.1 The Resistor

A resistor is a circuit element that has the property of ***resistance.*** The circuit symbol for a resistor is the "sawtooth" shown in Fig. 4.10. The R next to the resistor represents the

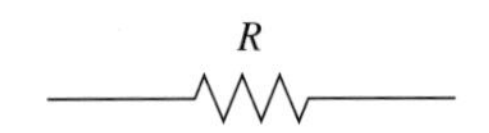

value of its ***resistance*** in ohms. The impedance of a resistor with resistance $R\ \Omega$ is given by Equation 4.22:

$$Z_R = R\angle 0^\circ\ \Omega \qquad \textbf{(Eq. 4.22)}$$

Equation 4.22 tells us two important things about the resistor:

1. The ***magnitude*** of its impedance is equal to the value of its resistance. Furthermore, this value is constant and ***independent*** of frequency.
2. The voltage across a resistor and the current through a resistor are always ***in phase*** ($\theta_R = 0^\circ$).

Resistors are utilized in many different types of circuits, which will be discussed in later chapters. For now, we will examine how to perform certain calculations involving resistors.

EXAMPLE 4.8

A resistor has a resistance value $R = 20\ \Omega$.
(a) Find its impedance.
(b) If the phasor voltage across the resistor is $V_R = 10\ \text{mV}\angle 45^\circ$, find the phasor current through the resistor, I_R.

Solution
(a) From Equation 4.22:

Answer $Z_R = 20\angle 0^\circ\ \Omega$.
(b) From Ohm's Law:

$$I = 10 \times 10^{-3}\angle 45^\circ/20\ \angle 0^\circ = 0.5 \times 10^{-3}\angle 45^\circ$$

Answer $I_R = 0.5\ \text{mA}\angle 45^\circ$.

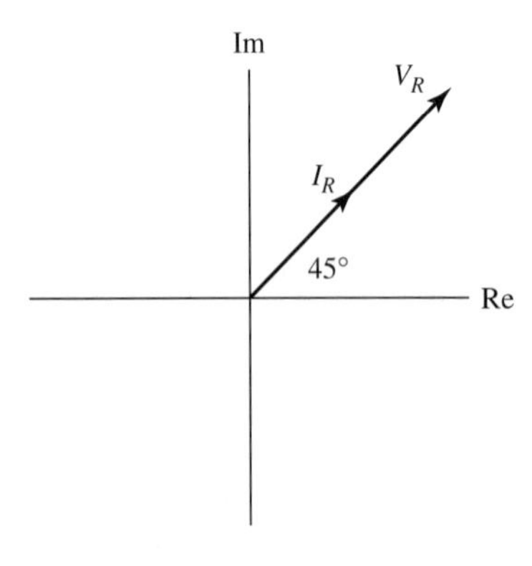

FIGURE 4.11
Phasor diagram for Example 4.8

Since the phasors V_R and I_R are in phase for a resistor, they will lie along the same straight line when plotted on the complex plane. These phasors are plotted in Fig. 4.11 for the resistor of Example 4.8.

The sinusoidal voltage and current corresponding to the phasors of Example 4.8 are:

$$V_R = 10\sin(\omega t + 45^\circ)\ \text{mV}$$
$$I_R = 0.5\sin(\omega t + 45^\circ)\ \text{mA}$$

The waveforms of these sinusoidal equations are plotted in Fig. 4.12. Note that the zeros and peaks of these two waveforms line up ***exactly*** with respect to time. This is the meaning of the statement that the voltage across a resistor and the current through a resistor are ***in phase*** (i.e., there is a 0° phase angle between these two parameters).

It is important to note that the ***length*** of a current phasor and the ***length*** of a voltage phasor bear no relation to one another because the units of current (amperes) and the units of voltage (volts) are different. These phasors are plotted on the complex plane together generally to enable us to examine their angular, or phase, relationship. As we will see later on, this provides us with valuable information regarding the characteristics of many different circuits.

As we are about to see, the resistor is the only one of the three basic passive circuit elements that exhibits the properties of constant impedance magnitude and constant impedance phase. The other two elements have impedances whose magnitudes are a function of frequency. We will see later on that this is a positive attribute because it enables us to obtain certain circuit characteristics that we could not obtain with only resistive impedances.

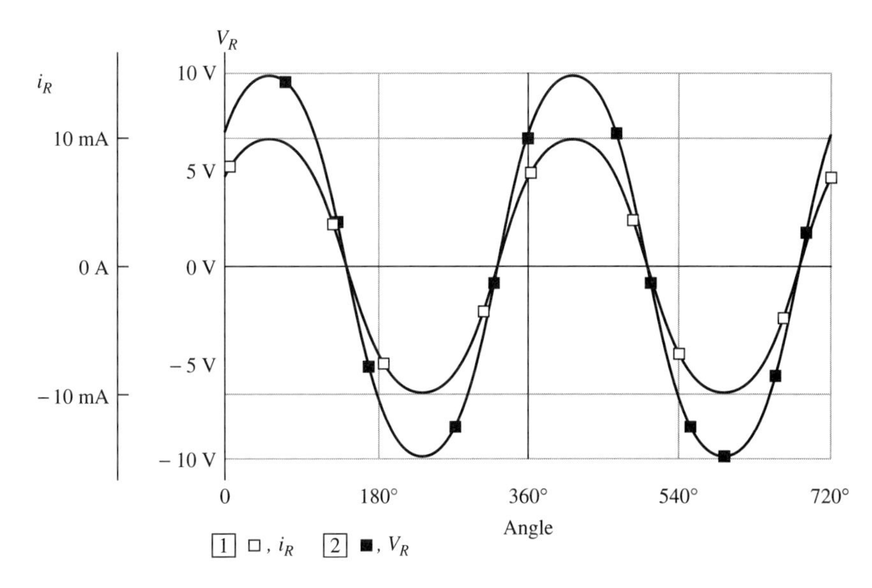

FIGURE 4.12
Voltage and current waveforms for Example 4.8

4.3.2 The Inductor

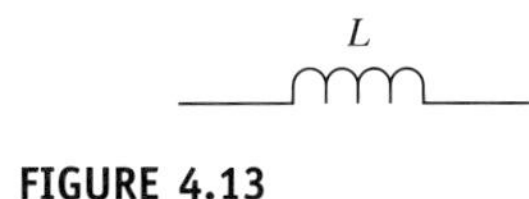

FIGURE 4.13
The inductor.

An inductor is a circuit element that has the property of ***inductance.*** The circuit symbol for the inductor is the "coil" shown in Fig. 4.13. The L next to the coil represents the value of the inductance in henries. The impedance of an inductor of inductance L henries is given by Equation 4.23.

$$Z_L = j\omega L\ \Omega = \omega L\angle 90°\ \Omega \qquad \textbf{(Eq. 4.23)}$$

Note that when the inductance L is given in henries and the radian frequency is given in radians per second, the unit for impedance is the ohm.

Equation 4.23 tells us two important things about the inductor:

1. The ***magnitude*** of the impedance is ωL.

Note that this magnitude is a function of frequency (ω). Recall that the impedance magnitude of the resistor was a constant equal to R ***independent*** of the frequency.

2. The voltage across the inductor ***leads*** the current by 90° ($\theta_L = 90°$).

Another way of saying the same thing is that the current through the inductor ***lags*** the voltage across it by 90°. The latter is the preferred way of describing the inductive current/voltage phase relationship because of its behavior from a power point of view. We will discuss this in more detail in Chapter 11.

When the phasor representing the voltage across an inductor and the phasor representing the current through an inductor are plotted on the same complex plane, they will have a 90° angle between them. In addition, the current ***lags*** the voltage. This is unlike the case of the resistor in which, as shown in Example 4.8, the current and voltage lie along the same straight line. This 90° relationship is demonstrated in the next example.

EXAMPLE 4.9

An inductance $L = 4$ mH has a current $i_L = 4\sin(2\pi ft)$ mA passing through it at a frequency of 100 kHz. Find the voltage across the inductor, v_L.

Solution From the given sinusoidal current, we can write the current phasor as:

$$I_L = 4\text{ mA}\angle 0°$$

From Ohm's Law:

$$V_L = Z_L I_L = \omega L\angle 90^\circ\, I = 2\pi f L\angle 90^\circ\, I, \qquad \text{since } \omega = 2\pi f$$

$$V_L = [2\pi(100 \times 10^3)(4 \times 10^{-3})\angle 90^\circ](4 \times 10^{-3}\angle 0^\circ) = 10.05\angle 90^\circ$$

$$v_L = 10.05\angle 90^\circ \text{ V}$$

Answer $v_L = 10.05 \sin(2\pi ft + 90^\circ)$ V.

As discussed earlier, θ_L is 90° because the voltage phasor V_L ***always*** leads the current phasor I_L by 90°. This is denoted by specifying $\theta_Z = 90^\circ$ for the inductor. These phasors are plotted in Fig. 4.14 for the inductor of Example 4.9.

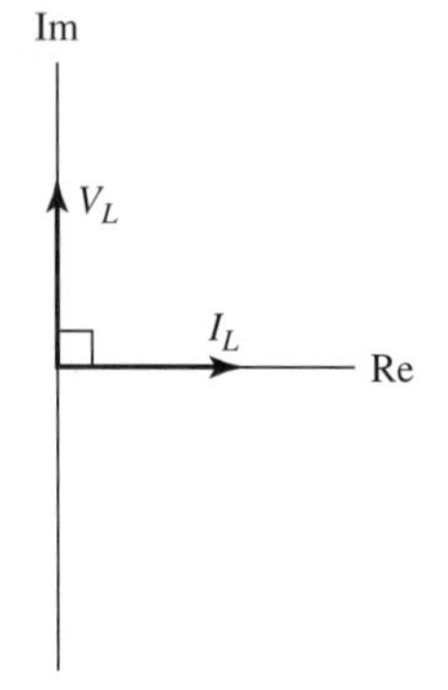

FIGURE 4.14
Phasor diagram for Example 4.9

The waveforms of the sinusoidal current and voltage of Example 4.9 are plotted in Fig. 4.15. Note how each point on the ***current*** waveform ***lags*** the corresponding point on the ***voltage*** waveform by 90°. (This may be easily seen by comparing the zeros and positive and negative peaks of the two waveforms.)

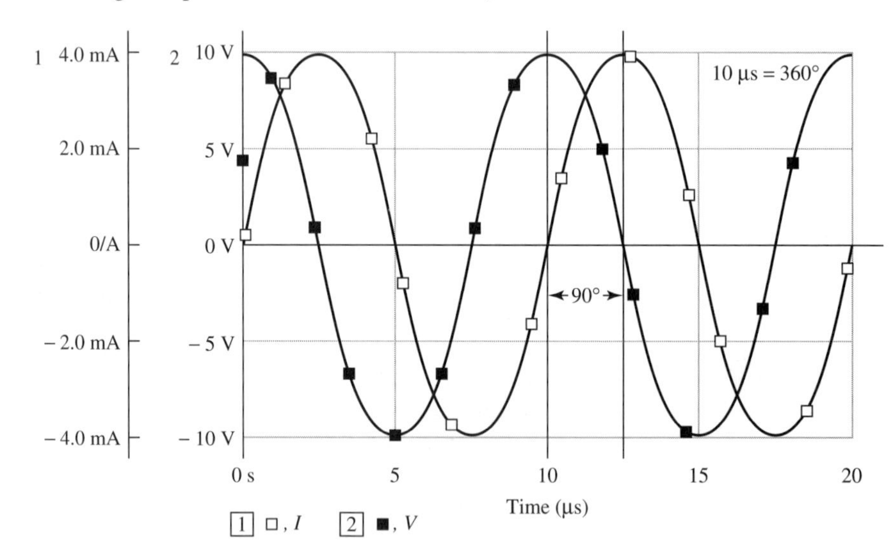

FIGURE 4.15
Voltage and current waveforms for Example 4.9

4.3.3 The Capacitor

A capacitor is a circuit element that has the property of ***capacitance.*** The circuit symbol used for the capacitor is generally two vertical lines, one straight and one curved. This symbol is shown in Fig. 4.16. The C represents the value of the capacitance in farads (F). The impedance of a capacitor of capacitance C farads is given by Equation 4.24.

$$Z_C = -j(1/\omega C)\,\Omega = (1/\omega C)\,\Omega\angle -90^\circ \qquad \textbf{(Eq. 4.24)}$$

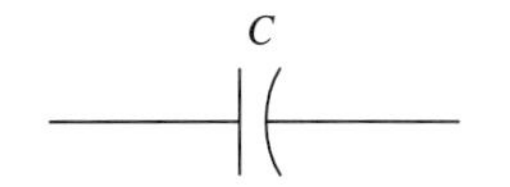

FIGURE 4.16
The capacitor

Note that when C is given in farads and ω is in rad/s, Z is in ohms.

Equation 4.24 tells us two important things about the capacitor:

1. The magnitude of the impedance is $1/\omega C$. This magnitude, like that of the inductor, also depends on frequency.
2. The voltage across the capacitor ***lags*** the current by 90° ($\theta_C = -90°$). Another way of saying the same thing is that the capacitive current ***leads*** the capacitive voltage by 90°. As was the case with the inductor, this is due to the behavior of a capacitor from a power viewpoint. This too will be discussed later in more detail.

EXAMPLE 4.10

The capacitor in Figure 4.16 has a phasor voltage across it of $V_C = 36\underline{/-28°}$ V. The phasor current through the capacitor is $I = 18\underline{/62°}$ A. The frequency of these parameters is 200 MHz.

(a) Find Z_C = impedance of the capacitor.
(b) Find C = value of the capacitance in farads.

Solution

(a) Using Ohm's Law:

$$Z_C = V_C/I_C = (36\underline{/-28°})/(18\underline{/62°}) = 2\underline{/-90°}$$

Answer $Z_C = 2\underline{/-90°}\ \Omega$.

We knew that the angle θ_C had to be $-90°$ before even going through the computations because θ_C, the angle by which the capacitive voltage leads the capacitive current, is always $-90°$.

(b) From Equation 4.24:

$$|Z_C| = 2 = 1/\omega C$$

or

$$C = 1/2\omega = 1/[(2)(2\pi f)] = 1/(2 \times 2 \times \pi \times 200 \times 10^6) = 3.98 \times 10^{-10}$$

Answer $C = 0.398$ nF

Since θ_C is always $-90°$, the voltage phasor will always lag the current phasor by 90° (or the current phasor will always lead the voltage phasor by 90°). These phasors are shown in Fig. 4.17 for the capacitor of Example 4.10.

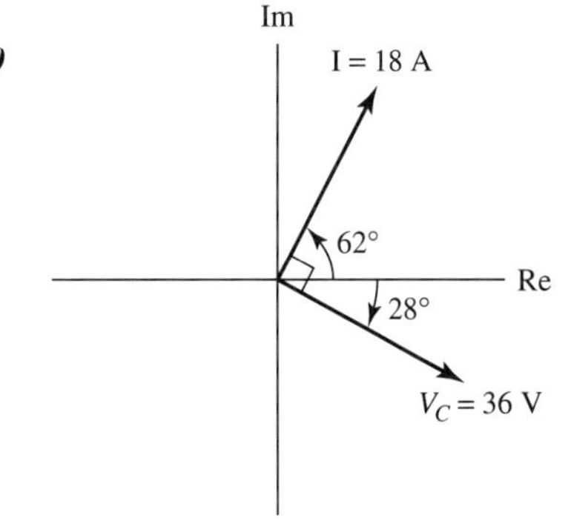

FIGURE 4.17
Phasor diagram for Example 4.10

The sinusoidal current and voltage waveforms for the capacitor of Example 4.10 are plotted in Fig. 4.18. Compare corresponding points on these two waveforms to demonstrate to yourself that the capacitive current always leads the capacitive voltage by 90°.

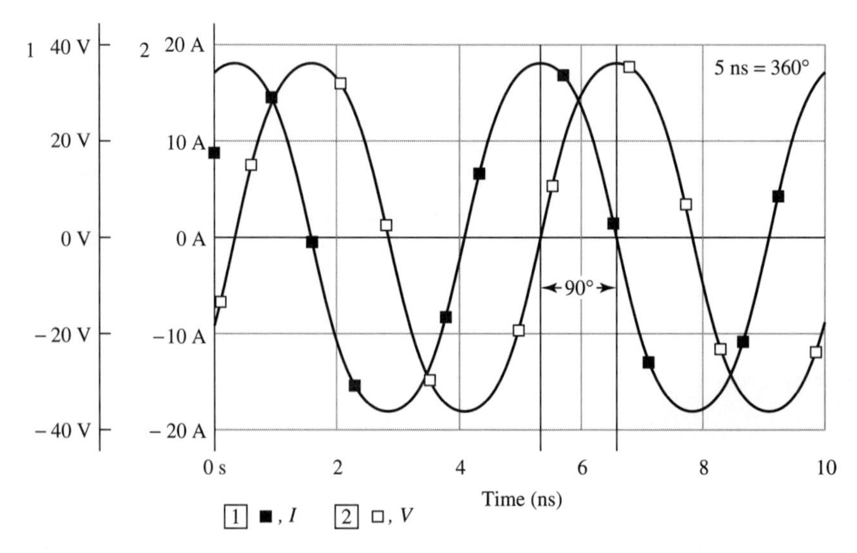

FIGURE 4.18
Voltage and current waveforms for Example 4.10

4.4 CIRCUIT ELEMENTS UNDER DC CONDITIONS

The behavior of circuit elements discussed so far has been under sinusoidal ac conditions. When the voltage and/or current source used is dc, the circuit elements behave somewhat differently. We will find, however, that having found expressions for the impedance of these elements under sinusoidal ac conditions, it will be relatively simple to understand their behavior under dc conditions.

Keep in mind during the following discussion that under ***dc*** conditions there is no meaning to the term ***phase angle.*** This term only has meaning when the voltage and current magnitudes involved are changing with time in a periodic fashion. Knowing the phase angle enables us to identify how one variable changes with respect to another variable. Since dc voltages and currents are constant with time, the term ***phase angle*** is meaningless. We will therefore be interested simply in impedance ***magnitudes.***

4.4.1 The Resistor under DC Conditions

Recall from Equation 4.22 that the ac impedance of a resistor is ***independent*** of frequency. This applies to both the magnitude and the phase angle of this impedance. This same equation is made applicable under dc conditions where the frequency is equal to 0 (except, of course, there is no phase angle involved). Therefore, the impedance of the resistor under dc conditions is simply given by Equation 4.25:

$$\text{Under DC Conditions} \quad Z_R = R\ \Omega \qquad \textbf{(Eq. 4.25)}$$

EXAMPLE 4.11

A resistor of value $R = 15\ \text{k}\Omega$ has a dc voltage of 60 V across it. Find the value of the dc current through the resistor.

Solution From Ohm's Law:

$$I = V/Z$$

From Equation 4.25:

$$Z = R = 15\ \text{k}\Omega$$

Therefore:

$I = 60/(15 \times 10^3) = 4 \times 10^{-3}$

Answer $I = 4$ mA.

Both the voltage and current are constant, unvarying values. A plot of these parameters as a function of time consists simply of straight, horizontal lines. This is shown in Fig. 4.19.

FIGURE 4.19
Current and voltage waveforms for Example 4.11

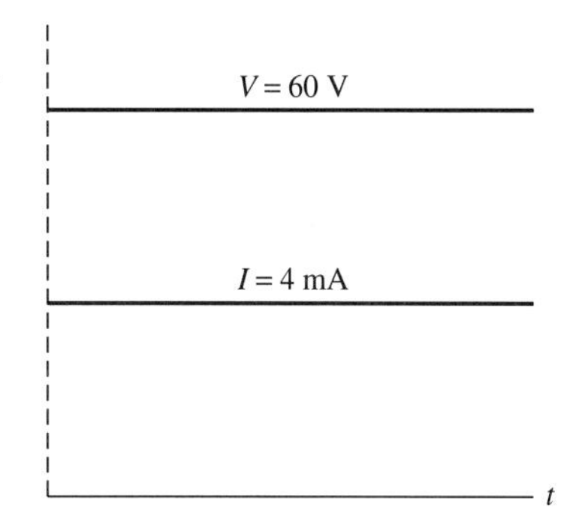

4.4.2 The Inductor under DC Conditions

Recall from Equation 4.23 that the impedance magnitude of the inductor is a direct function of frequency ($Z = \omega L$). Since under dc conditions the frequency $\omega = 0$, the impedance of an inductor in dc is given by Equation 4.26:

Under DC Conditions $Z_L = 0$ **(Eq. 4.26)**

Equation 4.26 tells us that under dc conditions, an inductor acts like a ***short circuit*** with 0 Ω impedance. This means that no matter how much ***dc current*** passes through an inductor, no voltage is developed across it. This is in accordance with Ohm's Law, which states that if $Z = 0$, $V = 0$. For an inductor in dc, therefore:

Under DC Conditions $V_L = 0$ **(Eq. 4.27)**

EXAMPLE 4.12

An inductance $L = 62$ mH is carrying a current of 100 Adc. How much voltage is developed across the inductor?

Solution From Equation 4.21, since this is a problem involving dc parameters, ***no voltage*** is developed across the inductor.

Answer $V_L = 0$ V.

4.4.3 The Capacitor under DC Conditions

Recall from Equation 4.24 that the magnitude of the impedance of a capacitor is an inverse function of frequency ($Z = 1/\omega C$). Again, under dc conditions we assume a zero frequency. The impedance of the capacitor then is given for dc conditions by Equation 4.28.

Under DC Conditions $Z_C = \infty$ **(Eq. 4.28)**

The infinite impedance of Equation 4.28 is interpreted to mean that under dc conditions, the capacitor acts as an ***open circuit.*** This means that no matter how much dc voltage is developed across a capacitor ***no current can pass through it.*** For a capacitor in dc, therefore:

Under DC Conditions $I_C = 0$ **(Eq. 4.29)**

EXAMPLE 4.13

A capacitor $C = 100\ \mu F$ has a voltage across it of $V_C = 200$ Vdc. How much current is flowing through the capacitor?

Solution From Equation 4.29, since this is a problem involving dc conditions, no current can flow through the capacitor.

Answer $I_C = 0$ A.

In summary, under dc steady state conditions:

(a) The inductor acts as a short circuit.
(b) The capacitor acts as an open circuit.
(c) The resistor acts exactly the same as under ac conditions in that the ratio of voltage to current is R under both conditions. In the case of dc, however, no phase angle is involved.

4.5 ADMITTANCE

In analyzing electric circuits, it is sometimes convenient to use a parameter known as ***admittance.*** Admittance is represented by the letter Y. It is defined as the reciprocal of impedance, as shown in Equation 4.30:

$$Y = 1/Z \qquad \textbf{(Eq. 4.30)}$$

When Z is in ohms, Y is in units of ***siemens,*** abbreviated by the letter S. Ohm's Law (Eq. 4.20) may be written in terms of admittance as shown in Equation 4.31.

$$Y = I/V \qquad \textbf{(Eq. 4.31)}$$

4.5.1 Admittance of Passive Elements

Each of the elements we have discussed (the resistor, capacitor, and inductor) can be analyzed using either its impedance or admittance. We have been using the impedance in all our examples so far. We will now look at some examples using the admittance concept.

EXAMPLE 4.13

A resistor of 20 kΩ is carrying a current of $i = 12 \sin(2000t - 20°)$ mA. Find:

(a) The admittance of the 20-kΩ resistor
(b) The voltage across the resistor

Solution From Equation 4.30:

(a) $Y = 1/Z = 1/20\text{ K} = 0.05 \times 10^{-3}$

Answer $Y = 0.05$ mS. From Equation 4.31:

(b) $V = I/Y = \dfrac{12 \times 10^{-3}\angle -20°}{0.05 \times 10^{-3}} = 240\angle -20°$

Answer $v = 240 \sin(2000t - 20°)$ V.

EXAMPLE 4.14

A capacitor of 0.3 μF is carrying a current of $i = 180\cos(200t + 50°)$ mA. Find:
(a) The admittance of the 0.3-μF capacitor
(b) The voltage across the capacitor

Solution

(a) $Z = (1/\omega C)\underline{/-90°} = \dfrac{1}{200 \times 0.3 \times 10^{-6}\underline{/-90°}} = 1.67 \times 10^4$

$$Y = 1/Z = \frac{1}{1.67 \times 10^4\underline{/-90°}} = 6 \times 10^{-5}\underline{/90°}$$

Answer $Y = 6 \times 10^{-5}\text{ S}\underline{/90°}$ or $Y = 60\ \mu\text{S}\underline{/90°}$.
(b) From Equation 4.31:

$$V = I/Y = \frac{180 \times 10^{-3}\underline{/140°}}{60 \times 10^{-6}\underline{/90°}} = 3 \times 10^3\underline{/50°}$$

Answer $V = 3\sin(200t + 50°)$ kV.

EXAMPLE 4.15

A 6-mH inductor has a voltage across it of $v = 10\sin(1000t - 60°)$ V. Find:
(a) The admittance of the inductor
(b) The current through the inductor

Solution
(a) From Equation 4.30:

$$Y = \frac{1}{1000 \times 6 \times 10^{-3}\underline{/90°}} = 0.167\underline{/-90°}$$

Answer $Y = 167\text{ mS}\underline{/-90°}$.
(b) From Equation 4.31:

$$I = YV = (167 \times 10^{-3}\underline{/-90°}) \times 10\underline{/-60°} = 1.67\underline{/-150°}$$

Answer $i = 1.67\sin(1000t - 150°)$ A.

4.5.2 Impedance and Admittance of Element Combinations

In the next chapter, we will see how the basic elements we have been discussing, the resistor, capacitor and inductor, can be combined in various ways. The resistor, for example, has a ***real*** impedance with no imaginary part. The inductor and capacitor each have an ***imaginary*** impedance only, with no real part. When connecting these elements in a circuit, we wind up with impedances having ***both*** real and imaginary parts. This results in complex impedances of the type we discussed in Section 4.2. We will now review how to handle complex impedances in general.

EXAMPLE 4.16

An impedance of $Z = 40\ \Omega\underline{/30°}$ has a current through it of $i = 60\sin(400t - 20°)$ mA. Find:
(a) Y = admittance of this impedance
(b) v = sinusoidal voltage across this impedance

Solution
(a) From Equation 4.30:

$$Y = \frac{1}{40\angle 30^\circ} = 0.025\angle -30^\circ$$

Answer $Y = 25 \text{ mS}\angle -30^\circ$.
(b) From Equation 4.31:

$$V = I/Y = \frac{60 \times 10^{-3}\angle -20^\circ}{25 \times 10^{-3}\angle -30^\circ} = 2.4\angle 10^\circ$$

Answer $v = 2.4 \sin(400t + 10^\circ)$ V.

EXAMPLE 4.17

An admittance of $Y = 32 \text{ mS}\angle -50^\circ$ carries a current of $i = 8 \cos(240t + 30^\circ)$ A.
(a) Find Z, the impedance of this admittance.
(b) Find v, the sinusoidal voltage across this admittance.

Solution
(a) From Equation 4.30:

$$Z = 1/Y = \frac{1}{32 \times 10^{-3}\angle -50^\circ} = 31.25\angle 50^\circ$$

Answer $Z = 31.25\ \Omega\angle 50^\circ$.
(b) From Equation 4.20:

$$V = IZ = 8\angle 120^\circ \times 31.25\angle 50^\circ = 250\angle 170^\circ$$

Answer $v = 250 \sin(240t + 170^\circ)$ V,

or

$$v = 250 \cos(240t + 80^\circ) \text{ V}.$$

EXAMPLE 4.18

An impedance carries a current of $i = 48 \cos(600t - 60^\circ)$ mA and has a voltage across it given by the equation $v = 24 \cos(600t + 15^\circ)$ V. Find:
(a) Z, the impedance
(b) Y, the admittance

Solution
(a) From Ohm's Law:

$$Z = V/I$$

where $V = 240 \text{ V}\angle 105^\circ$ and $I = 48 \times 10^{-3} \text{ A}\angle 30^\circ$.

$$Z = \frac{240\angle 105^\circ}{48 \times 10^{-3}\angle 30^\circ} = 5 \times 10^3\angle 75^\circ$$

Answer $Z = 5\angle 75^\circ$ kΩ.
(b) From the definition of admittance:

$$Y = 1/Z = \frac{1}{5 \times 10^3\angle 75^\circ} = 0.2 \times 10^3\angle -75^\circ$$

Answer $Y = 200\angle -75^\circ$ S.

4.6 CONDUCTANCE AND SUSCEPTANCE

4.6.1 Calculation of Conductance and Susceptance

As we have seen, impedances and admittances are generally complex numbers with both real and imaginary parts. Each part of these complex numbers has a special meaning, and we therefore assign a special name to each. The ***real*** part of an impedance is called the ***resistance*** and is assigned the letter R. The ***imaginary*** part of an impedance is called the ***reactance*** and is assigned the letter X. Using this notation, any complex impedance can be represented by Equation 4.32.

$$Z = R \pm jX \qquad \textbf{(Eq. 4.32)}$$

Note that for a ***pure resistance,*** $X = 0$ (see Eq. 4.22). For a ***pure inductance*** and a ***pure capacitance,*** $R = 0$ (see Eq. 4.23 and 4.24). The units for resistance and reactance are ***both*** ohms.

Similarly, the ***real*** part of an admittance is called the ***conductance*** and is assigned the letter G. The ***imaginary*** part of an admittance is called the ***susceptance*** and is assigned the letter B. Using this notation, any admittance can therefore be represented by Equation 4.33:

$$Y = G \pm jB \qquad \textbf{(Eq. 4.33)}$$

The units for conductance and susceptance are ***both*** siemens.

EXAMPLE 4.19

An impedance is given by $Z = 250\ \Omega\angle 30^\circ$. Find:
(a) R, the resistance part of this impedance
(b) X, the reactance part of this impedance
(c) G, the conductance
(d) B, the susceptance

Solution To find the resistance and reactance, we must express the impedance in rectangular form and then use Equation 4.32.

$$Z = (250 \cos 30^\circ + j250 \sin 30^\circ)\ \Omega = (216.5 + j125)\ \Omega$$

Answer
(a) $R = 216.5\ \Omega$
(b) $B = 125\ \Omega$

To find the conductance and susceptance, we must first find the admittance and express it in rectangular form. Then we may use Equation 4.33:

$$Y = 1/Z = \frac{1}{250\angle 30^\circ} = 0.004\angle -30^\circ = 4\ \text{mS}\angle -30^\circ$$

$$Y = [4 \times 10^{-3} \cos(-30^\circ) + 4 \times 10^{-3} \sin(-30^\circ)]\ \text{S}$$
$$= (3.46 \times 10^{-3} - j2 \times 10^{-3})\ \text{S}$$

Answer
(c) $G = 3.46$ mS
(d) $B = 2.0$ mS

Note that R, G, B, and X are magnitudes and therefore ***always*** positive numbers. The minus sign in front of the B of the last example means, as in any complex number, that if the susceptance were to be plotted on the complex plane, it would lie along the negative imaginary axis.

4.6.2 Plotting Z and Y

In plotting ***impedances*** on the complex plane it is instructive to remember that they must ***always*** lie in the right half of the plane because the real part of complex impedances represents resistance and there is no such thing, in electric circuits of common experience, as negative resistance. This is equally true for admittances. Since there are only real, positive conductances, admittances always appear in the right half of the complex plane.

It should also be pointed out that impedances and admittances, unlike phasors, ***do not*** represent sinusoidal functions of time. Since phasors and impedances are both complex numbers, beginning students often make the mistake of assuming that impedances represent time functions. Remember that impedances and admittances represent ***circuit elements,*** either single elements, as discussed in this chapter, or combinations of elements, as will be discussed in the remaining chapters.

EXAMPLE 4.20

Given the impedance $Z = 50\ \Omega\underline{/45^\circ}$, plot this impedance on the complex plane, showing both its polar and rectangular form. Find the resistance, R, and the reactance, X, as well.

Solution The impedance may be written in rectangular form as:

$$Z = (50 \cos 45^\circ + j50 \sin 45^\circ)\ \Omega = (35.35 + j35.35)\ \Omega$$

From this expression for the impedance

$$R = 35.34\ \Omega \quad \text{and} \quad X = 35.35\ \Omega$$

This impedance is plotted in Figure 4.20. Its resistance and reactance are shown as well.

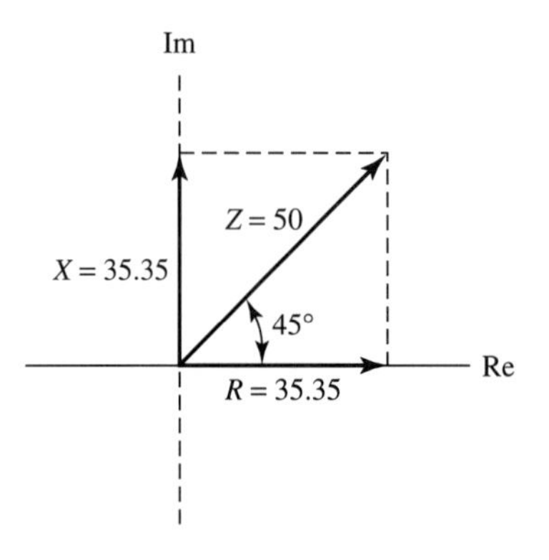

FIGURE 4.20
Phasor diagram for Example 4.19

■ SUMMARY

- A phasor is a complex number used to represent a sinusoidal function.
- The magnitude of the phasor is the amplitude of the sinusoid, and the angle of the phasor is the angle of the ***sine*** wave form of the function.
- Phasors are used to simplify the mathematical computations involving sinusoids.
- The impedance in ohms (Ω) between two points in a circuit is denoted by $Z = Z\underline{/\theta_Z}$.
- The impedance is defined as $Z = V/I$, where V is the phasor voltage across the two points and I is the phasor current through the two points (Ohm's Law).
- The impedance of a resistance value R is $Z_R = R\ \Omega\underline{/0^\circ}$.
- The impedance of an inductance value L is $Z_L = \omega L\ \Omega\underline{/90^\circ}$.
- The impedance of a capacitance value C is $Z_C = (1/\omega C)\ \Omega\underline{/-90^\circ}$.
- Combinations of the three basic circuit elements (R, L, and C) result in complex impedances with both real and imaginary parts.
- The real part of an impedance is called the resistance, R.

- The imaginary part of an impedance is called the reactance, X.
- The admittance between two points in an electric circuit is defined as $Y = 1/Z$.
- The real part of an admittance is called the conductance, G.
- The imaginary part of an admittance is called the susceptance, B.
- Impedance and admittance plots always appear in the right half of the complex plane.

PROBLEMS

4.1 Write each of the following sinusoidal equations in phasor form.

(a) $v = 200 \sin\omega t$ mV
(b) $i = 23.5 \sin(\omega t + 30°)$ A
(c) $v = 15 \sin(\omega t - 60°)$ V
(d) $i = 2 \times 10^{-6} \cos(300t + 45°)$ A
(e) $v = 4 \cos(100t - 85°)$ μA
(f) $v = 34 \cos(5000t + 150°)$ mV
(g) $i = 66 \sin(10t - 120°)$ μA

4.2 Plot each of the phasors of Problem 4.1 on the complex plane.

4.3 Perform each of the following mathematical computations using phasors and express the answer as a sinusoidal equation.

(a) $v = 6 \cos(1000t + 45°)$ V $+ 5 \sin(1000t + 60°)$ V
(b) $i = 100 \sin(377t - 30°)$ A $- 20 \cos(377t - 60°)$ A
(c) $v = 20 \sin(200t + 50°)$ mV $- 10 \sin(200t + 20°)$ mV
(d) $v = 30 \sin 400t$ V $+ 40 \cos 400t$ V
(e) $i = 4 \sin 1000t$ μA $- 2 \cos 1000t$ μA

4.4 Write a sentence that expresses Ohm's Law in English.

4.5 Given an impedance $Z_{ab} = 100\angle 0°$ Ω:

(a) Is this impedance a resistor, capacitor, or inductor?
(b) If the phasor voltage across this impedance is $V_{ab} = 100\angle 30°$ V, find the phasor current, I_{ab}, through this impedance.
(c) Draw a diagram showing the impedance with its proper symbol, the voltage polarity, and the current direction.
(d) If the frequency of the voltage is 1000 rad/s, write the sinusoidal equations for the voltage and current, v_{ab} and i_{ab}.
(e) Draw the phasors V_{ab} and I_{ab} on the same complex plane.
(f) Sketch the corresponding sinusoidal equations on the same set of axes.

4.6 Given an impedance $Z_{12} = 12\angle 90°$ Ω:

(a) Is this impedance a resistor, capacitor, or inductor?
(b) If the phasor current through this impedance is $I_{12} = 20\angle -30°$ mA, find the phasor voltage across this impedance, V_{12}.
(c) Draw a diagram showing the impedance with its proper symbol, the voltage polarity and the current direction.
(d) If the frequency of the current is 377 Hz, write the sinusoidal equations for the voltage and current, v_{12} and i_{12}.
(e) Draw the phasors V_{12} and I_{12} on the same complex plane.
(f) Sketch the corresponding sinusoidal equations on the same set of axes.
(g) Find the value of the element. That is, if the element is a resistor, find R. If it is an inductor find L. If it is a capacitor find C.

4.7 Given an impedance $Z = 40\angle -90°$ kΩ:

(a) Is this impedance a resistor, capacitor, or inductor?
(b) If the phasor voltage across this impedance is $V_{23} = 20\angle 45°$ mV, find the phasor current I_{23}, through this impedance.
(c) Draw a diagram showing the impedance with its proper symbol, the voltage polarity, and the current direction.
(d) If the frequency of the current is 100 Hz, write the sinusoidal equations for the voltage and current, v_{23} and i_{23}.
(e) Sketch the phasors V_{23} and I_{23} on the same complex plane.
(f) Sketch the corresponding sinusoidal equations on the same set of axes.
(g) Find the value of the element.

4.8 A dc voltage $V = 20$ V is impressed across a capacitor $C = 10$ μF. Find the current I flowing through this capacitor.

4.9 A dc current $I = 10$ mA flows through an inductor $L = 100$ mH. Find the voltage V across this inductor.

4.10 A dc current $I = 160$ mA flows through a resistor $R = 10$ kΩ. Find the voltage V across this resistor.

4.11 A voltage $v = 36 \sin(2000t + 45°)$ mV is impressed across an impedance, Z. The current through this impedance is $1.8 \sin(2000t + 45°)$ mA.

(a) Find the value of the impedance Z.
(b) What type of element is this impedance?
(c) Find the value of this element.

4.12 A voltage $v = 120 \sin(1000t + 60°)$ V is impressed across an impedance, Z. The current through this impedance is $12 \cos(1000t + 60°)$ A.

(a) Find the value of the impedance Z.
(b) What type of element is this impedance?
(c) Find the value of this element.

4.13 A current $i = 100 \cos(377t - 45°)$ mA flows through an impedance, Z. The voltage across this impedance is $v = 20 \cos(377t + 45°)$ V.

(a) Find the value of this impedance Z.
(b) What type of element is this impedance?
(c) Find the value of this element.

4.14 An impedance $Z = 50\angle 45°$ kΩ.

(a) Find R = the resistance of this impedance.
(b) Find X = the reactance of this impedance.

4.15 An impedance $Z = 100\angle -45°$ Ω.

(a) Find R = the resistance of this impedance.
(b) Find X = the reactance of this impedance.

4.16 An impedance $Z = 200\angle 60°$.

(a) Find the admittance, Y, of this impedance.
(b) Find G, the conductance of this admittance.
(c) Find B, the susceptance of this admittance.

4.17 An admittance $Y = 10\angle -90°$ mS.

(a) Find G, the conductance of this admittance.
(b) Find B, the susceptance of this admittance.
(c) What type of element is this (resistor, capacitor, or inductor)?

4.18 Write Ohm's Law using the admittance Y instead of the impedance Z.

4.19 What type of element is described by each of the following statements?

(a) The voltage across the element and the current through the element are always in phase.

(b) The current through the element always lags the voltage across the element by 90°.

(c) The current through the element always leads the voltage across the element by 90°.

(d) The voltage across the element in a dc circuit is always zero.

(e) The current through the element in a dc circuit is always zero.

4.20 **(a)** For impedances that consist of a single element, what are the only three impedance angles possible?

(b) For impedances that consist of combinations of elements, in what part of the complex plane must these impedances lie?

4.21 What type of element would you choose for each of the following conditions?

(a) You want the impedance to have a high value at low frequencies and a low value at high frequencies.

(b) You want the impedance to have the same value at all frequencies.

(c) You want the impedance to have a low value at low frequencies and a high value at high frequencies.

4.22 The circuit shown in Fig. P4.22 consists of a current source and a single resistor.

(a) Find the phasor voltage across this resistor.

(b) Find the sinusoidal voltage across this resistor.

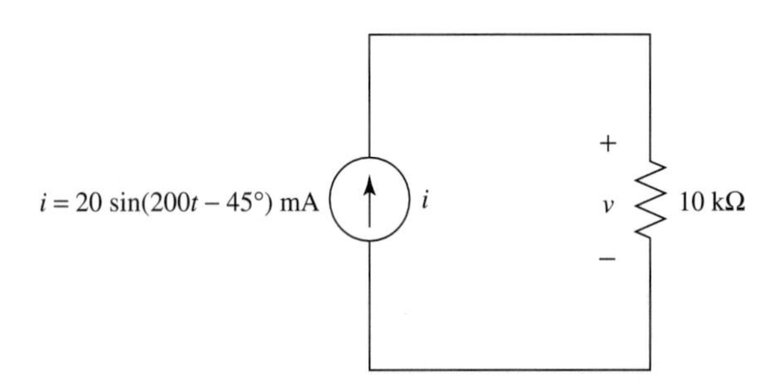

FIGURE P4.22

4.23 The circuit shown in Fig. P4.23 consists of a voltage source and a single capacitor.

(a) Find the phasor current through the capacitor.

(b) Find the sinusoidal current through this capacitor.

4.24 The circuit shown in Fig. P4.24 consists of a voltage source and a single capacitor. Find the current through this capacitor.

4.25 The circuit shown in Fig. P4.25 consists of a voltage source and a single resistor. Find the voltage across this resistor.

4.26 The circuit shown in Fig. P4.26 consists of a voltage source and a single inductor.

(a) Find the phasor current through this inductor.

(b) Find the sinusoidal current through this inductor.

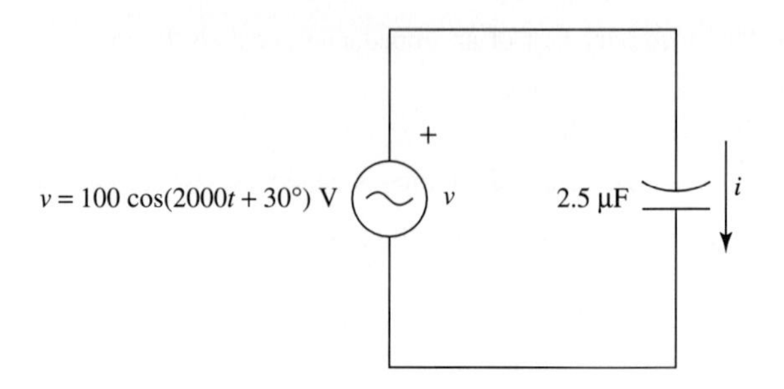

FIGURE P4.23

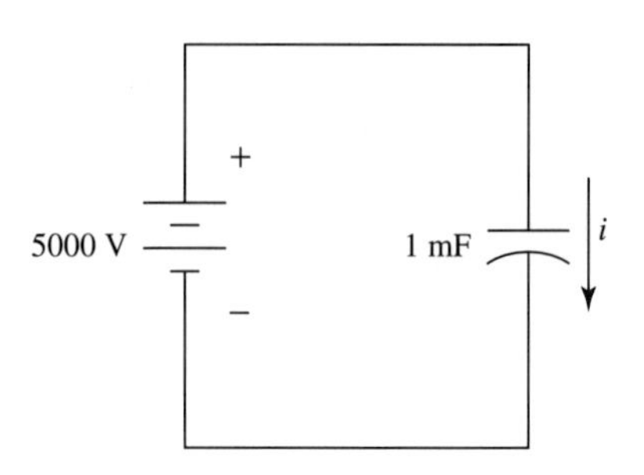

FIGURE P4.24

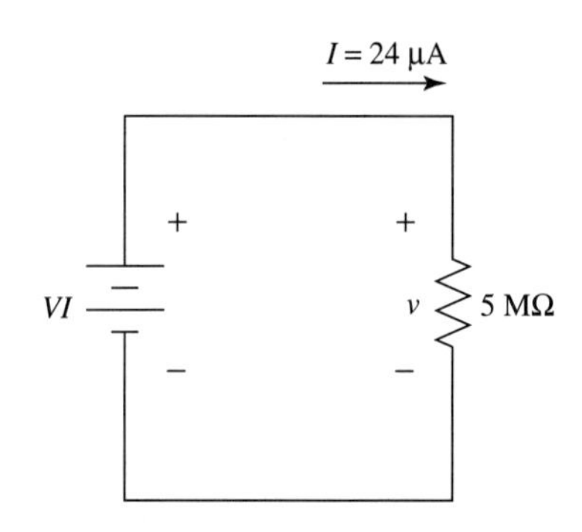

FIGURE P4.25

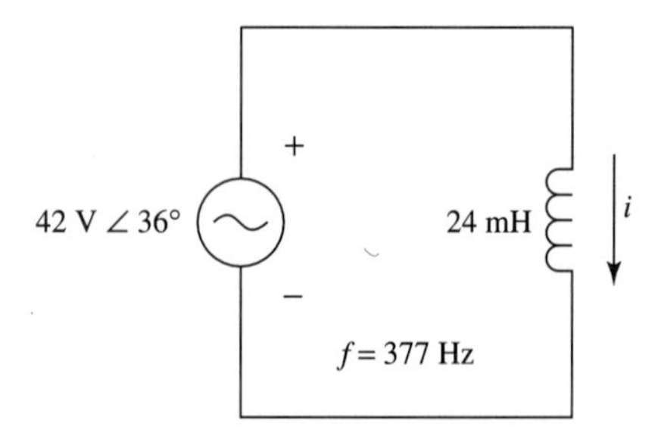

FIGURE P4.26

4.27 The circuit shown in Fig. P4.27 consists of a current source and a single inductor. Find the voltage across this inductor.

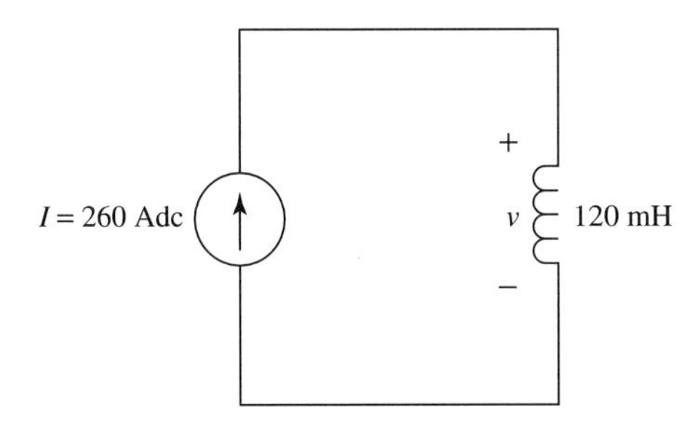

FIGURE P4.27

4.28 The circuit shown in Fig. P4.28 consists of a voltage source and a single impedance that represents some combination of the three basic elements.

(a) Find the phasor current through this impedance.

(b) Find the admittance corresponding to this impedance.

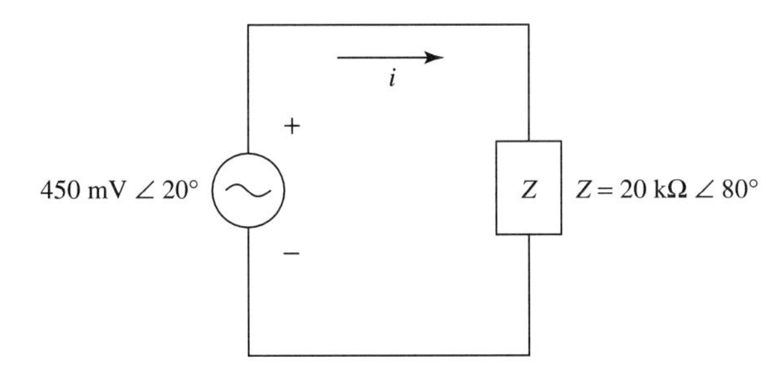

FIGURE P4.28

4.29 **(a)** Draw the impedance diagram of the circuit in problem 4.28.

(b) Draw the admittance diagram of the circuit in problem 4.28.

(c) Draw the voltage and current phasors on the same complex plane.

(d) Find the phase angle ***between*** the voltage phasor and the current phasor. Compare this value to the impedance angle. How do they compare? Is this result ***always*** true?

4.30 **(a)** Draw a graph showing the reactance of an inductor, X_L, on the vertical axis and the frequency, ω, on the horizontal axis. Use an inductor of $L = 10$ mH and let ω vary from 0 to 1000 rad/s.

(b) What type of graph did you get in part (a)?

(c) Did you expect this type of graph? Why?

4.31 **(a)** Draw a graph showing the reactance of a capacitor, X_C, on the vertical axis and the frequency, ω, on the horizontal axis. Use a capacitor $C = 10\ \mu$F and let ω vary from 0 to 1000 rad/s.

(b) What type of graph did you get in part (a)?

(c) Did you expect this type of graph? Why?

CHAPTER **5**

STEADY STATE ANALYSIS OF SERIES AND PARALLEL CIRCUITS

INTRODUCTION

In this chapter, we will examine simple series and parallel circuits. These circuits are analyzed using the response of passive circuit elements discussed in the last chapter together with additional circuit laws, which will be developed as we go along.

A ***circuit*** is a connection of two or more circuit elements. A ***closed*** circuit is the connection of two or more circuit elements in such a manner that a complete, unbroken path is made. An ***open*** circuit is the connection of two or more circuit elements such that no closed path exists.

Throughout our discussions of circuits, it is important to remember that current can only flow when a complete, unbroken path is made available to it. Therefore, current can only flow in a closed circuit and cannot flow in an open circuit.

We will start by looking at series circuits using the concepts of impedance and admittance discussed in the last chapter. We will then develop certain laws that simplify analysis of these circuits.

Next, we will form series circuits using combinations of the three basic passive circuit elements. We will then see how to apply the laws developed for series impedances and admittances to solve these circuits.

Following our analysis of series circuits, we will introduce parallel circuits. We will see how to analyze these circuits using the concepts of impedance and admittance together with the laws of parallel circuits, which we will develop.

Finally, we will form parallel circuits using combinations of the three basic passive circuit elements. We will analyze these circuits using the parallel circuit laws we have developed.

In any technical discussion it is necessary to be specific about the meanings of all terms used. We will therefore always define any term we introduce for the first time before using that term in any discussions that follow.

5.1 THE SERIES CIRCUIT

5.1.1 Definition

We will define the term ***series*** as follows:

> Two or more circuit impedances are in ***series*** if the ***same*** current flows through each impedance.

It is sometimes easier to visualize whether or not impedances are in series if one thinks of the current in the circuit as water flowing through a pipe system. If our impedances are connected so that the same water flows through each one, then these impedances are in series. If the water (current) encounters different pipes at any point connecting these impedances, then they are not in series.

Consider the circuits of Fig. 5.1. In the circuit of Fig. 5.1(a), the two impedances Z_1 and Z_2 are in series. As a matter of fact, the entire circuit including the voltage source V is a series circuit because the same current I flows through this source, as it does through each impedance in the circuit.

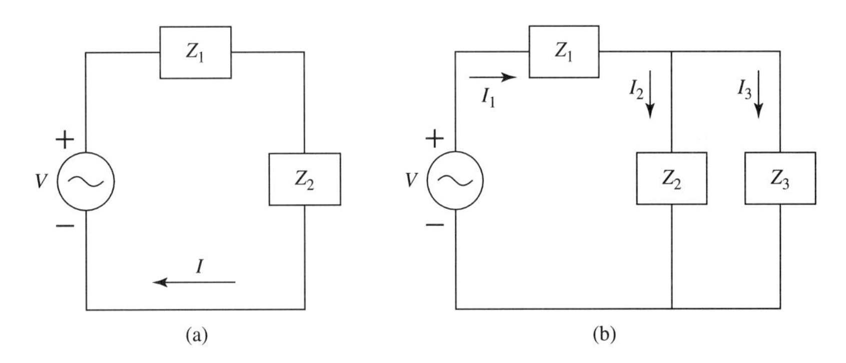

FIGURE 5.1

Now consider the circuit of Fig. 5.1(b). In this circuit, Z_1 and Z_2 are ***not*** in series because the current flowing through Z_1 is I_1, whereas the current flowing through Z_2 is I_2. These two currents are obviously not the same because I_2 is obtained by dividing the current (water) I_1 into one path (pipe) carrying I_2 and another path (pipe) carrying I_3. In this same circuit, however, note that the impedance Z_1 is in series with the voltage source, V. This is because both Z_1 and V carry the same current, I_1.

For now, we will not consider how the two impedances Z_2 and Z_3 in the circuit are connected. This will be the subject of our discussion on parallel circuits, which follows our discussion of series circuits.

Now consider the circuits of Fig. 5.2. In Fig. 5.2(a), the three impedances—Z_1, Z_2, and Z_3—are all in series. Here again, the entire circuit is a series circuit because the three impedances and the voltage source, V, all carry the same current I.

In the circuit of Fig. 5.2(b), observe that:

- V_1 is in series with Z_1.
- V_2 is in series with Z_4 and Z_5.
- Z_2 is in series with Z_3.

Note that no other series combination exists. The other circuit connections will be discussed later.

The circuits of Figs. 5.1 and 5.2 demonstrate that two or more impedances and voltage sources can all be in series provided they ***all carry the same current.*** These circuits also show that two or more impedances, or impedances and voltage sources, can be in series even though they are contained in a circuit with other impedances and voltage sources that are not in series.

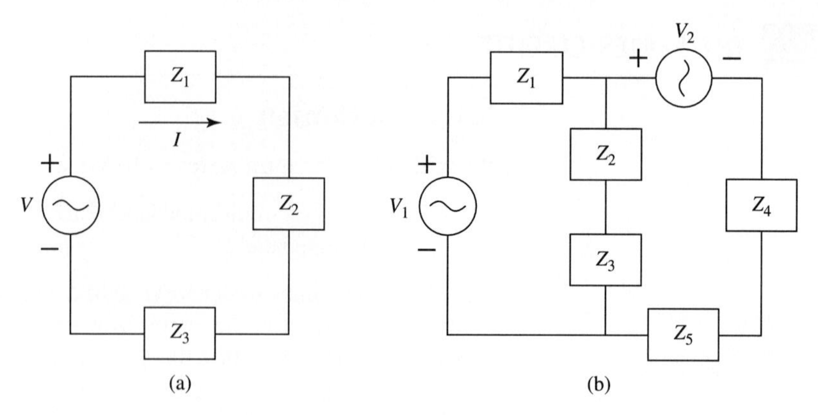

FIGURE 5.2

EXAMPLE 5.1

Identify all the series branches contained in the circuit of Fig. 5.3 by listing all the elements in each path.

FIGURE 5.3

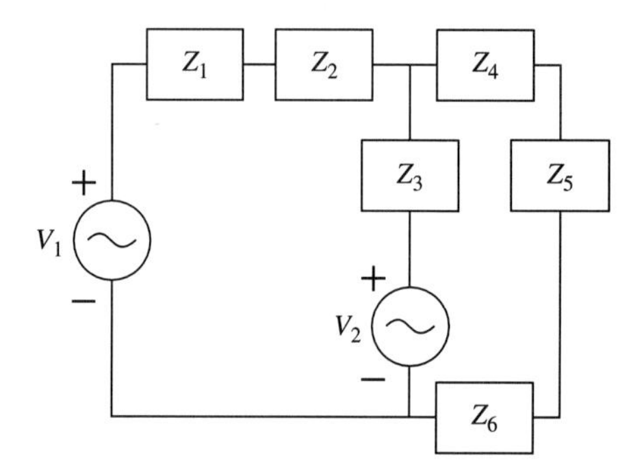

Solution There are three series branches as follows:
1. V_1, Z_1, and Z_2 **2.** V_2 and Z_3 **3.** Z_4, Z_5, and Z_6

EXAMPLE 5.2

In the circuit of Fig. 5.4, there are eight series branches. Identify each branch by listing all the components of the path.

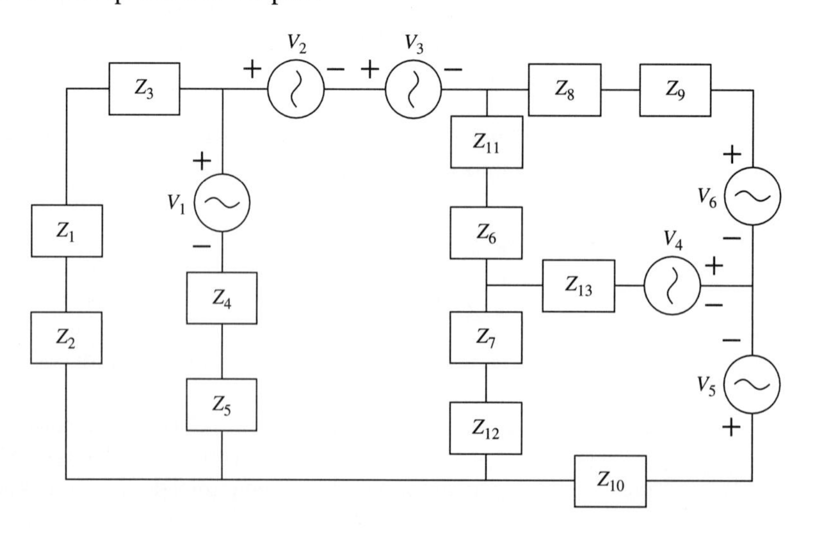

FIGURE 5.4

Solution The eight series branches are as follows:

1. Z_1, Z_2, and Z_3
2. V_1, Z_4, and Z_5
3. V_2 and V_3
4. Z_6 and Z_{11}
5. Z_7 and Z_{12}
6. Z_8, Z_9, and V_6
7. V_5 and Z_{10}
8. V_4 and Z_{13}

EXAMPLE 5.3

Find all the series branches of Fig. 5.5.

FIGURE 5.5

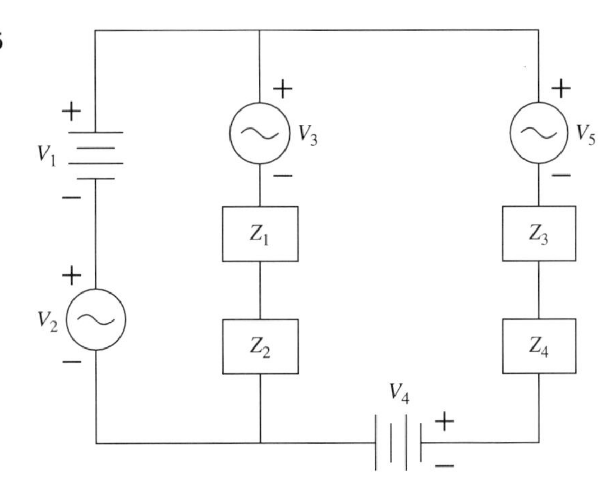

Solution

- V_1 and V_2
- V_3, Z_1, and Z_2
- V_5, Z_3, Z_4, and V_4

5.1.2 Kirchhoff's Voltage Law

Kirchhoff's Voltage Law, which will hereafter be abbreviated as KVL, is one of the basic laws of circuit analysis. This law is stated in words as follows:

KVL: The sum of the voltages around a closed loop is identically equal to zero.

Mathematically, this law is stated as Equation 5.1:

$$\Sigma V_{cl} = 0 \qquad \textbf{(Eq. 5.1)}$$

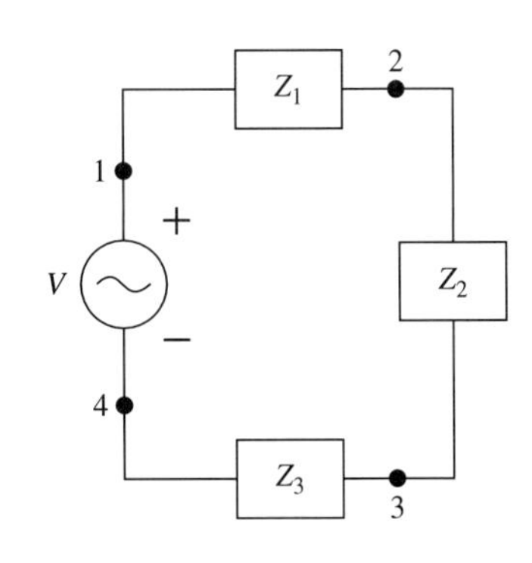

FIGURE 5.6

To understand this law, consider Fig. 5.6. Figure 5.6 is a series circuit consisting of a single voltage source and three impedances. KVL may be applied by starting at any point in the circuit of Fig. 5.6. If we start, for example, at point 1 and move clockwise (CW) around the loop until we return to point 1, we get Equation 5.2.

$$V_{12} + V_{23} + V_{34} + V_{41} = 0 \qquad \textbf{(Eq. 5.2)}$$

In evaluating an equation obtained by applying KVL, such as Equation 5.2, we must agree to a sign convention concerning voltage changes. We will use the convention that a voltage change will be considered ***positive*** across a voltage drop (i.e., when going from a positive terminal to a negative terminal) and ***negative*** across a voltage rise (i.e., when going from a negative terminal to a positive terminal).

EXAMPLE 5.4

In the circuit of Fig. 5.6, the following values are given: $V_{12} = 10\angle 60°$ V, $V_{23} = 12\angle 30°$ V, and $V_{34} = 8\angle 45°$ V. Find the voltage source, V, that is the same as V_{14}.

Solution From Equation 5.2, we may solve for V_{41} as follows:

$$V_{41} = -(V_{12} + V_{23} + V_{34}) \qquad \textbf{(Eq. 5.3)}$$

Since $V_{14} = -V_{41}$, we may write

$$V_{14} = V_{12} + V_{23} + V_{34} = V$$

Substituting the given values, we get:

$$V = 10\angle 60° + 12\angle 30° + 8\angle 45° = 29.26\angle 44.0°$$

Answer $V = 29.26\angle 44.0°$ V.

EXAMPLE 5.5

Figure 5.7 shows two dc voltage sources in series with three impedances. The following parameters are given: $V_1 = 100$ V, $V_{ab} = 10$ V, $V_2 = 40$ V, $V_{de} = 30$ V. Find V_{cd}, the voltage across impedance Z_2.

FIGURE 5.7

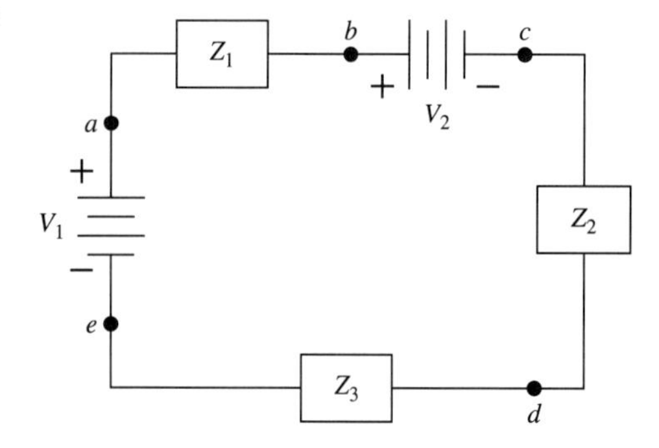

Solution Applying KVL to the circuit in Fig. 5.7 by moving in a CW direction from point a:

$$V_{ab} + V_{bc} + V_{cd} + V_{de} + V_{ea} = 0$$
$$10 + 40 + V_{cd} + 30 - 100 = 0 \qquad \textbf{(Eq. 5.4)}$$

Note that $V_{ea} = -100$ V because going from e to a represents a voltage rise.

$$-20 + V_{cd} = 0$$

Solving for V_{cd}, we get:

$$V_{cd} = 20$$

Answer $V_{cd} = 20$ V.

KVL may be applied moving in a CW direction or in a counterclockwise (CCW) direction. To demonstrate this, consider Example 5.5. We solved this example by moving in a CW direction from point a. We will now apply KVL by moving in a CCW direction from point a. From Equation 5.2 we get Equation 5.5:

$$V_{ae} + V_{ed} + V_{dc} + V_{cb} + V_{ba} = 0$$
$$100 - 30 - V_{cd} - 40 - 10 = 0 \qquad \text{since } V_{cd} = -V_{dc} \qquad \textbf{(Eq. 5.5)}$$
$$20 - V_{cd} = 0$$

Solving for V_{cd} we get:

$$V_{cd} = 20$$

Answer As before, $V_{cd} = 20$ V.

It also does not matter what point in the circuit we start with. The important thing is that after writing Equation 5.2 for a given loop we start and finish at the same point. This is because KVL is actually stating that $V_{xx} = 0$. In other words, the voltage of any point x with respect to itself is 0.

Another way of stating KVL is that the voltage between any two points is the same no matter what path we use to get from one point to the other. This is demonstrated by writing Equation 5.5 as Equation 5.6, remembering that $V_{xy} = -V_{yx}$:

$$V_{ae} = V_{ab} + V_{bc} + V_{cd} + V_{de} \qquad \textbf{(Eq. 5.6)}$$

Note that when we use this form of KVL, the first point and last point on the right side of the equation are the same as the two points on the left side of the equation in the same order.

5.1.3 Equivalent Series Impedance

Consider the circuit shown in Fig. 5.8(a). This circuit shows three impedances in series with a voltage source. From KVL, we may write Equation 5.7:

$$V = V_1 + V_2 + V_3 \qquad \textbf{(Eq. 5.7)}$$

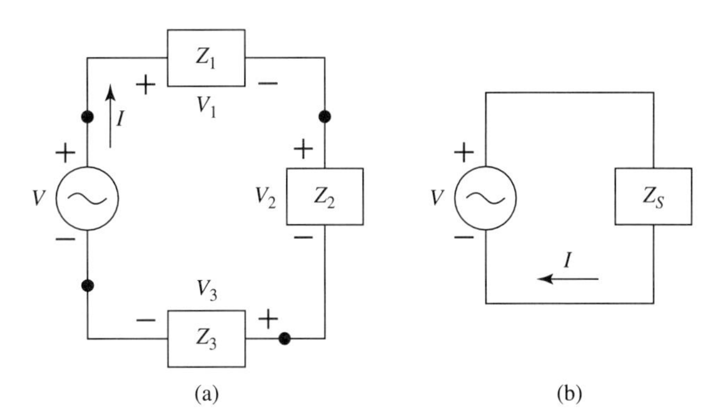

FIGURE 5.8

Now consider the circuit shown in Fig. 5.8(b). In this circuit, Z_s represents the ***equivalent impedance*** of the three series impedances shown in Fig. 5.8(a).

We define equivalent series impedance as follows:

> The equivalent impedance of two or more series impedances is defined as the single impedance that would have the same current flowing through it as the actual impedances in the circuit if it directly replaced those impedances.

Referring again to Fig. 5.8, Z_s in Fig. 5.8(b) is the equivalent impedance of the three series impedances in Fig. 5.8(a) if the current I flowing in Fig. 5.8(b) is the same as the current I flowing in Fig. 5.8(a). We will now find an expression that will enable us to determine equivalent impedances of series circuits.

From Ohm's Law, we may write each term in the right hand side of Equation 5.7 as follows:

$$V_1 = IZ_1 \qquad V_2 = IZ_2 \qquad V_3 = IZ_3 \qquad \textbf{(Eq. 5.8)}$$

Applying Ohm's Law to the circuit of Fig. 5.8(b):

$$V = IZ_s \qquad \textbf{(Eq. 5.9)}$$

Substituting Equations 5.8 and 5.9 into 5.7, we get:

$$IZ_s = IZ_1 + IZ_2 + IZ_3 \qquad \textbf{(Eq. 5.10)}$$

Factoring out I from the three terms on the right hand side of Equation 5.10:

$$IZ_s = I(Z_1 + Z_2 + Z_3) \qquad \textbf{(Eq. 5.11)}$$

Dividing both sides of Equation 5.11 by I, we get:

$$Z_s = Z_1 + Z_2 + Z_3 \qquad \textbf{(Eq. 5.12)}$$

Equation 5.12 states that when three impedances are connected in series, they may be replaced by a single impedance equal in value to the algebraic sum of the three impedances. In fact, this may be generalized to any number of impedances in series. The general statement of finding the equivalent impedance of any number of impedances in series is as follows:

> Any number of impedances connected in series may be replaced by a single equivalent impedance equal in value to the algebraic sum of all the impedances connected in series.

Mathematically, this law may be written as:

$$Z_s = \Sigma Z_n \quad (n = 1 \text{ to } N) \qquad \textbf{(Eq. 5.13)}$$

where N represents the number of impedances in series.

EXAMPLE 5.6

Figure 5.9(a) shows two impedances in series with a voltage source.

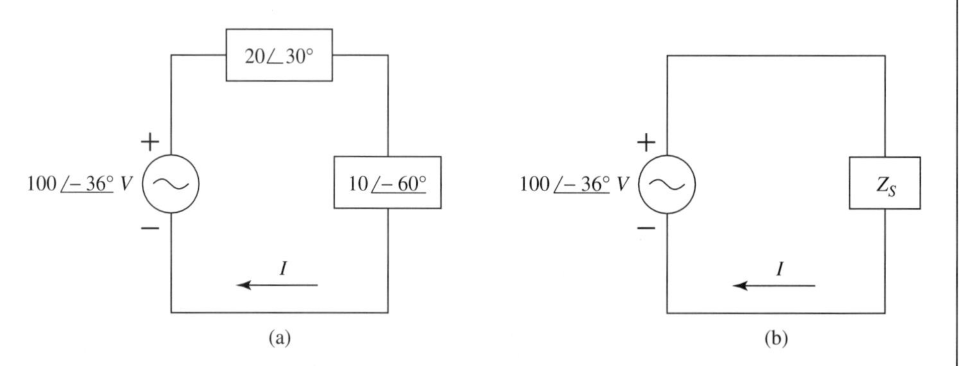

FIGURE 5.9

(a) Find Z_s, the equivalent impedance of those two series impedances.
(b) Redraw the circuit using the equivalent impedance in place of the two series impedances.
(c) Find the current flowing in the original and equivalent circuits.

Solution

(a) From Equation 5.13:

$$Z_s = 20\angle 30^\circ + 10\angle -60^\circ = 17.32 + j10.00 + 5.00 - j8.66$$
$$= 22.32 + j1.34$$

Answer $Z_s = (22.32 + j1.34)\ \Omega = 22.36\angle 3.44^\circ\ \Omega$.

(b) The circuit is redrawn in Fig. 5.9(b).
(c) Using Ohm's Law in Fig. 5.9(b):

$$I = V/Z_s = \frac{100\angle -36^\circ}{22.36\angle 3.44^\circ} = 4.47\angle -39.44^\circ$$

Answer $I = 4.47\angle -39.44^\circ$ A.

Note that this current flows in both the original circuit of Fig. 5.9(a) ***and*** the circuit of Fig. 5.9(b).

EXAMPLE 5.7

In the circuit of Fig. 5.10(a):
(a) Find Z_s, the equivalent impedance of the series combination of the 10-Ω resistor and the 5-mH inductor.
(b) Redraw the circuit using the equivalent impedance found in part (a) in place of the two elements.
(c) Find the circuit flowing in both the original and equivalent circuits.

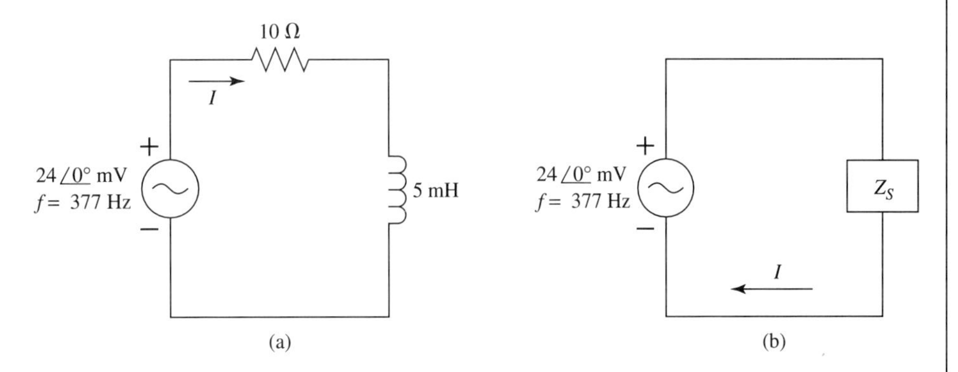

FIGURE 5.10

Solution
(a) Each of the elements in Fig. 5.10(a) represents an impedance. If Z_R represents the impedance of the resistor R and Z_L represents the impedance of the inductor L, then from the previous chapter:

$$Z_R = R = 10\ \Omega \quad Z_L = 2\pi fL\angle 90^\circ = 2\pi(377)(5)10^{-3}\angle 90^\circ = 11.84\angle 90^\circ\ \Omega$$

From Equation 5.13: $Z_s = Z_R + Z_L = (10 + j11.84)\ \Omega$.

Answer $Z_S = (10 + j11.84)\ \Omega$, or in polar form, $Z_S = 15.50\angle 49.8^\circ\ \Omega$.
(b) The redrawn circuit appears in Fig. 5.10(b).
(c) Applying Ohm's Law to Fig. 5.10(b):

$$I = \frac{24 \times 10^{-3}\angle 0^\circ}{15.50\angle 49.8^\circ} = 1.55 \times 10^{-3}\angle -49.8^\circ\ \text{A}$$

Answer $I = 1.59\angle -49.8^\circ$ mA.

As in the previous example, this current is the same for both circuits Fig. 5.10(a) and (b).

EXAMPLE 5.8

For the circuit of Fig. 5.11(a):

(a) Find Z_s, the equivalent impedance of the four series-connected elements.

(b) Redraw Fig. 5.11(a) using the equivalent impedance found in part (a) in place of the four series-connected elements.

(c) Find the current flowing in both circuits of Fig. 5.11.

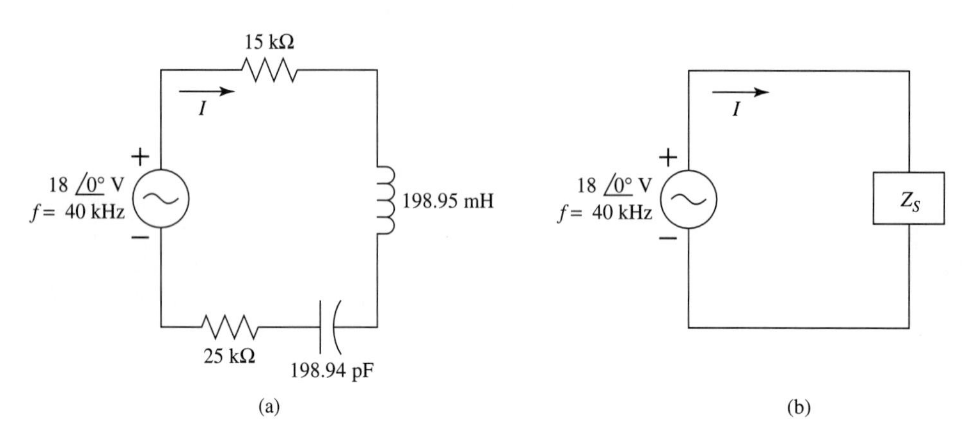

FIGURE 5.11

Solution

(a) We begin by finding the impedance of each of the elements in Fig. 5.11(a).
If Z_{R_1} represents the impedance of the 15-kΩ resistor R_1,
Z_{R_2} represents the impedance of the 25-kΩ resistor R_2,
Z_L represents the impedance of the 198.95 mH inductor L, and
Z_C represents the impedance of the 198.94 μF capacitor, C
then from Equation 5.13:

$$Z_s = Z_{R_1} + Z_{R_2} + Z_L + Z_C$$

$$Z_{R_1} = 15\text{ k}\Omega \qquad Z_{R_2} = 25\text{ k}\Omega,$$

$$Z_L = 2\pi fL = j2\pi(40 \times 10^3)(198.95 \times 10^{-3}) = j50\text{ k}\Omega$$

$$Z_C = -j(1/\omega C) = -j(1/2\pi fC)$$

$$= -j[1/(2\pi 40 \times 10^3 \times 198.94 \times 10^{-12})] = -j20\text{ k}\Omega$$

Substituting these values in the equation for Z_s, we get:

$$Z_s = 15\text{K} + 25\text{K} + j50\text{K} - j20\text{K} = (40\text{K} + j30\text{K})\ \Omega$$

Answer $Z_s = (40\text{K} + j30\text{K})\ \Omega = 50\text{K}\angle 36.9^\circ\ \Omega.$

(b) The circuit of Fig. 5.11(a) is redrawn in Fig. 5.11(b) using the equivalent impedance of the four elements in series.

(c) Applying Ohm's Law to the circuit of Fig. 5.11(b):

$$I = \frac{18\angle 0^\circ}{50\text{K}\angle 36.9^\circ} = .36 \times 10^{-3}\angle -36.9^\circ$$

Answer $I = 0.36\angle -36.9^\circ$ mA.

Note again that this current, I, is the same in both circuits of Fig. 5.11.

EXAMPLE 5.9

The circuit of Fig. 5.12(a) shows a dc voltage source in series with two elements, a resistor and an inductor.

(a) Find the equivalent impedance of the two elements in series, Z_s

(b) Redraw the circuit using the equivalent impedance found in part (a).

(c) Find the current flowing in both the original and equivalent circuits.

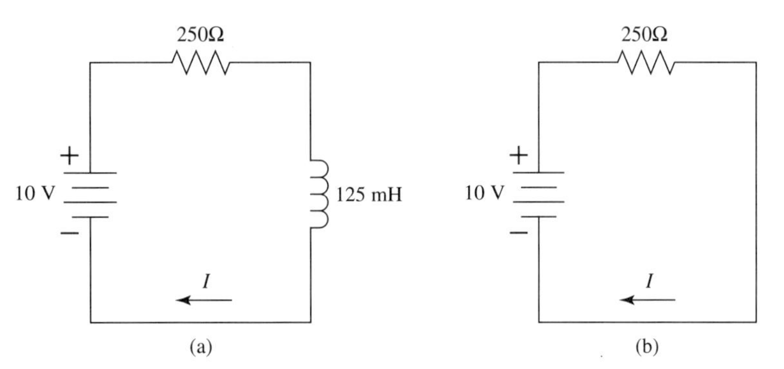

FIGURE 5.12

Solution

(a) The impedance of the resistor is $Z_R = 250\ \Omega$. The impedance of the inductor is $Z_L = 0\ \Omega$. (Recall that the impedance of an inductor in the case of dc power is $0\ \Omega$). Using these values in Equation 5.13, we get:

$$Z_s = Z_R + Z_L = 250 + 0 = 250$$

Answer $Z_s = 250\ \Omega$.

(b) The equivalent series circuit is drawn in Fig. 5.12(b).

(c) From Ohm's Law in Fig. 5.12(b):

$$I = 10/250 = 0.040$$

Answer $I = 40$ mA.

EXAMPLE 5.10

In the circuit of Fig. 5.13(a) two dc voltage sources are connected in series with three resistors.

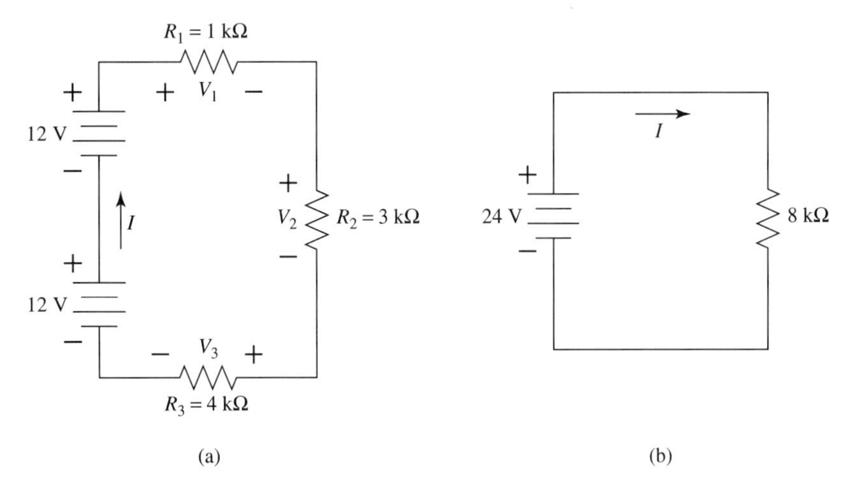

FIGURE 5.13

(a) Draw an equivalent circuit with a single voltage source and a single resistor.
(b) Find the current flowing in this circuit.
(c) Find the voltage across each resistor.

Solution

(a) The two series connected voltage sources may be replaced by a single voltage source V_t as follows:

$$V_t = 12 + 12 = 24 \text{ V}$$

The three resistors may be replaced by a single resistor having a resistance R_s:

$$R_s = 1\text{K} + 3\text{K} + 4\text{K} = 8\text{K}\Omega$$

The equivalent circuit is drawn in Fig. 5.13(b).

(b) Using Ohm's Law in Fig. 5.13(b):

$$I = V_t/R_s = 24/8\text{K} = 3 \text{ mA}$$

Answer $I = 3$ mA.

(c) The voltage across each resistor is found by applying Ohm's Law to each resistor as follows:

$$V_1 = R_1(I) = 1 \times 10^3(3 \times 10^{-3}) = 3$$
$$V_2 = R_2(I) = 3 \times 10^3(3 \times 10^{-3}) = 9$$
$$V_3 = R_3(I) = 4 \times 10^3(3 \times 10^{-3}) = 12$$

Answer $V_1 = 3$ V, $V_2 = 9$ V, $V_3 = 12$ V.

The polarity of the three resistor voltages is shown in Fig. 5.13(a). Note that this is consistent with our definitions of current through an impedance and voltage drop across an impedance as defined by Ohm's Law.

EXAMPLE 5.11

In the circuit of Fig. 5.14(a), two ac voltage sources are connected in series with three elements.

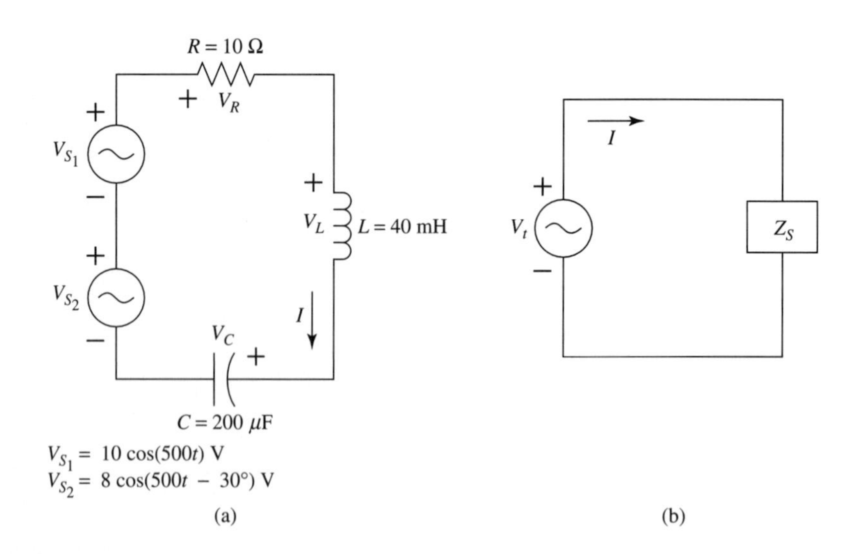

FIGURE 5.14

(a) Draw an equivalent circuit consisting of a single voltage source and a single impedance.
(b) Find the ***sinusoidal*** current flowing in this circuit.
(c) Find the voltage drop across each element.

Solution

(a) The two voltage sources, having the same frequency, may be added by phasor analysis. The two phasors are:

$$V_{s_1} = 10\underline{/90^\circ}\text{ V}, \qquad V_{s_2} = 8\underline{/60^\circ}\text{ V}$$
$$V_t = V_{s_1} + V_{s_2} = 10\underline{/90^\circ} + 8\underline{/60^\circ} = 17.39\text{ V }\underline{/76.7^\circ}$$

The equivalent impedance of the three elements is found by first finding the impedance of each element and then adding these three impedances as per Equation 5.13.

$$Z_R = R = 10\,\Omega \qquad Z_L = \omega L = 500(40 \times 10^{-3}) = j20\,\Omega$$
$$Z_C = 1/\omega C = 1/(500)(200 \times 10^{-6}) = -j10\,\Omega$$
$$Z_s = Z_R + Z_L + Z_C = 10 + j20 - j10 = (10 + j10)\,\Omega$$

Converting this to polar form, we get

$$Z_s = 14.14\underline{/45^\circ}\,\Omega$$

The equivalent circuit is redrawn in Fig. 5.14(b).

(b) We find the sinusoidal current flowing in this circuit by applying Ohm's Law to Fig. 5.14(b) using the phasors found in part (a) and then converting the phasor to a sinusoidal time function.

$$I = V_T/Z_s = \frac{17.39\underline{/76.7^\circ}}{14.14\underline{/45^\circ}} = 1.23\ \underline{/31.7^\circ}\text{ A}$$

Answer $i = 1.23\sin(500t + 31.7^\circ)$ A.

(c) Using Ohm's Law across each component:

$$V_R = RI = 10\underline{/0^\circ}(1.23\underline{/31.7^\circ}) = 12.3\underline{/31.7^\circ}\text{ V}$$
$$V_L = Z_L I = 20\underline{/90^\circ}(1.23\underline{/31.7^\circ}) = 24.6\underline{/121.7^\circ}\text{ V}$$
$$V_C = Z_C I = 10\underline{/-90^\circ}(1.23\underline{/31.7^\circ}) = 12.3\underline{/-58.3^\circ}\text{ V}$$

Answer $V_R = 12.3\sin(500t + 31.7^\circ)$ V,
$V_L = 24.6\sin(500t + 121.7^\circ)$ V,
$V_C = 12.3\sin(500t - 58.3^\circ)$ V.

EXAMPLE 5.12

In the circuit of Fig. 5.15(a), a dc voltage source is connected in series with three other elements.

(a) Draw an equivalent circuit.
(b) Find the current flowing in the circuit.
(c) Find the voltage across each element in Fig. 5.15(a).

Solution

(a) To draw an equivalent circuit we must first find the impedance

$$Z_L = 0\,\Omega$$

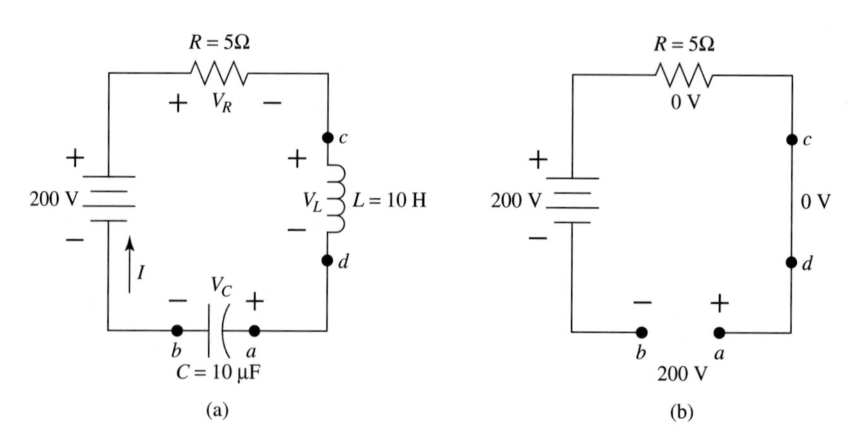

FIGURE 5.15

because the impedance of an inductor is 0 Ω (short circuit) in dc, and:

$$Z_C = \infty$$

because the impedance of a capacitor is ∞ (open circuit) in dc.

The equivalent circuit is drawn in Fig. 5.15(b). Note that the inductor has been replaced by a short and the capacitor has been replaced by an open.

(b) From Fig. 5.15(b), because the capacitor causes the circuit to be open, no current can flow. Therefore,

Answer $I = 0$.

(c) The voltage drop across each element is found as follows:

From Ohm's Law: $V_R = R(I) = R(0) = 0\text{ V}$

Recall that because an inductor acts like a short circuit in dc, there can be no dc voltage across it. Therefore $V_L = 0$ V.

From KVL, we may write:

$$V_R + V_L + V_C - 200 = 0 \quad \text{or} \quad 0 + 0 + V_C - 200 = 0$$

Solving for V_C we get: $V_C = 200$.

Answer $V_R = 0\text{ V}$, $V_L = 0\text{ V}$, $V_C = 200\text{ V}$.

Note from this example that in the case of a dc circuit, although a capacitor can have no current flowing through it, it can and often does have a dc voltage across it.

5.1.4 Voltage Division Law

The Voltage Division Law for series-connected impedances may be stated as follows:

> The voltage across any impedance in a series-connected circuit is equal to its impedance multiplied by the total voltage across all the series-connected impedances divided by the total (equivalent) series impedance.

To write this law mathematically, consider the circuit of Fig. 5.8, which is reproduced for convenience in Fig. 5.16. Equations 5.8 and 5.9, corresponding to this circuit, are reproduced as Equations 5.14 and 5.15:

For the circuit in Fig. 5.16(a):

$$V_1 = IZ_1 \qquad V_2 = IZ_2 \qquad V_3 = IZ_3 \qquad \textbf{(Eq. 5.14)}$$

For the circuit of Fig. 5.16(b):

$$V = IZ_s \qquad \textbf{(Eq. 5.15)}$$

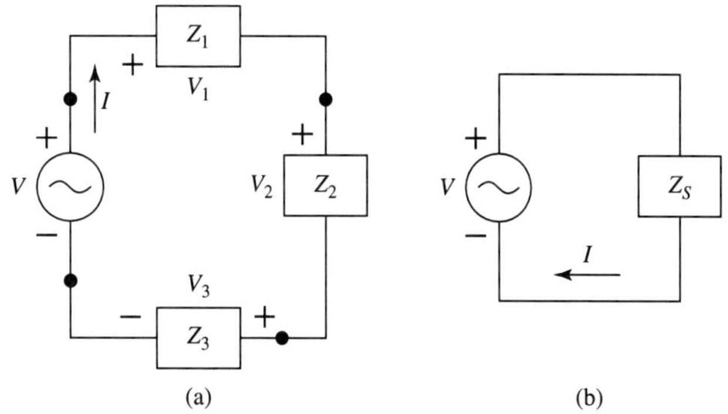

FIGURE 5.16

Dividing V_1 by V using Equations 5.14 and 5.15:

$$V_1/V = IZ_1/IZ_s = Z_1/Z_s \quad \textbf{(Eq. 5.16)}$$

Solving Equation 5.16 for V_1:

$$V_1 = (Z_1/Z_s)\ \mathrm{V} \quad \textbf{(Eq. 5.17)}$$

In a similar manner it can be shown that:

$$V_2 = (Z_2/Z_s)\ \mathrm{V} \quad \textbf{(Eq. 5.18)}$$

and

$$V_3 = (Z_3/Z_s)\ \mathrm{V} \quad \textbf{(Eq. 5.19)}$$

Equations 5.17, 5.18, and 5.19 may be summarized by Equation 5.20, in which it is assumed that the voltage V is the total voltage across n series impedances and V_n is the voltage across impedance Z_n:

$$\text{Voltage Division Law} \quad V_n = (Z_n/Z_s)\ \mathrm{V} \quad \textbf{(Eq. 5.20)}$$

EXAMPLE 5.13

Consider the series circuit of Fig. 5.17. The impedances of this circuit are given as follows: $Z_1 = 24\angle 30°\ \Omega$, $Z_2 = 10\angle 0°\ \Omega$, $Z_3 = 4\angle 45°\ \Omega$, and $Z_4 = 12\angle 90°\ \Omega$.

FIGURE 5.17

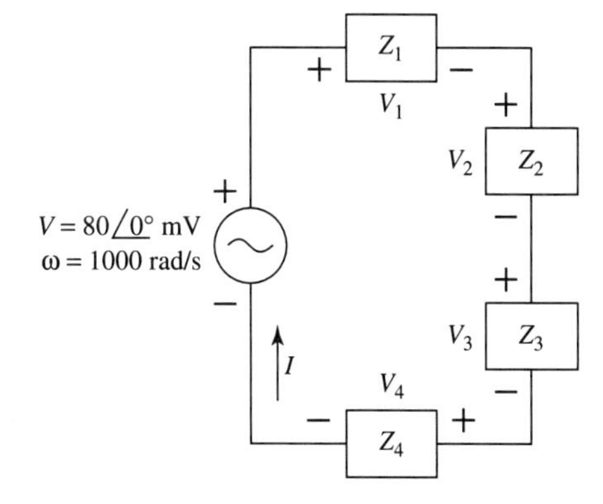

(a) Using the Voltage Division Law of Equation 5.20, find the sinusoidal voltage across each impedance in this circuit.

(b) Using Ohm's Law, find the sinusoidal voltage across each element in this circuit and compare each voltage to the corresponding value found in part (a).

Solution

(a) To use Equation 5.20, we must first find the total series equivalent impedance Z_s. From Equation 5.13:

$$Z_s = Z_1 + Z_2 + Z_3 + Z_4$$

or

$$Z_s = 24\underline{/30^\circ} + 10\underline{/0^\circ} + 4\underline{/45^\circ} + 12\underline{/90^\circ} = 43\underline{/38.6^\circ}\ \Omega$$

Using this value of Z_s in Equation 5.20 with the given impedance values, we find the following phasor voltages:

$$V_1 = \frac{24\underline{/30^\circ}}{43\underline{/38.6^\circ}} 80 \times 10^{-3} = 44.65 \times 10^{-3}\underline{/-8.6^\circ} = 44.65\underline{/-8.6^\circ}\ \text{mV}$$

$$V_2 = \frac{10\underline{/0^\circ}}{43\underline{/38.6^\circ}} 80 \times 10^{-3} = 18.61 \times 10^{-3}\underline{/-38.6^\circ} = 18.61\underline{/-38.6^\circ}\ \text{mV},$$

$$V_3 = \frac{4\underline{/45^\circ}}{43\underline{/38.6^\circ}} 80 \times 10^{-3} = 7.44 \times 10^{-3}\underline{/6.4^\circ} = 7.44\underline{/6.4^\circ}\ \text{mV}$$

$$V_4 = \frac{12\underline{/90^\circ}}{43\underline{/38.6^\circ}} 80 \times 10^{-3} = 22.33 \times 10^{-3}\underline{/51.40^\circ} = 23.33\underline{/51.40^\circ}\ \text{mV}$$

Answer $v_1 = 44.65 \sin(1000t - 8.6^\circ)$ mV, $v_2 = 18.61 \sin(1000t - 38.6^\circ)$ mV, $v_3 = 7.44 \sin(1000t + 6.4^\circ)$ mV, $v_4 = 22.33 \sin(1000t + 51.40^\circ)$ mV.

(b) We will now find these same voltages using Ohm's Law. To do this, we must first find the current in the circuit. Using the value of Z_s found in part (a):

$$I = V_s/Z_s = \frac{80 \times 10^{-3}\underline{/0^\circ}}{43\underline{/38.6^\circ}} = 1.86\underline{/-38.6^\circ}\ \text{mA}$$

Using this current with Ohm's Law for each impedance:

$$V_1 = IZ_1 = 1.86 \times 10^{-3}\underline{/-38.6^\circ}(24\underline{/30^\circ}) = 44.64\underline{/-8.6^\circ}\ \text{mV}$$
$$V_2 = IZ_2 = 1.86 \times 10^{-3}\underline{/-38.6^\circ}(10\underline{/0^\circ}) = 18.6\underline{/-38.6^\circ}\ \text{mV}$$
$$V_3 = IZ_3 = 1.86 \times 10^{-3}\underline{/-38.6^\circ}(4\underline{/45^\circ}) = 7.44\underline{/6.4^\circ}\ \text{mV}$$
$$V_4 = IZ_4 = 1.86 \times 10^{-3}\underline{/-38.6^\circ}(12\underline{/90^\circ}) = 22.32\underline{/51.4^\circ}\ \text{mV}$$

Comparing these voltages with their phasor counterparts in part (a) we find that they are exactly the same, which, of course, is what we expected.

EXAMPLE 5.14

In the circuit of Fig. 5.18, find the following phasor voltages by the Voltage Division Law.

(a) V_R **(b)** V_1 **(c)** V_C **(d)** V_2 **(e)** V_{ab} **(f)** I

Solution To use the Voltage Division Law (Eq. 5.20), we will find Z_s by adding up the individual impedances as per Equation 5.13:

$$Z_s = Z_R + Z_{L_1} + Z_{L_2} + Z_C$$
$$Z_R = R\underline{/0^\circ} = 2.2\text{K}\underline{/0^\circ}\ \Omega$$
$$Z_{L_1} = 2\pi f L_1\underline{/90^\circ} = 2\pi(8.76 \times 10^3)20 \times 10^{-3}\underline{/90^\circ} = 1.1\text{K}\underline{/90^\circ}\ \Omega$$
$$Z_{L_2} = 2\pi f L_2\underline{/90^\circ} = 2\pi(8.76 \times 10^3)40 \times 10^{-3}\underline{/90^\circ} = 2.2\text{K}\underline{/90^\circ}\ \Omega$$

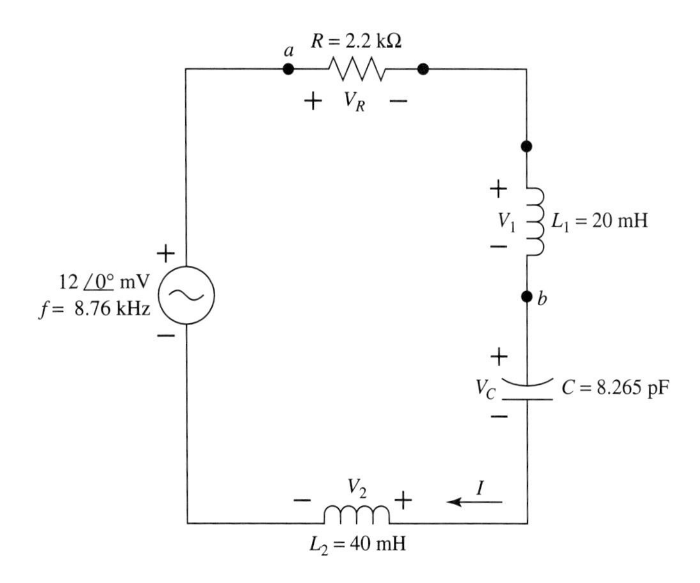

FIGURE 5.18

$$Z_C = \frac{1}{2\pi fC}\underline{/-90^\circ} = \frac{1}{2\pi 8.76 \times 10^3 \times 8.265 \times 10^{-9}}\underline{/-90^\circ}$$
$$= 2.2\text{K}\underline{/-90^\circ}\ \Omega$$
$$Z_s = 2.2\text{K}\underline{/0^\circ} + 1.1\text{K}\underline{/90^\circ} + 2.2\text{K}\underline{/90^\circ} + 2.2\text{K}\underline{/-90^\circ}$$
$$Z_s = 2.46\text{K}\underline{/26.57^\circ}\ \Omega$$

(a) Using these impedances in Equation 5.20:

$$V_R = (R/Z_s)V_s = \frac{2.2\text{K}\underline{/0^\circ}}{2.46\text{K}\underline{/26.57^\circ}}12 \times 10^{-3} = 10.73 \times 10^{-3}\underline{/-26.57^\circ}\text{ V}$$
$$V_1 = (Z_{L_1}/Z_s)V_s = \frac{1.1\text{K}\underline{/90^\circ}}{2.46\text{K}\underline{/26.57^\circ}}12 \times 10^{-3} = 5.37 \times 10^{-3}\underline{/63.43^\circ}\text{ V}$$
$$V_C = (Z_C/Z_s)V_s = \frac{2.2\text{K}\underline{/-90^\circ}}{2.46\text{K}\underline{/26.57^\circ}}12 \times 10^{-3} = 10.73 \times 10^{-3}\underline{/-116.57^\circ}\text{ V}$$
$$V_2 = (Z_{L_2}/Z_s)V_s = \frac{2.2\text{K}\underline{/90^\circ}}{2.46\text{K}\underline{/26.57^\circ}}12 \times 10^{-3} = 10.73 \times 10^{-3}\underline{/63.43^\circ}\text{ V}$$

From KVL,

$$V_{ab} = V_R + V_1 = 10.73 \times 10^{-3}\underline{/-26.57^\circ} + 5.37 \times 10^{-3}\underline{/63.43^\circ}\text{ V}$$
$$= (9.60 - j4.80 + 2.40 + j4.80) \times 10^{-3} = 12.0 \times 10^{-3}\underline{/0^\circ}\text{ V}$$

The current is found by using Ohm's Law with the total series equivalent impedance:

$$I = V_s/Z_s = \frac{12 \times 10^{-3}\underline{/0^\circ}}{2.46 \times 10^3\underline{/26.57^\circ}} = 4.88 \times 10^{-6}\underline{/-26.57^\circ}\text{ A}$$

Answer

(a) $V_R = 10.73\underline{/-26.57^\circ}$ mV
(b) $V_1 = 5.37\underline{/63.43^\circ}$ mV
(c) $V_C = 10.73\underline{/-116.57^\circ}$ mV
(d) $V_2 = 10.73\underline{/63.43^\circ}$ mV
(e) $V_{ab} = 12.00\underline{/0^\circ}$ mV
(f) $I = 4.88\underline{/-26.57^\circ}$ μA

5.1.5 Impedance and Phasor Diagrams of Series Circuits

A phasor diagram of a given circuit will show the relationship between the voltages and currents in that circuit. The phasor diagram of the series circuit in Example 5.14 is drawn in Fig. 5.19.

FIGURE 5.19

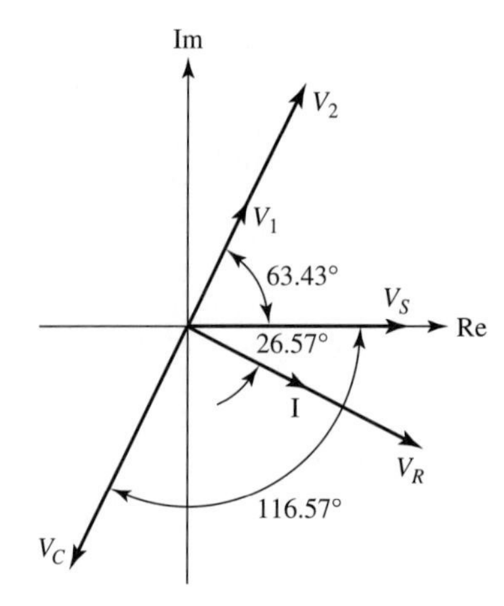

When drawing such a diagram, it is important to remember that it is the angular relationship ***between*** phasors that is important rather than the absolute value of each phasor angle. For example, note the following in the phasor diagram of Fig. 5.19:

- The current phasor I is ***in phase*** with the voltage phasor V_R. This is expected because the angle of a resistive impedance is 0°.
- The current phasor I ***leads*** the voltage phasor, V_C, by 90°. This is expected because the angle of a capacitive impedance is −90°.
- The current phasor I ***lags*** each of the voltages V_1 and V_2 by 90°. This is expected because the angle of an inductive impedance is 90°.
- The current phasor I lags the input voltage phasor by 26.57°. This is expected because the angle of the total equivalent series impedance is 26.57°.

From the preceding discussion, we can see that an impedance diagram of a circuit has the same angular relationship as the corresponding phasor diagram. This is demonstrated by drawing the impedance diagram in Fig. 5.20.

FIGURE 5.20

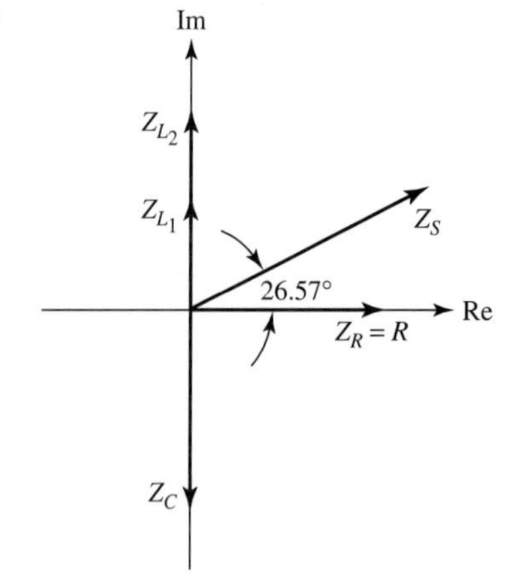

Comparing the impedance diagram of Fig. 5.20 with the phasor diagram of Fig. 5.19 we see that:

1. The angle between the resistance and inductive impedances is 90°, the same as the angle between the current phasor and the inductive voltages V_1 and V_2.

2. The angle between the resistance and capacitive impedance is $-90°$, the same as the angle between the current phasor and the capacitive voltage, V_C.
3. The angle between the resistance and the total series equivalent impedance is 26.57°, the same as the angle between the current phasor and the total voltage, V_s.

Since in a series circuit the current is the same for each element it is often convenient to plot the current phasor as the reference (i.e., along the positive real axis). If we did this in Example 5.14, the phasor diagram of Fig. 5.19 would lie along the exact same lines as the impedance diagram of Fig. 5.20. As noted previously, it is the angular relationship between phasors that is of importance and not the absolute angles of each phasor. The current phasor I would then have an angle of 0°, V_1 and V_2 would each have an angle of 90°, V_C would have an angle of $-90°$, and V_s would have an angle of 26.57°.

5.1.6 Equivalence of Like Passive Series Elements

Just as impedances in series may be combined to yield a total equivalent series impedance, ***like elements*** in series may also be combined to yield a single equivalent element. This can often simplify a circuit problem because we would be dealing with a single element of each type for much of the problem.

5.1.6.1 Resistors in Series

Consider the circuit of Fig. 5.21(a), which shows three resistors in series with a voltage source. Since the impedance of a resistor is equal to its resistance we can find the total series equivalent impedance as:

$$Z_s = R_1 + R_2 + R_3$$

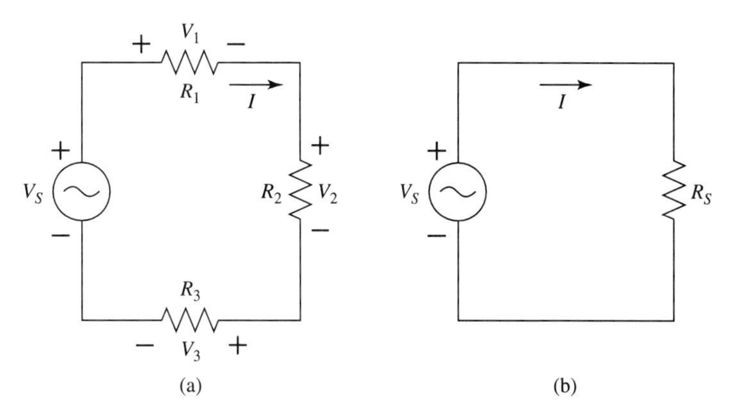

FIGURE 5.21

or, since the equivalent series impedance is a resistor:

$$R_s = R_1 + R_2 + R_3$$

This demonstrates the general law of series resistors:

> Any number of series connected resistors may be replaced by a single resistor whose resistance is given by the sum of all the series resistance values.

This law may be stated mathematically as follows:

$$R_s = \Sigma R_n \ (n = 1 \text{ to } N) \qquad \textbf{(Eq. 5.21)}$$

where N represents the number of resistors in series.

The circuit of Fig. 5.21(a) has been redrawn in Fig. 5.21(b) with the three series resistors replaced by the single equivalent resistor, R_s.

EXAMPLE 5.15

Consider the series circuit of Fig. 5.22(a), which consists of four resistors and two voltage sources in series.

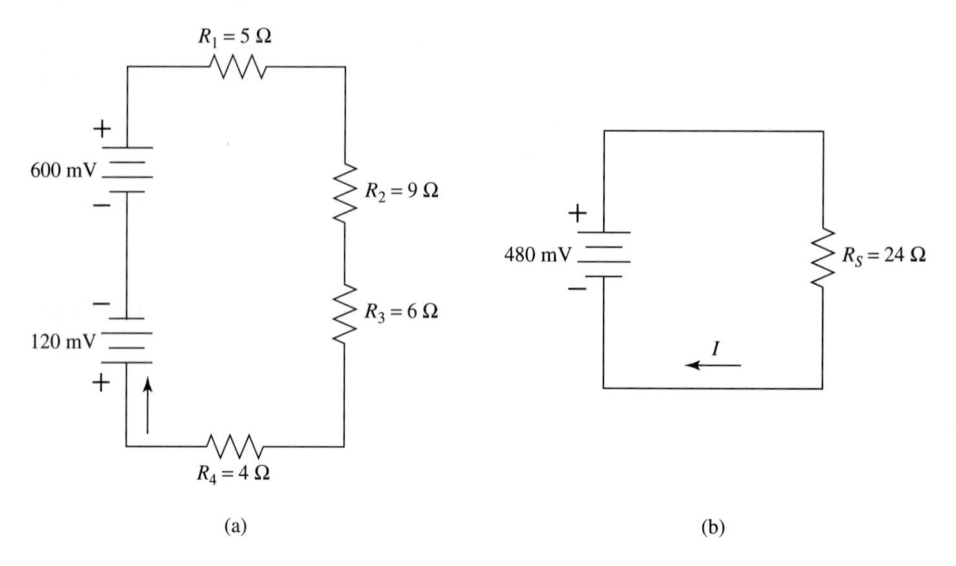

FIGURE 5.22

(a) Find a single resistor to replace the four series resistors.
(b) Draw an equivalent circuit using the single equivalent resistor found in part (a) and a single voltage source.
(c) Find the current flowing in the circuit of Fig. 5.21(a).

Solution
(a) From Equation 5.21, $R_s = 5 + 9 + 6 + 4 = 24\ \Omega$.

Answer $R_s = 24\ \Omega$.
(b) The two series voltage sources are algebraically added to yield the equivalent circuit shown in Fig. 5.22(b).
(c) The current flowing in the circuit of Fig. 5.22(a) is the same as the current flowing in the circuit of Fig. 5.22(b). From Ohm's Law applied to Fig. 5.22(b):

$$I = V_s/R_s = 480 \times 10^{-3}/24 = 20 \times 10^{-3}\ \text{A}$$

Answer $I = 20$ mA.

5.1.6.2 Inductors in Series

Consider the circuit of Fig. 5.23(a), which includes three inductors in series. The impedance of each inductor is as follows: $Z_1 = j2\pi fL_1$, $Z_2 = j2\pi fL_2$, and $Z_3 = j2\pi fL_3$. The single impedance that is equivalent to the three series impedances of these inductors is found from Equation 5.13:

$$\begin{aligned} Z_s &= Z_1 + Z_2 + Z_3 = j2\pi fL_1 + j2\pi fL_2 + j2\pi fL_3 \\ &= j2\pi f(L_1 + L_2 + L_3) \end{aligned}$$

The sum of the inductances in the previous equation can be considered a single inductance, L_s, such that

$$Z_s = j2\pi fL_s \quad \text{and} \quad L_s = L_1 + L_2 + L_3$$

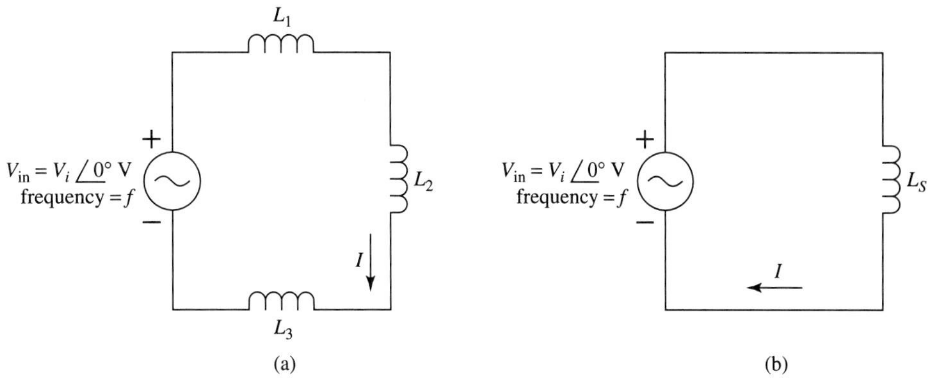

FIGURE 5.23

This result may be stated as the general law of series inductors:

> Any number of series connected inductors may be replaced by a single inductor whose inductance is given by the sum of all the series inductance values.

This law may be stated mathematically as follows:

$$L_s = \Sigma L_n \ (n = 1 \text{ to } N) \qquad \textbf{(Eq. 5.22)}$$

where N represents the number of inductors in series.

The circuit of Fig. 5.23(a) is redrawn in Fig. 5.23(b), where the single equivalent inductance, L_s, has replaced the three series inductors.

EXAMPLE 5.16

In the circuit of Fig. 5.24(a), two inductors are in series with a voltage source, V_i.

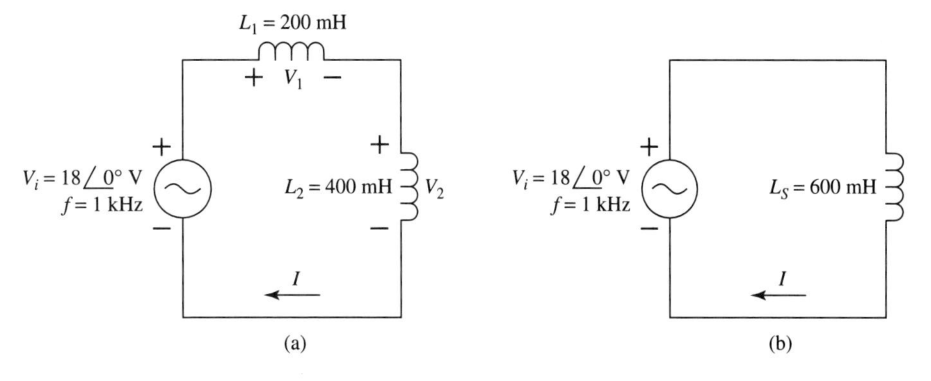

FIGURE 5.24

(a) Find the single inductor that is equivalent to the two series inductors.
(b) Redraw the circuit of Fig. 5.24(a) with the two series inductors replaced by the single equivalent inductor of part (a).
(c) Calculate the current flowing in the circuit of Fig. 5.24(a).
(d) Find V_1 and V_2, the phasor voltages across the inductors L_1 and L_2 respectively.

Solution

(a) From Equation 5.22 the single inductor that is equivalent to the two series inductors of Fig. 5.24(a) is:

$$L_s = L_1 + L_2 = 200 \times 10^{-3} + 400 \times 10^{-3} = 600 \times 10^{-3}$$

Answer $L_s = 600$ mH.

(b) The equivalent circuit is drawn in Fig. 5.24(b), with the single equivalent inductor replacing the two original inductors.

(c) Since the currents flowing in the two circuits of Fig. 5.24 are exactly the same, we may find this current by applying Ohm's Law to the circuit of Fig. 5.24(b):

$$I = V_i/Z_s$$

$$Z_s = 2\pi f L_s \angle 90^\circ = 2\pi(10^3)600 \times 10^{-3}\angle 90^\circ = 3.77 \times 10^3 \angle 90^\circ\ \Omega$$

$$I = \frac{18\angle 0^\circ}{3.77 \times 10^3 \angle 90^\circ} = 4.77 \times 10^{-3}\angle -90^\circ\ \text{A}$$

Answer $I = 4.77\angle -90^\circ$ mA.

(d) Since we now know the current in the circuit of Fig. 5.24(a), we can find the voltages V_1 and V_2 using Ohm's Law across each inductor:

$$\begin{aligned} V_1 &= 2\pi f L_1 \angle 90^\circ\, I \\ &= 2\pi(10^3)(200 \times 10^{-3})\angle 90^\circ(4.77 \times 10^{-3}\angle -90^\circ) = 6.00\angle 0^\circ\ \text{V} \\ V_2 &= 2\pi f L_2 \angle 90^\circ\, I \\ &= 2\pi(10^3)(400 \times 10^{-3})\angle 90^\circ(4.77 \times 10^{-3}\angle -90^\circ) = 12.00\angle 0^\circ\ \text{V} \end{aligned}$$

As a check, we will calculate the KVL around the circuit. Starting at the positive terminal of V_i and moving CW around Fig. 5.24(a):

$$V_1 + V_2 - V_i = 0 \qquad 6.00\angle 0^\circ + 12\angle 0^\circ - 18\angle 0^\circ = 0$$

The fact that our voltage calculations satisfy KVL is an indication that our figures are correct.

5.1.6.3 Capacitors in Series

A series circuit consisting of three capacitors in series with a voltage source is shown in Fig. 5.25(a). The impedance of each capacitor is given by $Z_1 = -j(1/2\pi f C_1)$, $Z_2 = -j(1/2\pi f C_2)$, $Z_3 = -j(1/2\pi f C_3)$. From Equation 5.13:

$$\begin{aligned} Z_s &= Z_1 + Z_2 + Z_3 = -j(1/2\pi f C_1) - j(1/2\pi f C_2) - j(1/2\pi f C_3) \\ &= (-j/2\pi f)\left(\frac{1}{C_1} + \frac{1}{C_2} + \frac{1}{C_3}\right) \end{aligned}$$

(a) (b)

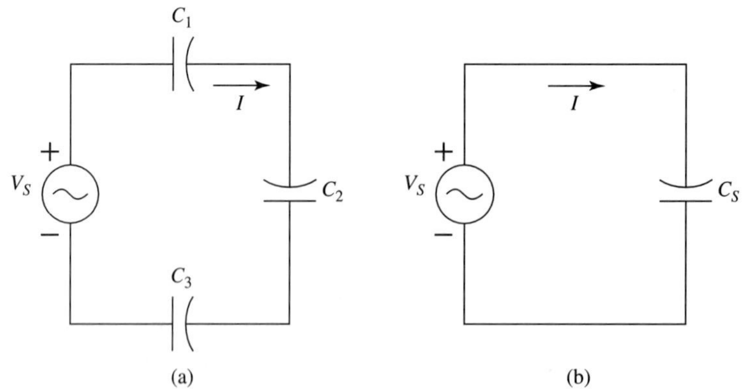

FIGURE 5.25

The sum of the terms in parentheses may be considered an equivalent capacitance such that:

$$Z_s = -j(1/2\pi f C_s)$$

where $\dfrac{1}{C_s} = \dfrac{1}{C_1} + \dfrac{1}{C_2} + \dfrac{1}{C_3}$

This result may be stated in general terms as follows:

Any number of capacitors in series may be replaced by a single capacitor whose capacitance is equal to the reciprocal of the sum obtained by adding the reciprocals of all individual capacitances.

This law may be stated mathematically as follows:

$$1/C_s = \Sigma(1/C_n) \quad (n = 1 \text{ to } N) \qquad \textbf{(Eq. 5.23)}$$

where N represents the number of capacitors in series.

Note that this is unlike the cases of the resistors and inductors in series, where the equivalent series resistance was a simple sum of all the resistances and the equivalent series inductance was a simple sum of all the inductances.

EXAMPLE 5.17

The circuit of Fig. 5.26(a) contains a voltage source and three capacitors in series.

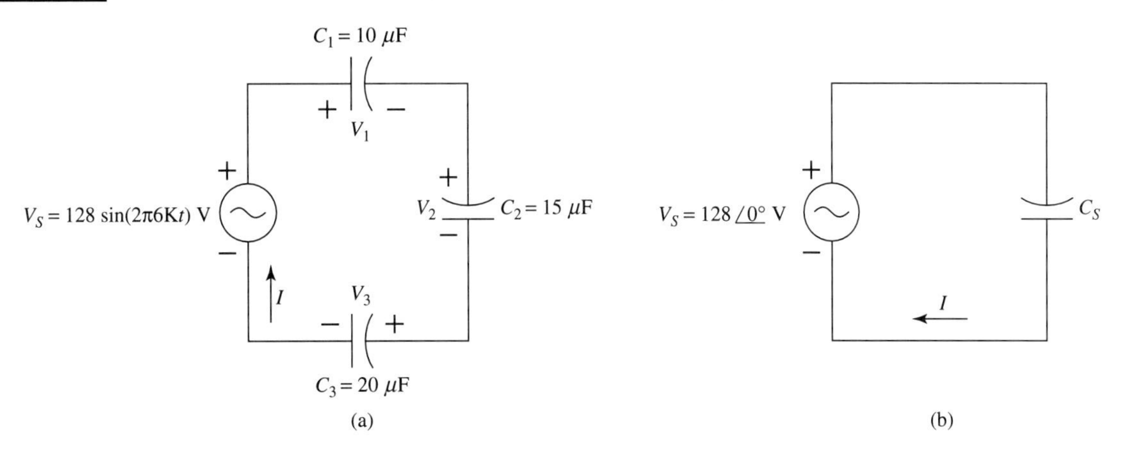

FIGURE 5.26

(a) Find a single capacitor that is the equivalent of the three series capacitors.
(b) Redraw the circuit of Fig. 5.26(a) with the three capacitors replaced by their single equivalent capacitor.
(c) Find the sinusoidal current flowing in the circuit of Fig. 5.26(a).
(d) Find the sinusoidal voltage across each of the capacitors in Fig. 5.26(a).

Solution

(a) From Equation 5.23 the single capacitance, C_s, that is equivalent to the three series capacitances is given by:

$$\frac{1}{C_s} = \frac{1}{C_1} + \frac{1}{C_2} + \frac{1}{C_3} = \frac{1}{10 \times 10^{-6}} + \frac{1}{15 \times 10^{-6}} + \frac{1}{20 \times 10^{-6}}$$
$$= 10^5 + 0.67 \times 10^5 + 0.5 \times 10^5 = 10^5(1 + 0.67 + 0.50) = 2.17 \times 10^5$$

Taking the reciprocal of both sides of the previous equation:

$$C_s = 1/(2.17 \times 10^5) = 4.61 \times 10^{-6}$$

Answer $C_s = 4.61\ \mu$F.

(b) The circuit of Fig. 5.26(a) is redrawn in Fig. 5.26(b) with the three capacitors replaced by their single equivalent capacitor.

(c) The current in Fig. 5.26(a) may be found easily by using the equivalent circuit of Fig. 5.26(b) with the voltage source expressed as a phasor. The impedance of Z_s is:

$$Z_s = -j1/(2\pi f C_s) = -j[1/(2\pi 6 \times 10^3 \times 4.61 \times 10^{-6})] = 5.75\angle -90°\ \Omega$$

Using Z_s in Ohm's Law:

$$I = V_s/Z_s \qquad V_s = 128\angle 0^\circ$$

$$I = \frac{128\angle 0^\circ}{5.75\angle -90^\circ} = 22.26\angle 90^\circ \text{ A}$$

Answer $i = 22.26 \sin(2\pi 6\text{K}t + 90^\circ)$ A.

(d) The voltage across each capacitor in Fig. 5.26(a) may be found by applying Ohm's Law to each capacitor. We first find the impedance of each capacitor as follows:

$$Z_1 = -j[1/(2\pi f C_1) = -j[1/(2\pi 6 \times 10^3 \times 10 \times 10^{-6}) = -j2.65\ \Omega$$
$$Z_2 = -j[1/(2\pi f C_2) = -j[1/(2\pi 6 \times 10^3 \times 15 \times 10^{-6}) = -j1.77\ \Omega$$
$$Z_3 = -j[1/(2\pi f C_3) = -j[1/(2\pi 6 \times 10^3 \times 20 \times 10^{-6}) = -j1.33\ \Omega$$

We now apply Ohm's Law to each capacitive impedance using the phasor current found in part (c):

$$V_1 = IZ_1 = 22.26\angle 90^\circ(2.65\angle -90^\circ) = 59.00\angle 0^\circ \text{ V}$$
$$V_2 = IZ_2 = 22.26\angle 90^\circ(1.77\angle -90^\circ) = 39.40\angle 0^\circ \text{ V}$$
$$V_3 = IZ_3 = 22.26\angle 90^\circ(1.33\angle -90^\circ) = 29.60\angle 0^\circ \text{ V}$$

Converting each of these voltage phasors to sinusoidal time functions we get:

Answer $v_1 = 59.0 \sin(2\pi 6\text{K}t)$ V, $v_2 = 39.4 \sin(2\pi 6\text{K}t)$ V, $v_3 = 29.6 \sin(2\pi 6\text{K}t)$ V.

Equation 5.23, which enables us to calculate the single capacitance, C_s, that is equivalent to any number of capacitances in series, may be simplified for the case where only ***two*** capacitors are in series. In this case, if C_1 and C_2 are the two series capacitors, Equation 5.23 reduces to:

$$\frac{1}{C_s} = \frac{1}{C_1} + \frac{1}{C_2} \qquad \textbf{(Eq. 5.24)}$$

Taking the reciprocal of both sides of equation 5.24:

$$C_s = 1/[(1/C_1) + (1/C_2)]$$

or

$$C_s = C_1C_2/(C_1 + C_2) \qquad \textbf{(Eq. 5.25)}$$

Equation 5.25 may be expressed as follows:

> Two capacitors in series may be replaced by a single capacitor whose capacitance is equal to the product of the two capacitances divided by the sum of the two capacitances.

EXAMPLE 5.18

In the circuit of Fig. 5.27(a), two capacitors are in series with a voltage source, V_s.

(a) Find the single capacitance that will be equivalent to the two series capacitances of Fig. 5.27(a).

(b) Find the phasor current flowing in the circuit of Fig. 5.27(a).

(c) Find the voltage across each capacitor of Fig. 5.27(a).

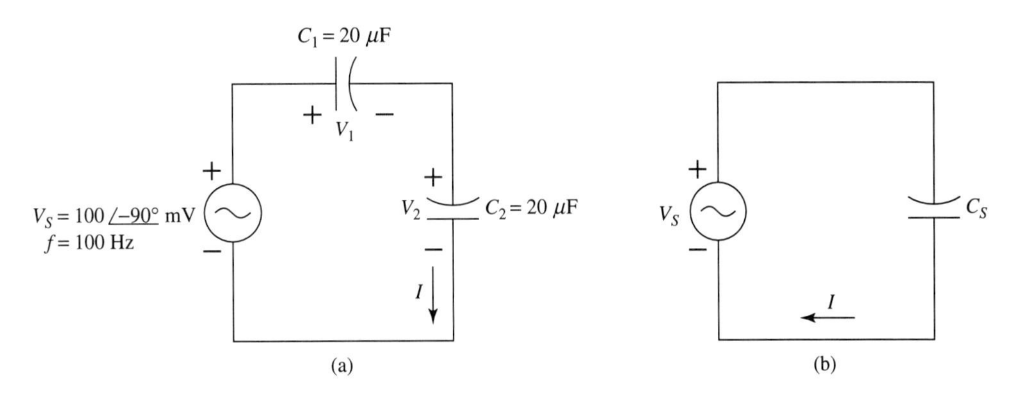

FIGURE 5.27

Solution

(a) From Equation 5.25:

$$C_s = C_1C_2/(C_1 + C_2)$$
$$= 20 \times 10^{-6} \times 20 \times 10^{-6}/(20 \times 10^{-6} + 20 \times 10^{-6}) = 10 \times 10^{-6}\ F$$

Answer $C_s = 10\ \mu F$.

(b) To find the current flowing in the circuit of Fig. 5.27(a), we first redraw the circuit as in Fig. 5.27(b) replacing the two series capacitors by their single equivalent capacitor. Applying Ohm's Law to Fig. 5.27(b):

$$I = V_s/Z_s$$
$$Z_s = -j[1/(2\pi f C_s)] = -j[1/(2\pi 100 \times 10 \times 10^{-6})] = -j159.15\ \Omega$$
$$I = \frac{100 \times 10^{-3}\angle{-90°}}{159.15\angle{-90°}} = 0.63 \times 10^{-3}\angle{0°}\ A$$

Answer $I = 0.63\angle{0°}$ mA

(c) The voltage across each capacitor may be found by Ohm's Law applied to each capacitive impedance as follows:

$$Z_1 = -j[1/(2\pi f C_1)] = -j[1/(2\pi 100 \times 20 \times 10^{-6})] = -j79.6\ \Omega$$

Since $C_2 = C_1$, then $Z_2 = Z_1$, and therefore $Z_2 = -j79.6\ \Omega$

$$V_1 = IZ_1 = (0.63 \times 10^{-3}\angle{0°}) \times (79.6\angle{-90°}) = 50.1 \times 10^{-3}\angle{-90°}\ V$$

Since the two capacitances have equal impedances:

$$V_2 = V_1 = 50.1 \times 10^{-3}\angle{-90°}\ V$$

Answer $V_1 = 50.1\angle{-90°}$ mV, $V_2 = 50.1\angle{-90°}$ mV.

Note from the previous example that the equivalent capacitance of the ***two equal*** series capacitances turned out to be equal to ***one half*** of each individual capacitance. This may be generalized by the following rule:

If ***n equal*** capacitances of value C are in series, then they may be replaced by a single equivalent capacitance C_s equal in value to C/n.

This may be mathematically stated as follows:

$$C_s = C/n \qquad \textbf{(Eq. 5.26)}$$

where C = the value of each series connected capacitance
n = the number of capacitors connected in series
C_s = the single equivalent capacitance

EXAMPLE 5.19

Eight capacitors are connected in series. Each capacitor has a capacitance $C = 12\ \mu F$. Find the single capacitance that is equivalent to these eight series-connected capacitors.

Solution From Equation 5.26:

$$C_s = 12\ \mu F/8 = 1.5\ \mu F$$

Answer $C_s = 1.5\ \mu F$

5.2 THE PARALLEL CIRCUIT

5.2.1 Definition

We will define the term ***parallel*** as follows:

> Two or more circuit impedances are in ***parallel*** if the ***same voltage*** exists across each impedance.

Do not confuse the ***same*** voltage with ***equal*** voltage. Two or more impedances can have equal voltages but not be connected in parallel. This definition implies that each parallel impedance must be ***physically connected*** between the ***same*** two voltage points.

Consider the circuit of Figure 5.28. In this circuit, Z_1 and Z_2 are in parallel because they are both connected between the same two voltage points a and b. Since they are connected in parallel, then the voltage across impedance Z_2 must equal the voltage across Z_1, which is given as $12\ V\angle 40°$.

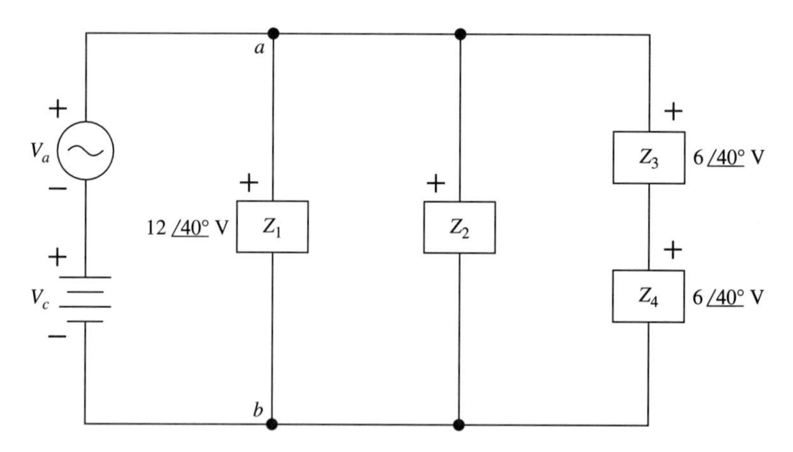

FIGURE 5.28

In this same circuit, impedances Z_3 and Z_4 both have equal voltages of $6\ V\angle 60°$ across them but they are ***not*** in parallel because they are each connected between two different voltage points. Z_3 and Z_4 are in fact connected in series.

Note that if impedances Z_3 and Z_4 were replaced by their single equivalent impedance Z_s, then ***that*** equivalent impedance ***would*** be in parallel with other two impedances Z_1 and Z_2 because it would be connected between voltage points a and b.

EXAMPLE 5.20

Find all the parallel impedance combinations in Fig. 5.29.

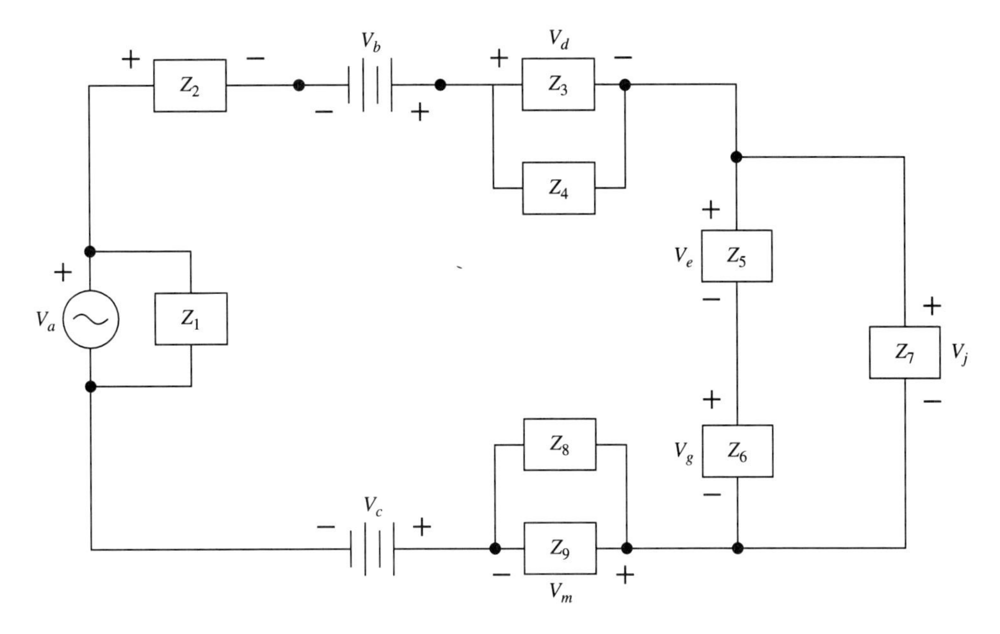

FIGURE 5.29

Solution The following parallel impedances appear in Fig. 5.29:

- Z_3 is in parallel with Z_4.
- Z_1 is in parallel with V_a.
- Z_8 is in parallel with Z_9.
- Z_7 is in parallel with the ***series combination*** of Z_5 and Z_6.

5.2.2 Kirchhoff's Current Law

Kirchhoff's Current Law, which will hereafter be abbreviated as KCL, is another basic law of circuit analysis. This law may be stated as follows:

> KCL: The algebraic sum of currents entering or leaving a node is equal to zero.

A ***node*** will be defined as a voltage point at which three or more impedances and/or power sources (i.e., voltage and/or current sources) are connected. KCL may be mathematically stated as follows:

$$\Sigma I_e = \Sigma I_l = 0 \qquad \textbf{(Eq. 5.27)}$$

where ΣI_e = the sum of currents ***entering*** the node
ΣI_l = the sum of currents ***leaving*** the node

In any given problem, either form of Equation 5.27 may be used.

EXAMPLE 5.21

In the circuit of Fig. 5.30, the following currents are given: $I_1 = 120\underline{/20^\circ}$ mA, $I_2 = 60\underline{/45^\circ}$ mA, and $I_5 = 38\underline{/50^\circ}$ mA. Find the currents I_3 and I_6.

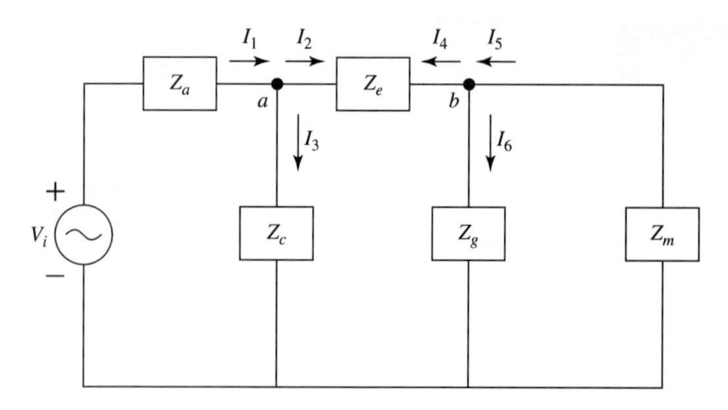

FIGURE 5.30

Solution To solve this problem, we first identify two nodes in Fig. 5.30 as nodes a and b. At node a we use the following form of KCL:

$$(\Sigma I_e)_a = 0$$
$$(\Sigma I_e)_a = I_1 - I_2 - I_3 = 0$$

Note that I_1 is positive because it is ***entering*** node a and we are using the form of KCL in which the sum of currents entering a node is equal to zero. Solving this equation for I_3, we get:

$$\begin{aligned} I_3 = I_1 - I_2 &= (120 \times 10^{-3}\underline{/20^\circ}) - (60 \times 10^{-3}\underline{/45^\circ})\text{ A} \\ &= 112.76 \times 10^{-3} + j41.04 \times 10^{-3} - (42.43 \times 10^{-3} + j42.43 \times 10^{-3})\text{ A} \\ &= (112.76 - 42.43) \times 10^{-3} + j(41.04 - 42.43) \times 10^{-3}\text{ A} \\ &= (70.33 - j1.39) \times 10^{-3}\text{ A} \end{aligned}$$

Converting this to polar form, we get:

Answer $I_3 = 70.34\underline{/-1.13^\circ}$ mA.

To find the current I_6 we apply KCL to node b. This time, we will use the form of KCL that states that the sum of currents ***leaving*** a node is equal to zero:

$$(\Sigma I_l)_b = 0$$

At node b:

$$(\Sigma I_l) = I_4 - I_5 + I_6 = 0$$

Solving this equation for I_6, we get:

$$I_6 = I_5 - I_4$$

Note from the circuit of Fig. 5.30 that $I_4 = -I_2$ because both currents flow in the same leg but are opposite in direction. Therefore,

$$I_4 = -60\underline{/45^\circ}\text{ mA}$$

Using this value of I_4 and the given value of I_5 in the preceding equation for I_6 we may write:

$$\begin{aligned} I_6 &= (38 \times 10^{-3}\underline{/50^\circ}) - (-60 \times 10^{-3}\underline{/45^\circ})\text{ A} \\ &= (38\underline{/50^\circ} + 60\underline{/45^\circ}) \times 10^{-3}\text{ A} \\ &= (24.43 + j29.10 + 42.43 + j42.43) \times 10^{-3}\text{ A} \\ &= (66.86 + j71.53) \times 10^{-3}\text{ A} \end{aligned}$$

Converting this to polar form we get:

Answer $I_6 = 97.91\underline{/46.93^\circ}$ mA.

EXAMPLE 5.22

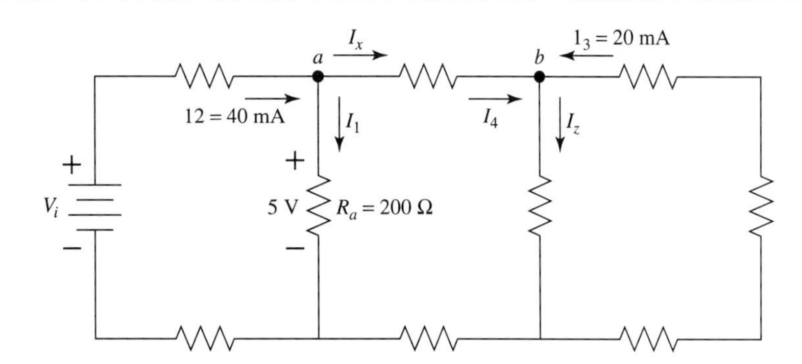

FIGURE 5.31

(a) Find the current I_x.
(b) Find the current I_z.

Solution

(a) Applying KCL to node a:

$$(\Sigma I_e)_a = I_2 - I_x - I_1 = 0$$

Solving for I_x:

$$I_x = I_2 - I_1$$

I_2 is given. To find I_1, we apply Ohm's Law to resistor R_a:

$$I_1 = 5/200 = 25 \text{ mA}$$

Using the values of I_1 and I_2 in the equation for I_x, we get:

$$I_x = 40 \times 10^{-3} - 25 \times 10^{-3} = 15 \times 10^{-3} \text{ A}$$

Answer $I_x = 15$ mA.

(b) Applying KCL to node b:

$$(\Sigma I_e)_b = I_4 + I_3 - I_z = 0$$

Solving for I_z:

$$I_z = I_4 + I_3$$

I_3 is given and $I_4 = I_x$ because they are both in the same series leg. Therefore:

$$I_z = 15 \times 10^{-3} + 20 \times 10^{-3} = 35 \times 10^{-3} \text{ A}$$

Answer $I_z = 35$ mA.

5.2.3 Equivalent Parallel Impedance

Consider the circuit shown in Fig. 5.32(a). This circuit shows three impedances in parallel with a current source. From KCL:

$$I_s = I_1 + I_2 + I_3 \qquad \textbf{(Eq. 5.28)}$$

Since these impedances are all in parallel, the same voltage, V, exists across each impedance. Applying Ohm's Law to each impedance:

$$I_1 = V/Z_1 \qquad I_2 = V/Z_2 \qquad I_3 = V/Z_3 \qquad \textbf{(Eq. 5.29)}$$

Now consider the circuit in Fig. 5.32(b). In this circuit, Z_p represents the single impedance that is equivalent to the three impedances in parallel. By equivalent we mean that

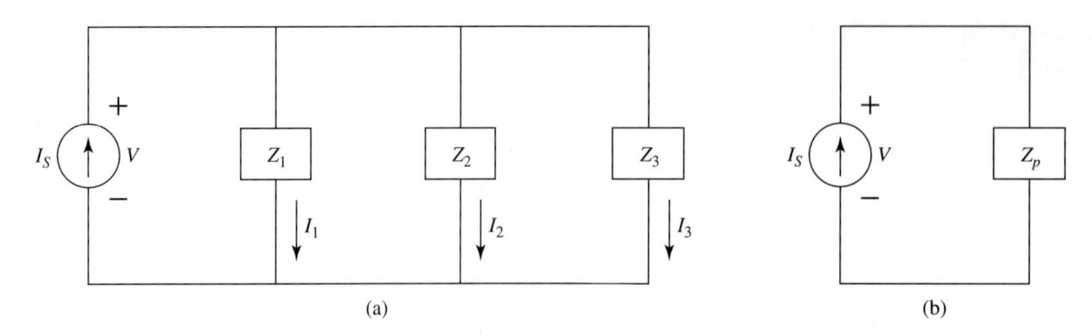

FIGURE 5.32

the same current source placed across Z_p will result in the same voltage across Z_p as the voltage across the three parallel impedances in Fig. 5.32(a). Therefore:

$$I_s = V/Z_p \qquad \textbf{(Eq. 5.30)}$$

Using Equations 5.29 and 5.30 in Equation 5.28:

$$V/Z_p = V/Z_1 + V/Z_2 + V/Z_3 \qquad \textbf{(Eq. 5.31)}$$

Dividing both sides of Equation 5.31 by V, we get Equation 5.32.

$$1/Z_p = 1/Z_1 + 1/Z_2 + 1/Z_3 \qquad \textbf{(Eq. 5.32)}$$

Equation 5.32 states that three impedances in parallel may be replaced by a single impedance such that the reciprocal of the equivalent impedance equals the sum of the reciprocals of each of the three impedances.

This result may be generalized for any number of impedances in parallel as follows:

> Any number of impedances in parallel may be replaced by a single impedance such that the reciprocal of the equivalent impedance is equal to the sum of the reciprocals of each of the individual impedances.

Mathematically, this may be stated as follows:

$$1/Z_p = \Sigma(1/Z_n) \quad (n = 1 \text{ to } N) \qquad \textbf{(Eq. 5.33)}$$

where N represents the number of impedances in parallel.

EXAMPLE 5.23

In the circuit of Fig. 5.32(a): $Z_1 = 20\underline{/36.9^\circ}\ \Omega$, $Z_2 = 15\underline{/53.1^\circ}\ \Omega$, $Z_3 = 25\underline{/60^\circ}\ \Omega$, and $I_s = 10\underline{/0^\circ}$ A. Find:

(a) The single impedance, Z_p, that is equivalent to the three parallel impedances
(b) The phasor voltage across each impedance
(c) The sinusoidal voltage across each impedance if the frequency of the current source is $f = 1$ MHz.

Solution

(a) From Equation 5.33:

$$\begin{aligned} 1/Z_p &= \frac{1}{20\underline{/36.9^\circ}} + \frac{1}{15\underline{/53.1^\circ}} + \frac{1}{25\underline{/60^\circ}} \\ &= 0.05\underline{/-36.9^\circ} + 0.067\underline{/-53.1^\circ} + 0.04\underline{/-60^\circ} \\ &= 0.04 - j0.03 + 0.04 - j0.05 + 0.02 - j0.03 \\ &= 0.10 - j0.11 = 0.15\underline{/-49.71^\circ} \end{aligned}$$

Taking the reciprocal of both sides of this equation:

$$Z_p = \frac{1}{0.15\underline{/-49.71^\circ}} = 6.67\underline{/49.71^\circ}\ \Omega$$

Answer $Z_p = 6.67\angle 47.73°\ \Omega$.

(b) The circuit of Fig. 5.32(b) is the equivalent of the circuit of Fig. 5.32(a) with the three impedances replaced by their single equivalent impedance, Z_p. Using Ohm's Law in Fig. 5.32(b):

$$V = I_s Z_p = (10\angle 0°)(6.67\angle 47.73°) = 66.67\angle 47.73°$$

Answer $V = 66.67\angle 47.73°$ V.

(c) Converting the phasor voltage to a sinusoidal voltage:

Answer $v = 66.67 \sin(2\pi 10^6 t + 47.73°)$ V

A special case of Equation 5.33 occurs when there are only ***two*** impedances in parallel. In this case, Equation 5.33 becomes

$$\frac{1}{Z_p} = \frac{1}{Z_1} + \frac{1}{Z_2} \qquad \textbf{(Eq. 5.34)}$$

If we take the reciprocal of both sides of Equation 5.34 we get:

$$Z_p = \frac{1}{(1/Z_1) + (1/Z_2)} \qquad \textbf{(Eq. 5.35)}$$

Multiplying the right side of Equation 5.35 by Z_1Z_2/Z_1Z_2 (this fraction is equal to 1, so it does not change anything) we get:

$$Z_p = Z_1Z_2/(Z_1 + Z_2) \qquad \textbf{(Eq. 5.36)}$$

Equation 5.36 may be stated as follows:

> Two impedances in parallel may be replaced by a single impedance equal in value to the product of the two impedances divided by the sum of the two impedances.

EXAMPLE 5.24

Fig. 5.33(a) shows a circuit consisting of two impedances in parallel with a voltage source. Given $Z_a = 1.2\text{K}\angle 30°\ \Omega$, $Z_c = 0.8\text{K}\angle -60°\ \Omega$, and $V_s = 120\angle 0°$ V, find:

(a) Z_p = the single impedance that is equivalent to the two parallel impedances
(b) I_s = the current supplied by the voltage source
(c) I_a = the current flowing through Z_a
(d) I_c = the current flowing through Z_c

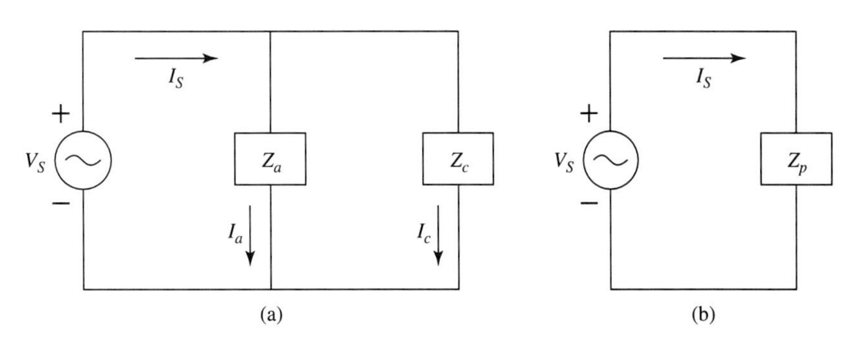

FIGURE 5.33

Solution

(a) Applying Equation 5.36 to the two parallel impedances of Fig. 5.33(a), the single equivalent impedance is given by:

$$Z_p = \frac{(1.2\text{K}\angle 30^\circ)(0.8\text{K}\angle -60^\circ)}{(1.2\text{K}\angle 30^\circ + 0.8\text{K}\angle -60^\circ)}$$
$$= \frac{(1.2\text{K})(0.8\text{K})\angle -30^\circ}{(1.04\text{K} + j0.60\text{K} + 0.4\text{K} - j0.69\text{K})}$$
$$= 665.64\angle -26.3^\circ$$

Answer $Z_p = 665.64\angle -26.3^\circ\ \Omega$.

(b) The circuit of Fig. 5.33(a) is redrawn in Fig. 5.33(b) with the two parallel impedances replaced by their single equivalent impedance. We may calculate I_s by applying Ohm's Law to Z_p in Fig. 5.33(b).

$$I_s = V_s/Z_p = (120\angle 0^\circ)/(665.64\angle -26.3^\circ) = 180\angle 26.3^\circ \text{ mA}$$

Answer $I_s = 180\angle 26.30^\circ$ mA.

(c) The current I_a is found by applying Ohm's Law to Z_a in Fig. 5.33(a):

$$I_a = V_s/Z_a = (120\angle 0^\circ)/(1.2\text{K}\angle 30^\circ) = 100\angle -30^\circ \text{ mA}$$

Answer $I_a = 100\angle -30^\circ$ mA.

(d) The current I_c is found by applying Ohm's Law to Z_c in Fig. 5.33(a):

$$I_c = V_s/Z_c = (120\angle 0^\circ)/(0.8\text{K}\angle -60^\circ) = 150\angle 60^\circ \text{ mA}$$

Answer $I_c = 150\angle 60^\circ$ mA.

5.2.4 Admittances in Parallel

When a circuit contains parallel impedances, it may be easier to perform certain calculations using the admittance values. This may be demonstrated by considering Equation 5.33, which is reproduced here as Equation 5.37.

$$1/Z_p = \Sigma(1/Z_n) \quad (n = 1 \text{ to } N) \qquad \textbf{(Eq. 5.37)}$$

where Z_p is the single equivalent impedance of N impedances in parallel.

From the definition of admittance in Equation 4.30:

$$Y_p = 1/Z_p \quad \text{and} \quad Y_n = 1/Z_n \qquad \textbf{(Eq. 5.38)}$$

Using Equation 5.38 in Equation 5.37:

$$Y_p = \Sigma Y_n \quad (n = 1 \text{ to } N) \qquad \textbf{(Eq. 5.39)}$$

Equation 5.39 may be stated as follows:

> Any number of admittances in parallel may be replaced by a single equivalent admittance equal to the sum of all the individual admittances.

EXAMPLE 5.25

For the three parallel admittances of Fig. 5-34(a), $V_s = 420\angle 30^\circ$ mV, $Y_1 = 10\angle 60^\circ$ S, $Y_2 = 14\angle 45^\circ$ S, and $Y_3 = 18\angle 20^\circ$ S. Find:

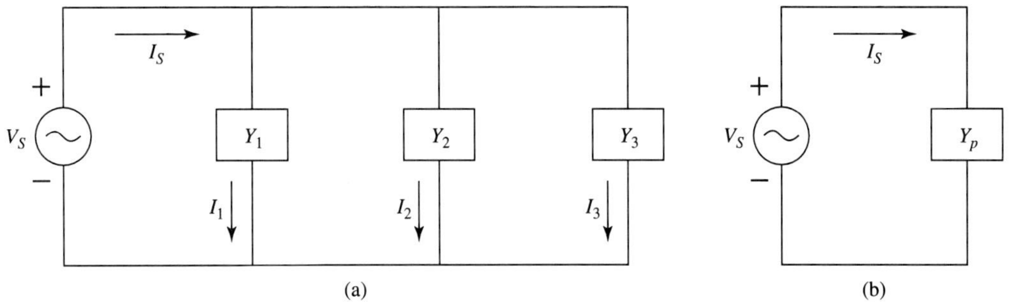

FIGURE 5.34

(a) The single equivalent admittance
(b) The single equivalent impedance
(c) The phasor current through each admittance
(d) The total phasor current delivered by the voltage source

Solution

(a) To find the single equivalent admittance we will use Equation 5.39:

$$\begin{aligned} Y_p = Y_1 + Y_2 + Y_3 &= (10\angle 60^\circ) + (14\angle 45^\circ) + (18\angle 20^\circ) \\ &= 5 + j8.66 + 9.90 + j9.90 + 16.91 + j6.16 \\ &= (31.81 + j24.72)\text{ S} \end{aligned}$$

In polar form: $Y_p = 40.26\angle 37.85^\circ$

Answer $Y_p = 40.26\angle 37.85^\circ$ S.

(b) The single equivalent impedance may easily be found from the admittance using the basic definition of admittance as the reciprocal of the impedance:

$$Z_p = 1/Y_p = 1/(40.26\angle 37.85^\circ) = 24.8\angle -37.85^\circ\text{ m}\Omega$$

(c) The phasor current through each admittance may be found using Ohm's Law and recognizing that the same voltage, V_s, appears across each admittance. Using the admittance form of Ohm's Law:

$$\begin{aligned} I_1 &= Y_1V_s = (10\angle 60^\circ)(420 \times 10^{-3}\angle 30^\circ) = 4.2\angle 90^\circ \\ I_2 &= Y_2V_s = (14\angle 45^\circ)(420 \times 10^{-3}\angle 30^\circ) = 5.88\angle 75^\circ \\ I_3 &= Y_3V_s = (18\angle 20^\circ)(420 \times 10^{-3}\angle 30^\circ) = 7.56\angle 50^\circ \end{aligned}$$

Answer $I_1 = 4.2\angle 90^\circ$ A, $I_2 = 5.88\angle 75^\circ$ A, $I_3 = 7.56\angle 50^\circ$ A.

(d) The total phasor current may be found in either of two ways. First, using Ohm's Law in the circuit of Fig. 5.34(b)

$$I_s = Y_pV_s = (40.26\angle 37.85^\circ)(420 \times 10^{-3}\angle 30^\circ) = 16.91\angle 67.85^\circ$$

Answer $I_s = 16.91\angle 67.85^\circ$ A.

Second, using KCL in the circuit of Fig. 5.34(a):

$$I_s = I_1 + I_2 + I_3 = 4.2\angle 90^\circ + 5.88\angle 75^\circ + 7.56\angle 50^\circ$$

As an exercise, convert the polar form of these phasors into their rectangular counterparts so that the addition may be performed. Then convert the rectangular sum back into polar form and compare it to the answer obtained using Ohm's Law. They should, of course, be the same.

Answer $I_s = 6.38 + j15.67 = 16.91\angle 67.85^\circ$ A.

EXAMPLE 5.26

In Example 5.25, find the total equivalent conductance and susceptance of the circuit.

Solution Since Y_p represents the total equivalent admittance, we may find the total equivalent conductance and susceptance by expressing Y_p in rectangular form as follows:

$$Y_p = 40.26\angle 37.85° = 31.79 + j24.70 \text{ S}$$

Recall that the conductance, G, is defined as the real part of the admittance and the susceptance, B, is defined as the imaginary part of the admittance. Therefore:

Answer $G = 31.79$ S, $B = 24.70$ S.

5.2.5 Current Division Law

Consider a circuit of three parallel admittances as in Fig. 5.34(a) of the previous example. From KCL:

$$I_s = I_1 + I_2 + I_3 \qquad \textbf{(Eq. 5.40)}$$

From Ohm's Law:

$$I_s = Y_p V_s \qquad \textbf{(Eq. 5.41)}$$

and

$$I_1 = Y_1 V_s \qquad I_2 = Y_2 V_s \qquad I_3 = Y_3 V_s \qquad \textbf{(Eq. 5.42)}$$

Dividing each term in Equation 5.42 by I_s of Equation 5.41:

$$\begin{aligned} I_1/I_s &= Y_1 V_s / Y_p V_s = Y_1/Y_p \\ I_2/I_s &= Y_2 V_s / Y_p V_s = Y_2/V_p \\ I_3/I_s &= Y_3 V_s / Y_p V_s = Y_3/V_p \end{aligned} \qquad \textbf{(Eq. 5.43)}$$

Rearranging terms, we get the following:

$$I_1 = (Y_1/Y_p)I_s \qquad I_2 = (Y_2/Y_p)I_s \qquad I_3 = (Y_3/Y_p)I_s \qquad \textbf{(Eq. 5.44)}$$

The result of Equation 5.44 may be generalized in words as follows:

> For a circuit consisting of any number of admittances in parallel, the current through any individual admittance is equal to that individual admittance divided by the total equivalent admittance of the parallel circuit multiplied by the total current entering the parallel circuit.

This statement is known as the ***Current Division Law for parallel admittances.*** This law may be mathematically stated as follows:

$$I_n = (Y_n/Y_p)I_s \qquad \textbf{(Eq. 5.45)}$$

where I_s = total current entering the parallel circuit
I_n = current through admittance Y_n
Y_p = total equivalent admittance of the parallel circuit

EXAMPLE 5.27

The circuit of Fig. 5.35 shows three admittances in parallel with a current source. In this circuit: $I_s = 25\angle 0°$ mA, $Y_a = 32\angle 50°$ S, $Y_c = 40\angle 36.9°$ S, and $Y_e = 50\angle 25°$ S. Using the Current Division Law of parallel admittances, calculate the current through each admittance.

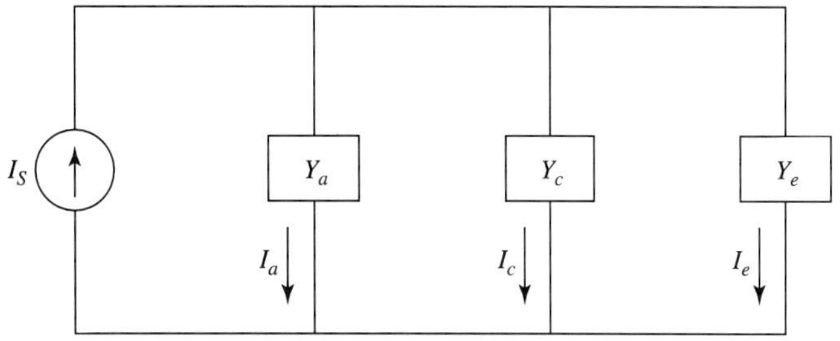

FIGURE 5.35

Solution To use Equation 5.45 we must first calculate the total equivalent admittance for this circuit as follows:

$$\begin{aligned} Y_p &= Y_a + Y_c + Y_e = (32\angle 50^\circ) + (40\angle 36.9^\circ) + (50\angle 25^\circ) \\ &= 20.57 + j24.51 + 32.00 + j24.00 + 45.32 + j21.13 \\ &= 97.89 + j69.64 = 120.13\angle 35.43^\circ \text{ S} \end{aligned}$$

We can now use this value of Y_p together with the given values of the individual admittances in Equation 5.45.

$$\begin{aligned} I_a &= (Y_a/Y_p)I_s = [(32\angle 50^\circ)/(120.13\angle 35.43^\circ)]25\angle 0^\circ = 6.67\angle 14.57^\circ \text{ mA} \\ I_c &= (Y_c/Y_p)I_s = [(40\angle 36.9^\circ)/(120.13\angle 35.43^\circ)]25\angle 0^\circ = 8.32\angle 1.47^\circ \text{ mA} \\ I_e &= (Y_e/Y_p)I_s = [(50\angle 25^\circ)/(120.13\angle 35.43^\circ)]25\angle 0^\circ = 10.41\angle -10.43^\circ \text{ mA} \end{aligned}$$

Answer $I_a = 6.67\angle 14.57^\circ$ mA, $I_c = 8.32\angle 1.47^\circ$ mA, $I_e = 10.41\angle -10.43^\circ$ mA.

The Current Division Law may be employed whenever there are admittances in parallel even when the parallel admittances are part of a larger circuit containing other elements that are differently connected.

EXAMPLE 5.28

In the circuit of Fig. 5.36:

$$I_2 = 156\angle -60^\circ \text{ mA} \qquad Y_a = 260\angle 30^\circ \text{ mS} \qquad Y_c = 120\angle 45^\circ \text{ mS}$$

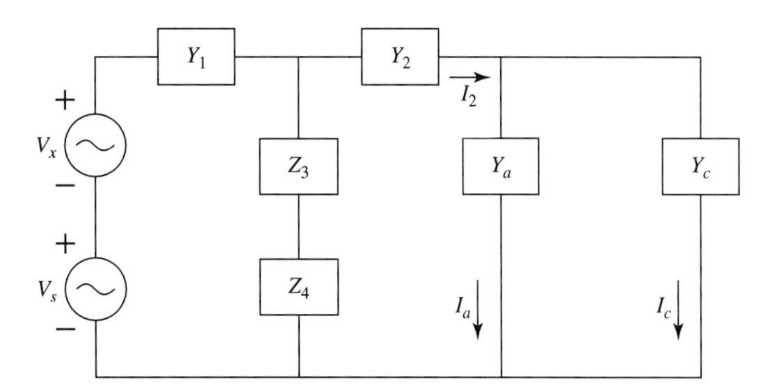

FIGURE 5.36

Find the current through Y_a and through Y_c.

Solution The two admittances Y_a and Y_c are in parallel. We are given I_2, which is the total current into the parallel combination of Y_a and Y_c. Despite the fact that these two admittances are part of a larger circuit of differently connected admittances, the known quantities are sufficient to allow us to use Equation 5.45 as follows:

$$I_a = (Y_a/Y_p)I_2 \qquad I_c = (Y_c/Y_p)I_2$$
$$Y_p = Y_a + Y_c = (260\angle 30^\circ) + (120\angle 45^\circ)$$
$$= 225.17 + j130.00 + 84.85 + j84.85 = 310 + j214.85 \text{ mS}$$

Converting this rectangular value for Y_p to polar form:

$$Y_p = 377.17\angle 34.72^\circ \text{ mS}$$

Substituting this value of Y_p and the given values of Y_a, Y_c, and I_2 into the preceding equations for I_a and I_c, we get:

$$I_a = [(260\angle 30^\circ)/(377.17\angle 34.72^\circ)]156\angle -60^\circ = 107.54\angle -64.72^\circ \text{ mA}$$
$$I_c = [(120\angle 45^\circ)/(377.17\angle 34.72^\circ)]156\angle -60^\circ = 49.63\angle -49.72^\circ \text{ mA}$$

Answer $I_a = 107.54\angle -64.72^\circ$ mA, $I_c = 49.63\angle -49.72^\circ$ mA.

When ***two*** admittances are in parallel, the Current Division Law of Equation 5.45 may be expressed just as easily in terms of the impedances. If Z_1 and Z_2 represent two impedances in parallel, then

$$Y_1 = 1/Z_1 \quad \text{and} \quad Y_2 = 1/Z_2 \qquad \textbf{(Eq. 5.46)}$$

Equation 5.45 reduces to the following for two admittances:

$$I_1 = (Y_1/Y_s)I_s \qquad I_2 = (Y_2/Y_s)I_s \qquad \textbf{(Eq. 5.47)}$$

From Equation 5.36, the two parallel impedances may be written as:

$$Z_p = Z_1Z_2/(Z_1 + Z_2) = 1/Y_p \qquad \textbf{(Eq. 5.48)}$$

Substituting Equations 5.48 and 5.47 into Equation 5.45:

$$I_1 = [Z_2/(Z_1 + Z_2)]I_s \qquad I_2 = [Z_1/(Z_1 + Z_2)]I_s \qquad \textbf{(Eq. 5.49)}$$

Equation 5.49 may be stated in words as follows:

> If two impedances are in parallel, the current through any one impedance is equal to the opposite impedance divided by the sum of the two impedances multiplied by the total current through the parallel combination of the two impedances.

EXAMPLE 5.29

In the circuit of Fig. 5.37 two impedances are in parallel with a voltage source. If $I_s = 15\angle 53.1^\circ$ mA, $Z_a = 12\angle 40^\circ$ kΩ, and $Z_c = 16\angle 60^\circ$ kΩ, find the current through each impedance.

FIGURE 5.37

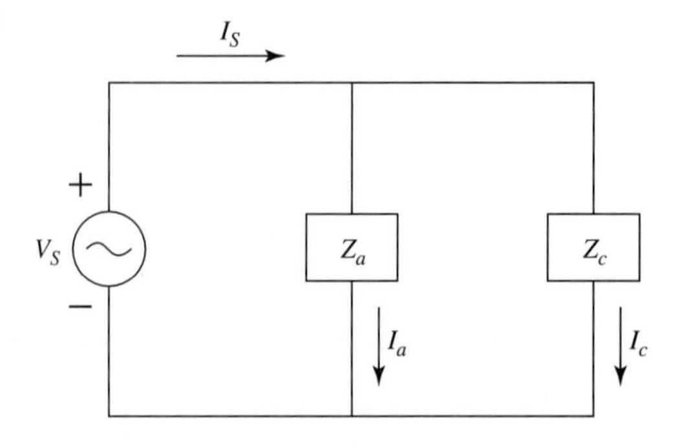

Solution This problem may be solved using either the Current Division Law of Equation 5.45 or Equation 5.49. Since we are given the impedances in this example, it is easier to use Equation 5.49. Before solving this example, we will add Z_a and Z_c so that we can place this sum in the denominator of the right side of Equation 5.49.

$$Z_a + Z_c = (12\angle 40°) + (16\angle 60°) = 9.19 + j7.71 + 8.00 + j13.86$$
$$= 17.19 + j21.57 \text{ k}\Omega$$

Converting this rectangular form to is polar form equivalent:

$$Z_a + Z_c = 27.58\angle 51.44° \text{ k}\Omega$$

Substituting this value and the given values for I_s, Z_a, and Z_c into Equations 5.49:

$$I_a = [Z_c/(Z_a + Z_c)]I_s$$
$$= [(16\angle 60°)/(27.58\angle 51.44°)]15\angle 53.1° = 8.70\angle 61.55° \text{ mA}$$
$$I_c = [Z_a/(Z_a + Z_c)]I_s$$
$$= [(12\angle 40°)/(27.58\angle 51.44°)]15\angle 53.1° = 6.53\angle 41.66° \text{ mA}$$

Answer $I_a = 8.70\angle 61.55°$ mA, $I_c = 6.53\angle 41.66°$ mA.

As an exercise, repeat this problem using the admittance form of the Current Division Law, Equation 5.45.

EXAMPLE 5.30

In the circuit of Fig. 5.38(a) find:
(a) The source current I_s
(b) The current through each resistor

$R_a = 36$ kΩ $R_c = 54$ kΩ $R_e = 18$ kΩ $V_s = 108$ V

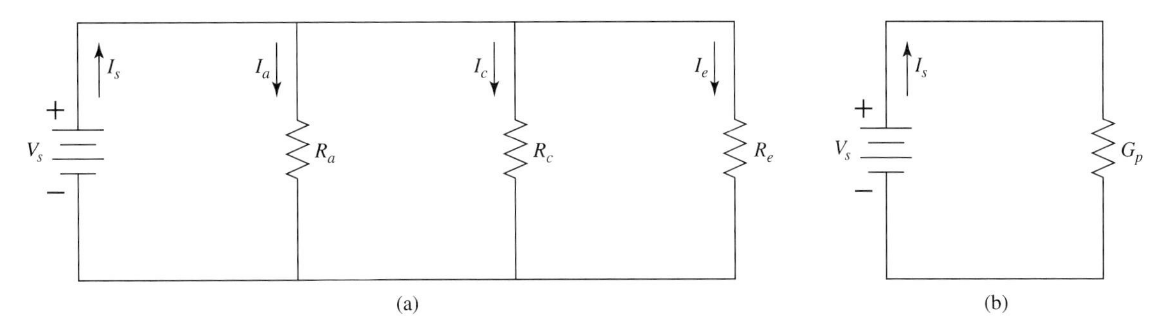

FIGURE 5.38

Solution

(a) Recall that the admittance of a resistor is equal to its conductance, G. The current division equation (5.45) for resistances may be written as:

$$I_n = (G_n/G_p)I_s \qquad \textbf{(Eq. 5.50)}$$

In the circuit of Fig. 5.38(a) Equation 5.50 becomes:

$$I_a = (G_a/G_p)I_s \qquad I_c = (G_c/G_p)I_s \qquad I_e = (G_e/G_p)I_s \qquad \textbf{(Eq. 5.51)}$$

$$G_a = 1/R_a = 1/36\text{K} = 0.028 \text{ mS}$$
$$G_c = 1/R_c = 1/54\text{K} = 0.019 \text{ mS}$$
$$G_e = 1/R_e = 1/18\text{K} = 0.060 \text{ mS}$$
$$G_p = G_a + G_c + G_e = 0.028 + 0.019 + 0.060 = 0.107 \text{ mS}$$

Fig. 5.38(a) is redrawn as Fig. 5.38(b) with the three resistances replaced by their single equivalent conductance, G_p. The source current may be calculated from this circuit using Ohm's Law:

$$I_s = G_p V_s = 0.107(108) = 11.56 \text{ mA}$$

Answer $I_s = 11.56$ mA.

(b) We may now substitute this value of I_s and all the other calculated and given values into Equation 5.51.

$$I_a = (0.028/0.107)11.56 = 3.03 \text{ mA}$$
$$I_c = (0.019/0.107)11.56 = 2.05 \text{ mA}$$
$$I_e = (0.060/0.107)11.56 = 6.48 \text{ mA}$$

5.2.6 Admittance and Phasor Diagrams of Parallel Circuits

Just as in the case of series circuits, it is often helpful to draw the phasor diagram of a parallel circuit. Keep in mind when drawing these diagrams that it is always the angular relationship ***between*** the phasors that is important rather than the absolute phasor angles.

EXAMPLE 5.31

Consider the circuit diagram of Fig. 5.39, consisting of a resistor, inductor, and capacitor, all in parallel with a voltage source. In this circuit, $V = 36 \text{ V}\angle 0°$, $R = 10 \text{ K}\Omega$, $C = 0.167\ \mu\text{F}$, $L = 8$ H, and $\omega = 1000$ rad/s.

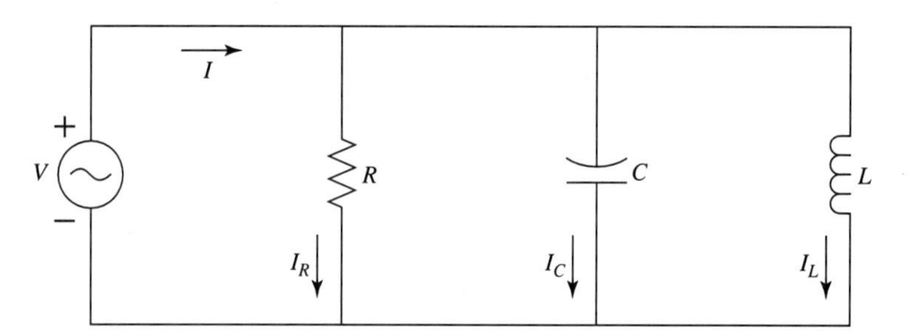

FIGURE 5.39

(a) Calculate the current through each element.
(b) Calculate the source current.
(c) Draw the phasor diagram for this circuit, showing V, I_R, I_C, I_L, and I.

Solution

(a) To perform the calculations requested, we will first calculate the impedance of each element.

$$Z_R = R = 10 \text{ k}\Omega$$
$$Z_C = -j[1/\omega C] = -j[1/1\text{K}(0.167\ \mu\Omega)] = -j6 \text{ k}\Omega$$
$$Z_L = j\omega L = j1\text{K}(8) = j8 \text{ k}\Omega$$

The voltage across each element is equal to the source voltage, V. We may therefore find the current through each element by Ohm's Law as follows:

$$I_R = V/R = (36\angle 0°)/10\text{K} = 3.6\angle 0° \text{ mA}$$
$$I_C = V/Z_C = (36\angle 0°)/(6\text{K}\angle -90°) = 6\angle 90° \text{ mA}$$
$$I_L = V/Z_L = (36\angle 0°)/(8\text{K}\angle 90°) = 4.5\angle -90° \text{ mA}$$

Answer $I_R = 3.6\underline{/0°}$ mA, $I_C = 6.0\underline{/90°}$ mA, $I_L = 4.5\underline{/-90°}$ mA.

(b) From KCL:

$$I = I_R + I_C + I_L = 3.6 \text{ mA} + j6.0 \text{ mA} - j4.5 \text{ mA} = 3.6 + j1.5 \text{ mA}$$

Converting this to polar form: $I = 3.9\underline{/22.6°}$ mA

Answer $I = 3.9\underline{/22.60}$ mA.

(c) The phasor diagram is drawn in Fig. 5.40.

FIGURE 5.40

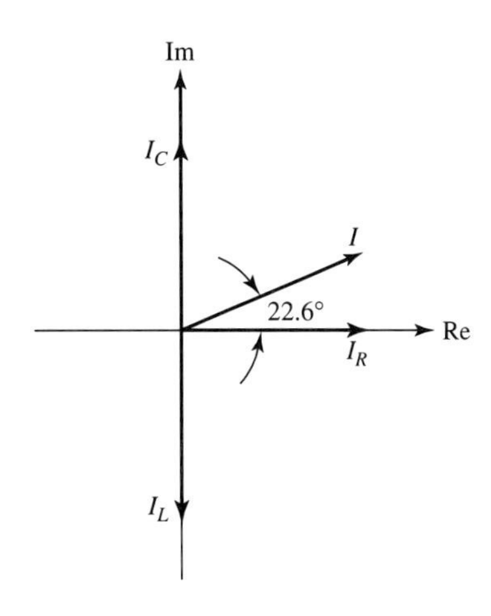

Note the following from the phasor diagram.

- The current through the resistor is in phase with the voltage across it.
- The current through the capacitor ***leads*** the voltage across it by 90°.
- The current through the inductor ***lags*** the voltage across it by 90°.

These results will of course always be true. They are the same no matter how the elements are connected. Compare these results with those obtained for a circuit where these same element types are connected in series (e.g., Example 5.14 and Fig. 5.19).

The admittance diagram for the parallel circuit of Fig. 5.39 will have the same angular relationship as the voltages and currents of the phasor diagram. Note that:

$$G_R = 1/R = 1/10\text{K} = 0.10 \text{ mS}$$
$$G_C = 1/Z_C = 1/(6\text{K}\underline{/-90°}) = 0.16\underline{/90°} \text{ mS}$$
$$G_L = 1/Z_L = 1/(8\text{K}\underline{/90°}) = 0.125\underline{/-90°} \text{ mS}$$

Note that G_R, G_C, and G_L have the same angles as I_R, I_C, and I_L, respectively.

5.2.7 Equivalence of Like Passive Parallel Elements

Just as impedances in parallel may be combined to yield a total equivalent impedance, ***like elements*** in parallel may also be combined to yield a single equivalent element. This can often help simplify the analysis of a circuit.

5.2.7.1 Resistors in Parallel

Consider the circuit of Fig. 5.41(a), which shows three resistors in parallel with a voltage source. We wish to find an expression for the single resistor, R_s, that is equivalent to the three parallel resistors as shown in Fig. 5.41(b). From Equation 5.39 for three admittances in parallel:

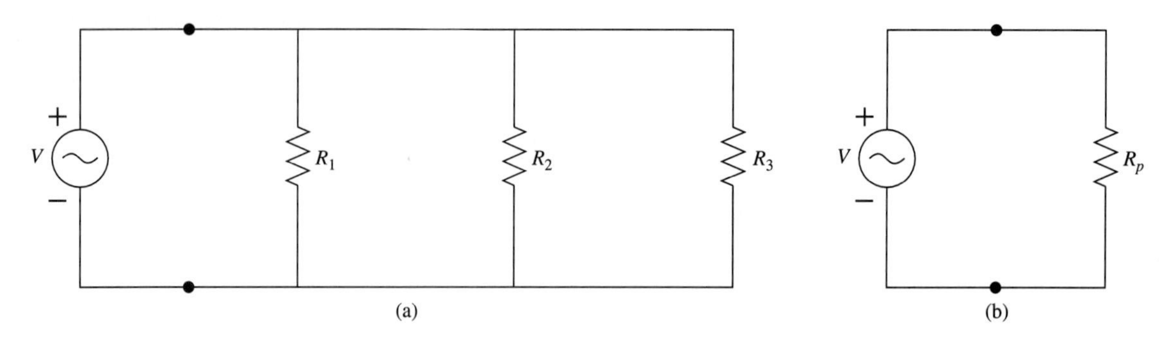

FIGURE 5.41

$$Y_p = Y_1 + Y_2 + Y_3 \quad \textbf{(Eq. 5.51)}$$

Since $Y = G$ for a resistor, Equation 5.51 becomes:

$$G_p = G_1 + G_2 + G_3 \quad \textbf{(Eq. 5.52)}$$

Since, by definition, $G = 1/R$, we may write Equation 5.52 as:

$$\frac{1}{R_p} = \frac{1}{R_1} + \frac{1}{R_2} + \frac{1}{R_3} \quad \textbf{(Eq. 5.53)}$$

Equation 5.53 may be generalized as follows:

$$1/R_p = \Sigma(1/R_n)\ (n = 1 \text{ to } N) \quad \textbf{(Eq. 5.54)}$$

Equation 5.54 may be stated as follows:

> Any number (N) of resistors in parallel may be replaced by a single equivalent resistor such that the reciprocal of the equivalent resistance (R_p) is equal to the sum of the reciprocals of each individual parallel resistance.

EXAMPLE 5.32

For the circuit of Fig. 5.42(a): $V_s = 36\angle 40°$ mV, $R_a = 2$ kΩ, $R_c = 4$ kΩ, $R_e = 5$ kΩ, and $R_m = 8$ kΩ.

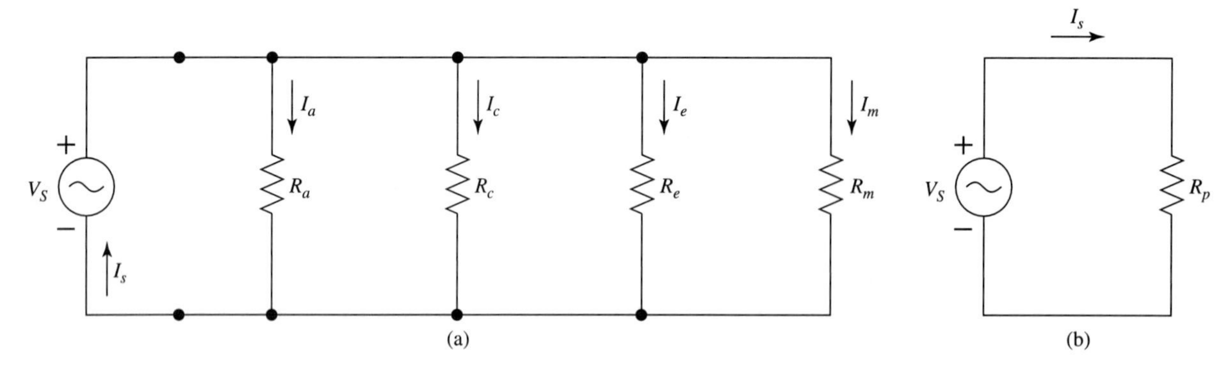

FIGURE 5.42

(a) Find the single equivalent resistor that can replace the four parallel resistors.
(b) Draw an equivalent circuit to Fig. 5.42(a) using the single equivalent resistor found in part (a).

(c) Find the current flowing in each resistor.
(d) Find the source current.

Solution

(a) Using Equation 5.54:

$$\frac{1}{R_p} = \frac{1}{R_a} + \frac{1}{R_c} + \frac{1}{R_e} + \frac{1}{R_m} = \frac{1}{2\text{K}} + \frac{1}{4\text{K}} + \frac{1}{5\text{K}} + \frac{1}{8\text{K}}$$

$$1/R_p = (0.5 \times 10^{-3}) + (0.25 \times 10^{-3}) + (0.2 \times 10^{-3}) + (0.125 \times 10^{-3})$$
$$= 1.075 \times 10^{-3}$$

Taking the reciprocal of both sides of this equation:

$$R_p = 1/(1.075 \times 10^{-3}) = 0.93\ \text{k}\Omega$$

Answer $R_p = 930\ \Omega$

(b) The equivalent circuit is drawn in Fig. 5.42(b).

(c) The current through each resistor is found by Ohm's Law.

$$I_a = \frac{V_s}{R_a} = \frac{36 \times 10^{-3}\angle 40^\circ}{2\text{K}} = 18\angle 40^\circ\ \mu\text{A}$$

$$I_c = \frac{V_s}{R_c} = \frac{36 \times 10^{-3}\angle 40^\circ}{4\text{K}} = 9\angle 40^\circ\ \mu\text{A}$$

$$I_e = \frac{V_s}{R_e} = \frac{36 \times 10^{-3}\angle 40^\circ}{5\text{K}} = 7.2\angle 40^\circ\ \mu\text{A}$$

$$I_m = \frac{V_s}{R_m} = \frac{36 \times 10^{-3}\angle 40^\circ}{8\text{K}} = 4.5\angle 40^\circ\ \mu\text{A}$$

Answer $I_a = 18\angle 40^\circ\ \mu\text{A}, I_c = 9\angle 40^\circ\ \mu\text{A}, I_e = 7.2\angle 40^\circ\ \mu\text{A}$, and $I_m = 4.5\angle 40^\circ\ \mu\text{A}$.

(d) The current supplied by the source can be found in two ways. We may use KCL and add up each of the currents found in part (c) ***or*** we may use the fact that Fig. 5.42(b) is equivalent to Fig. 5.42(a) and use Ohm's Law in Fig. 5.42(b). It is obviously simpler to use the latter method, and this is done next.

$$I_s = \frac{V_s}{R_p} = \frac{36 \times 10^{-3}\angle 40^\circ}{930} = 38.7 \times 10^{-6}\angle 40^\circ\ \text{A}$$

Answer $I_s = 38.7\angle 40^\circ\ \mu\text{A}$.

As an exercise, do part (d) using KCL and verify that the same answer is obtained.

A simplified form of Equation 5.54 is obtained when only ***two*** resistors are in parallel. This is the same as Equation 5.36 for resistors:

$$R_p = R_1R_2/(R_1 + R_2) \qquad \textbf{(Eq. 5.55)}$$

Equation 5.55 may be stated as follows:

> Two resistors in parallel may be replaced by a single resistor whose resistance is equal in value to the product of the two resistances divided by the sum of the two resistances.

EXAMPLE 5.33

The circuit in Fig. 5.43(a) consists of two resistors in parallel with a dc current source. If $I_s = 24$ mA, $R_a = 2\ \Omega$, and $R_c = 8\ \Omega$, find:

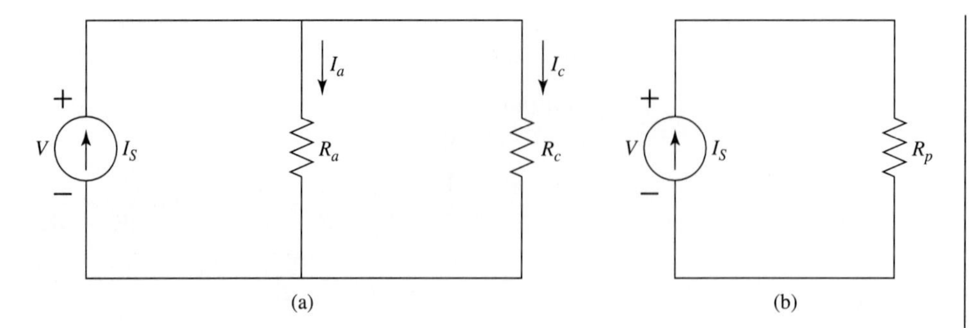

FIGURE 5.43

(a) The voltage across each resistor
(b) The current through each resistor

Solution

(a) The two resistors may be replaced by a single equivalent resistor as shown in Fig. 5.43(b).

$$R_p = R_a R_c/(R_a + R_c) = 2(8)/(2 + 8) = 1.6\ \Omega$$

Since the resistor R_p is equivalent to the parallel combination of R_a and R_c, we may use Ohm's Law to find the voltage across R_p, and this will be the same as the voltage across each resistor.

$$V = I_s R_p = 24 \times 10^{-3}(1.6) = 38.4\ \text{mV}$$

Answer $V = 38.4$ mV.

(b) The current through each resistor may be found in two ways. We may use Ohm's Law because we found the voltage across each resistor in part (a) ***or*** we may use the Current Division Law because we know the total current flowing into the parallel combination of the two resistors.

Using the Current Division Law in the circuit of Fig. 5.43(a):

$$I_a = [R_c/(R_a + R_c)]I_s = [8/10]24\ \text{mA} = 19.2\ \text{mA}$$
$$I_c = [R_a/(R_a + R_c)]I_s = [2/10]24\ \text{mA} = 4.8\ \text{mA}$$

Answer $I_a = 19.2$ mA, $I_c = 4.8$ mA.

Note that the sum of I_a and I_c is 24 mA, which, as dictated by KCL, is equal to I_s.

As an exercise, solve for I_a and I_c using Ohm's Law and confirm that the answers are the same as we obtained using the Current Division Law.

Another simplified form of Equation 5.54 is obtained for the special case when a number of equal resistances are in parallel. This may be stated as follows:

Any number (N) of ***equal*** resistors in parallel may be replaced by a single equivalent resistor (R_p) whose resistance equals an individual resistance (R) divided by the number of resistors.

This may be expressed mathematically as:

$$R_p = R/N \qquad \textbf{(Eq. 5.56)}$$

5.2.7.2 Inductors in Parallel

Consider the circuit of Fig. 5.44(a), consisting of three inductors in parallel with a current source, I_s. We wish to find an expression for a single inductor that is equivalent to the three inductors in parallel as shown in Fig. 5.44(b). To do this, we will apply Equation 5.39 for admittances in parallel to our circuit in Fig. 5.44(a):

$$Y_p = Y_1 + Y_2 + Y_3 \qquad \textbf{(Eq. 5.57)}$$

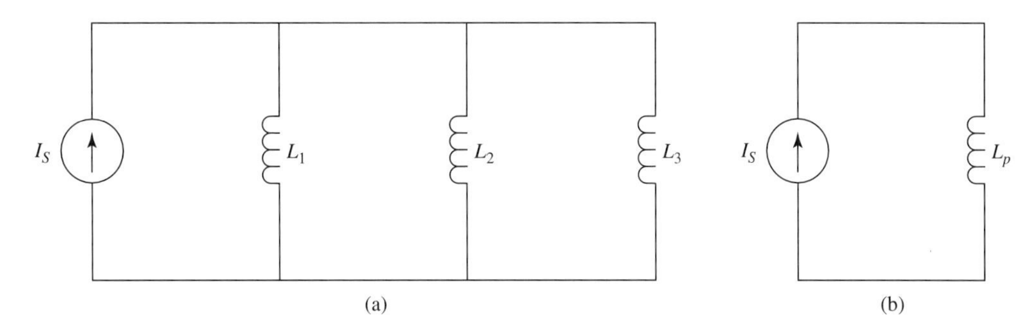

FIGURE 5.44

where Y_p represents the admittance of the equivalent inductance, L_p. Using the expression for admittance of an inductor, Equation 5.57 becomes:

$$-j(1/2\pi f L_p) = -j(1/2\pi f L_1) - j(1/2\pi f L_2) - j(1/2\pi f L_3) \qquad \textbf{(Eq. 5.58)}$$

Canceling out the common factor of $-j(1/2\pi f)$ we get

$$\frac{1}{L_p} = \frac{1}{L_1} + \frac{1}{L_2} + \frac{1}{L_3} \qquad \textbf{(Eq. 5.59)}$$

Equation 5.59 may be generalized as follows:

$$1/L_p = \Sigma(1/L_n) \quad (n = 1 \text{ to } N) \qquad \textbf{(Eq. 5.60)}$$

where L_p is the equivalent inductance of N inductances in parallel.

Equation 5.60 may be stated as follows:

> Any number (N) of parallel inductors may be replaced by a single inductor (L_p) such that the reciprocal of the equivalent inductance is equal to the sum of the reciprocals of each of the individual inductances.

Notice the similarity between Equation 5.54, which gives the ***equivalent resistance*** of any number of resistances in parallel, and Equation 5.60, which gives the ***equivalent inductance*** of any number of inductances in parallel.

EXAMPLE 5.34

Consider the circuit of Fig. 5.45(a), consisting of three inductors in parallel with a voltage source. If $V_s = 50\text{ V}\angle 0°$, $L_a = 10$ mH, $L_c = 5$ mH, $L_e = 20$ mH, and $f = 200$ Hz:

(a) Find the single inductance that is equivalent to the three parallel inductances.
(b) Find the source phasor current.
(c) Find the current through each inductor.

Solution

(a) From Equation 5.60:

$$\frac{1}{L_p} = \frac{1}{L_a} + \frac{1}{L_c} + \frac{1}{L_e} = \frac{1}{10 \times 10^{-3}} + \frac{1}{5 \times 10^{-3}} + \frac{1}{20 \times 10^{-3}}$$
$$= 0.1\text{K} + 0.2\text{K} + 0.05\text{K} = 0.35\text{K}$$

Taking the reciprocal of both sides of this equation:

$$L_p = 1/0.35\text{K} = 2.86 \text{ mH}$$

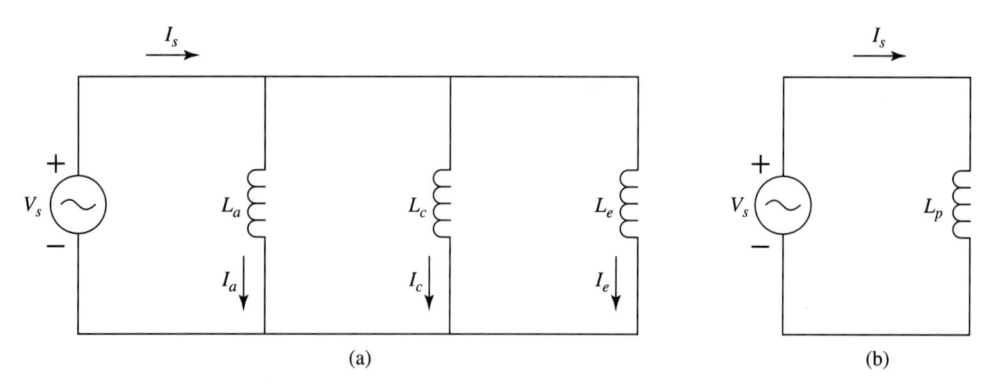

FIGURE 5.45

Answer $L_p = 2.86$ mH

(b) From the fact that L_p is equivalent to the three inductors in parallel, we may use the circuit in Fig. 5.45(b) to calculate the source current using Ohm's Law:

$$I_s = V_s/Z_s = V_s/j2\pi fL_p = (50\angle 0^\circ)/j2\pi(200)2.86 \times 10^{-3} = 13.9\angle -90^\circ \text{ A}$$

Answer $I_s = 13.9\angle -90^\circ$ A.

(c) The current through each inductor may be found by Ohm's Law ***or*** by current division. Using Ohm's Law, the current through each inductor is:

$$\begin{aligned} I_a &= V_s/Z_a = (50\angle 0^\circ)/j2\pi fL_a \\ &= (50\angle 0^\circ)/(j2\pi(200)10 \times 10^{-3}) = 3.98\angle -90^\circ \text{ A} \\ I_c &= V_s/Z_c = 50\angle 0^\circ/j2\pi fL_c \\ &= 50\angle 0^\circ/(j2\pi(200)5 \times 10^{-3}) = 7.96\angle -90^\circ \text{ A} \\ I_e &= V_s/Z_e = 50\angle 0^\circ/j2\pi fL_e \\ &= 50\angle 0^\circ/(j2\pi(200)20 \times 10^{-3} = 1.99\angle -90^\circ \text{ A} \end{aligned}$$

Answer $I_a = 3.98\angle -90^\circ$ A, $I_c = 7.96\angle -90^\circ$ A, $I_e = 1.99\angle -90^\circ$ A.

As an exercise, repeat part (c) using current division and confirm that the same answers are obtained. In either case, check the answer using KCL.

For the special case of ***two*** inductors in parallel, it can be shown that Equation 5.60 reduces to the following:

$$L_p = L_1L_2/(L_1 + L_2) \qquad \textbf{(Eq. 5.61)}$$

Equation 5.61 may be stated as follows:

> Two inductors in parallel may be replaced by a single equivalent inductor whose inductance (L_p) is equal to the product of the two inductances divided by the sum of the two inductances.

Note the similarity between Equation 5.61 for ***two inductances*** in parallel and Equation 5.55 for ***two resistances*** in parallel.

EXAMPLE 5.35

For the circuit of Fig. 5.46(a):

(a) Find the single inductance that is equivalent to the two parallel inductances.
(b) Find the phasor voltage across each inductor.
(c) Find the phasor current through each inductor.

$$I_s = 50\angle 0^\circ \text{ mA} \qquad L_a = 100 \text{ mH} \qquad L_c = 60 \text{ mH} \qquad f = 120 \text{ Hz}$$

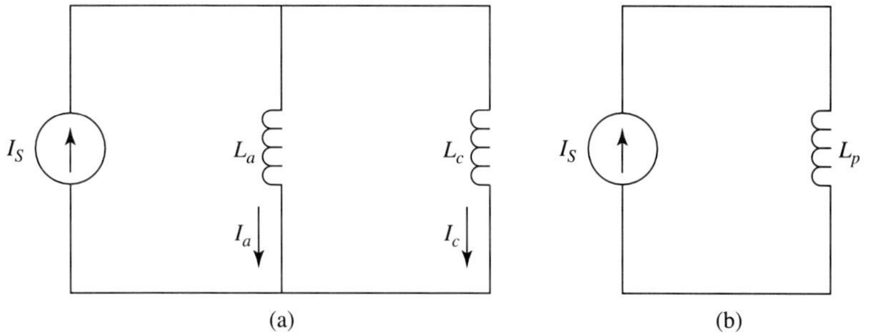

(a) (b)

FIGURE 5.46

Solution

(a) Using Equation 5.61:

$$L_p = L_a L_c/(L_a + L_c) = 100 \times 10^{-3}(60 \times 10^{-3})/(160 \times 10^{-3}) = 37.5 \text{ mH}$$

Answer $L_p = 37.5$ mH

(b) Since L_p represents the equivalent inductance of the two parallel inductors, we may redraw the circuit of Fig. 5.46(a) as shown in Fig. 5.46(b). In this circuit, we may use Ohm's Law to calculate the voltage across L_p and recognize that this is the same voltage across each of the two parallel inductors.

$$V_s = I_s Z_s = I_s(j2\pi f L_s) = 50\text{m}\angle 0°(j2\pi(120)(37.5 \text{ m}) = 1.41\angle 90° \text{ V}$$

Answer $V_a = V_c = V_s = 1.41\angle 90°$ V

The current through each inductor may be found by using Ohm's Law in the circuit of Fig. 5.46(a).

$$I_a = \frac{V_s}{Z_a} = \frac{1.41\angle 90°}{j2\pi f L_a} = \frac{1.41\angle 90°}{j2\pi(120)100 \times 10^{-3}} = 18.7\angle 0° \text{ mA}$$

$$I_c = \frac{V_s}{Z_c} = \frac{1.41\angle 90°}{j2\pi f L_c} = \frac{1.41\angle 90°}{j2\pi(120)60 \times 10^{-3}} = 31.2\angle 0° \text{ mA}$$

Answer $I_a = 18.7\angle 0°$ mA, $I_c = 31.2\angle 0°$ mA.

Another simplified form of Equation 5.60 occurs when a number of ***equal*** inductors are in parallel. This may be stated as follows:

> Any number (N) of ***equal*** inductances in parallel may be replaced by a single equivalent inductance (L_p) equal to an individual inductance (L) divided by the number of parallel inductors.

This may be expressed mathematically as follows:

$$L_p = L/N \qquad \textbf{(Eq. 5.62)}$$

5.2.7.3 Capacitors in Parallel

Consider the circuit of Fig. 5.47(a), which includes three capacitors in parallel. We wish to find a single capacitor, C_p, that is equivalent to the three parallel capacitors. To do this, we will apply Equation 5.39 for admittances in parallel to our circuit.

$$Y_p = Y_1 + Y_2 + Y_3 \qquad \textbf{(Eq. 5.63)}$$

Using the expression for admittance of a capacitor in Equation 5.63:

$$Y_p = j2\pi f C_1 + j2\pi f C_2 + j2\pi f C_3 = j2\pi f C_s \qquad \textbf{(Eq. 5.64)}$$

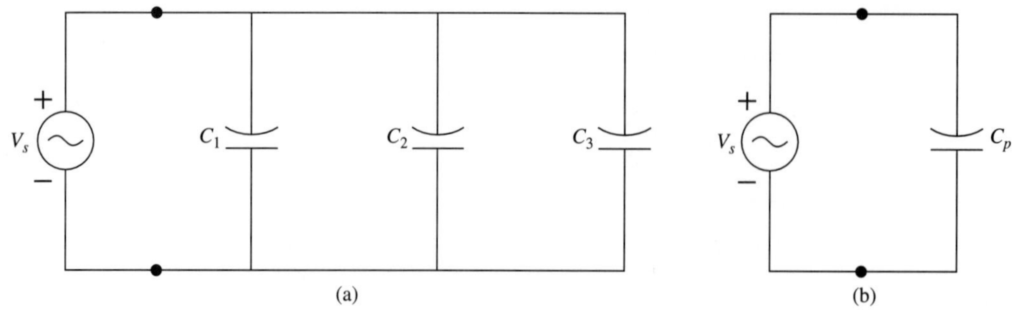

(a) (b)

FIGURE 5.47

Canceling out the common factor of $j2\pi f$, we get the following:

$$C_p = C_1 + C_2 + C_3 \qquad \textbf{(Eq. 5.65)}$$

The circuit of Fig. 5.47(a) may now be redrawn as shown in Fig. 5.47(b), where the single capacitor, C_p, replaces the three parallel capacitors.

Equation 5.65 may be generalized for any number of capacitors in parallel as in the following equation:

$$C_s = \Sigma C_n \quad (n = 1 \text{ to } N) \qquad \textbf{(Eq. 5.66)}$$

Equation 5.66 may be stated as follows:

> Any number (N) of capacitors in parallel may be replaced by a single equivalent capacitor whose capacitance (C_p) is equal to the sum of all the individual capacitances.

Note that this expression for ***parallel capacitors*** is similar in form to resistances and inductances in ***series.***

EXAMPLE 5.36

For the circuit of Fig. 5.48(a):

(a) Find the single capacitor that is equivalent to the four capacitors in parallel.

(b) Redraw this circuit using the equivalent capacitor in place of the four parallel capacitors.

(c) Find the phasor current through each capacitor.

(d) Find the phasor source current.

$$V_s = 40\text{ V}\angle 0^\circ \qquad C_a = 200\ \mu\text{F} \qquad C_c = 250\ \mu\text{F} \qquad C_e = 400\ \mu\text{F}$$
$$C_m = 500\ \mu\text{F} \qquad f = 50\text{ Hz}$$

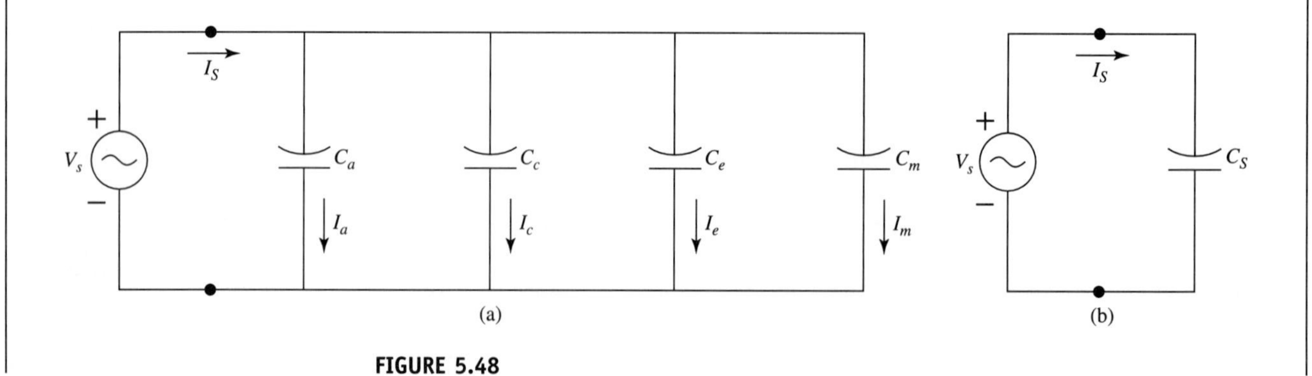

FIGURE 5.48

Solution

(a) Using Equation 5.66 for four capacitors in parallel:

$$C_p = C_a + C_c + C_e + C_m = (200 \times 10^{-6}) + (250 \times 10^{-6}) + (400 \times 10^{-6}) + (500 \times 10^{-6}) = 1350 \times 10^{-6}$$

Answer $C_s = 1350\ \mu\text{F}$

(b) The circuit is redrawn in Fig. 5.48(b). Note the junction points in this circuit. We will use these points to indicate points of equivalence in more complicated circuits in the following chapters.

(c) The individual capacitor phasor currents are found using Ohm's Law for each capacitor.

$$I_a = Y_aV_s = j2\pi fC_aV_s = j2\pi(50)200 \times 10^{-6}(40\angle 0°) = j2.5\text{ A}$$
$$I_c = Y_cV_s = j2\pi fC_cV_s = j2\pi(50)250 \times 10^{-6}(40\angle 0°) = j3.1\text{ A}$$
$$I_e = Y_eV_s = j2\pi fC_eV_s = j2\pi(50)400 \times 10^{-6}(40\angle 0°) = j5.0\text{ A}$$
$$I_m = Y_mV_s = j2\pi fC_mV_s = j2\pi(50)500 \times 10^{-6}(40\angle 0°) = j6.3\text{ A}$$

Answer $I_a = 2.5\angle 90°$ A, $I_c = 3.1\angle 90°$ A, $I_e = 5.0\angle 90°$ A, $I_m = 6.3\angle 90°$ A.

(d) The phasor source current may be found by Ohm's Law using Fig. 5.48(b).

$$I_s = Y_sV_s = j2\pi C_sV_s = j2\pi(50)1350\ \mu\text{A}\ (40\angle 0°) = 16.9\text{ A}$$

As an exercise, use KCL to obtain this same result.

■ SUMMARY

- A ***circuit*** is a connection of two or more elements.
- A ***closed circuit*** is one in which a complete, unbroken path for current exists.
- An ***open circuit*** is one in which no current can flow due to the lack of an unbroken path.
- A ***series*** circuit is one in which the ***same current*** flows through each impedance.
- Kirchhoff's Voltage Law (KVL) states that the sum of the voltages around a closed loop is identically equal to zero ($\Sigma V_{cl} = 0$).
- Any number (N) of series connected impedances may be replaced by an equivalent single impedance (Z_s) equal to the sum of all the individual impedances [$Z_s = \Sigma Z_n \quad (n = 1 \text{ to } N)$].
- ***Voltage Division Law:*** The voltage across any impedance in a series-connected circuit is equal to the circuit's impedance multiplied by the total voltage across all the series-connected impedances divided by the total (equivalent) impedance [$V_n = (Z_n/Z_s)V$].
- Any number (N) of series-connected resistors may be replaced by a single resistor whose resistance (R_s) is given by the sum of all the series resistance values [$R_s = \Sigma R_n \quad (n = 1 \text{ to } N)$].
- Any number (N) of series-connected inductors may be replaced by a single inductor whose inductance (L_s) is given by the sum of all the series inductance values [$L_s = \Sigma L_n \quad (n = 1 \text{ to } N)$].
- Any number (N) of series connected capacitors may be replaced by a single capacitor whose capacitance (C_s) is equal to the reciprocal of the sum obtained by adding the reciprocals of all the individual capacitances [$1/C_s = \Sigma(1/C_n) \quad (n = 1 \text{ to } N)$].
- If n capacitors of ***equal capacitance,*** C, are in series then the equivalent capacitance, C_s, is given by $C_s = C/n$.
- A ***parallel*** circuit is one in which all the impedances have the ***same voltage*** across them.
- A ***node*** is a voltage point at which three or more impedances and/or power sources are commonly connected.
- Kirchhoff's Current Law states that the algebraic sum of currents entering or leaving a node is equal to zero.
- Any number (N) of impedances in parallel may be replaced by a single equivalent impedance (Z_p) such that the reciprocal of the equivalent impedance is equal to the sum of the reciprocals of each of the individual impedances [$1/Z_p = \Sigma(1/Z_n)$ $(n = 1 \text{ to } N)$].
- ***Two*** impedances in parallel may be replaced by a single equivalent impedance (Z_p) equal in value to the product of the two impedances divided by their sum [$Z_p = Z_1Z_2/(Z_1 + Z_2)$].
- Any number (N) of admittances in parallel may be replaced by a single equivalent admittance (Y_p) equal to the sum of all the individual admittances [$Y_p = \Sigma Y_n \quad (n = 1 \text{ to } N)$].
- ***Current Division Law:*** The current (I_n) through any admittance (Y_n) in a parallel circuit equals that admittance divided by the single equivalent admittance (Y_s) of the parallel circuit multiplied by the total current (I_s) through the parallel circuit [$I_n = (Y_n/Y_s)I_s$].
- If ***two*** impedances are in parallel, the current through any one impedance is equal to the opposite impedance divided by the sum of the two impedances multiplied by the total current (I_s) through the parallel combination of the two impedances [$I_1 = Z_1I_s/(Z_1 + Z_2)$].

- Any number (N) of resistors in parallel may be replaced by a single equivalent resistor (R_p) such that the reciprocal of the equivalent resistance is equal to the sum of the reciprocals of each individual parallel resistance $[1/R_p = \Sigma(1/R_n)$ $(n = 1 \text{ to } N)]$.
- Two resistors in parallel may be replaced by a single resistor whose resistance is equal in value to the product of the two resistances divided by the sum of the two resistances $[R_p = R_1R_2/(R_1 + R_2)]$.
- Any number (N) of ***equal*** resistances in parallel may be replaced by a single equivalent resistance (R_p) equal in value to an individual resistance (R) divided by the number of resistors in parallel ($R_p = R/N$).
- Any number (N) of inductors in parallel may be replaced by a single equivalent inductor whose inductance (L_p) is such that its reciprocal is equal to the sum of the reciprocals of each of the individual inductances $[1/L_p = \Sigma(1/L_n) \quad (n = 1 \text{ to } N)]$.
- Two inductors in parallel may be replaced by a single equivalent inductor whose inductance (L_p) is equal to the product of the two inductances divided by their sum $[L_p = L_1L_2/(L_1 + L_2)]$.
- Any number (N) of capacitors in parallel may be replaced by a single equivalent capacitor whose capacitance (C_p) is equal to the sum of all the individual capacitances $[C_p = \Sigma C_n \quad (n = 1 \text{ to } N)]$.
- A phasor diagram is a plot of the voltage and current phasors of a circuit on the complex plane. It is used to analyze the angular relationship between these phasors.
- It is useful to plot the ***current phasor*** along the real axis in a ***series circuit*** because it is common to all elements in that circuit.
- It is useful to plot the ***voltage phasor*** along the real axis in a ***parallel circuit*** because it is common to all elements.

■ PROBLEMS

5.1 Define each of the following terms:
 (a) Series circuit
 (b) Parallel circuit
 (c) Closed circuit
 (d) Open circuit

5.2 Write KVL in your own words.

5.3 Find the phasor voltage V_{ab} in the circuit of Fig. P5.3.

$V_s = 20\text{ V}\angle 30°$ $\quad V_{bc} = 10\text{ V}\angle 60°$
$V_{ce} = 12\text{ V}\angle 36.9°$

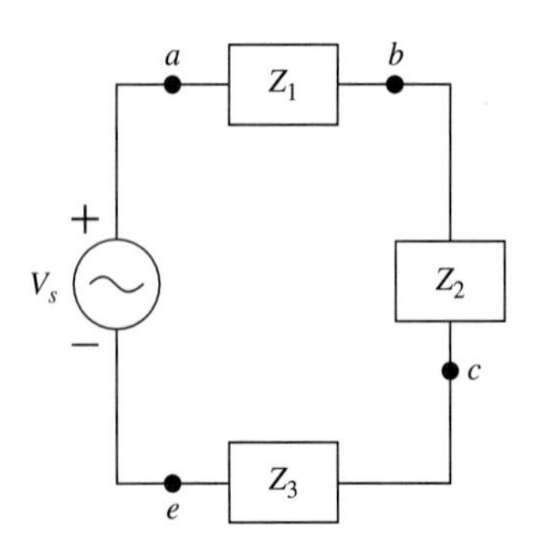

FIGURE P5.3

5.4 Find the phasor voltage V_s in the circuit of Fig. P5.4.

$V_1 = 32\angle 45°\text{ mV}$ $\quad V_2 = 16\angle 53.1°\text{ mV}$
$V_3 = 24\angle 60°\text{ mV}$

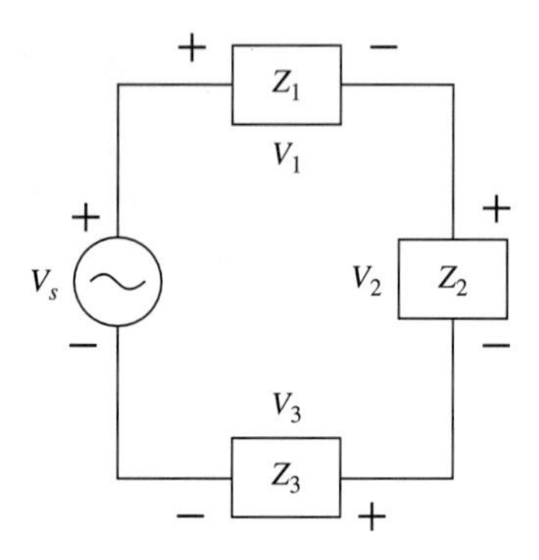

FIGURE P5.4

5.5 Find the phasor voltage V_{ac} in the circuit of Fig. P5.5.

$V_s = 42\angle 0°\text{ mV}$ $\quad V_{cd} = 24\angle 0°\text{ mV}$
$V_{de} = 18\angle 0°\text{ mV}$

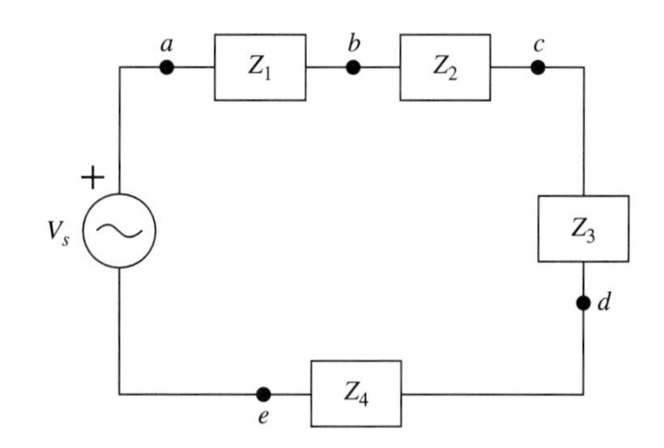

FIGURE P5.5

5.6 Find the voltage E in the circuit of Fig. P5.6.

$V_{12} = 50\text{ V}$ $\quad V_{24} = 30\text{ V}$ $\quad V_{45} = 10\text{ V}$

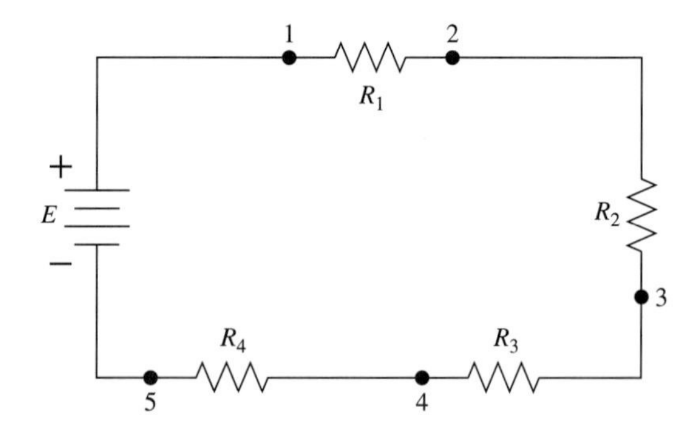

FIGURE P5.6

5.7 **(a)** Find the single equivalent impedance that can replace the two impedances of Fig. P5.7.
 (b) Redraw this circuit using the equivalent impedance in place of the two series impedances.

$V_s = 18\text{ mV}\angle 36.9°$ $\quad Z_1 = 20\angle 45°\ \Omega$
$Z_2 = 12\angle 30°\ \Omega$

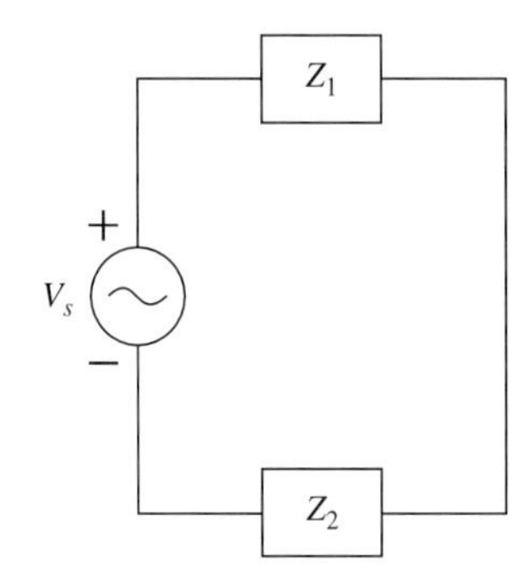

FIGURE P5.7

5.8 **(a)** Find the single equivalent impedance that can replace the four resistances of Fig. P5.8.
(b) Redraw this circuit using the equivalent impedance in place of the four resistances.

$E = 200$ mV $\quad R_1 = 10$ kΩ $\quad R_2 = 14$ kΩ
$R_3 = 8$ kΩ $\quad R_4 = 20$ kΩ

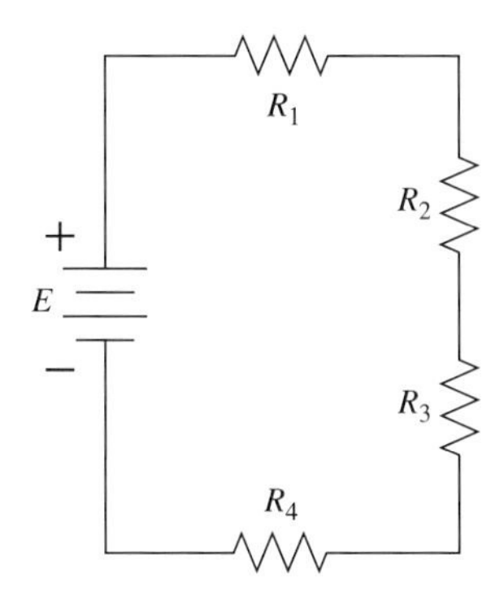

FIGURE P5.8

5.9 **(a)** Find the single equivalent impedance that can replace the three elements of Fig. P5.9.
(b) Redraw this circuit using the equivalent impedance in place of the three elements.

$V_s = 100\angle 20°$ V $\quad R_a = 50\ \Omega$ $\quad R_c = 200\ \Omega$
$L = 10$ mH $\quad f = 1.5$ kHz

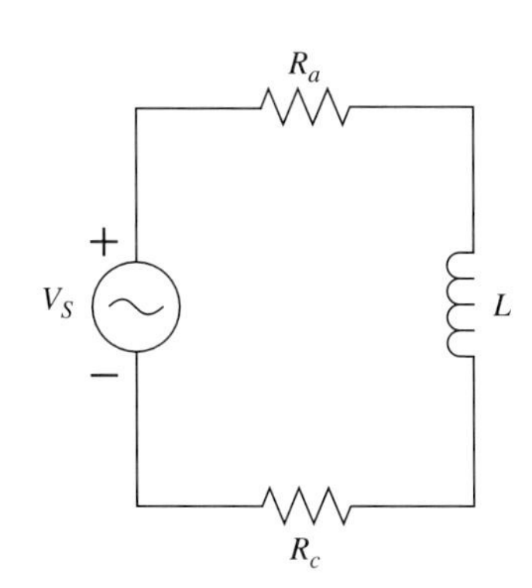

FIGURE P5.9

5.10 **(a)** Find the single equivalent impedance that can replace the three elements of Fig. P5.10.
(b) Redraw this circuit using the equivalent impedance in place of the three elements.

$V_s = 42\angle 53.1°$ V $\quad R = 4$ kΩ $\quad L = 4$ mH
$C = 2$ nF $\quad f = 150$ kHz

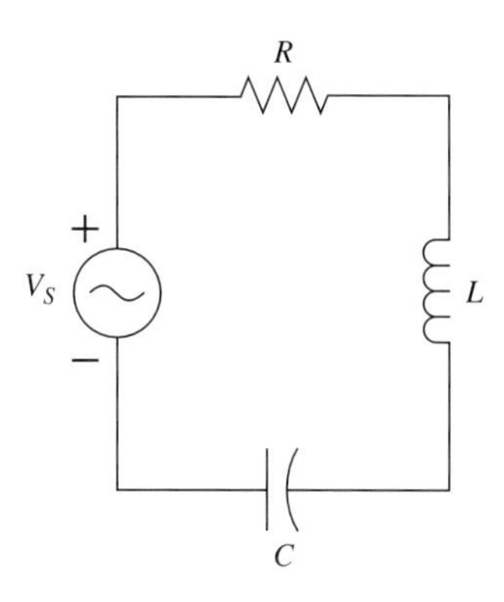

FIGURE P5.10

5.11 Use the Voltage Division Law to find each of the following phasor voltages in the circuit of Fig. P5.11.
(a) V_R
(b) V_L
(c) V_C
(d) Check these answers by applying KVL to this circuit.
(e) Write the sinusoidal time function corresponding to each phasor of parts (a) through (c).

$V_s = 36\angle 26°$ mV $\quad R = 200\ \Omega$ $\quad L = 100$ mH
$C = 2\ \mu$F $\quad f = 400$ Hz

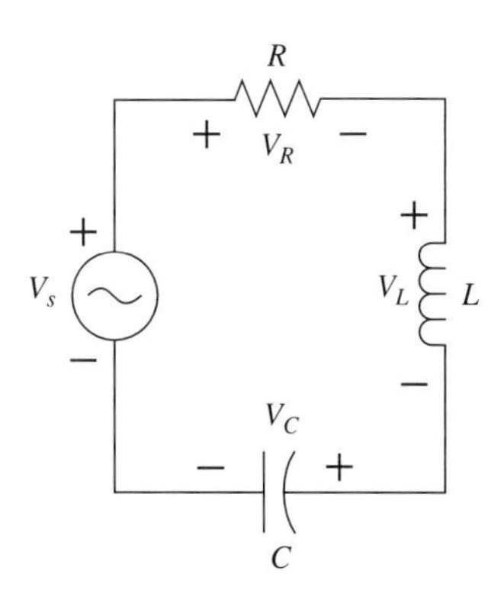

FIGURE P5.11

5.12 Use the Voltage Division Law to find each of the following voltages in the circuit of Fig. P5.12.
(a) V_a
(b) V_c
(c) V_e
(d) Check these answers by applying KVL to this circuit.

$E = 360$ V $\quad R_a = 12\ \Omega$ $\quad R_c = 24\ \Omega$
$R_e = 18\ \Omega$

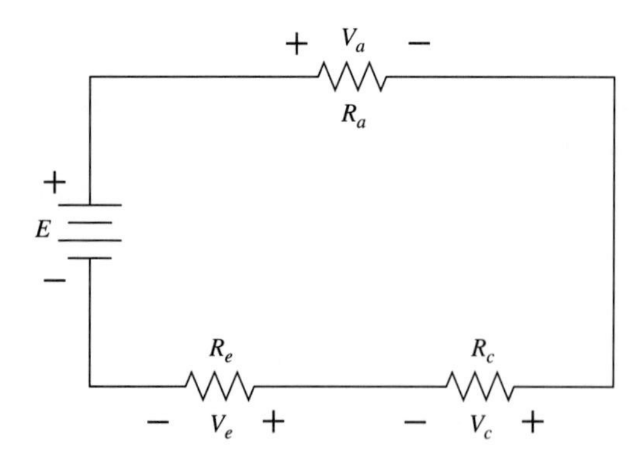

FIGURE P5.12

5.13 In the circuit of Fig. P5.13:

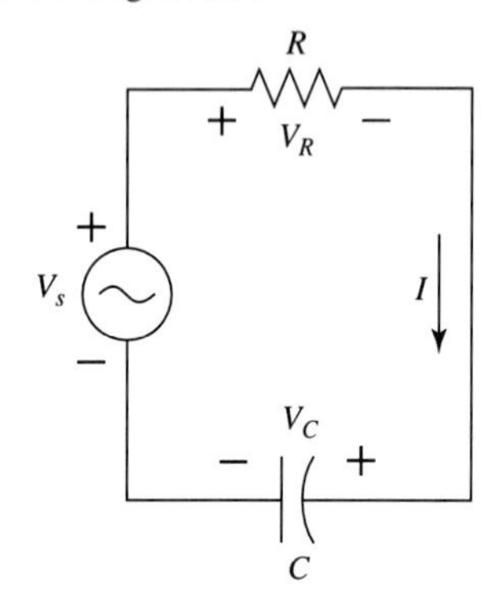

FIGURE P5.13

(a) Find the voltage V_R by voltage division.
(b) Find the voltage V_C using KVL.
(c) Find the current in this circuit by first solving for the single equivalent impedance that can replace the two passive elements in the circuit and then applying Ohm's Law to this impedance.
(d) Verify your answer in part (c) by applying Ohm's Law to the resistor alone.
(e) Explain why the answers to parts (c) and (d) must be the same.

$V_s = 82\underline{/40°}$ mV $R = 400\ \Omega$ $C = 0.8\ \mu$F
$f = 650$ Hz

5.14 Draw a phasor diagram for the circuit of Fig. P5.13 showing the current I, V_C, V_R and V_s.

5.15 In the circuit of Fig. P5.15:

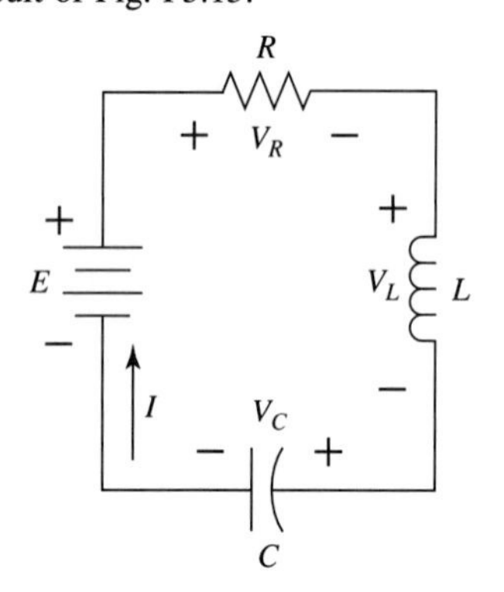

FIGURE P5.15

(a) Find the current I and explain your answer.
(b) Find the voltage V_R and explain your answer.
(c) Find the voltage V_L and explain your answer.
(d) Find the voltage V_C and explain your answer.

$E = 60$ V $R = 10$ KΩ $L = 1$ mH
$C = 4\ \mu$F

5.16 Draw a phasor diagram for the circuit of Fig. P5.11 showing the current I, V_C, V_R, V_L, and V_s.

5.17 For the circuit of Fig. P5.17, find each of the following in phasor form:
(a) I
Using voltage division find:
(b) V_x, the voltage across the two series inductors
(c) V_y, the voltage across the two series resistors
(d) V_z, the voltage across the two series capacitors
(e) V_q, the voltage across one resistor and one capacitor
(f) Verify your answers to (b) and (e) by calculating V_x and V_q using Ohm's Law.

$V_s = 14\underline{/36.9°}$ V $R_e = 1$ kΩ $R_m = 3$ kΩ
$w = 1$ K Rad/sec $L_a = 1.5$ H $L_c = 1$ H
$C_o = 0.125\ \mu$F $C_u = 0.083\ \mu$F

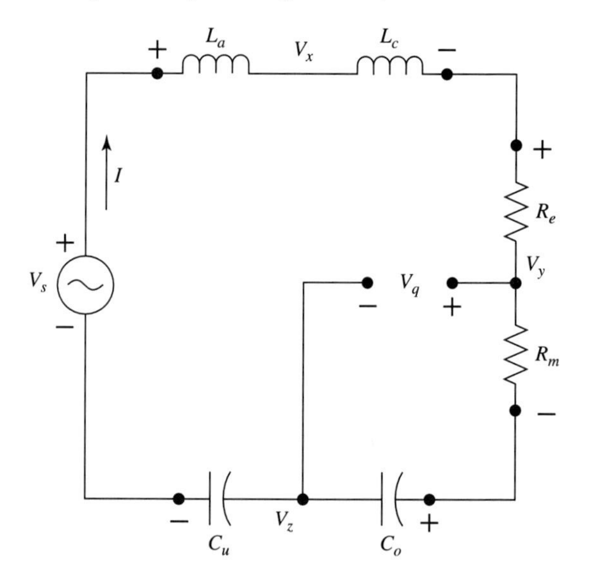

FIGURE P5.17

5.18 Write KCL in your own words.

5.19 Find the phasor current I_a in the circuit of Fig. P5.19.

$I_s = 10\underline{/45°}$ mA $I_b = 4\underline{/60°}$ mA

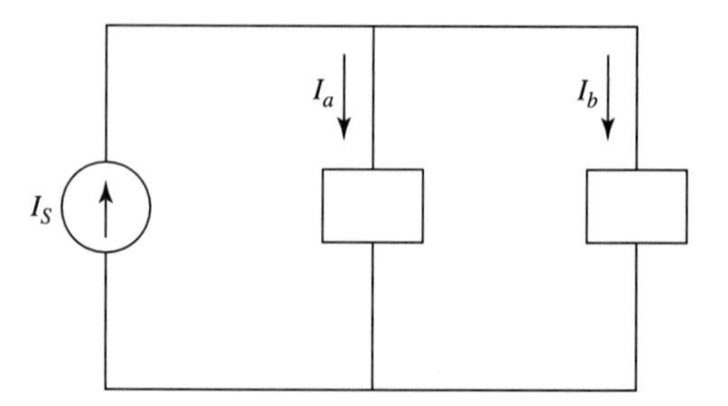

FIGURE P5.19

5.20 In the circuit of Fig. P5.20, find the following phasors:
(a) I_a
(b) I_c

$I_s = 20\angle 30^\circ$ A $\quad I_e = 10\angle 36.9^\circ$ A
$I_m = 5\angle 40^\circ$ A

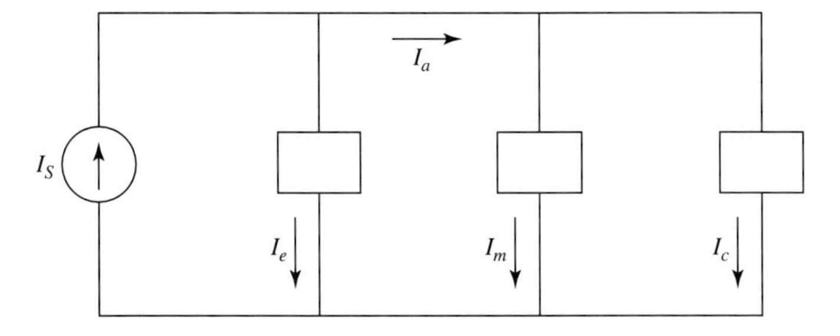

FIGURE P5.20

5.21 Find the following currents in the circuit of Fig. P5.21:
(a) I_4
(b) I_3
(c) I

$I_6 = 200\ \mu$A $\quad I_5 = 50\ \mu$A $\quad I_2 = 350\ \mu$A
$I_1 = 60\ \mu$A

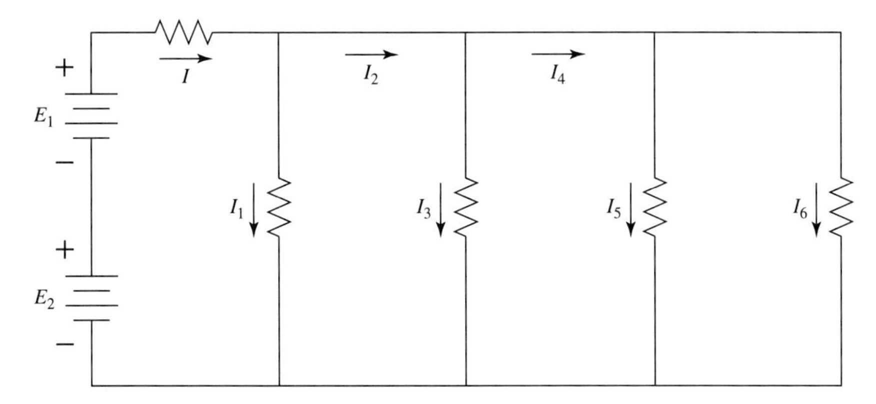

FIGURE P5.21

5.22 For the circuit of Fig. P5.22:
(a) Find the single equivalent impedance that can replace the three impedances shown.
(b) Redraw the circuit using the single equivalent impedance in place of the three impedances shown.

$Z_1 = 20\angle 60^\circ\ \Omega \quad Z_2 = 16\angle 50^\circ\ \Omega$
$Z_3 = 24\angle 30^\circ\ \Omega$

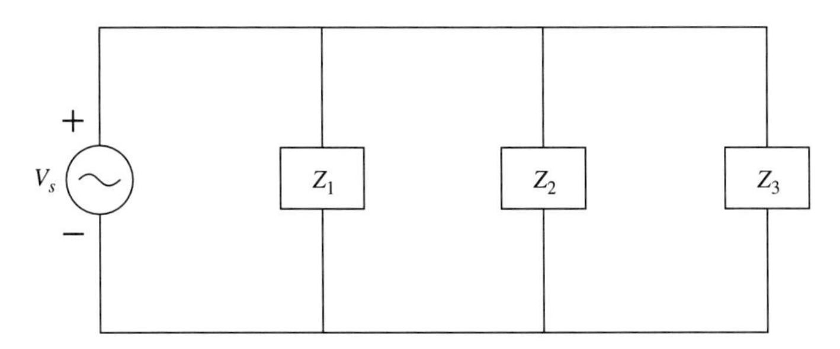

FIGURE P5.22

5.23 **(a)** Find the single impedance that is equivalent to the two parallel impedances shown in Fig. P5.23.
(b) Redraw the circuit using the single equivalent impedance in place of the two impedances shown.

$Z_a = 100\angle 45°\ \Omega \qquad Z_c = 60\angle 36.9°\ \Omega$

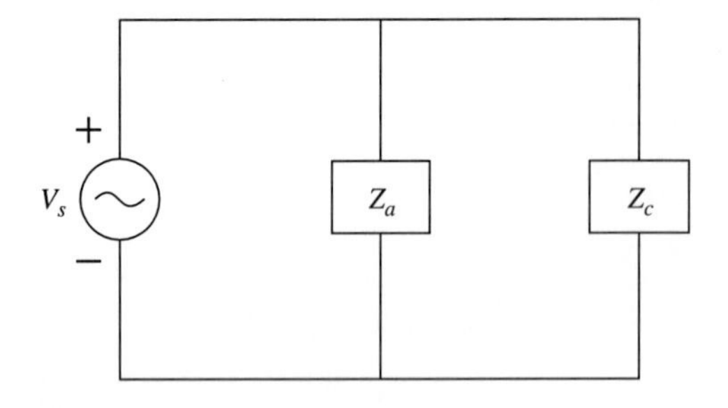

FIGURE P5.23

5.24 **(a)** Find the single impedance that is equivalent to the three parallel resistors shown in Fig. P5.24.
(b) Redraw the circuit using the single equivalent impedance in place of the three resistors.

$R_1 = 4\ \text{k}\Omega \qquad R_2 = 5\ \text{k}\Omega \qquad R_3 = 2\ \text{k}\Omega$

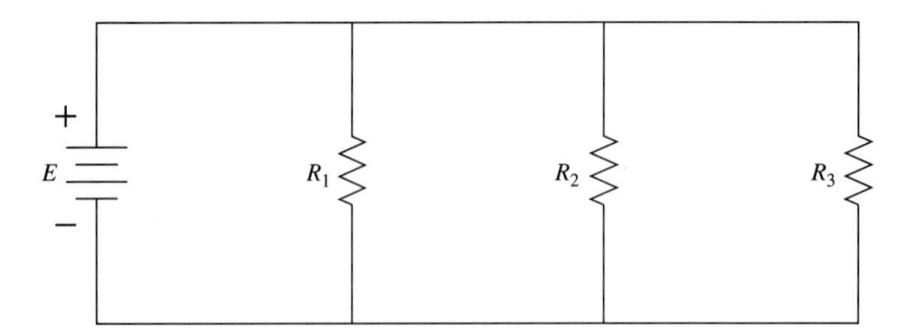

FIGURE P5.24

5.25 **(a)** Find the single admittance that is equivalent to the four parallel admittances shown in Fig. P5.25.
(b) Find the single impedance that is equivalent to the four parallel admittances shown.
(c) Redraw this circuit using the single equivalent admittance (or impedance) in place of the four admittances.

$Y_a = 40\angle 36.9°\ \text{mS} \qquad Y_c = 30\angle 60°\ \text{mS}$

$Y_e = 50\angle 30°\ \text{mS} \qquad Y_m = 60\angle -53.1°\ \text{mS}$

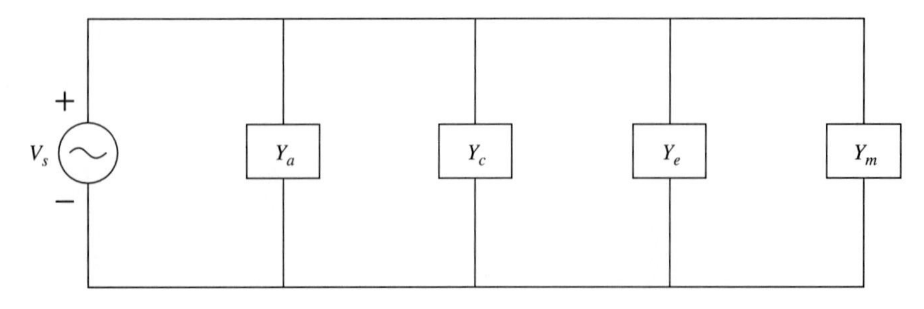

FIGURE P5.25

5.26 For the circuit shown in Fig. P5.26:

(a) Find the single equivalent capacitor that can replace the three capacitors shown.

(b) Redraw this circuit using the single equivalent capacitor in place of the three capacitors.

$C_1 = 100\ \mu F \qquad C_2 = 120\ \mu F \qquad C_3 = 180\ \mu F$

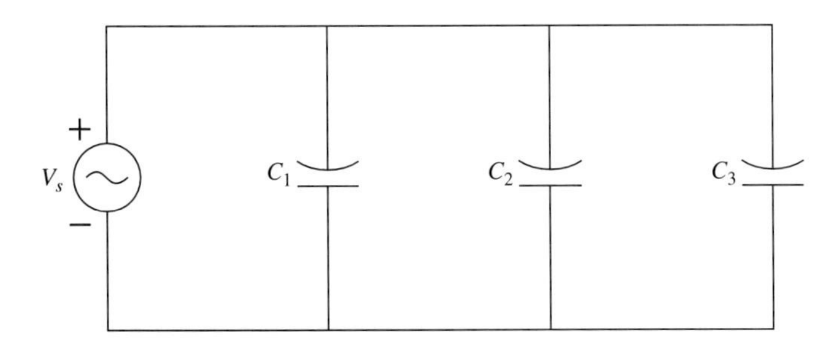

FIGURE P5.26

5.27 For the circuit shown in Fig. P5.27:

(a) Find the single equivalent inductor that can replace the three inductors shown.

(b) Redraw the circuit using this equivalent inductor in place of the three inductors.

$L_1 = 24$ mH $\qquad L_2 = 26$ mH $\qquad L_3 = 30$ mH

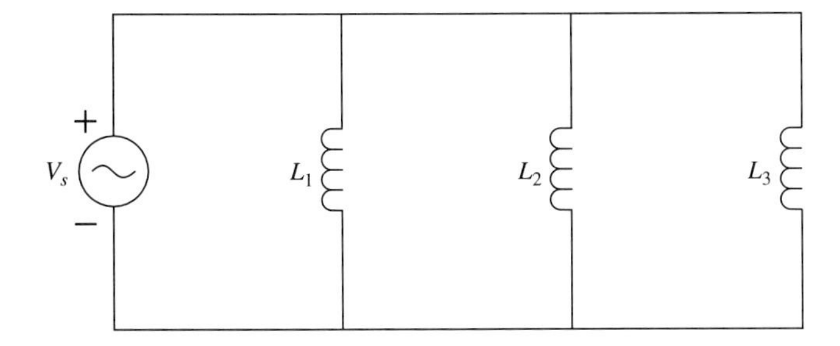

FIGURE P5.27

5.28 Eight resistors are connected in parallel. Each resistor has a resistance of 320 kΩ. Find the single equivalent resistance.

5.29 Twelve inductors are connected in parallel. Each inductor has an inductance of 54 mH. Find the single equivalent inductance.

5.30 Ten capacitors are connected in parallel. Each capacitor has a capacitance of 12 μF. Find the single equivalent capacitance.

5.31 For the R–L–C circuit shown in Fig. P5.31:

(a) Find the single equivalent admittance that can replace the three elements shown.

(b) Find the phasor current I.

(c) Find the phasor current I_R using current division.

(d) Find the phasor current I_C using current division.

(e) Find the phasor current I_L using current division.

(f) Express each phasor as its corresponding sinusoidal time function.

$V_s = 360\underline{/-60^\circ}$ mV $\qquad R = 200\ \Omega \qquad L = 5$ mH

$C = 0.067\ \mu F \qquad f = 9.55$ kHz

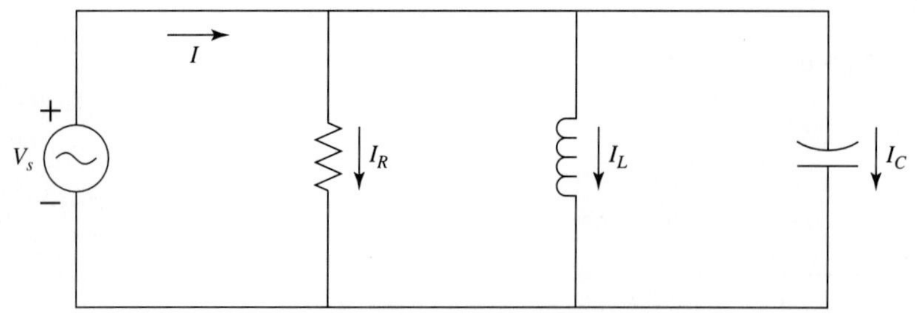

FIGURE P5.31

5.32 Repeat parts (c) through (e) of Problem 5.31 using Ohm's Law. Should these answers be the same? Why?

5.33 **(a)** Draw the phasor diagram for the circuit of Fig. P5.31. Include the following: V_s, I, I_R, I_C, I_L.

(b) What is the phase relationship (magnitude and angle) between each of the following: I_R and V_s, I_C and V_s, I_L and V_s, I and V_s, I_L and I_C. Explain why these phase relationships are expected.

5.34 For the circuit of Fig. P5.34:

(a) Find the phasor current I_R.

(b) Find the phasor current I_C.

(c) Find the phasor voltage V_R.

(d) Find the phasor voltage V_C.

(e) Express each of these phasors as their sinusoidal time function counterparts.

$I_s = 20\angle 0°$ mA $\quad R = 2$ kΩ $\quad C = 0.05\ \mu$F

$\omega = 10$ krad/s

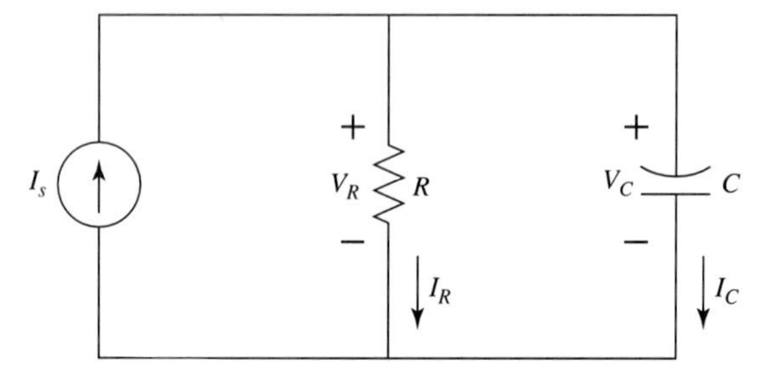

FIGURE P5.34

CHAPTER 6

STEADY STATE ANALYSIS OF SERIES–PARALLEL CIRCUITS

INTRODUCTION

In the last chapter we developed the circuit laws applicable to standalone series circuits and standalone parallel circuits. Many circuits, however, consist of combinations of series and parallel circuits and are appropriately known as "series–parallel" circuits.

The analysis of series–parallel circuits does not involve any new theories or any new circuit laws over and above those covered in previous chapters. It basically involves the proper application of those circuit laws already discussed. We will therefore see that the first step in such analysis usually involves identifying which impedances are connected in series and which are connected in parallel.

Since series–parallel circuits are generally more complex than individual series or parallel circuits, we will find ourselves performing more extensive mathematical calculations. All the required mathematical techniques have been covered in previous chapters, so refer to those chapters whenever the need for a review arises.

At this point we will introduce the following shorthand notation:

$$Z_1 \| Z_2$$

This notation is read as "Z_1 is in parallel with Z_2."

The use of two vertical parallel lines to represent parallel connection of the specified impedances is common and should be committed to memory.

6.1 DEFINITION

We define the series–parallel circuit as one that consists of ***both*** impedances connected in series ***and*** impedances connected in parallel. As we will see in later chapters, some circuits (e.g., bridge circuits) have to be modified before they can be treated as series–parallel circuits. In this chapter, we limit ourselves to circuits that are obviously series–parallel configurations.

6.2 EQUIVALENT IMPEDANCE

We define the equivalent impedance of a series–parallel circuit as follows:

> The equivalent impedance of a series–parallel circuit is the single impedance that can replace the circuit and result in the same source current when the same source voltage is applied. (It is equally accurate to say that replacement will result in the same source voltage when the same source current is applied).

6.2.1 Finding Equivalent Impedances

To find the equivalent impedance of a series–parallel circuit we will use the equations developed in the last chapter for equivalent impedance of a series circuit, and equivalent impedance of a parallel circuit. As noted earlier, it is the proper application of these equations that is important.

EXAMPLE 6.1

In the circuit of Fig. 6.1(a), the following values are given: $Z_1 = 3\angle 53.1^\circ\ \Omega$, $Z_2 = 4\angle 36.9^\circ\ \Omega$, $Z_3 = 6\angle 30^\circ\ \Omega$. Find Z_e = the equivalent impedance of the circuit looking into terminals a and b.

FIGURE 6.1

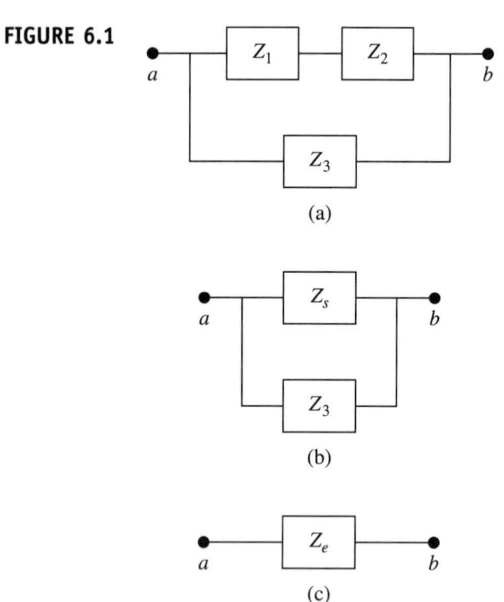

Solution The first step in the solution of these types of problems is identifying which elements are connected in series and which in parallel. In Fig. 6.1(a) Z_1 and Z_2 are connected in series. They may therefore be replaced by a single equivalent impedance Z_s using Equation 5.13 as follows:

$$Z_s = Z_1 + Z_2 = 3\angle 53.1^\circ + 4\angle 36.9^\circ = 1.8 + j2.4 + 3.2 + j2.4$$
$$= (5 + j4.8)\ \Omega = 6.9\angle 43.8^\circ\ \Omega$$

This series equivalent impedance, Z_s, is in parallel with impedance Z_3 as shown in Fig. 6.1(b). We now have a simple parallel circuit with Z_s in parallel with Z_3. The equivalent impedance of the ***circuit,*** Z_e, is therefore the equivalent parallel impedance of Z_s and Z_3 as shown in Fig. 6.1(c). This may be found using Equation 5.36 as follows:

$$Z_e = Z_s Z_3/(Z_s + Z_3) = (6.9\angle 43.8^\circ)(6\angle 30^\circ)/[(6.9\angle 43.8^\circ) + (6\angle 30^\circ)]\ \Omega$$

Answer $Z_e = 3.23\angle 36.4^\circ\ \Omega$.

Example 6.1 involved a simple series–parallel circuit with only two impedances in series and a single impedance in parallel with those two. The procedure followed in this example can be employed regardless of the number of impedances involved. The next series of examples demonstrates how this is accomplished.

EXAMPLE 6.2

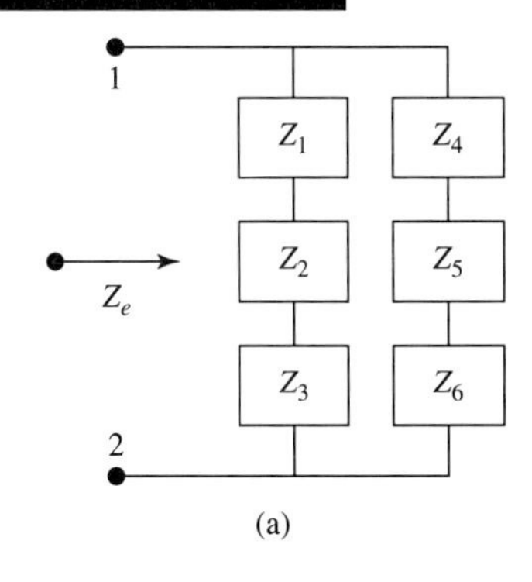

(a)

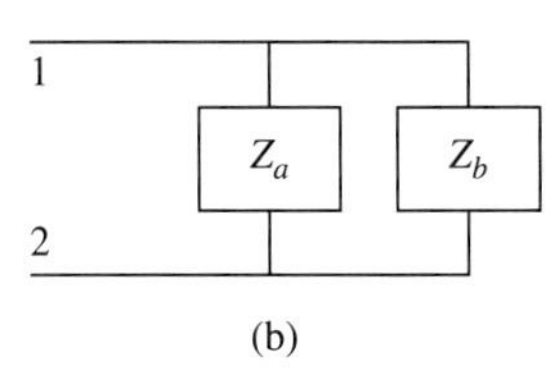

(b)

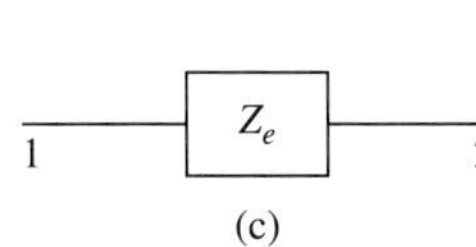

(c)

FIGURE 6.2

Find the equivalent impedance, Z_e, of the series–parallel circuit shown in Fig. 6.2(a) with respect to terminals 1 and 2. The values of the impedances are $Z_1 = 12\text{K}\angle 45^\circ\ \Omega$, $Z_2 = 8\text{K}\angle 60^\circ\ \Omega$, $Z_3 = 15\text{K}\angle 26^\circ\ \Omega$, $Z_4 = 10\text{K}\angle 36.9^\circ\ \Omega$, $Z_5 = 14\text{K}\angle 53.1^\circ\ \Omega$, $Z_6 = 6\text{K}\angle 30^\circ\ \Omega$.

Solution The three impedances, Z_1, Z_2, and Z_3, are in series. They may be replaced by a single equivalent impedance Z_x as follows:

$$Z_x = Z_1 + Z_2 + Z_3 = 12\text{K}\angle 45^\circ + 8\text{K}\angle 60^\circ + 15\text{K}\angle 26^\circ$$
$$= 26\text{K} + j22\text{K} = 34.1\text{K}\angle 40.2^\circ\ \Omega$$

The three impedances, Z_4, Z_5, and Z_6 are in series. These three impedances may be replaced by a single equivalent impedance Z_y as follows:

$$Z_y = Z_4 + Z_5 + Z_6 = 10\text{K}\angle 36.9^\circ + 14\text{K}\angle 53.1^\circ + 6\text{K}\angle 30^\circ$$
$$= 21.6\text{K} + j20.2\text{K} = 29.6\text{K}\angle 43.1^\circ\ \Omega$$

The original series–parallel circuit may now be redrawn as shown in Fig. 6.2(b). The equivalent impedance of our circuit, Z_e, may now be expressed as the equivalent impedance of the two parallel impedances Z_x and Z_y as follows:

$$Z_e = Z_x Z_y/(Z_x + Z_y) = \frac{(34.1\text{K}\angle 40.2^\circ)(29.6\text{K}\angle 43.1^\circ)}{(34.1\text{K}\angle 40.2^\circ + 29.6\text{K}\angle 43.1^\circ)}\ \Omega$$

Answer $Z_e = 15.9\text{K}\angle 41.7^\circ\ \Omega$.

Fig. 6.2(c) shows the single equivalent impedance, Z_e, replacing the original series–parallel circuit of Fig. 6.2(a).

EXAMPLE 6.3

Find the equivalent impedance, Z_e, of the series–parallel circuit in Fig. 6.3(a) with respect to terminals c and d. The values of the impedances are $Z_1 = 100\angle -60^\circ\ \Omega$, $Z_2 = 140\angle 30^\circ\ \Omega$, $Z_3 = 180\angle -36.9^\circ\ \Omega$, $Z_4 = 200\angle 70^\circ\ \Omega$, $Z_5 = 150\angle -45^\circ\ \Omega$, $Z_6 = 120\angle -30^\circ\ \Omega$, $Z_7 = 140\angle 60^\circ\ \Omega$, and $Z_8 = 130\angle 45^\circ\ \Omega$.

Solution In the circuit of Fig. 6.3(a), we will represent individual series and parallel combinations as follows:

$$Z_a = Z_1 \| Z_2 = Z_1 Z_2/(Z_1 + Z_2) = \frac{(100\angle -60^\circ)(140\angle 30^\circ)}{(100\angle -60^\circ) + (140\angle 30^\circ)}\ \Omega$$
$$= 81.4\angle -24.5^\circ\ \Omega = (74 - j33.8)\ \Omega$$

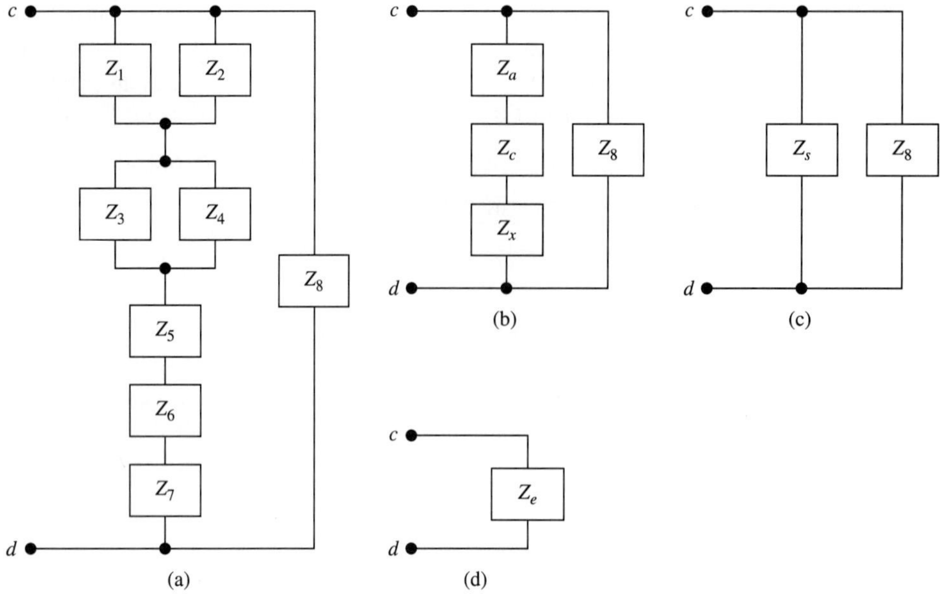

(a) (b) (c) (d)

FIGURE 6.3

$$Z_c = Z_3 \| Z_4 = Z_3Z_4/(Z_3 + Z_4) = \frac{(180\angle -36.9^\circ)(200\angle 70^\circ)}{(180\angle -36.9^\circ) + (200\angle 70^\circ)}\ \Omega$$
$$= 158.7\angle 12.5^\circ\ \Omega = (154.9 + j34.3)\ \Omega$$
$$Z_x = Z_5 + Z_6 + Z_7 = [(150\angle -45^\circ) + (120\angle -30^\circ) + (140\angle 60^\circ)]\ \Omega$$
$$= (280 - j44.9)\ \Omega = 283.6\angle 9.1^\circ\ \Omega$$

The circuit is redrawn in Fig. 6.3(b) using the equivalent impedances found here. In Fig. 6.3(b), impedances Z_a, Z_c, and Z_e are in series. Their single series equivalent impedance is Z_s as found by:

$$Z_s = Z_a + Z_c + Z_x = 74 - j33.8 + 154.9 + j34.3 + 280 - j44.9$$
$$= (508.9 - j44.4)\ \Omega = 510.8\angle -5^\circ\ \Omega$$

The impedances Z_s and Z_8 are in parallel as shown in Fig. 6.3(c). The equivalent impedance of these two in parallel is the total circuit equivalent impedance, Z_e:

$$Z_e = Z_s \| Z_8 = \frac{Z_sZ_8}{(Z_s + Z_8)} = \frac{(510.8\angle -5^\circ)(130\angle 45^\circ)}{(510.8\angle -5^\circ) + (130\angle 45^\circ)}\ \Omega$$

Answer $Z_e = 110\angle 35.5^\circ\ \Omega = (89.6 + j63.9)\ \Omega$.

The equivalent impedance, Z_e, is shown in Fig. 6.3(d).

We will now look at some examples involving discrete components.

EXAMPLE 6.4

Find the equivalent impedance of the circuit of Fig. 6.4(a) with respect to terminals 1 and 2. The values of the components are $R_1 = 400\ \Omega$, $R_2 = 300\ \Omega$, $C_1 = 1\ \mu$F, $C_2 = 1\ \mu$F. Assume a frequency $f = 200$ Hz.

Solution Since R_1 and C_1 are in series, we may add their impedances to find their equivalent impedance Z_1 as follows:

$$Z_1 = Z_{R_1} + Z_{C_1} = R_1 - j(1/2\pi fC_1) = 400 - j[1/(2\pi(200)1 \times 10^{-6})]$$
$$= (400 - j796)\ \Omega = 891\angle -63.3^\circ\ \Omega$$

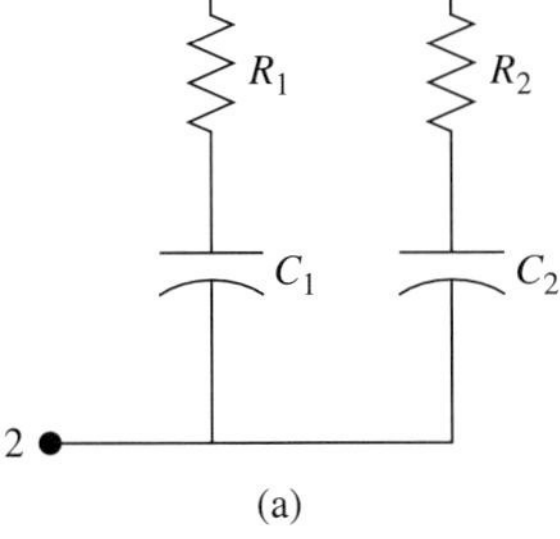
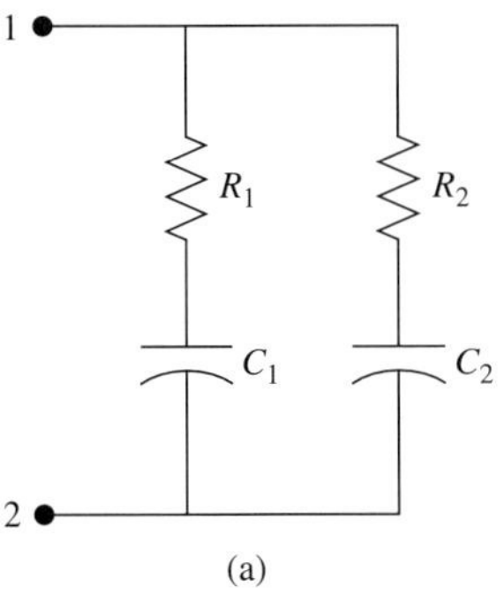

FIGURE 6.4

Similarly, R_2 and C_2 are in series. If their equivalent impedance is represented by Z_2, then:

$$Z_2 = Z_{R_2} + Z_{C_2} = R_2 - j(1/2\pi f C_2) = 300 - j[1/(2\pi(200)1 \times 10^{-6})]$$
$$= (300 - j796)\ \Omega = 849\underline{/-69.3^\circ}\ \Omega$$

The circuit of Fig. 6.4(a) is redrawn in Fig. 6.4(b) with Z_1 and Z_2 replacing their discrete components. These two impedances are in parallel and their equivalent impedance is Z_e, the equivalent impedance of the circuit we are looking for as shown in Fig. 6.4(c).

$$Z_e = Z_1 \| Z_2 = \frac{Z_1 Z_2}{(Z_1 + Z_2)} = \frac{(891\underline{/-63.3^\circ})(849\underline{/-69.3^\circ})}{400 - j796 + 300 - j796}\ \Omega$$

Answer $Z_e = 435\underline{/-66.3^\circ}\ \Omega$.

Note that the impedance angle is negative and less than 90° in magnitude. Recall that this is expected because a capacitive circuit will always have a phase angle lying between 0 and −90°.

EXAMPLE 6.5

The circuit in Fig. 6.5(a) consists of resistors and inductors. Find Z_e, the equivalent impedance of this circuit between terminals a and b. Assume $f = 4$ kHz, $R_a = 2\ \text{k}\Omega$, $R_c = 4\ \text{k}\Omega$, $R_u = 3\ \text{k}\Omega$, $R_n = 5\ \text{k}\Omega$, $R_o = 6\ \text{k}\Omega$, $L_u = 100$ mH, and $L_n = 120$ mH.

Solution Since R_a and R_c are in series, we may find their equivalent impedance, Z_a, as follows:

$$Z_a = R_a + R_c = 2\text{K} + 4\text{K} = 6\text{K}\underline{/0^\circ}\ \Omega$$

R_u and L_u are similarly in series. We may find their equivalent impedance, Z_u, as follows:

$$Z_u = R_u + j2\pi f L_u = 3\text{K} + j2\pi(4000)(100)10^{-3} = 3\text{K} + j2.5\text{K}$$
$$= 3.9\text{K}\underline{/39.8^\circ}\ \Omega$$

The series combination of R_n and L_n may be replaced by their equivalent impedance, Z_n.

$$Z_n = R_n + j2\pi f L_n = 5\text{K} + j2\pi(4000)(120)10^{-3} = 5\text{K} + j3\text{K}$$
$$= 5.8\text{K}\underline{/31^\circ}\ \Omega$$

The single resistor R_o may be replaced by its impedance, Z_o:

$$Z_o = R_o = 6\ \text{k}\Omega$$

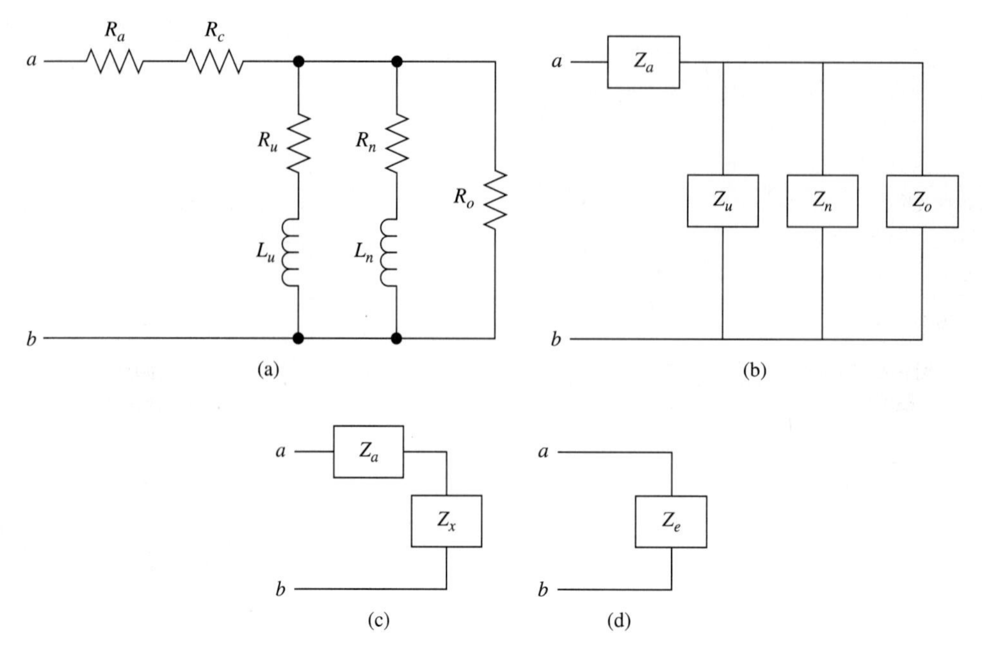

FIGURE 6.5

The circuit of Fig. 6.5(a) is redrawn in Fig. 6.5(b) using the equivalent impedances found here. The impedances Z_u, Z_n, and Z_o are in parallel. If we denote their single equivalent impedance by Z_x, then:

$$Z_x = 1/Y_x$$

where Y_x is the equivalent admittance of these three parallel impedances. We find Y_x as follows:

$$Y_x = Y_u + Y_n + Y_o$$

$$\text{where } Y_u = 1/Z_u = 1/(3.9\text{K}\underline{/39.8°}) = 0.26 \times 10^{-3}\underline{/-39.8°}$$
$$= (0.20 - j0.17)\text{ mS}$$
$$Y_n = 1/Z_n = 1/(5.8\text{K}\underline{/31°}) = 0.17 \times 10^{-3}\underline{/-31°}$$
$$= (0.15 - j0.09)\text{ mS}$$
$$Y_o = 1/Z_o = 1/(6\text{K}\underline{/0°}) = 0.17\text{ mS}$$

Substituting these values for Y_u, Y_n, and Y_o into the equation for Y_x:

$$Y_x = .26 \times 10^{-3}\underline{/-39.8°} + .17 \times 10^{-3}\underline{/-31°} + .17 \times 10^{-3}$$
$$= .57\underline{/-26.2°}\text{ mS}$$
$$Z_x = 1/.57\text{ m}\underline{/-26.2°} = 1.74\text{K}\underline{/26.2°}$$

Fig. 6.5(b) is redrawn in Fig. 6.5(c) with Z_x replacing the three parallel impedances, Z_u, Z_n, and Z_o. Here we see that Z_a and Z_x are in series. The single equivalent impedance of these two in series is the same as the total circuit equivalent impedance, Z_e. This is shown in Fig. 6.5(d), where the single impedance Z_e replaces the entire series–parallel circuit of Fig. 6.5(a).

$$Z_e = Z_a + Z_x = 6\text{K}\underline{/0°} + 1.74\text{K}\underline{/26.2°} = 7.6\text{K}\underline{/5.8°}$$

Answer $Z_e = 7.6\underline{/5.8°}\text{ k}\Omega$

6.2.2 Series–Parallel Circuit Problems

We will now look at several examples involving the determination of voltages and/or currents in series–parallel circuits.

EXAMPLE 6.6

In the circuit of Fig. 6.6(a), $v = 200\sin(500t)$ V, $R_1 = 200\ \Omega$, $R_2 = 500\ \Omega$, $R_3 = 400\ \Omega$, $L_1 = 200$ mH, $L_2 = 800$ mH, and $C = 5\ \mu$F.

Find the following:

(a) i_s, the sinusoidal source current
(b) i_2, the sinusoidal current through the R_2–L_2 string
(c) i_3, the sinusoidal current through the R_3–C string
(d) v_{cb}, the sinusoidal voltage across terminals cb
(e) v_3, the sinusoidal voltage across the R_3–C series string.

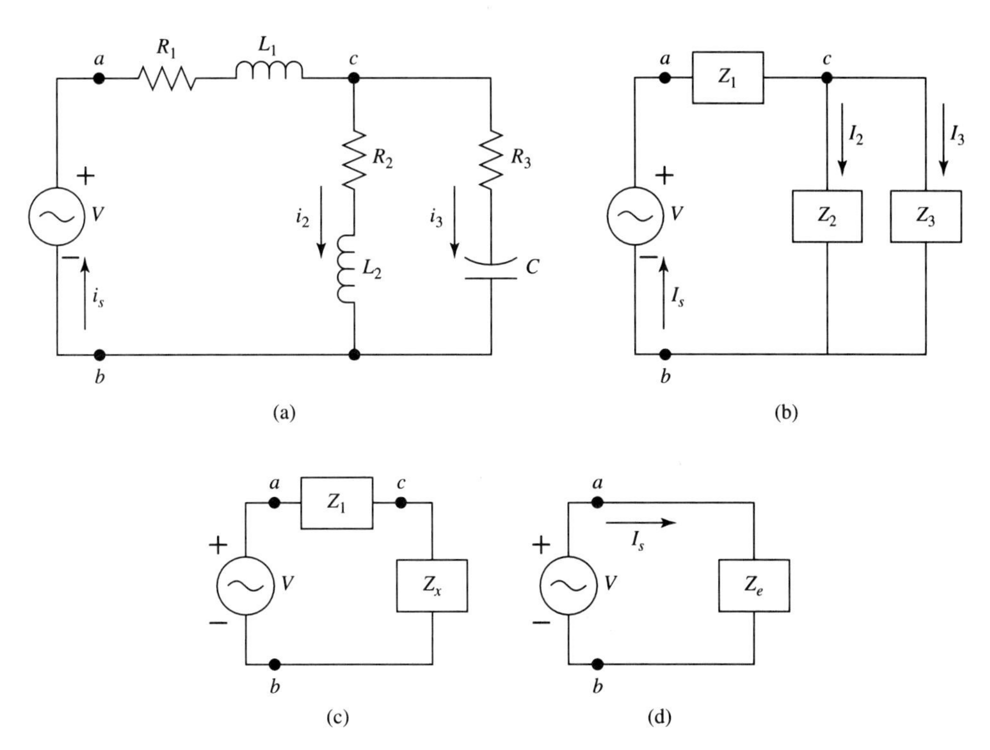

FIGURE 6.6

Solution The first step in the solution of most of these problems is to redraw the original circuit while substituting phasors for their sinusoidal counterparts and impedances for their discrete component counterparts. This is done in Fig. 6.6(b), where:

$V = 200\angle 0^\circ$ V

I_s is the phasor representing the sinusoidal current i_s

I_2 is the phasor representing the sinusoidal current i_2

I_3 is the phasor representing the sinusoidal current i_3

$$Z_1 = R_1 + j\omega L_1 = 200 + j(500)(200)10^{-3} = (200 + j100)\ \Omega = 223.6\angle 26.6^\circ\ \Omega$$

$$Z_2 = R_2 + j\omega L_2 = 500 + j(500)(500)10^{-3} = (500 + j250)\ \Omega = 559\angle 26.6^\circ\ \Omega$$

$$\begin{aligned} Z_3 &= R_3 - j(1/\omega C) = 400 - j[1/500(5)10^{-6}] \\ &= (400 - j400)\ \Omega = 565.7\angle -45^\circ\ \Omega \end{aligned}$$

(a) To find the source current we will first find Z_e, the equivalent impedance of the circuit between terminals a and b, and then use Ohm's Law. Accordingly, if we let $Z_x = Z_2 \| Z_3$ in Fig. 6.6(b):

$$\begin{aligned} Z_x &= \frac{Z_2 Z_3}{Z_2 + Z_3} = \frac{(559\angle 26.6^\circ)(565.7\angle -45^\circ)}{500 + j250 + 400 - j400} \\ &= 346.6\angle -8.9^\circ = (342.4 - j53.6)\ \Omega \end{aligned}$$

The circuit is redrawn in Fig. 6.6(c) with Z_x replacing $Z_2 \| Z_3$. From this circuit we can see that

$$Z_e = Z_1 + Z_x = 200 + j100 + 342.4 - j53.6 = (542.4 + j46.4)\ \Omega$$
$$= 544.4\angle 4.9°\ \Omega$$

Fig. 6.6(d) shows Z_e replacing the entire series parallel impedance connected across the voltage source. From Ohm's Law

$$I_s = V/Z_e = (200\angle 0°)/(544.4\angle 4.9°) = 0.37\angle -4.9° = 370\angle -4.9°\ \text{mA}$$

Answer $i_s = 370\ \sin(500t - 4.9°)$ mA.

(b) We may note from Fig. 6.6(b) that the source current, I_s, divides between the two parallel impedances Z_2 and Z_3. We may therefore find the current I_2 by current division.

$$I_2 = \frac{Z_3 I_s}{Z_2 + Z_3} = \frac{(565.7\angle -45°)(0.37\angle -4.9°)}{900 - j150} = 229.4\angle -40.44°\ \text{mA}$$

Answer $i_2 = 229.4\ \sin(500t - 40.44°)$ mA.

(c) The phasor current I_3 may be found either by voltage division ***or*** by using KCL at node c. Since we used voltage division in part (b), let us use KCL to find I_3.

$$I_3 = I_s - I_2 = (370\angle -4.9°) - (229.4\angle -40.44°) = 226.7\angle 31.1°\ \text{mA}$$

Answer $i_3 = 226.7\ \sin(500t + 31.1°)$ mA.

(d) The voltage V_{cb} may be found by noting in Fig. 6.6(c) that this is the voltage across Z_x. Since Z_x is in series with Z_1 we will use voltage division as follows:

$$V_{cb} = Z_x \times V/(Z_x + Z_1)$$
$$= (346.6\angle -8.9°)(200\angle 0°))/(342.4 - j53.6 + 200 + j100)$$
$$= (346.6)(200)\angle -8.9°/544.4\angle 4.9° = 127.3\angle -13.8°\ \text{V}$$

Answer $v_{cb} = 127.3\ \sin(500t - 13.8°)$ V.

(e) Since $Z_x = Z_3 \| Z_2$, the voltage across Z_3 is the same as the voltage across Z_x, which is the voltage found in part (d). Accordingly, $v_3 = v_{cb}$ and $v_3 = 127.3\ \sin(500t - 13.8°)$ V.

EXAMPLE 6.7

The components of Fig. 6.7(a) have the following values: $v = 36\ \sin(10{,}000t)$ V, $R_1 = 4\ \text{k}\Omega$, $R_2 = 6\ \text{k}\Omega$, $R_3 = 10\ \text{k}\Omega$, $R_4 = 12\ \text{k}\Omega$, $R_5 = 2\ \text{k}\Omega$, $L_1 = 200$ mH, $L_3 = 1$ H, $C_2 = 0.02\ \mu$F, $C_4 = 0.01\ \mu$F, $C_5 = 0.04\ \mu$F. Find the sinusoidal voltage v_{cd}.

Solution We will first convert the voltage source to its phasor form and all the components to their corresponding impedances.

$$Z_1 = R_1 + j\omega L_1 = 4\text{K} + j(10\text{K})(200)10^{-3} = (4 + j2)\ \text{k}\Omega = 4.5\angle 26.6°\ \text{k}\Omega$$

$$Z_2 = R_2(1/\omega C_2)\angle -90°/[R_2 - (j/\omega C_2)]$$
$$= 6\ \text{K}\frac{\{1/[(10\text{K})(0.02)10^{-6}]\}\angle -90°}{\{(6\text{K} - j)/[(10\text{K})(0.02)10^{-6}]\}} = 3.8\angle -50.2°\ \text{k}\Omega = (2.4 - j2.9)\ \text{k}\Omega$$

$$Z_3 = \frac{R_3(\omega L_3)\angle 90°}{R_3 + j\omega L_3} = \frac{(10\text{K})(10\text{K})(1)\angle 90°}{10\text{K} + j(10\text{K})(1)} = 7\angle 45°\ \text{k}\Omega = (5 + j5)\ \text{k}\Omega$$

$$Z_4 = R_4 - j(1/\omega C_4) = 12\text{K} - j\frac{1}{(10\text{K})(0.01)10^{-6}} = (12 - j10)\ \text{k}\Omega$$
$$= 15.6\angle -39.8°\ \text{k}\Omega$$

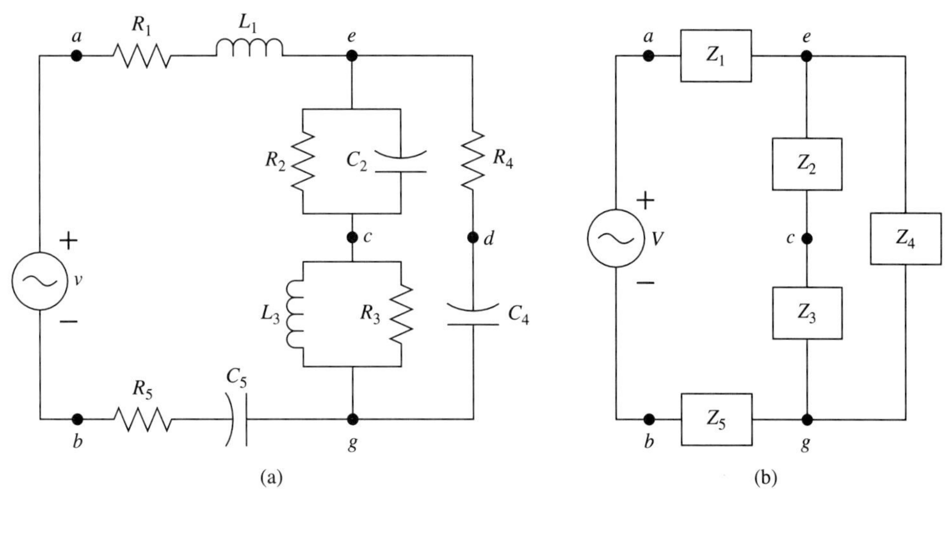

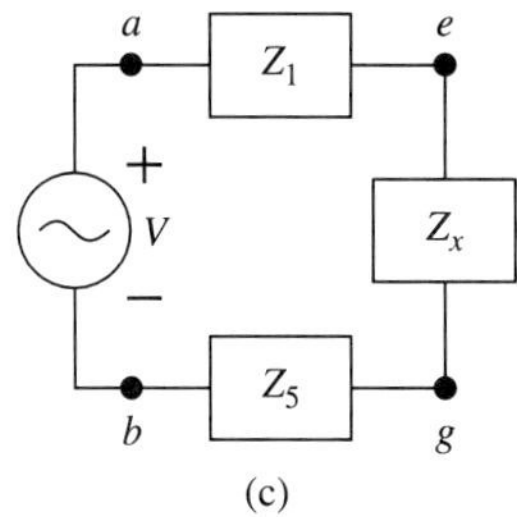

FIGURE 6.7

$$\begin{aligned} Z_5 &= R_5 - j(1/\omega C_5) = 2\text{K} - j\frac{1}{(10\text{K})(0.04)10^{-6}} \\ &= (2 - j2.5)\ \text{k}\Omega = 3.2\underline{/-51.3^\circ}\ \text{k}\Omega \\ \text{V} &= 36\underline{/0^\circ}\ \text{V} \end{aligned}$$

The circuit is redrawn in Fig. 6.7(b) using the impedances and voltage phasor just found.

From Fig. 6.7(a) and KVL, we may write:

$$V_{cd} = V_{ce} + V_{ed}$$

Note that if we find V_{eg}, we can then find V_{ce} and V_{ed} by voltage division: V_{ce} is found by voltage division between the series impedances Z_2 and Z_3 in Fig. 6.7(b) and V_{ed} by voltage division between the series components R_4 and C_4 in Fig. 6.7(a).

Accordingly, we will first find V_{eg}. To do this, we may further simplify the circuit by combining Z_2, Z_3, and Z_4 into a single impedance Z_x as shown in Fig. 6.7(c).

$$\begin{aligned} Z_x &= Z_4 \| (Z_2 + Z_3) \\ Z_2 + Z_3 &= (2.4 - j2.9 + 5 + j5) \times 10^3 = (7.4 + j2.1)\ \text{k}\Omega = 7.7\underline{/15.4^\circ}\ \text{k}\Omega \\ Z_x &= \frac{(15.6\underline{/-39.8^\circ})(7.7\underline{/15.4^\circ})}{12 - j10 + 7.4 + j2.1} \times 10^3 = 5.73\underline{/-2.1^\circ}\ \text{k}\Omega \end{aligned}$$

In Fig. 6.7(c) the three impedances are in series with the voltage source. From the voltage division rule:

$$V_{eg} = \frac{Z \times V}{Z_1 + Z_x + Z_5} = \frac{(5.73\text{K}\underline{/-2.1^\circ})(36\underline{/0^\circ})}{(4.5\text{K}\underline{/26.6^\circ}) + (5.73\text{K}\underline{/-2.1^\circ}) + (3.2\text{K}\underline{/-51.3^\circ})}$$

$$V_{eg} = 17.52\underline{/1.3^\circ}\ \text{V}$$

From Fig. 6.7(b), using voltage division:

$$V_{ec} = \frac{Z_2 V_{eg}}{Z_2 + Z_3} = \frac{(3.8\text{K}\underline{/-50.2^\circ})(17.52\underline{/1.3^\circ})}{7.7\text{K}\underline{/15.4^\circ})} = 8.65\underline{/-64.3^\circ}\text{ V}$$

$$V_{ce} = -V_{ec} = 8.65\underline{/115.7^\circ}\text{ V}$$

From Fig. 6.7(a), using voltage division:

$$V_{ed} = R_4 V_{eg}/Z_4 = (12\text{K}\underline{/0^\circ})(17.52\underline{/1.3^\circ})/(15.6\text{K}\underline{/-39.8^\circ})$$

$$V_{ed} = 13.48\underline{/41.1^\circ}$$

From KVL, as previously noted:

$$V_{cd} = V_{ce} + V_{ed} = 8.65\underline{/115.7^\circ} + 13.48\underline{/41.1^\circ}$$

$$V_{cd} = 17.8\underline{/69^\circ}\text{ V}$$

Answer $V_{cd} = 17.8\sin(10{,}000t + 69^\circ)$ V.

The circuits in Fig. 6.8 are known as voltage divider circuits. Using series-connected resistors as shown, a voltage source of a fixed value may be used to supply a different value according to the rule of voltage division.

In the circuit of Fig. 6.8(a) a single output voltage is available as shown across resistor R_2 in addition to the source voltage E. In Fig. 6.8(c), two output voltages are available, one at point a and a second at point b, in addition to the source voltage E. In both cases, the output voltages may be calculated using the rules of voltage division.

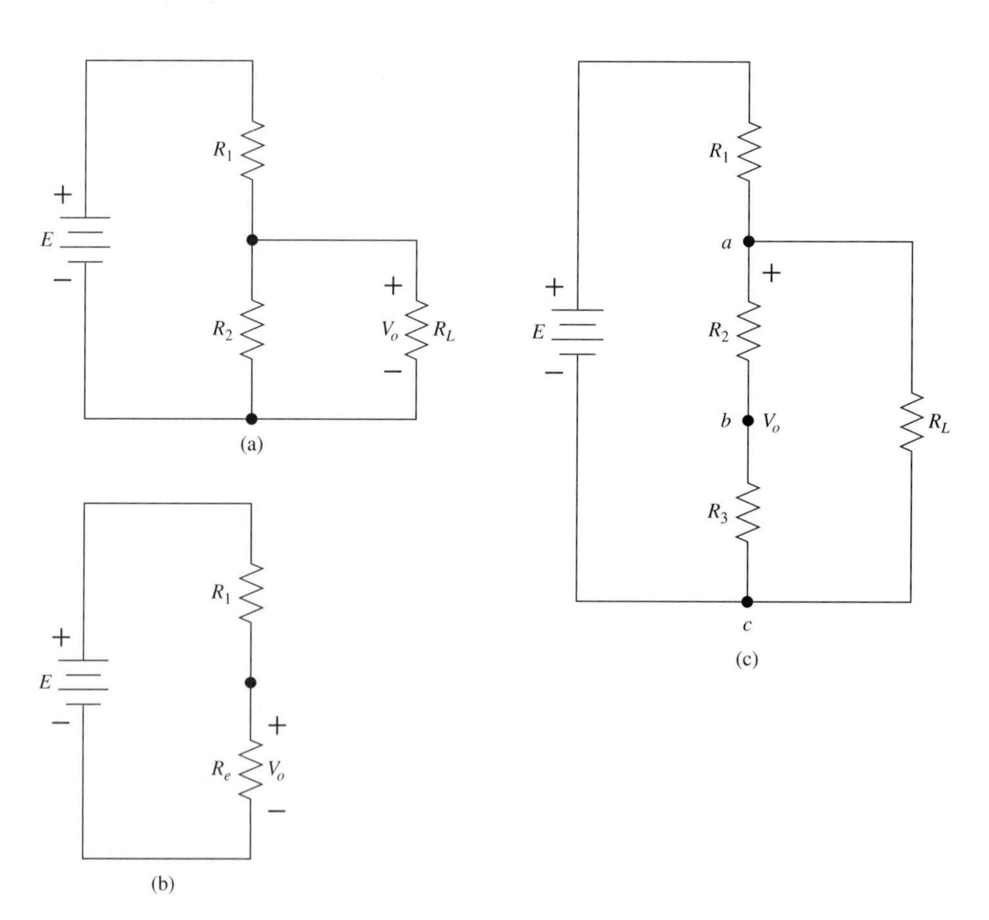

FIGURE 6.8

EXAMPLE 6.8

(a) Find the voltage across the load resistor R_L in Fig. 6.8(a) given the following values: $E = 120$ V, $R_1 = 100\ \Omega$, $R_2 = 30\ \Omega$, and $R_L = 10\ \text{M}\Omega$.
(b) Repeat part (a) if $R_L = 30\ \Omega$.

Solution

(a) The two parallel resistors, R_2 and R_L, may be replaced by a single equivalent resistor, R_e, given by Equation 5.36, where

$$R_e = R_2R_L/(R_2 + R_L)$$

In performing this calculation, it is apparent that the R_e is effectively the same as R_2, 30 Ω. This is because of the general rule: ***Two parallel, unequal resistors are always equivalent to the smaller resistor whenever the larger value resistor is at least 40 times the value of the smaller resistor.***

These two resistors are replaced by their single equivalent R_e in Fig. 6.8(b). In this figure, V_o is calculated by voltage division between R_e and R_1 as follows:

$$V_o = R_eE/(R_e + R_1) = 30(120)/(30 + 100) = 30(120)/130 = 27.69 \text{ V}$$

Answer $V_o = 27.7$ V.

(b) The two parallel resistors, R_1 and R_L, may again be replaced by their equivalent single resistor, R_e. This time, however, the two resistors are of the same order of magnitude. When Equation 5.36 is applied, a value of R_e equal to 15 Ω is found. This is because of the general rule that when two ***equal*** resistors are in parallel, the result is equal to half the value of one resistor.

V_o may now be found by voltage division as:

$$V_o = R_eE/(R_e + R_1) = 15(120)/(15 + 100) = 15(120)/115 = 15.65 \text{ V}$$

Answer $V_o = 15.7$ V.

This example points out that in designing a voltage divider circuit, it is desirable to make sure that the resistors in the voltage divider circuit are at least 40 times smaller than any load resistor that will be used. In this way, the voltage across the various points in the voltage divider circuit can be calculated and assumed to be constant, independent of the value of the load resistor used.

EXAMPLE 6.9

In the circuit of Fig. 6.9(a), $R_a = 12\ \text{k}\Omega$, $R_c = 16\ \text{k}\Omega$, $R_m = 8\ \text{k}\Omega$, $E_a = 40$ V, and $E_m = 30$ V. Find:
(a) I, the current through the resistor R_a
(b) V_c, the voltage across the capacitor

Solution

(a) Since the circuit is dc, we must replace all the components by their dc impedances. Accordingly, the capacitor becomes an open circuit, the inductors become short circuits, and the resistors remain unchanged. The equivalent circuit is redrawn in Fig. 6.9(b). In this circuit, R_a and R_m are in series and may be replaced by a single equivalent resistor, R_e:

$$R_e = R_a + R_m = 12\text{K} + 8\text{K} = 20\ \text{k}\Omega$$

The voltage sources E_a and E_m are also in series and may be replaced by a single equivalent source, E:

$$E = E_a + E_m = 40 + 30 = 70 \text{ V}$$

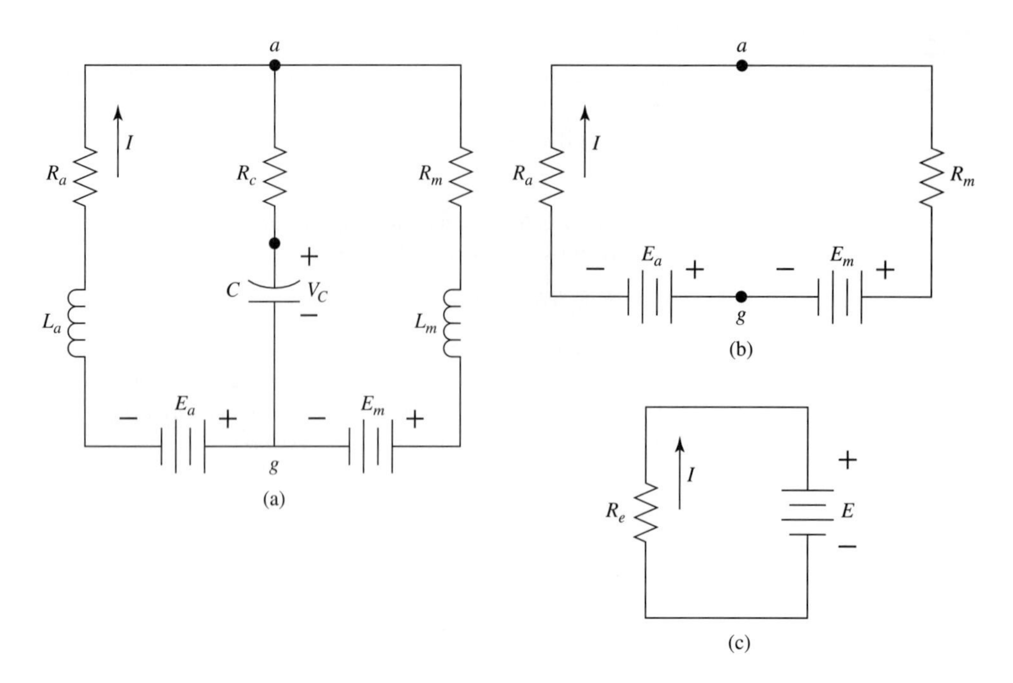

FIGURE 6.9

The circuit is redrawn in Fig. 6.9(c) using R_e and E. From this circuit:

$$I = -E/R_e = -70/20\text{K} = -3.5 \text{ mA}$$

Answer $I = -3.5$ mA.

(b) Since the capacitor is an open circuit there is no current through the resistor R_C. This means that the voltage across R_C is zero and

$$V_C = V_{ag}$$

V_{ag} may be found using KVL in Fig. 6.9 (a) or (b) as follows:

$$V_{ag} = -IR_a - E_a = -(-3.5 \times 10^{-3})(12 \times 10^3) - 40 \text{ V} = 42 - 40 = 2 \text{ V}$$

Answer $V_c = 2$ V.

6.3 CIRCUIT REFERENCE POINTS

We know that two points are always required to specify a voltage. The two subscripts used with each voltage indicate that the voltage specified is the voltage of the first subscript (point) with respect to the second subscript (point).

In many circuits, a single point is selected as a reference point. Often, this reference point is referred to as a "ground," because that is a common reference. When such a reference point exists, the voltages in the circuit may be written using a single subscript. The subscript indicates the point at which we are finding the voltage, and the second point is assumed to be the reference.

In the circuit of Fig. 6.10(a) for example, the ground point g is used as the reference. This means that the voltage V_{ag} may be written as V_a with the g assumed as the reference. Similarly, the voltage V_{ng} may be written as V_n. Of course, if we are considering the voltage between two points that do not include the reference, then we must specify both points as usual.

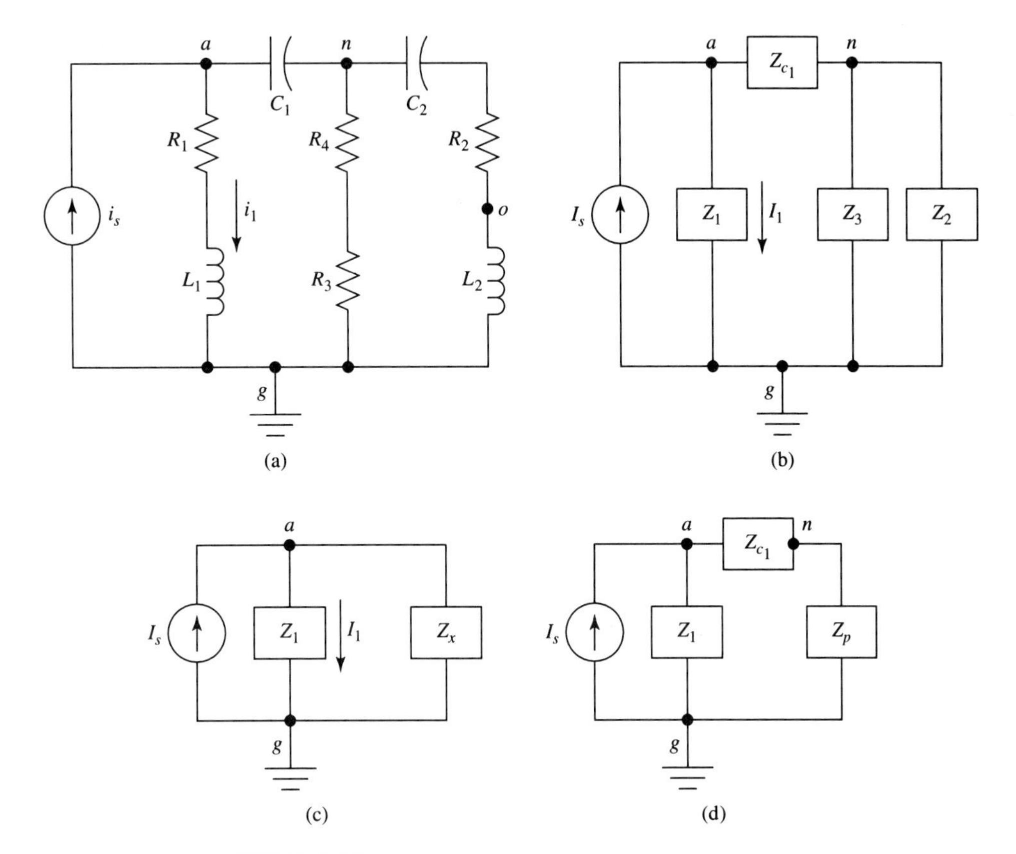

FIGURE 6.10

EXAMPLE 6.10

In the circuit of Fig. 6.10(a), $i_s = 10\cos(24{,}000t)$ mA, $R_1 = 6\text{ k}\Omega$, $R_2 = 4\text{ k}\Omega$, $R_3 = 5\text{ k}\Omega$, $R_4 = 3\text{ k}\Omega$, $L_1 = 250$ mH, $L_2 = 210$ mH, $C_1 = 0.01\ \mu$F, $C_2 = 0.0046\ \mu$F. Find:

(a) i_1, the sinusoidal current through R_1
(b) v_a, the sinusoidal voltage across the R_1–L_1 series string
(c) v_n, the sinusoidal voltage across the R_4–R_3 series string
(d) v_{an}, the sinusoidal voltage across capacitor C_1

Solution We first redraw the circuit replacing the components by their impedances and the current source by its phasor representation. This is done in Fig. 6.10(b), where:

$$Z_1 = R_1 + j\omega L_1 = 6\text{K} + j(24\text{K})(250 \times 10^{-3})$$
$$= (6 + j6)\text{ k}\Omega = 8.49\text{K}\underline{/45°}\ \Omega$$
$$Z_2 = R_2 + j[\omega L_2 - (1/\omega C_2)]$$
$$= 4\text{K} + j(24\text{K})(210 \times 10^{-3}) - [1/(24\text{K})(0.0046 \times 10^{-6})]$$
$$= 4\text{K} + j(5\text{K} - 9\text{K}) = (4 - j4)\text{ k}\Omega = 5.66\text{K}\underline{/-45°}\ \Omega$$
$$Z_3 = R_3 + R_4 = 5\text{K} + 3\text{K} = 8\text{K}\underline{/0°}\ \Omega$$
$$Z_{C_1} = -j(1/\omega C_1) = -j1/[(24\text{K})(0.01 \times 10^{-6})] = 4.17\text{K}\underline{/-90°}\ \Omega$$
$$I_s = 10\underline{/90°}\text{ mA}$$

(a) To find I_1, we recognize that Z_1 is in parallel with the single impedance that is equivalent to Z_{C_1} in series with the parallel combination of Z_3 and Z_2. If we represent this single impedance by Z_x, then:

$$Z_x = Z_3 \| Z_2 + Z_{C_1}$$

$$Z_3 \| Z_2 = \frac{Z_3 Z_2}{Z_3 + Z_2} = \frac{(8\text{K})(5.66\text{K}\angle -45^\circ)}{(8 + 4 - j4)\text{K}} = \frac{8\text{K}(5.66\text{K})\angle -45^\circ}{(12 - j4)\text{K}}$$

$$= \frac{8\text{K}(5.66\text{K})\angle -25^\circ}{12.65\text{K}\angle -18.4^\circ} = 3.58\text{K}\angle -26.6^\circ\ \Omega$$

$$Z_x = 3.58\text{K}\angle -26.6^\circ + 4.17\text{K}\angle -90^\circ = 6.6\text{K}\angle -61^\circ$$

The circuit is once again redrawn in Fig. 6.10(c). I_1 may now be found by current division between Z_1 and Z_x:

$$I_1 = \frac{Z_x I_s}{Z_x + Z_1} = \frac{(6.6\text{K}\angle -61^\circ)(10 \times 10^{-3}\angle 90^\circ)}{(3.2 - j5.77 + 6 + j6)\text{K}} = 7.17\angle 27.57^\circ \text{ mA}$$

Answer $i_1 = 7.17 \sin(24{,}000t + 27.57^\circ)$.

(b) From Ohm's Law:

$$V_a = I_1 Z_1 = (7.17 \times 10^{-3}\angle 27.57^\circ)(8.45\text{K}\angle 45^\circ) = 60.6\angle 72.6^\circ \text{ V}$$

Answer $v_a = 60.6 \sin(24{,}000t + 72.6^\circ)$ V.

(c) To find V_n, note that the parallel combination of Z_2 and Z_3 may be represented by a single impedance, Z_p. This single impedance is in series with Z_{C_1} as in Fig. 6.10(d). We may therefore find V_n by voltage division of V_a between Z_{C_1} and Z_p.

The parallel combination of Z_2 and Z_3 was found in part (a). Therefore:

$$Z_p = (3.2 - j1.6)\ \text{k}\Omega = 3.58\text{K}\angle -26.6^\circ\ \Omega$$

$$V_n = Z_p V_a / (Z_p + Z_{C_1})$$

The series combination of $Z_p + Z_{C_1}$ was also found in part (a) as Z_x. Therefore:

$$V_n = Z_p V_a / Z_x = (3.58\text{K}\angle -26.6^\circ)(60.6\angle 72.6^\circ)/6.6\text{K}\angle -61^\circ$$

$$= 32.87\angle 107^\circ \text{ V}$$

Answer $v_n = 32.87 \sin(24{,}000t + 107^\circ)$ V.

(d) From KVL:

$$V_{an} = V_a - V_n = (60.6\angle 72.6^\circ) - (32.87\angle 107^\circ) = 38.28\angle 43.6^\circ \text{ V}$$

Answer $v_{an} = 38.28 \sin(24{,}000t + 43.6^\circ)$ V

The circuit of Fig. 6.10(a) is known as a ***ladder*** circuit because it resembles a ladder on its side with each series "rung" separated by an impedance. The technique for solving these circuits is simply consecutive combinations of impedances as demonstrated in the Example 6.10.

6.4 SERIES TO PARALLEL CIRCUIT CONVERSION

It is sometimes convenient to convert a series circuit to a parallel circuit and vice versa. The procedure is simply to apply the circuit laws and impedance equations to one configuration and determine which other configuration will yield the same result.

6.4.1 Impedance Conversion

EXAMPLE 6.11

The circuit of Fig. 6.11(a) consists of a resistor and capacitor in series. Find a combination of parallel elements that will result in the same impedance between

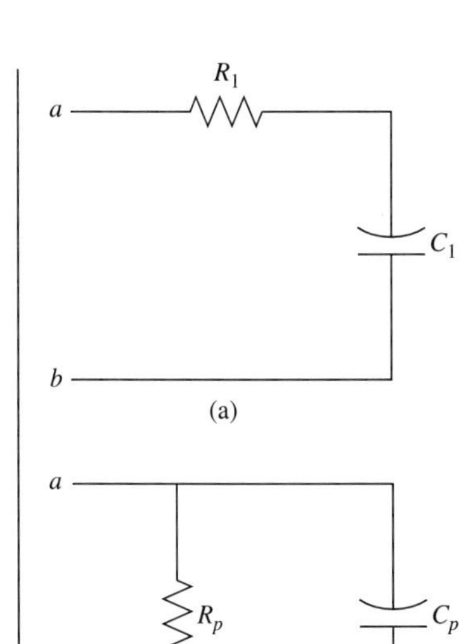

FIGURE 6.11

terminals *a* and *b*.

$$R_1 = 10\ \text{k}\Omega \qquad C_1 = 0.1\ \mu\text{F} \qquad f = 100\ \text{Hz}$$

Solution The equivalent impedance of R_1 and C_1 in series is as follows:

$$\begin{aligned} Z_{ab} &= R_1 - j(1/2\pi f C_1) = 10\text{K} - j[1/2\pi 100(0.1)(10^{-6})] \\ &= 10\text{K} - j15.9\text{K} = 18.78\text{K}\underline{/-57.8^\circ}\ \Omega \end{aligned}$$

To find the combination of parallel elements that will result in the same equivalent impedance, we first find the admittance of this impedance as Y_{ab}.

$$Y_{ab} = 1/Z_{ab} = 1/(18.78\text{K}\underline{/-57.8^\circ}) = 0.05\underline{/57.8^\circ}\ \text{mS}$$

Since parallel admittances add, we convert the polar form to rectangular form as follows:

$$Y_{ab} = (0.027 - j0.042)\ \text{mS} = G + jB$$

The conductance G represents a resistance R_p such that

$$R_p = 1/G_p = 1/0.027(10^{-3}) = 37\ \text{k}\Omega$$

The susceptance B represents a capacitance C_p such that $B = \omega C_p$.

$$C_p = 0.042(10^{-3})/2\pi(100) = 0.067 \times 10^{-6}$$

Answer The equivalent circuit is shown in Fig. 6.11(b), where

$$R_p = 37\ \text{k}\Omega \quad \text{and} \quad C_p = 0.067\ \mu\text{F}.$$

To convert a parallel combination of elements to a series combination of elements, the procedure is similar. We simply find the admittance of the parallel combination, take the reciprocal to obtain the impedance, write the impedance in rectangular form, and determine which elements will result in the required impedance.

EXAMPLE 6.12

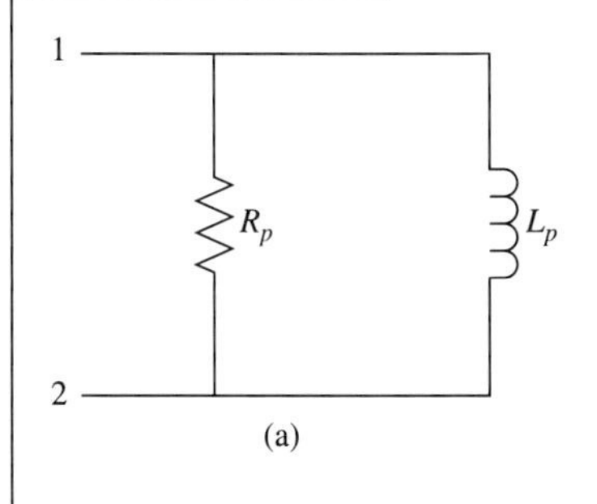

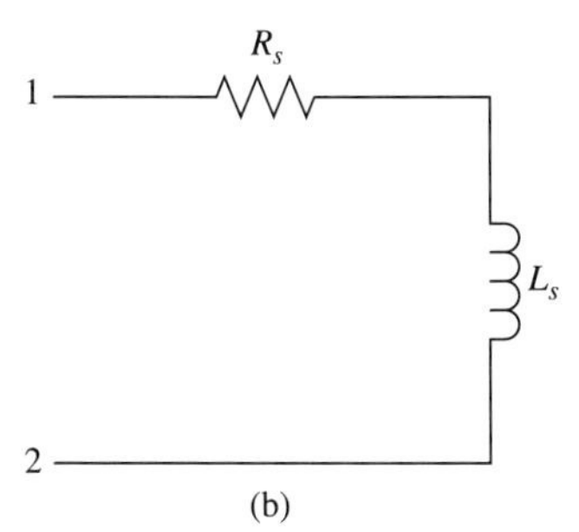

FIGURE 6.12

The parallel circuit of Fig. 6.12(a) is to be converted to a series combination of elements that yields the same impedance between terminals 1 and 2. Find the series elements to accomplish this.

$$R_p = 600\ \Omega \qquad L_p = 20\ \text{mH} \qquad f = 6.37\ \text{kHz}$$

Solution The admittance of the parallel combination is Y_p:

$$\begin{aligned} Y_p &= (1/R_p) - j(1/2\pi f L_p) = (1/600) - j[1/2\pi(6.37)10^3(20)10^{-3}] \\ &= (1.67 - j1.25)\ \text{mS} = 2.09\underline{/-36.8^\circ}\ \text{mS} \end{aligned}$$

The equivalent impedance is found by taking the reciprocal of the admittance:

$$Z_p = 1/(2.09\ \text{mS}\underline{/-36.8^\circ}) = 479\underline{/36.8^\circ}\ \Omega$$

The impedance is now changed to rectangular form:

$$Z_p = (383.6 + j286.9)\ \Omega$$

The real part represents a resistance $R_s = 383.6\ \Omega$. The imaginary part represents an inductive reactance, X_s. The corresponding inductance is found using the equation for the inductive reactance:

$$L_s = X_s/2\pi f = 286.9/2\pi(6.37)10^3 = 7.2\ \text{mH}$$

Answer The equivalent circuit is shown in Fig. 6.12(b), where $R_s = 383.6\ \Omega$, and $L_s = 7.2$ mH.

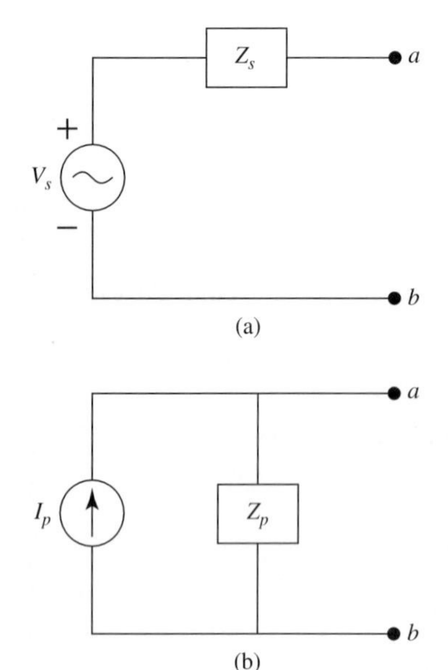

FIGURE 6.13

6.4.2 Source Conversion

Converting from a voltage source with a series impedance to a current source with a parallel impedance is sometimes advisable to facilitate analysis of a given circuit. The criteria for equivalence are the voltage and current that result at designated terminals for a given load impedance.

Consider the circuit of Fig. 6.13(a). This circuit includes a voltage source, V_s, in series with an impedance, Z_s. We wish to determine the current source, I_p, and the parallel impedance, Z_p, which will make the circuit of Fig. 6.13(b) equivalent to that of Fig. 6.13(a) with respect to terminals a and b.

To do this, we recognize that the short circuit current and open circuit voltage of both circuits must be the same. This results in the following equivalent equations:

$$I_p = V_s/Z_s \qquad \textbf{(Eq. 6.1)}$$

$$Z_p = Z_s \qquad \textbf{(Eq. 6.2)}$$

These equivalence equations can of course be used whether converting ***from*** a voltage source with series impedance ***to*** a current source with parallel impedance or vice versa.

EXAMPLE 6.13

Consider the circuit of Fig. 6.14(a). The voltage source and its series impedance are $V_s = 100\angle 0°$ V, $Z_s = 25\text{K}\angle 36.9°\ \Omega$, and $f = 1000$ Hz. Find a circuit consisting of a sinusoidal current source and a parallel impedance equivalent to Fig. 6.14(a) with respect to terminals 1 and 2.

FIGURE 6.14

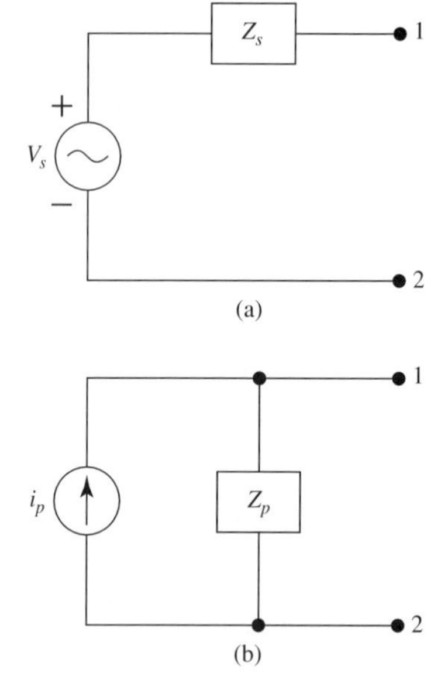

Solution The equivalent circuit is drawn in Fig. 6.14(b). The values of Z_p and I_p are determined using Equations 6.1 and 6.2.

$$I_p = V_s/Z_s = (100\angle 0°)/[25(10^3)\angle 36.9°] = 4\angle -36.9° \text{ mA}$$
$$Z_p = Z_s = 25\text{K}\angle 36.9°\ \Omega$$

Answer The circuit of Fig. 6.14(b) such that $i_p = 4\sin(2\pi 1000t - 36.9°)$ mA and $Z_p = 25\angle 36.9°\ \Omega$.

EXAMPLE 6.14

Determine the series circuit that is equivalent to the circuit of Fig. 6.15(a) with respect to terminals 1 and 2.

$$I_p = 240 \text{ mAdc} \qquad R_p = 2 \text{ k}\Omega$$

FIGURE 6.15

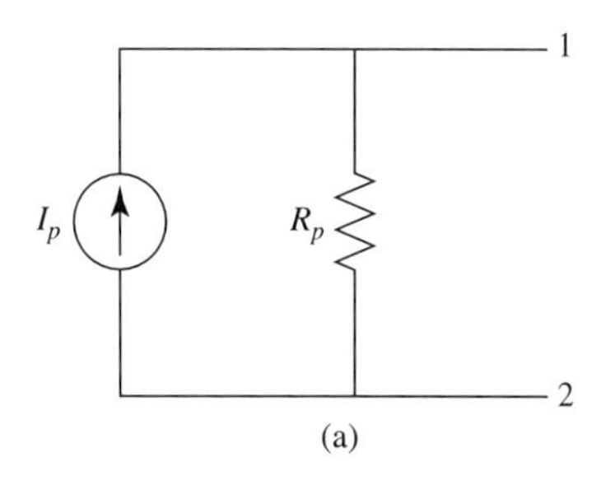

(a)

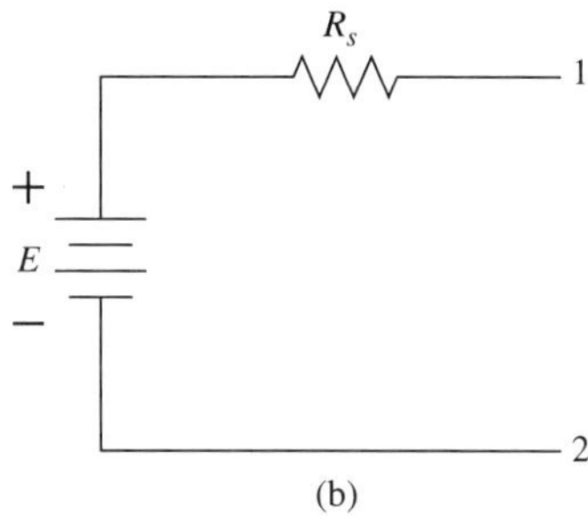

(b)

Solution Using Equations 6.1 and 6.2 for a dc circuit:

$$E = I_p R_p = 240 \times 10^{-3} \times 2 \times 10^3 = 480 \text{ V}$$
$$R_s = R_p = 2 \text{ k}\Omega$$

Answer The equivalent circuit is shown in Fig. 6.15(b), where

$$E = 480 \text{ V} \quad \text{and} \quad R_s = 2 \text{ k}\Omega.$$

■ SUMMARY

- A series–parallel circuit is defined as one that consists ***both*** of impedances connected in series ***and*** impedances connected in parallel.
- Series–parallel circuits require no new theory for their solution. It is the proper application of the theory studied for simple series and parallel circuits that is important.
- Many circuits contain a "reference" point (such as a "ground"). In such cases, voltages may be written with a single subscript and it is understood that the second subscript is the reference point.
- Ladder circuits are solved by consecutive combinations of impedances into their equivalent single impedances.
- A voltage source in series with an impedance is equivalent to a current source in parallel with the same impedance if the two sources are related by $I_p = V_p/Z_p$.

■ PROBLEMS

6.1 Find the equivalent impedance of the circuit in Fig. P6.1 looking into terminals a and b. The source frequency is 1 kHz, and $R_1 = 12 \text{ k}\Omega$, $R_2 = 14 \text{ k}\Omega$, $C_1 = 0.012\ \mu\text{F}$, and $C_2 = 0.008\ \mu\text{F}$.

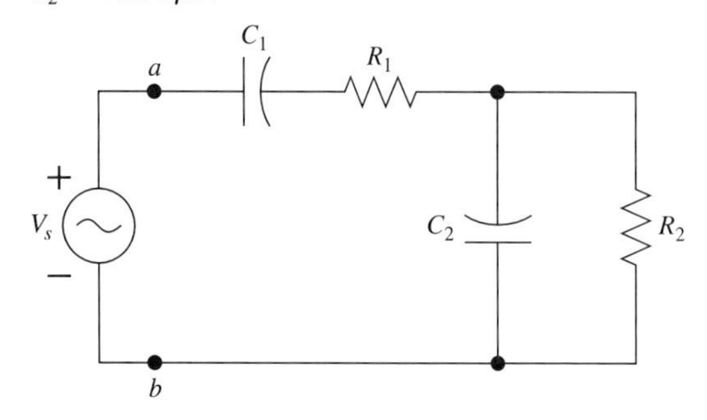

FIGURE P6.1

6.2 Find the equivalent impedance of the circuit in Fig. P6.2 looking into terminals 1 and 2. The source frequency is 20 kHz and $R_1 = 800\ \Omega$, $R_2 = 1200\ \Omega$, $R_3 = 1600\ \Omega$, $L = 10$ mH, $C_1 = 0.006\ \mu$F, and $C_3 = 0.01\ \mu$F.

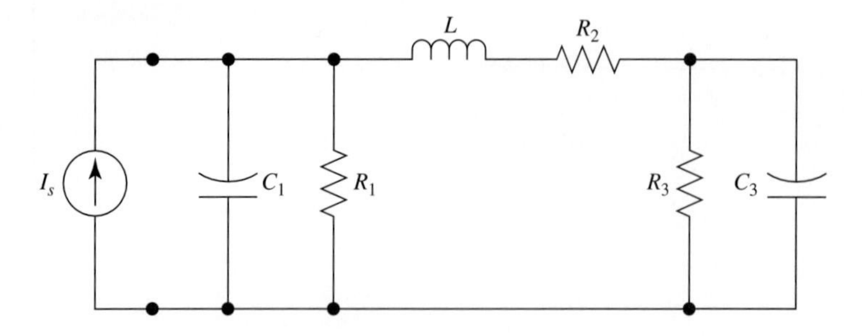

FIGURE P6.2

6.3 Find the equivalent impedance of the dc circuit shown in Fig. P6.3 looking into terminals a and b given $R_a = 6$ kΩ, $R_c = 4$ kΩ, $R_m = 8$ kΩ, $R_o = 10$ kΩ, $R_u = 12$ kΩ, $C_a = 4\ \mu$F, $C_c = 0.1\ \mu$F, $C_m = 80$ nF, $L_a = 10$ mH, and $L_c = 4$ mH.

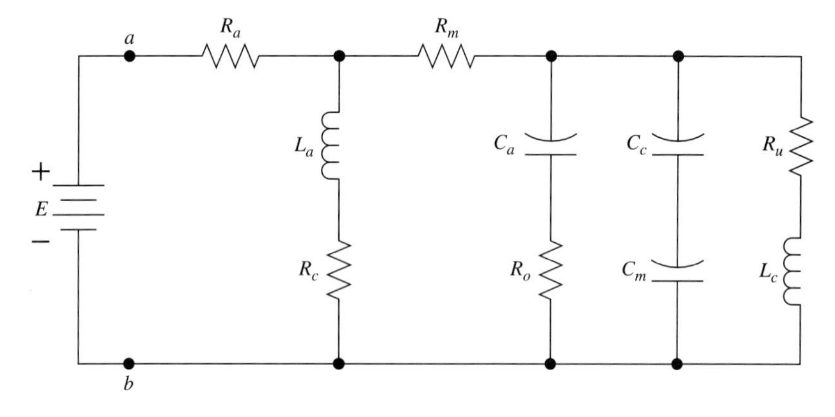

FIGURE P6.3

6.4 Find the equivalent impedance of the dc circuit of Fig. P6.4 looking into terminals 1 and 2 given $R_1 = 40\ \Omega$, $R_2 = 60\ \Omega$, $R_3 = 50\ \Omega$, $R_4 = 80\ \Omega$, $R_5 = 100\ \Omega$, $L_2 = 24$ H, $L_5 = 60$ H, and $C_1 = 0.1\ \mu$F.

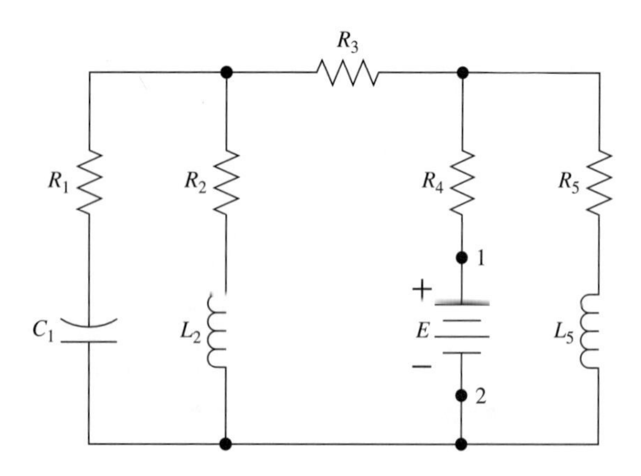

FIGURE P6.4

6.5 Find the equivalent impedance of the circuit of Fig. P6.5 looking into terminals a and b. The source frequency is 500 Hz and $R_1 = 600\ \Omega$, $R_2 = 400\ \Omega$, $R_3 = 100\ \Omega$, $L_1 = 270$ mH, $L_3 = 30$ mH, and $C_2 = 1\ \mu$F.

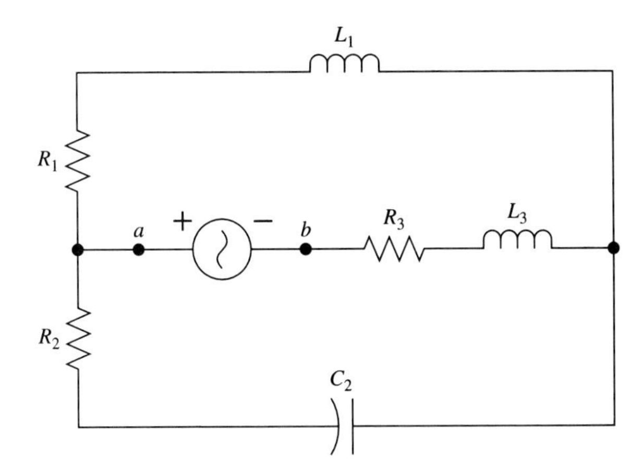

FIGURE P6.5

6.6 In the circuit of Fig. P6.6, $e = 28\sin(2\pi 6000t)$ V, $R_a = 36\ \Omega$, $R_c = 54\ \Omega$, $R_o = 100\ \Omega$, $C_o = 3\ \mu$F, $L_a = 1.5$ mH, and $L_c = 1.2$ mH. Find each of the following sinusoidal quantities:
(a) i **(b)** i_1 **(c)** i_2 **(d)** v_c

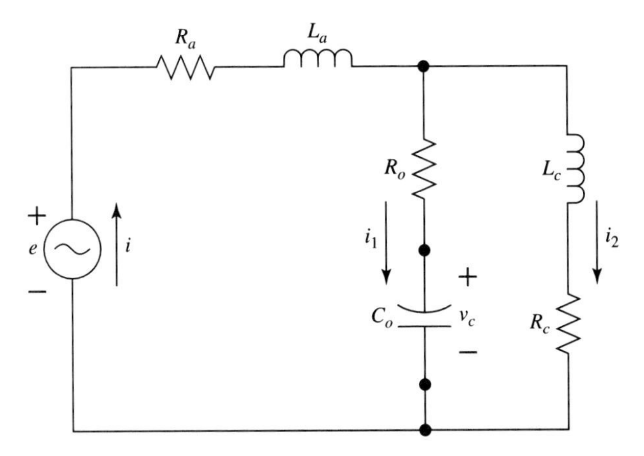

FIGURE P6.6

6.7 In the dc circuit of Fig. P6.7, $E = 48$ V, $R_a = 2.4\ \text{k}\Omega$, $R_c = 1.2\ \text{k}\Omega$, $R_o = 9.6\ \text{k}\Omega$, $R_m = 1.6\ \text{k}\Omega$, $L_a = 20\ \mu$H, $L_c = 40$ mH, $C_o = 20$ F, and $C_m = 10\ \mu$F. Find each of the following:
(a) I **(b)** V_{ac} **(c)** V_{cb}

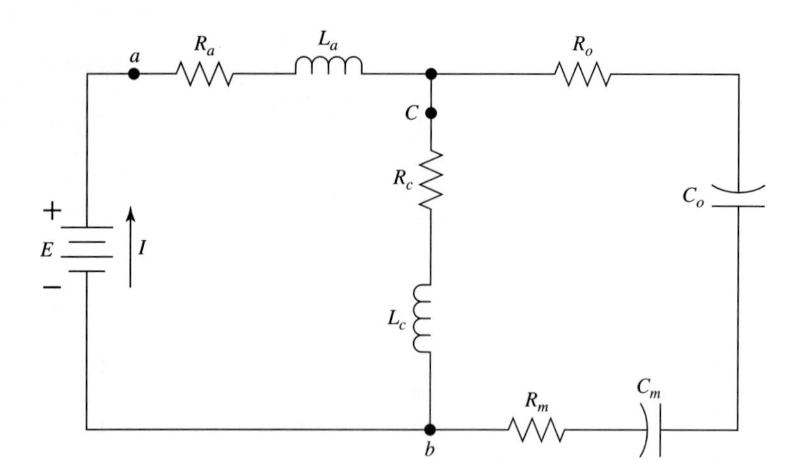

FIGURE P6.7

6.8 The circuit of Fig. P6.8 contains two current sources of the same frequency, $f = 2400$ Hz. The values of the components and source currents are $R_1 = 100\ \Omega$, $R_2 = 400\ \Omega$, $R_3 = 200\ \Omega$, $L_1 = 8$ mH, $L_2 = 30$ mH, $C = 0.2\ \mu$F, $I_{S_1} = 20\underline{/45^\circ}$ mA, and $I_{S_2} = 40\underline{/53.1^\circ}$ mA. Find each of the following sinusoidal quantities:

(a) i_x **(b)** i_y **(c)** v_{ab} **(d)** v_{cb}

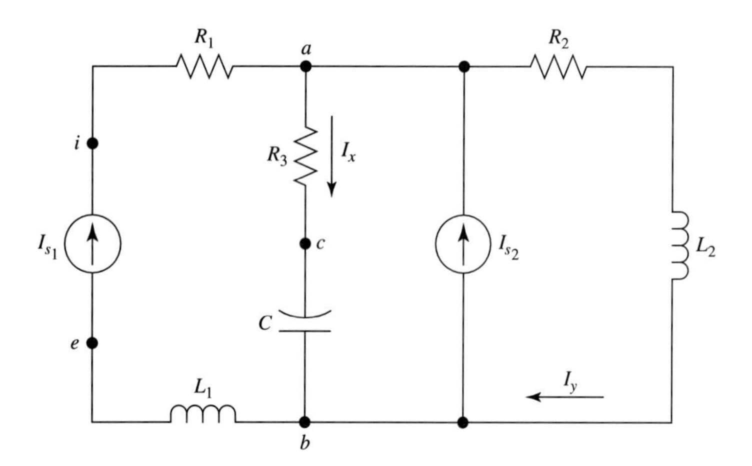

FIGURE P6.8

6.9 In the series voltage source circuit of Fig. P6.9, $v_s = 50 \sin(2000t + 45^\circ)$ V and $Z_s = 250\underline{/53.1^\circ}\ \Omega$. Find the equivalent current source circuit with respect to terminals a and b.

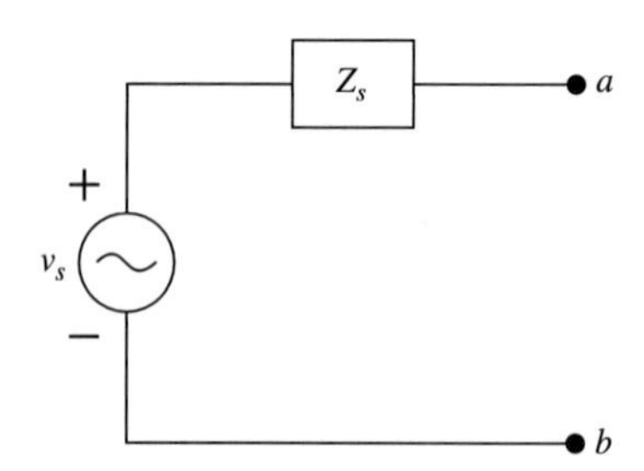

FIGURE P6.9

6.10 The current source and parallel impedance of Fig. P6.10 are given as $i_p = 60 \sin(400t - 75°)$ mA and $Z_p = (10 + j10)\ \Omega$. Find the equivalent voltage source circuit with respect to terminals a and b.

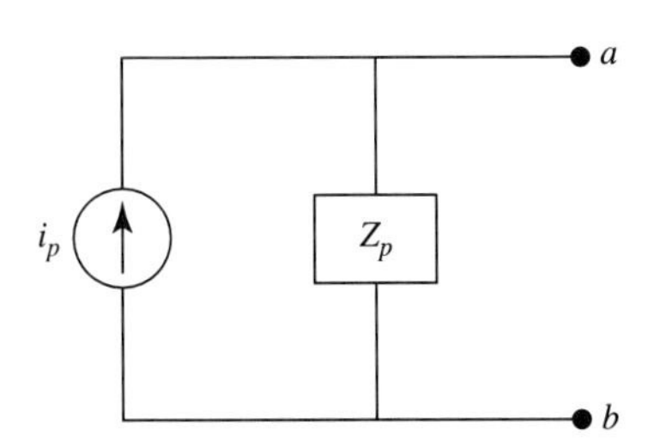

FIGURE P6.10

6.11 Convert the dc voltage source circuit of Fig. P6.11 to a current source circuit that is equivalent with respect to terminals a and b given $V_s = 24$ V and $R_s = 18$ kΩ.

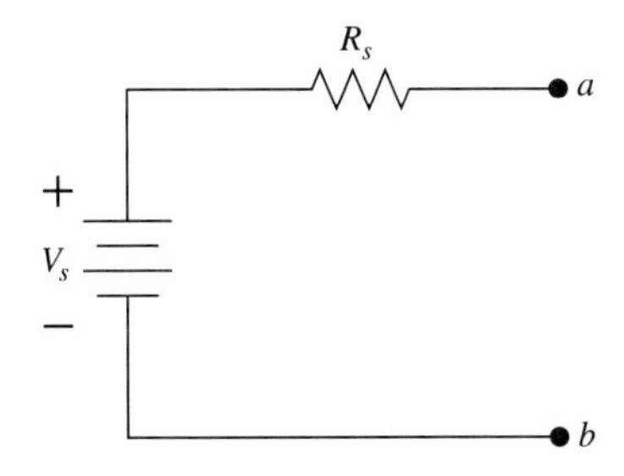

FIGURE P6.11

6.12 In the circuit of Fig. P6.12, $v_s = 480 \sin(4800t)$ mV, $R_s = 20\ \Omega$, and $L_s = 1$ mH. Find the current source circuit that is equivalent to this voltage source circuit with respect to terminals 1 and 2.

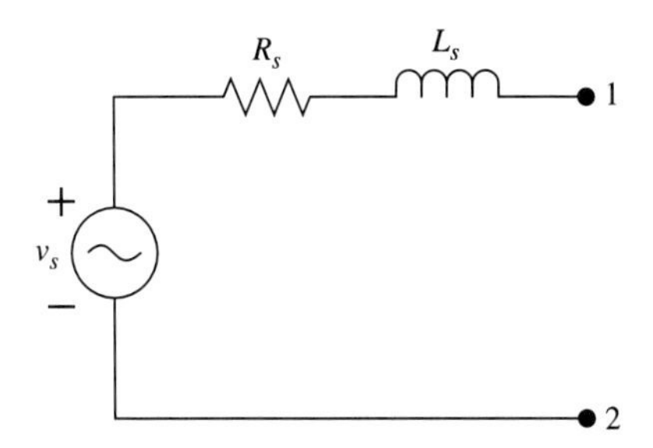

FIGURE P6.12

CHAPTER 7

FORMAL STEADY STATE CIRCUIT ANALYSIS TECHNIQUES AND THEOREMS

INTRODUCTION

In all the previous chapters, the circuits analyzed contained a single voltage source or a single current source. Alternatively, those circuits that contained more than one voltage source had those voltage sources in series so that they could be added to produce a single voltage source. If more than one current source are contained in a circuit, they can be added provided they are connected in parallel. (Refer to Chapter 2 for the combining of voltage sources in series and current sources in parallel).

In many cases, circuits contain more than one source and those sources are not connected in series or in parallel. In such cases, more formal techniques may be employed to analyze those circuits. These techniques (mesh analysis and nodal analysis) result in simultaneous equations involving the voltages and currents of the circuit being analyzed.

Such equations can generally be solved more easily if one uses the method of determinants. Accordingly, we will review the procedures involved when using determinants to solve simultaneous equations before introducing the techniques of mesh and nodal analysis.

7.1 DETERMINANTS

Determinants are defined as square matrices. A matrix is basically an array of numbers, and matrices are covered in a number of mathematical texts on advanced algebra. The reader is encouraged to review this topic for a more complete understanding of determinants.

Fortunately, it is not necessary to acquire a knowledge of matrices in order to use the method of determinants. We will therefore describe the method of determinants in the following sections. The reader should become familiar with this method before proceeding to the sections on mesh and nodal analysis.

It is emphasized that although simultaneous equations can be solved using methods other than the method of determinants, this method is recommended when three or more variables are involved. It is more straightforward than other methods and, once mastered, can actually be easier than the others.

7.1.1 2 × 2 Determinants

When the simultaneous equations involved include two variables, the determinants that result are known as 2 × 2 determinants because the determinants contain 2 rows and 2 columns. To illustrate the use of such determinants, consider the following general two-variable set of equations:

$$ax + by = c \qquad \textbf{(Eq. 7.1)}$$

$$dx + ey = r \qquad \textbf{(Eq. 7.2)}$$

In these equations, y and x are the variables whose values we wish to find. The remaining letters (a, b, c, d, e, and r) all represent constants. These constants may be positive, negative, or complex numbers.

The method of determinants first sets each of the variables equal to the ratio of two determinants as follows:

$$x = N_1/D \qquad \textbf{(Eq. 7.3)}$$

$$y = N_2/D \qquad \textbf{(Eq. 7.4)}$$

In these equations, N_1, N_2, and D are all determinants. These determinants are formed as follows: D is formed by using the coefficients of the variables x and y and placing them as follows:

$$D = \begin{vmatrix} a & b \\ d & e \end{vmatrix} \qquad \textbf{(Eq. 7.5)}$$

N_1 is formed by replacing the coefficients of x (a and d) in D by the constants on the other side of the equal sign. The coefficients of y are left as they are in the D determinant.

$$N_1 = \begin{vmatrix} c & b \\ r & e \end{vmatrix} \qquad \textbf{(Eq. 7.6)}$$

N_2 is formed by replacing the coefficients of y (b and e) in D by the constants on the other side of the equal sign. The coefficients of x are left as they are in the D determinant.

$$N_2 = \begin{vmatrix} a & c \\ d & r \end{vmatrix} \qquad \textbf{(Eq. 7.7)}$$

Each determinant is evaluated in the same manner. The product formed by the diagonal going from the upper left term to the lower right term is decreased by the product formed by the diagonal going from the upper right term to the lower left term. This is expressed by the following three equations:

$$D = ae - bd \qquad \textbf{(Eq. 7.8)}$$

$$N_1 = ce - br \quad \textbf{(Eq. 7.9)}$$

$$N_2 = ar - cd \quad \textbf{(Eq. 7.10)}$$

The constants obtained using Equations 7.8, 7.9, and 7.10 are then substituted into Equations 7.3 and 7.4 to yield the solutions for the unknowns x and y.

EXAMPLE 7.1

Solve the following two simultaneous equations for x and y using the method of determinants:

$$4x + 2y = 28$$
$$6x - 3y = 18$$

Solution From Equation 7.5:

$$D = \begin{vmatrix} 4 & 2 \\ 6 & (-3) \end{vmatrix}$$

Evaluating this determinant as described in Equation 7.8:

$$D = (4)(-3) - (2)(6) = -12 - 12 = -24$$

From Equation 7.6:

$$N_1 = \begin{vmatrix} 28 & 2 \\ 18 & (-3) \end{vmatrix}$$

Evaluating this determinant as described in Equation 7.9:

$$N_1 = (28)(-3) - (2)(18) = -84 - 36 = -120$$

From Equation 7.7:

$$N_2 = \begin{vmatrix} 4 & 28 \\ 6 & 18 \end{vmatrix}$$

Evaluating this determinant as described in Equation 7.10:

$$N_2 = (4)(18) - (6)(28) = 72 - 168 = -96$$

We now substitute the values of D, N_1, and N_2 into Equations 7.3 and 7.4 to find the values of the unknowns x and y.

$$x = -120/-24 = 5 \qquad y = -96/-24 = 4$$

Answer $x = 5, y = 4$.

EXAMPLE 7.2

The set of two simultaneous equations shown here contains complex numbers. Find the values of the unknowns, x and y, that will satisfy these simultaneous equations.

$$(5\angle 53.1°)x - (10\angle 36.9°)y = 12\angle 0°$$
$$(8\angle 45°)x + (12\angle 30°)y = 10\angle 60°$$

Solution Using Equation 7.5:

$$D = \begin{vmatrix} (5\angle 53.1°) & (-10\angle 36.9°) \\ (8\angle 45°) & (12\angle 30°) \end{vmatrix}$$

Evaluating this determinant as described in Equation 7.8:

$$D = (5\angle 53.1^\circ)(12\angle 30^\circ) - (8\angle 45^\circ)(-10\angle 36.9^\circ)$$
$$= (60\angle 83.1^\circ) + (80\angle 81.9^\circ) = 140\angle 82.4^\circ$$

Using Equation 7.6:

$$N_1 = \begin{vmatrix} (12\angle 0^\circ) & (-10\angle 36.9^\circ) \\ (10\angle 60^\circ) & (12\angle 30^\circ) \end{vmatrix}$$

Evaluating this determinant as described in Equation 7.9:

$$N_1 = (12\angle 0^\circ)(12\angle 30^\circ) - (-10\angle 36.9^\circ)(10\angle 60^\circ)$$
$$= (144\angle 30^\circ) + (100\angle 96.9^\circ) = 205\angle 56.7^\circ$$

Using Equation 7.7:

$$N_2 = \begin{vmatrix} (5\angle 53.1^\circ) & (12\angle 0^\circ) \\ (8\angle 45^\circ) & (10\angle 60^\circ) \end{vmatrix}$$

Evaluating this determinant as described in Equation 7.10:

$$N_2 = (5\angle 53.1^\circ)(10\angle 60^\circ) - (8\angle 45^\circ)(12\angle 0^\circ)$$
$$= (50\angle 113.1^\circ) - (96\angle 45^\circ) = 90.2\angle -166^\circ$$

Substituting the values just found for D, N_1 and N_2 into Equations 7.3 and 7.4, we get the following solutions for the unknowns:

$$x = (205\angle 56.7^\circ)/(140\angle 82.4^\circ) = 1.46\angle -25.7^\circ$$
$$y = (90.2\angle -166^\circ)/(140\angle 82.4^\circ) = 0.64\angle 111.6^\circ$$

Answer $x = 1.46\angle -25.7^\circ$, $y = 0.64\angle 111.6^\circ$.

The previous examples used the technique of determinants to solve a pair of simultaneous equations with two unknowns. This was primarily to demonstrate the procedure to be followed when using this technique. The usefulness of the method of determinants becomes more compelling when more than two simultaneous equations are involved.

In the next section, we will investigate the use of determinants to solve three simultaneous equations with three unknowns. Once the procedure is mastered, it will be readily apparent that it is more desirable than using other techniques such as substitution.

7.1.2 3 × 3 Determinants

When there are three simultaneous equations to be solved involving three unknowns, the determinants that result are known as 3 × 3 determinants. This is because the determinants include 3 rows and 3 columns.

To illustrate the use of these determinants, consider the following general set of three simultaneous equations:

$$ax + by + cz = r \qquad \textbf{(Eq. 7.11)}$$

$$dx + ey + mz = s \qquad \textbf{(Eq. 7.12)}$$

$$nx + oy + vz = u \qquad \textbf{(Eq. 7.13)}$$

In these equations, x, y, and z are the three unknowns whose values we wish to find. The letters a, b, c, d, e, m, n, o, v, r, s, and u are all constants. These constants, as always, may be positive, negative or complex numbers.

Using the method of determinants, each unknown is set to equal the ratio of two determinants as follows:

$$x = N_1/D \quad \textbf{(Eq. 7.14)}$$

$$y = N_2/D \quad \textbf{(Eq. 7.15)}$$

$$z = N_3/D \quad \textbf{(Eq. 7.16)}$$

N_1, N_2, N_3, and D are all determinants. These determinants are formed in a manner similar to that used in the solution of two simultaneous equations with two unknowns.

D is formed by using the coefficients of the x, y, and z as follows:

$$D = \begin{vmatrix} a & b & c \\ d & e & m \\ n & o & v \end{vmatrix} \quad \textbf{(Eq. 7.17)}$$

N_1 is formed by replacing the column in D that includes the coefficients of x with the constants on the opposite side of the equal sign in the simultaneous equations.

$$N_1 = \begin{vmatrix} r & b & c \\ s & e & m \\ u & o & v \end{vmatrix} \quad \textbf{(Eq. 7.18)}$$

N_2 is formed by replacing the column in D that includes the coefficients of y with the constants on the opposite side of the equal sign in the simultaneous equations.

$$N_2 = \begin{vmatrix} a & r & c \\ d & s & m \\ n & u & v \end{vmatrix} \quad \textbf{(Eq. 7.19)}$$

N_3 is formed by replacing the column in D that includes the coefficients of z with the constants on the opposite side of the equal sign in the simultaneous equations.

$$N_3 = \begin{vmatrix} a & b & r \\ d & e & s \\ n & o & u \end{vmatrix} \quad \textbf{(Eq. 7.20)}$$

To evaluate each determinant, several methods are used. One such method involves reducing the 3×3 determinants to a set of 2×2 determinants and evaluating the 2×2 determinants in the same manner as discussed in the last section.

To illustrate this procedure, consider the determinant D in Equation 7.17. Each term in the top row is used as a coefficient for a set of three 2×2 determinants. Each constant in the 2×2 determinants is determined by crossing out the row and column containing the coefficient of that 2×2 determinant and using the constants that remain. This is demonstrated in Equation 7.21:

$$D = a\begin{vmatrix} e & m \\ o & v \end{vmatrix} - b\begin{vmatrix} d & m \\ n & v \end{vmatrix} + c\begin{vmatrix} d & e \\ n & o \end{vmatrix} \quad \textbf{(Eq. 7.21)}$$

The determinants N_1, N_2, and N_3 are similarly evaluated as shown in Equations 7.22, 7.23 and 7.24.

$$N_1 = r\begin{vmatrix} e & m \\ o & v \end{vmatrix} - b\begin{vmatrix} s & m \\ u & v \end{vmatrix} + c\begin{vmatrix} s & e \\ u & o \end{vmatrix} \quad \textbf{(Eq. 7.22)}$$

$$N_2 = a\begin{vmatrix} s & m \\ u & v \end{vmatrix} - r\begin{vmatrix} d & m \\ n & v \end{vmatrix} + c\begin{vmatrix} d & s \\ n & u \end{vmatrix} \quad \textbf{(Eq. 7.23)}$$

$$N_3 = a\begin{vmatrix} e & s \\ o & u \end{vmatrix} - b\begin{vmatrix} d & s \\ n & u \end{vmatrix} + r\begin{vmatrix} d & e \\ n & o \end{vmatrix} \quad \textbf{(Eq. 7.24)}$$

Each 2×2 determinant is evaluated and multiplied by its coefficient. They are all then combined as in Equations 7.21 through 7.24.

Note that each coefficient has a positive sign except for the middle (second) coefficient, which has a negative sign. The sign of any coefficient is determined by adding the numbers of the row and column containing that term. If the sum is an ***even*** number the sign of the coefficient is positive, and if the sum is ***odd*** the sign of the coefficient is negative.

EXAMPLE 7.3

Solve the following three simultaneous equations for x, y, and z using the method of determinants.

$$2x - 2y + 4z = 28$$
$$3x + y - 3z = 16$$
$$x + y + z = 8$$

Solution From Equation 7.17:

$$D = \begin{vmatrix} 2 & (-2) & 4 \\ 3 & 1 & (-3) \\ 1 & 1 & 1 \end{vmatrix}$$

From Equation 7.21:

$$D = 2\begin{vmatrix} 1 & (-3) \\ 1 & 1 \end{vmatrix} - (-2)\begin{vmatrix} (3) & (-3) \\ 1 & 1 \end{vmatrix} + 4\begin{vmatrix} 3 & 1 \\ 1 & 1 \end{vmatrix}$$
$$= 2(1 + 3) + 2(3 + 3) + 4(3 - 1) = 8 + 12 + 8 = 28 \qquad \mathbf{D = 28}$$

From Equation 7.18:

$$N_1 = \begin{vmatrix} 28 & (-2) & 4 \\ 16 & 1 & (-3) \\ 8 & 1 & 1 \end{vmatrix}$$

From Equation 7.22:

$$N_1 = 28\begin{vmatrix} 1 & (-3) \\ 1 & 1 \end{vmatrix} - (-2)\begin{vmatrix} 16 & (-3) \\ 8 & 1 \end{vmatrix} + 4\begin{vmatrix} 16 & 1 \\ 8 & 1 \end{vmatrix}$$
$$= 28(1 + 3) + 2(16 + 24) + 4(16 - 8) = 112 + 80 + 32 \qquad \mathbf{N_1 = 224}$$

From Equation 7.19:

$$N_2 = \begin{vmatrix} 2 & 28 & 4 \\ 3 & 16 & (-3) \\ 1 & 8 & 1 \end{vmatrix}$$

From Equation 7.23:

$$N_2 = 2\begin{vmatrix} 16 & (-3) \\ 8 & 1 \end{vmatrix} - 28\begin{vmatrix} 3 & (-3) \\ 1 & 1 \end{vmatrix} + 4\begin{vmatrix} 3 & 16 \\ 1 & 8 \end{vmatrix}$$
$$= 2(16 + 24) - 28(3 + 3) + 4(24 - 16) = 80 - 168 + 32 \qquad \mathbf{N_2 = -56}$$

From Equation 7.20:

$$N_3 = \begin{vmatrix} 2 & (-2) & 28 \\ 3 & 1 & 16 \\ 1 & 1 & 8 \end{vmatrix}$$

From Equation 7.24:

$$N_3 = 2\begin{vmatrix}1 & 16\\ 1 & 8\end{vmatrix} - (-2)\begin{vmatrix}3 & 16\\ 1 & 8\end{vmatrix} + 28\begin{vmatrix}3 & 1\\ 1 & 1\end{vmatrix}$$

$$= 2(8-16) + 2(24-16) + 28(3-1) = -16 + 16 + 56 \qquad \mathbf{N_3 = 56}$$

Using the values found for D, N_1, N_2, and N_3 in Equations 7.14, 7.15, and 7.16, we may now solve for the three unknowns:

$$x = 224/28 \qquad y = -56/28 \qquad z = 56/28$$

Answer $x = 8, y = -2, z = 2.$

This procedure is equally valid with complex numbers.

EXAMPLE 7.4

Solve the following three complex simultaneous equations for the unknown phasor quantities V_1, V_2, and V_3 using the method of determinants.

$$(2\angle 45^\circ)V_1 + (3\angle 90^\circ)V_2 + (2\angle -30^\circ)V_3 = 24.5\angle 119.2^\circ$$

$$(3\angle 90^\circ)V_1 - (8\angle -45^\circ)V_2 - (4\angle 60^\circ)V_3 = 100.7\angle -177.4^\circ$$

$$(2\angle -30^\circ)V_1 - (4\angle -45^\circ)V_2 + (12\angle 36.9^\circ)V_3 = 108.6\angle -8.8^\circ$$

Solution From Equation 7.17:

$$D = \begin{vmatrix}(2\angle 45^\circ) & (3\angle 90^\circ) & (2\angle -30^\circ)\\ (3\angle 90^\circ) & -(8\angle -45^\circ) & -(4\angle 60^\circ)\\ (2\angle -30^\circ) & -(4\angle -45^\circ) & (12\angle 36.9^\circ)\end{vmatrix}$$

From Equation 7.21:

$$D = (2\angle 45^\circ)\begin{vmatrix}-(8\angle -45^\circ) & -(4\angle 60^\circ)\\ (-4\angle -45^\circ) & (12\angle 36.9^\circ\end{vmatrix} - (3\angle 90^\circ)\begin{vmatrix}(3\angle 90^\circ) & -(4\angle 60^\circ)\\ (2\angle -30^\circ) & (12\angle 36.9^\circ)\end{vmatrix}$$

$$+ (2\angle -30^\circ)\begin{vmatrix}(3\angle 90^\circ) & -(8\angle -45^\circ)\\ (2\angle -30^\circ) & -(4\angle -45^\circ)\end{vmatrix}$$

$$= 2\angle 45^\circ[(96\angle 171.9^\circ) - (16\angle 15^\circ)] - 3\angle 90^\circ[(36\angle 126.9^\circ) + (8\angle 30^\circ)]$$

$$+ 2\angle -30^\circ[(12\angle -135^\circ) + (16\angle -75^\circ)]$$

$$= 170.4\angle -127^\circ \qquad \mathbf{D = 170.4\angle -127^\circ}$$

From Equation 7.18:

$$N_1 = \begin{vmatrix}24.5\angle 119.2^\circ & (3\angle 90^\circ) & (2\angle -30^\circ)\\ 100.7\angle -177.4^\circ & (-8\angle -45^\circ) & (-4\angle 60^\circ)\\ 108.6\angle -8.8^\circ & (-4\angle -45^\circ) & (12\angle 36.9^\circ)\end{vmatrix}$$

From Equation 7.22:

$$N_1 = (24.5\angle 119.2^\circ)\begin{vmatrix}(-8\angle -45^\circ) & (-4\angle 60^\circ)\\ (-4\angle -45^\circ) & (12\angle 36.9^\circ)\end{vmatrix}$$

$$- (3\angle 90^\circ)\begin{vmatrix}(100.7\angle -177.4^\circ) & (-4\angle 60^\circ)\\ (108.6\angle -8.8^\circ) & (12\angle 36.9^\circ)\end{vmatrix}$$

$$+ (2\angle -30^\circ)\begin{vmatrix}(100.7\angle -177.4^\circ) & (-8\angle -45^\circ)\\ (108.6\angle -8.8^\circ) & (-4\angle -45^\circ)\end{vmatrix}$$

$$= 24.5\angle 119.2^\circ[(96\angle 171.9^\circ) - (16\angle 15^\circ)] - (3\angle 90^\circ)[(1208.4\angle -140.5^\circ) + (434.4\angle 51.2^\circ)] + (2\angle -30^\circ)[(402.8\angle -42.4^\circ) + (868.8\angle -53.8^\circ)]$$

$$= 3001.3\angle -85.0^\circ \qquad \mathbf{N_1 = 3001.3\angle -85^\circ}$$

From Equation 7.19:

$$N_2 = \begin{vmatrix} (2\angle 45^\circ) & (24.5\angle 119.2^\circ) & (2\angle -30^\circ) \\ (3\angle 90^\circ) & (100.7\angle -177.4^\circ) & (-4\angle 60^\circ) \\ (2\angle -30^\circ) & (108.6\angle -8.8^\circ) & (12\angle 36.9^\circ) \end{vmatrix}$$

From Equation 7.23:

$$N_2 = (2\angle 45^\circ)\begin{vmatrix} (100.7\angle -177.4^\circ) & (-4\angle 60^\circ) \\ (108.6\angle -8.8^\circ) & (12\angle 36.9^\circ) \end{vmatrix} - (24.5\angle 119.2^\circ)\begin{vmatrix} (3\angle 90^\circ) & (-4\angle 60^\circ) \\ (2\angle -30^\circ) & (12\angle 36.9^\circ) \end{vmatrix} + (2\angle -30^\circ)\begin{vmatrix} (3\angle 90^\circ) & (100.7\angle -177.4^\circ) \\ (2\angle -30^\circ) & (108.6\angle -8.8^\circ) \end{vmatrix}$$

$$= 2\angle 45^\circ\,[(1208.4\angle -140.5^\circ) + (434.4\angle 51.2^\circ)] - (24.5\angle 119.2^\circ)[(36\angle 126.9^\circ) + (8\angle 30^\circ)] + 2\angle -30^\circ[(325.8\angle 81.2^\circ) - (201.4\angle 152.6^\circ)] = 940.0\angle -39.7^\circ \qquad \mathbf{N_2 = 1061.6\angle -39.0^\circ}$$

From Equation 7.20:

$$N_3 = \begin{vmatrix} (2\angle 45^\circ) & (3\angle 90^\circ) & (24.5\angle 119.2^\circ) \\ (3\angle 90^\circ) & (-8\angle -45^\circ) & (100.7\angle -177.4^\circ) \\ (2\angle -30^\circ) & (-4\angle -45^\circ) & (108.6\angle -8.8^\circ) \end{vmatrix}$$

From Equation 7.24:

$$N_3 = (2\angle 45^\circ)\begin{vmatrix} (-8\angle -45^\circ) & (100.7\angle -177.4^\circ) \\ (-4\angle -45^\circ) & (108.6\angle -8.8^\circ) \end{vmatrix} - (3\angle 90^\circ)\begin{vmatrix} (3\angle 90^\circ) & (100.7\angle -177.4^\circ) \\ (2\angle -30^\circ) & (108.6\angle -8.8^\circ) \end{vmatrix} + (24.5\angle 119.2^\circ)\begin{vmatrix} (3\angle 90^\circ) & (-8\angle -45^\circ) \\ (2\angle -30^\circ) & (-4\angle -45^\circ) \end{vmatrix}$$

$$= 2\angle 45^\circ[(868.8\angle 126.2^\circ) + (402.8\angle 137.6^\circ)] - (3\angle 90^\circ)[(325.8\angle 81.2^\circ) - (201.4\angle 152.6^\circ)] + (24.5\angle 119.2^\circ)[(12\angle -135^\circ) + (16\angle -75^\circ)]$$

$$= 1297.2\angle -168.3^\circ \qquad \mathbf{N_3 = 1297.2\angle -168.3^\circ}$$

Placing these values for D, N_1, N_2, and N_3 in Equations 7.14, 7.15, and 7.16 yields the final answers for the unknown phasors:

$$V_1 = (3001.3\angle -85^\circ)/(170.4\angle -127^\circ) = 17.61\angle 42^\circ$$

$$V_2 = (1061.6\angle -39^\circ)/(170.4\angle -127^\circ) = 6.23\angle 88^\circ$$

$$V_3 = (1297.2\angle -168.3^\circ)/(170.4\angle -127^\circ) = 7.6\angle -41.3^\circ$$

Answer $V_1 = 17.61\angle 42^\circ$ V, $V_2 = 6.23\angle 88^\circ$ V, $V_3 = 7.6\angle -41^\circ$ V.

7.2 MESH ANALYSIS

7.2.1 Definition

A ***mesh*** may be defined as a ***complete*** circuit loop. Circuits can often be completely divided into two or more independent meshes. (An independent mesh does not ***contain*** any other meshes.)

Normal circuit analysis usually involves the writing of Kirchhoff's Voltage Law around each closed loop, generally together with the use of Ohm's Law. The resulting equations must then be solved for the unknown quantities.

A close look at the resulting equations led to the understanding that these equations always followed a pattern. This pattern was formalized into the mesh analysis approach to solving circuits. Mesh analysis is a very useful technique for solving circuit problems, especially when more than one voltage source is involved.

7.2.2 Two-Mesh Circuits

There is no limit to the number of meshes that may be included in this type of analysis. To illustrate the procedure, however, the two-mesh circuit shown in Fig. 7.1 will be used.

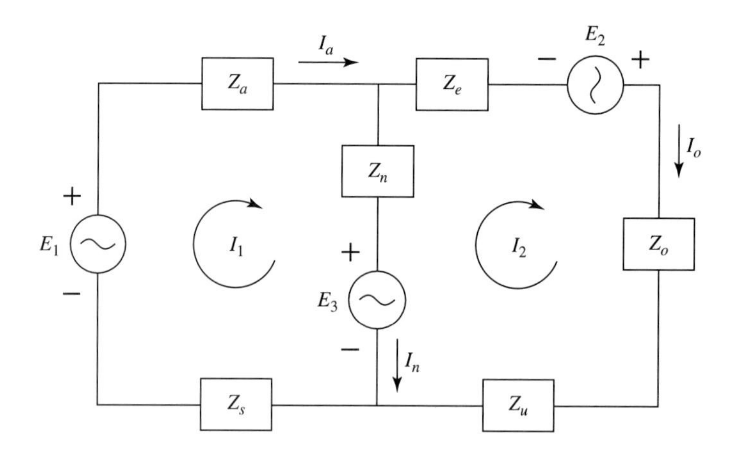

FIGURE 7.1

Step 1. Assign a mesh current to each mesh in a clockwise (CW) direction.

In Fig. 7.1, the mesh currents I_1 and I_2 have been assigned. Each current is assumed to flow in a CW direction around the entire loop. These currents are fictitious in nature and merely serve as convenient concepts in our analysis.

The real currents, I_a, I_n, and I_o may each be expressed in terms of the mesh currents as follows:

$$I_a = I_1 \quad I_o = I_2 \quad I_n = I_1 - I_2$$

Therefore, once the mesh currents have been found the real currents have been determined.

Step 2. Write the generalized mesh equations for this circuit.

The generalized mesh equations for a two-mesh circuit are as follows:

$$Z_{11}I_1 - Z_{12}I_2 = \Sigma E_{S_1} \quad \text{(mesh 1 equation)} \qquad \textbf{(Eq. 7.25)}$$

$$-Z_{21}I_1 + Z_{22}I_2 = \Sigma E_{S_2} \quad \text{(mesh 2 equation)} \qquad \textbf{(Eq. 7.26)}$$

In Equations 7.25 and 7.26, I_1 and I_2 have already been defined as the mesh currents. Note that the signs of the coefficients Z_{11} and Z_{22} are positive whereas the signs of the coefficients Z_{12} and Z_{21} are negative.

$$Z_{11} = \text{Sum of all impedances included in mesh 1} \quad \textbf{(Eq. 7.27)}$$

$$Z_{22} = \text{Sum of all impedances included in mesh 2} \quad \textbf{(Eq. 7.28)}$$

$$Z_{12} = Z_{21} = \text{Sum of all impedances that are common to both mesh 1 and mesh 2} \quad \textbf{(Eq. 7.29)}$$

$$\Sigma E_{S_1} = \text{Sum of all voltage sources in mesh 1} \quad \textbf{(Eq. 7.30)}$$

$$\Sigma E_{S_2} = \text{Sum of all voltage sources in mesh 2} \quad \textbf{(Eq. 7.31)}$$

Z_{11}, Z_{22}, and Z_{12} are all determined by simply adding all the impedances as defined above. In Fig. 7.1:

$$Z_{11} = Z_a + Z_n + Z_s \qquad Z_{22} = Z_e + Z_o + Z_u + Z_n \qquad Z_{12} = Z_{21} = Z_n$$

The terms ΣE_{S_1} and ΣE_{S_2} are determined by adding all the voltage sources as defined above. The sign of each source in the addition process is determined as follows:

> In finding ΣE_{S_1}, a voltage source is positive if the mesh current I_1 enters the voltage source at the negative terminal. Otherwise, it is negative.
>
> In finding ΣE_{S_2} a voltage source is positive if the mesh current I_2 enters the voltage source at the negative terminal. Otherwise, it is negative.

Using these rules with Equations 7.30 and 7.31 in Fig. 7.1 gives:

$$\Sigma E_{S_1} = E_1 - E_3 \quad \text{and} \quad \Sigma E_{S_2} = E_3 + E_2$$

The values found are then substituted into the generalized Equations (7.25 and 7.26). This results in two simultaneous equations with two unknowns I_1 and I_2. These may then be solved using the method of determinants discussed in previous sections.

EXAMPLE 7.5

In Fig. 7.1 $Z_a = 10\angle 36.9°\ \Omega$, $Z_s = 12\angle 53.1°\ \Omega$, $Z_n = 8\angle -45°\ \Omega$, $Z_e = 14\angle -30°\ \Omega$, $Z_o = 18\angle 60°\ \Omega$, $Z_u = 24\angle 20°\ \Omega$, $E_1 = 36\angle 0°$ V, $E_2 = 28\angle 60°$ V, and $E_3 = 30\angle 30°$ V. Using mesh analysis, find the currents I_a, I_n, and I_o.

Solution From Equations 7.27, 7.28, and 7.29:

$$Z_{11} = (10\angle 36.9°) + (8\angle -45°) + (12\angle 53.1°) = 23.1\angle 25.5°\ \Omega$$
$$Z_{22} = (14\angle -30°) + (18\angle 60°) + (24\angle 20°) + (8\angle -45°) = 50.6\angle 12.7°\ \Omega$$
$$Z_{12} = Z_{21} = 8\angle -45°\ \Omega$$

From Equations 7.30 and 7.31:

$$\Sigma E_{S_1} = (36\angle 0°) - (30\angle 30°) = 18\angle -56.26°$$
$$\Sigma E_{S_2} = (28\angle 60°) + (30\angle 30°) = 56\angle 44.47°$$

We next substitute the above values into the mesh Equations 7.25 and 7.26:

$$(23.1\angle 25.5°)I_1 - (8\angle -45°)I_2 = 18\angle -56.26°$$
$$-(8\angle -45°)I_1 + (50.6\angle 12.7°)I_2 = 56\angle 44.47°$$

These two simultaneous equations may now be solved by recognizing that they are of the form of Equations 7.1 and 7.2. I_1 and I_2 are the two unknowns represented by x and y. The coefficients of I_1 and I_2 are represented by the coefficients a, b, d, and e of these equations. From Equations 7.3 and 7.4:

$$I_1 = N_1/D \qquad I_2 = N_2/D$$

The determinants N_1, N_2, and D may now be found by using Equations 7.5, through 7.9 as follows:

$$D = \begin{vmatrix} (23.1\angle 25.5^\circ) & (-8\angle -45^\circ) \\ (-8\angle -45^\circ) & (50.6\angle 12.7^\circ) \end{vmatrix}$$

$$= (23.1\angle 25.5^\circ)(50.6\angle 12.7^\circ) - (-8\angle -45^\circ)(-8\angle -45^\circ) = \mathbf{1209.48\angle 40.6^\circ}$$

$$N_1 = \begin{vmatrix} (18\angle -56.26^\circ) & (-8\angle -45^\circ) \\ (56\angle 44.47^\circ) & (50.6\angle 12.7^\circ) \end{vmatrix}$$

$$= (18\angle -56.26^\circ)(50.6\angle 12.7^\circ) - (-8\angle -45^\circ)(56\angle 44.47^\circ)$$

$$= \mathbf{1275.46\angle -29.7^\circ}$$

$$N_2 = \begin{vmatrix} (23.1\angle 25.5^\circ) & (18\angle -56.26^\circ) \\ (-8\angle -45^\circ) & (56\angle 44.47^\circ) \end{vmatrix}$$

$$= (23.1\angle 25.5^\circ)(56\angle 44.47^\circ) - (18\angle -56.26^\circ)(-8\angle -45^\circ)$$

$$= \mathbf{1151.37\angle 68.9^\circ}$$

We may now substitute these values for N_1, N_2, and D into the preceding expressions for I_1 and I_2:

$$I_1 = (1275.46\angle -29.7^\circ)/(1209.48\angle 40.6^\circ) = \mathbf{1.05\angle -70.3^\circ}$$

$$I_2 = (1151.37\angle 68.9^\circ)/(1209.48\angle 40.6^\circ) = \mathbf{0.95\angle 28.3^\circ}$$

The real currents in the circuit may now be found as follows:

$$I_a = I_1 = 1.05\angle -70.3^\circ$$

$$I_o = I_2 = 0.95\angle 28.3^\circ$$

$$I_n = I_1 - I_2 = (1.05\angle -70.3^\circ - 0.95\angle 28.3^\circ) = 1.52\angle -108.5^\circ$$

Answer $I_a = 1.05\angle -70.3^\circ$ A, $I_o = 0.95\angle 28.3^\circ$ A, $I_n = 1.52\angle -108.5^\circ$ A.

EXAMPLE 7.6

Using mesh analysis, find the sinusoidal currents i_a and i_c in Fig. 7.2.

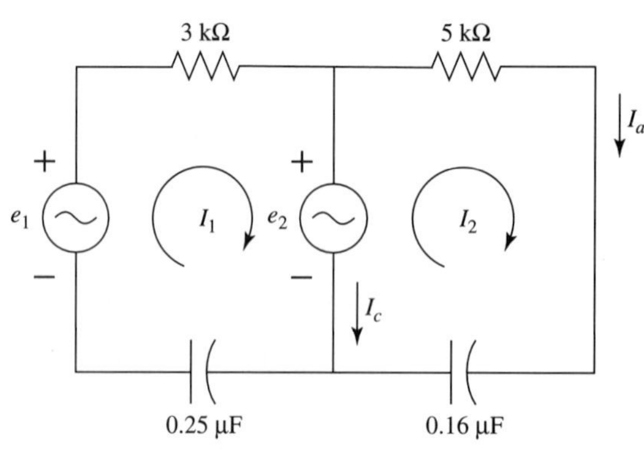

FIGURE 7.2

Solution From Equations 7.27, 7.28 and 7.29:

$$Z_{11} = \{3000 - [j1/(2000)(0.25 \times 10^{-6})]\}\Omega$$
$$= (3000 - j2000)\ \Omega = 3606\underline{/-33.7^\circ}\ \Omega$$
$$Z_{22} = \{5000 - [j1/(2000)(0.16 \times 10^{-6})]\}\ \Omega$$
$$= (5000 - j3125)\ \Omega = 5896\underline{/-32^\circ}\ \Omega$$
$$Z_{12} = Z_{21} = 0\ \Omega$$

From Equations 7.30 and 7.31:

$$\Sigma E_{S_1} = E_1 - E_2 = (24\underline{/0^\circ}) - (12\underline{/30^\circ}) = 14.87\underline{/-23.80}\ \text{V}$$
$$\Sigma E_{S_2} = E_2 = 12\underline{/30^\circ}\ \text{V}$$

Substituting these values in Equations 7.25 and 7.26:

$$(3606\underline{/-33.7^\circ})I_1 - 0I_2 = 14.87\underline{/-23.80^\circ}$$
$$0I_1 + (5896\underline{/-32^\circ})I_2 = 12\underline{/30^\circ}$$

The zero coefficients in these two equations make their solution simple because determinants need not be used.

From the first equation:

$$I_1 = (14.87\underline{/-23.8^\circ})/(3606\underline{/-33.7^\circ}) = 4.12\underline{/9.9^\circ}\ \text{mA}.$$

From the second equation:

$$I_2 = (12\underline{/30^\circ})/(5896\underline{/-32^\circ}) = 2.04\underline{/62^\circ}\ \text{mA}.$$

The real currents, I_a and I_c may now be found.

$$I_a = I_2 = 2.04\underline{/62^\circ}\ \text{mA}$$
$$I_c = I_1 - I_2 = (4.12\underline{/9.9^\circ}) - (2.04\underline{/62^\circ})\ \text{mA} = 3.29\underline{/-19.4^\circ}\ \text{mA}$$

Answer $i_a = 2.04 \sin(2000t + 62^\circ)$ mA, $i_c = 3.29 \sin(2000t - 19.4^\circ)$ mA.

7.2.3 Three-Mesh Circuits

The procedure used to solve a three-mesh circuit is similar to the one used in Examples 7.5 and 7.6. The difference is that in a three-mesh circuit we wind up with three simultaneous equations instead of two. We therefore use the same method of determinants as described in Examples 7.3 and 7.4.

Consider the three-mesh circuit shown in Fig. 7.3.

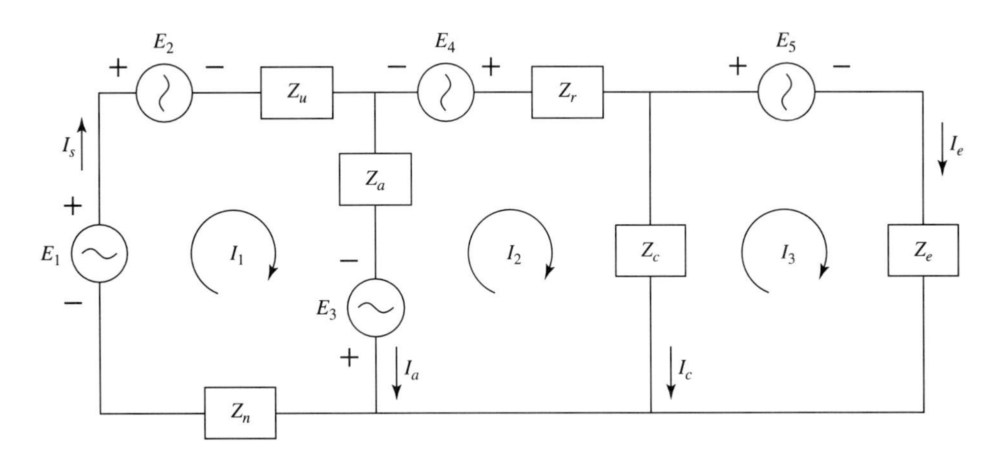

FIGURE 7.3

Step 1. Assign a mesh current to each mesh in a CW direction.

In Fig. 7.3, the mesh currents I_1, I_2, and I_3 have been assigned. The "real" branch currents I_a, I_c, I_s, and I_e will be found in terms of the mesh currents in the same manner as in the two mesh circuits.

The generalized mesh equations for a three-mesh circuit are as follows:

$$Z_{11}I_1 - Z_{12}I_2 - Z_{13}I_3 = \Sigma E_{S_1} \quad \text{(mesh 1 equation)} \qquad \textbf{(Eq. 7.32)}$$

$$-Z_{21}I_1 + Z_{22}I_2 - Z_{23}I_3 = \Sigma E_{S_2} \quad \text{(mesh 2 equation)} \qquad \textbf{(Eq. 7.33)}$$

$$-Z_{31}I_1 - Z_{32}I_2 + Z_{33}I_3 = \Sigma E_{S_3} \quad \text{(mesh 3 equation)} \qquad \textbf{(Eq. 7.34)}$$

The coefficients of the three mesh currents in these equations are defined in a similar manner to those of two mesh equations. The following equations define these quantities.

$$Z_{11} = \text{Sum of all impedances included in mesh 1} \qquad \textbf{(Eq. 7.35)}$$

$$Z_{22} = \text{Sum of all impedances included in mesh 2} \qquad \textbf{(Eq. 7.36)}$$

$$Z_{33} = \text{Sum of all impedances included in mesh 3} \qquad \textbf{(Eq. 7.37)}$$

$$Z_{12} = Z_{21} = \text{Sum of all impedances common to mesh 1 and mesh 2} \qquad \textbf{(Eq. 7.38)}$$

$$Z_{13} = Z_{31} = \text{Sum of all impedances common to mesh 1 and mesh 3} \qquad \textbf{(Eq. 7.39)}$$

$$Z_{23} = Z_{32} = \text{Sum of all impedances common to mesh 2 and mesh 3} \qquad \textbf{(Eq. 7.40)}$$

Note: In all mesh equations, the sign of the impedance coefficients is ***positive*** if the impedance subscripts are both the same and ***negative*** if the impedance subscripts are different.

The remaining terms are also defined in a similar manner to those of their two mesh counterparts.

$$\Sigma E_{S_1} = \text{Sum of all voltage sources in mesh 1} \qquad \textbf{(Eq. 7.41)}$$

$$\Sigma E_{S_2} = \text{Sum of all voltage sources in mesh 2} \qquad \textbf{(Eq. 7.42)}$$

$$\Sigma E_{S_3} = \text{Sum of all voltage sources in mesh 3} \qquad \textbf{(Eq. 7.43)}$$

In adding the voltage sources, the same criteria are used to determine the sign of the source as in the two-mesh case.

> In finding ΣE_{S_1} a voltage source is positive if the mesh current I_1 enters the voltage source at the negative terminal. Otherwise, it is negative.
>
> In finding ΣE_{S_2} a voltage source is positive if the mesh current I_2 enters the voltage source at the negative terminal. Otherwise, it is negative.
>
> In finding ΣE_{S_3} a voltage source is positive if the mesh current I_3 enters the voltage source at the negative terminal. Otherwise, it is negative.

EXAMPLE 7.7

In Fig. 7.3 the following values apply: $E_1 = 300\underline{/36^\circ}$ V, $E_2 = 260\underline{/0^\circ}$ V, $E_3 = 240\underline{/45^\circ}$ V, $E_4 = 340\underline{/60^\circ}$ V, $E_5 = 200\underline{/50^\circ}$ V, $Z_a = 8\text{K}\underline{/0^\circ}\ \Omega$, $Z_c = 15\text{K}\underline{/45^\circ}\ \Omega$, $Z_e = 12\text{K}\underline{/-60^\circ}\ \Omega$, $Z_n = 20\text{K}\underline{/-30^\circ}\ \Omega$, $Z_r = 6\text{K}\underline{/80^\circ}\ \Omega$, $Z_u = 10\text{K}\underline{/70^\circ}\ \Omega$. Using mesh analysis, find the branch currents I_a, I_c, I_e, and I_s.

Solution From Equations 7.35 through 7.40, we calculate the impedance coefficients of the generalized mesh equations:

$$Z_{11} = Z_a + Z_n + Z_u = (8\text{K}\underline{/0^\circ}) + (20\text{K}\underline{/-30^\circ}) + (10\text{K}\underline{/70^\circ})$$
$$= 28.7\text{K}\underline{/-1.2^\circ}\ \Omega$$

$$Z_{22} = Z_a + Z_r + Z_c = (8\text{K}\angle 0°) + (6\text{K}\angle 80°) + (15\text{K}\angle 45°)$$
$$= 25.7\text{K}\angle 40°\ \Omega$$
$$Z_{33} = Z_c + Z_e = (15\text{K}\angle 45°) + (12\text{K}\angle -60°)$$
$$= 16.6\text{K}\angle -0.74°\ \Omega$$
$$Z_{12} = Z_{21} = Z_a = 8\text{K}\angle 0°\ \Omega$$
$$Z_{13} = Z_{31} = 0$$
$$Z_{23} = Z_{32} = Z_c = 15\text{K}\angle 45°\ \Omega$$

The voltage source summations are determined from Equations 7.41 through 7.43 as follows:

$$\Sigma E_{S_1} = E_1 - E_2 + E_3$$
$$= (300\angle 36°) - (260\angle 0°) + (240\angle 45°) = 378.1\angle 66.2°\ \text{V}$$
$$\Sigma E_{S_2} = E_4 - E_3 = (340\angle 60°) - (240\angle 45°)$$
$$= 124.7\angle 89.9°\ \text{V}$$
$$\Sigma E_{S_3} = -E_5 = -200\angle 50°\ \text{V}$$

Substituting these values in Equations 7.32 through 7.34 gives us the following three simultaneous equations for this circuit:

$$(28.7\text{K}\angle -1.2°)\,I_1 - (8\text{K}\angle 0°)\,I_2 - (0)\,I_3 = 378.1\angle 66.2°$$
$$-(8\text{K}\angle 0°)\,I_1 + (25.7\text{K}\angle 40°)\,I_2 - (15\text{K}\angle 45°)\,I_3 = 124.7\angle 89.9°$$
$$(0)\,I_1 - (15\text{K}\angle 45°)\,I_2 + (16.6\text{K}\angle -0.74°)\,I_3 = -200\angle 50°$$

From Equations 7.14 through 7.16:

$$I_1 = N_1/D \qquad I_2 = N_2/D \qquad I_3 = N_3/D$$

where the determinants N_1, N_2, N_3, and D are calculated from Equations 7.17 through 7.20, with:

1. The coefficients of x, y, and z replaced by the impedance coefficients of I_1, I_2, and I_3
2. The constants r, s, and u replaced by the voltage source summations ΣE_{S_1}, ΣE_{S_2}, and ΣE_{S_3}

This gives us the following determinant calculations:

$$D = \begin{vmatrix} (28.7\text{K}\angle -1.2°) & (-8\text{K}\angle 0°) & (0) \\ (-8\text{K}\angle 0°) & (25.7\text{K}\angle 40°) & (-15\text{K}\angle 45°) \\ (0) & (-15\text{K}\angle 45°) & (16.6\text{K}\angle -0.74°) \end{vmatrix}$$

$$= (28.7\text{K}\angle -1.2°)\begin{vmatrix} (25.7\text{K}\angle 40°) & (-15\text{K}\angle 45°) \\ (-15\text{K}\angle 45°) & (16.6\text{K}\angle -0.74°) \end{vmatrix}$$

$$- (-8\text{K}\angle 0°)\begin{vmatrix} (-8\text{K}\angle 0°) & (-15\text{K}\angle 45°) \\ (0) & (16.6\text{K}\angle -0.74°) \end{vmatrix} + (0)$$

$$= (28.7\text{K}\angle -1.2°)(333.37 \times 10^6\angle 7.75°)$$
$$- (-8\text{K}\angle 0°)(132.8 \times 10^6\angle 179.26°) \qquad \mathbf{D = 8.52 \times 10^{12}\angle 7.46°}$$

$$N_1 = \begin{vmatrix} (378.1\angle 66.2°) & (-8\text{K}\angle 0°) & (0) \\ (124.7\angle 89.9°) & (25.7\text{K}\angle 40°) & (-15\text{K}\angle 45°) \\ (-200\angle 50°) & (-15\text{K}\angle 45°) & (16.6\text{K}\angle -0.74°) \end{vmatrix}$$

$$= (378.1\angle 66.2°)\begin{vmatrix} (25.7\text{K}\angle 40°) & (-15\text{K}\angle 45°) \\ (-15\text{K}\angle 45°) & (16.6\text{K}\angle -0.74°) \end{vmatrix}$$

$$- (-8\text{K}\angle 0°)\begin{vmatrix} (124.7\angle 89.9°) & (-15\text{K}\angle 45°) \\ (-200\angle 50°) & (16.6\text{K}\angle -0.74°) \end{vmatrix} + (0)$$

$$= (378.1\angle 66.2^\circ)(333.37 \times 10^6\angle 7.75^\circ) - (-8\text{K}\angle 0^\circ)(0.96 \times 10^6\angle -72.38^\circ) \qquad \mathbf{N_1 = 1.2 \times 10^{11}\angle 71.9^\circ}$$

$$N_2 = \begin{vmatrix} (28.7\text{K}\angle -1.2^\circ) & (378.1\angle 66.2^\circ) & (0) \\ (-8\text{K}\angle 0^\circ) & (124.7\angle 89.9^\circ) & (-15\text{K}\angle 45^\circ) \\ (0) & (-200\angle 50^\circ) & (16.6\text{K}\angle -0.74^\circ) \end{vmatrix}$$

$$= (28.7\text{K}\angle -1.2^\circ)\begin{vmatrix} (124.7\angle 89.9^\circ) & (-15\text{K}\angle 45^\circ) \\ (-200\angle 50^\circ) & (16.6\text{K}\angle -0.74^\circ) \end{vmatrix}$$

$$- (378.1\angle 66.2^\circ)\begin{vmatrix} (-8\text{K}\angle 0^\circ) & (-15\text{K}\angle 45^\circ) \\ (0) & (16.6\text{K}\angle -0.74^\circ) \end{vmatrix} + (0)$$

$$= (28.7\text{K}\angle -1.2^\circ)(9.64 \times 10^5\angle -72.40^\circ) - (378.1\angle 66.2^\circ)(1.32 \times 10^8\angle 179.26^\circ) \qquad \mathbf{N_2 = 3.42 \times 10^{10}\angle 33.46^\circ}$$

$$N_3 = \begin{vmatrix} (28.7\text{K}\angle -1.2^\circ) & (-8\text{K}\angle 0^\circ) & (378.1\angle 66.2^\circ) \\ (-8\text{K}\angle 0^\circ) & (25.7\text{K}\angle 40^\circ) & (124.7\angle 89.9^\circ) \\ (0) & (-15\text{K}\angle 45^\circ) & (-200\angle 50^\circ) \end{vmatrix}$$

To evaluate N_3 we will use the first column, instead of the first row, to make use of the zero in that column. The same procedure and rules apply as when we are using the first row.

$$N_3 = (28.7\text{K}\angle -1.2^\circ)\begin{vmatrix} (25.7\text{K}\angle 40^\circ) & (124.7\angle 89.9^\circ) \\ (-15\text{K}\angle 45^\circ) & (-200\angle 50^\circ) \end{vmatrix}$$

$$- (-8\text{K}\angle 0^\circ)\begin{vmatrix} (-8\text{K}\angle 0^\circ) & (378.1\angle 66.2^\circ) \\ (-15\text{K}\angle 45^\circ) & (-200\angle 50^\circ) \end{vmatrix}$$

$$= (28.7\text{K}\angle -1.2)(4.04 \times 10^6\angle -109.1^\circ) - (-8\text{K}\angle 0^\circ)(6.59 \times 10^6\angle 98.9^\circ)$$

$$\mathbf{N_3 = 7.45 \times 10^{10}\angle -130.5^\circ}$$

We now use these determinant values to solve for the three mesh currents:

$$I_1 = (1.2 \times 10^{11}\angle 71.9^\circ)/(8.52 \times 10^{12}\angle 7.46^\circ) = 1.48 \times 10^{-2}\angle 64.4^\circ$$

$$\mathbf{I_1 = 14.8\angle 64.4^\circ\ mA}$$

$$I_2 = (3.42 \times 10^{10}\angle 33.46^\circ)/(8.52 \times 10^{12}\angle 7.46^\circ) = 4.01 \times 10^{-3}\angle 26^\circ$$

$$\mathbf{I_2 = 4.01\angle 26^\circ\ mA}$$

$$I_3 = (7.45 \times 10^{10}\angle -135^\circ)/(8.52 \times 10^{12}\angle 7.46^\circ) = 8.74 \times 10^{-3}\angle -142.5^\circ$$

$$\mathbf{I_3 = 8.74\angle -142.5^\circ\ mA}$$

We now evaluate the desired branch currents as follows:

$$I_s = I_1 = 14.8\angle 64.4^\circ \text{ mA} \qquad I_e = I_3 = 8.74\angle -142.5^\circ \text{ mA}$$

$$I_a = I_1 - I_2 = (14.8\angle 64.4^\circ) - (4.01\angle 26^\circ) = 11.92\angle 76.5^\circ \text{ mA}$$

$$I_c = I_2 - I_3 = (4.01\angle 26^\circ) - (8.74\angle -142.5^\circ) = 12.69\angle 33.89^\circ \text{ mA}$$

Answer $I_s = 14.8\angle 64.4^\circ$ mA, $I_e = 8.74\angle -142.5^\circ$ mA, $I_a = 11.9\angle 76.5^\circ$ mA, $I_c = 12.7\angle 33.9^\circ$ mA.

Mesh analysis with the method of determinants becomes more useful as the number of meshes increases. The emergence of calculators capable of evaluating determinants makes this method even more useful. The student is cautioned, however, that correct setup of the determinants is crucial to assuring that the results obtained using any computing device will be accurate.

7.3 NODAL ANALYSIS

7.3.1 Definition

Recall that our definition of a node is a point where three or more branches are joined. The use of Kirchhoff's Current Law at the different nodes in electric circuits containing current sources, together with Ohm's Law, results in several simultaneous equations, which form patterns that are similar to those previously noted in the section on mesh analysis.

This leads to another useful type of analysis, called ***nodal analysis.*** In this type of analysis current sources are employed and the resulting equations are called ***nodal equations.*** The coefficients of the nodal voltages are admittances, and the resulting simultaneous equations are solved for these ***nodal voltages*** using the method of determinants discussed in Section 7.2.

Before setting up the nodal equations, it is necessary to define a ***reference node,*** which is the node with respect to which all nodal voltages apply. This is done by inspection and, after some practice, the reference node is generally identified easily.

7.3.2 Two-Node Circuit

We will first consider a circuit with two independent nodes (in addition to the reference node). The two nodes are labeled 1 and 2, and the two voltages associated with these two nodes are V_1 and V_2. It is understood that the two voltages are with respect to the reference node.

The generalized nodal equations for a two-node circuit are:

$$Y_{11}V_1 - Y_{12}V_2 = \Sigma I_{S_1} \qquad \textbf{(Eq. 7.44)}$$

$$-Y_{21}V_1 + Y_{22}V_2 = \Sigma I_{S_2} \qquad \textbf{(Eq. 7.45)}$$

The coefficients of the two nodal voltages are defined as follows:

Y_{11} = Sum of all admittances connected between node 1 and all other nodes **(Eq. 7.46)**

Y_{22} = Sum of all admittances connected between node 2 and all other nodes **(Eq. 7.47)**

$Y_{12} = Y_{21}$ = Sum of all admittances connected between nodes 1 and 2 **(Eq. 7.48)**

Note: In all nodal equations, the sign of the admittance coefficient is ***positive*** if the subscripts are both the same and ***negative*** if the subscripts are different.

ΣI_{S_1} = Sum of all current sources connected between node 1 and all other nodes **(Eq. 7.49)**

ΣI_{S_2} = Sum of all current sources connected between node 2 and all other nodes **(Eq. 7.50)**

The sign of each current source is found using the following criteria:

In finding ΣI_{S_1}, a current source is ***positive*** if its current is entering node 1 and ***negative*** if its current is leaving node 1.

In finding ΣI_{S_2}, a current source is ***positive*** if its current is entering node 2 and ***negative*** if its current is leaving node 2.

EXAMPLE 7.8

In Fig. 7.4 the following values apply: $I_1 = 100\angle 0°$ mA, $I_2 = 80\angle 30°$ mA, $I_3 = 120\angle -60°$ mA, $Y_1 = 200\angle 45°$ S, $Y_2 = 150\angle 36.9°$ S, and $Y_3 = 180\angle -53.1°$ S. Using nodal analysis find V_{12}, the voltage ***between*** the two nodes, and I_x, the current through the admittance Y_1.

FIGURE 7.4

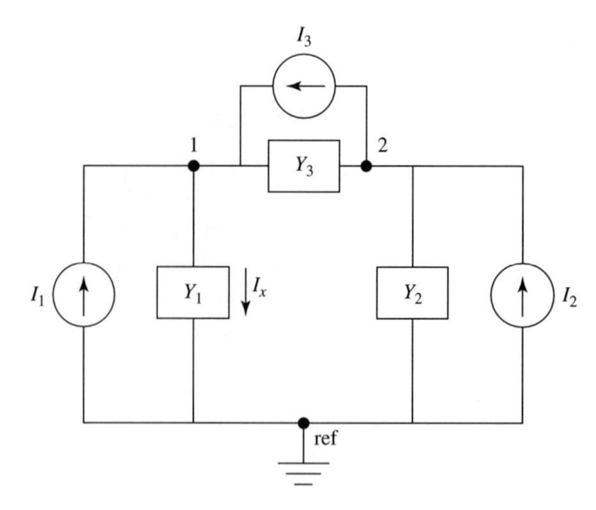

Solution We first find the admittance coefficients using Equations 7.46 through 7.48.

$$Y_{11} = Y_1 + Y_3 = 200\underline{/45^\circ} + 180\underline{/-53.1^\circ} = 249.5\underline{/-0.56^\circ}\text{ S}$$
$$Y_{22} = Y_2 + Y_3 = 150\underline{/36.9^\circ} + 180\underline{/-53.1^\circ} = 234.3\underline{/-13.3^\circ}\text{ S}$$
$$Y_{12} = Y_{21} = Y_3 = 180\underline{/-53.1^\circ}\text{ S}$$

The current source summations are found as follows, using Equations 7.49 and 7.50 and the sign conventions discussed earlier:

$$\Sigma I_{S_1} = I_1 + I_3 = (100\underline{/0^\circ} + 120\underline{/-60^\circ})\text{ mA} = 190.8\underline{/-33^\circ}\text{ mA}$$
$$\Sigma I_{S_2} = I_2 - I_3 = (80\underline{/30^\circ} - 120\underline{/-60^\circ})\text{ mA} = 144.2\underline{/86.3^\circ}\text{ mA}$$

Substituting these values of admittance coefficients and current source summations in Equations 7.44 and 7.45 yields the following two simultaneous equations:

$$(249.5\underline{/-0.56^\circ})V_1 - (180\underline{/-53.1^\circ})V_2 = 190.8 \times 10^{-3}\underline{/-33^\circ}$$
$$-(180\underline{/-53.1^\circ})V_1 + (234.3\underline{/-13.3^\circ})V_2 = 144.2 \times 10^{-3}\underline{/86.3^\circ}$$

These two simultaneous equations can now be solved for the two unknown nodal voltages using determinants. Recall that the unknowns can be written as the ratio of two determinants as follows: $V_1 = N_1/D$ and $V_2 = N_2/D$.

where

$$D = \begin{vmatrix} (249.5\underline{/-0.56^\circ}) & (-180\underline{/-53.1^\circ}) \\ (-180\underline{/-53.1^\circ}) & (234.3\underline{/-13.3^\circ}) \end{vmatrix}$$
$$= (249.5\underline{/-0.56^\circ})(234.3\underline{/-13.3^\circ}) - (-180\underline{/-53.1^\circ})(-180\underline{/-53.1^\circ})$$

$$\mathbf{D = 6.80 \times 10^4\underline{/14.6^\circ}}$$

$$N_1 = \begin{vmatrix} (190.8 \times 10^{-3}\underline{/-33^\circ}) & (-180\underline{/-53.1^\circ}) \\ (144.2 \times 10^{-3}\underline{/86.3^\circ}) & (234.3\underline{/-13.3^\circ}) \end{vmatrix}$$
$$= (190.8 \times 10^{-3}\underline{/-33^\circ})(234.3\underline{/-13.3^\circ}) - (180\underline{/-53.1^\circ})(144.2 \times 10^{-3}\underline{/86.3^\circ}) \qquad \mathbf{N_1 = 55.63\underline{/-19.0^\circ}}$$

$$N_2 = \begin{vmatrix} (249.5\underline{/-0.56^\circ}) & (190.8 \times 10^{-3}\underline{/-33^\circ}) \\ (-180\underline{/-53.1^\circ}) & (144.2 \times 10^{-3}\underline{/86.3^\circ}) \end{vmatrix}$$
$$= (249.5\underline{/-0.56^\circ})(144.2 \times 10^{-3}\underline{/86.3^\circ}) - (190.8 \times 10^{-3}\underline{/-33^\circ})(-180\underline{/-53.1^\circ}) \qquad \mathbf{N_2 = 5.26\underline{/17.9^\circ}}$$

V_1 and V_2 may now be found using these values for N_1, N_2, and D in the appropriate ratio.

$$V_1 = N_1/D = (55.63\underline{/-19°})/(6.8 \times 10^4\underline{/14.6°}) = 8.18 \times 10^{-4}\underline{/-33.6°}\text{ V}$$
$$V_2 = N_2/D = (5.26\underline{/17.9°})/(6.8 \times 10^4\underline{/14.6°}) = 7.73 \times 10^{-5}\underline{/3.3°}\text{ V}$$

From KVL: $V_{12} = V_1 - V_2$
$= (8.18 \times 10^{-4}\underline{/-33.6°}) - (7.73 \times 10^{-5}\underline{/3.3°})$
$= 7.58 \times 10^{-4}\underline{/-37.1°}\text{ V}$

From Ohm's Law: $I_x = Y_1V_1$
$= (200\underline{/45°})(8.18 \times 10^{-4}\underline{/-33.6°})$
$= 1.64 \times 10^{-1}\underline{/11.4°}$

Answer $V_{12} = 0.758\underline{/-37.1°}$ mV, $I_x = 164\underline{/11.4°}$ mA.

EXAMPLE 7.9

In Fig. 7.5, use nodal analysis to find:
(a) The two nodal voltages, V_1 and V_2
(b) The sinusoidal current i_x
(c) The sinusoidal current i_z

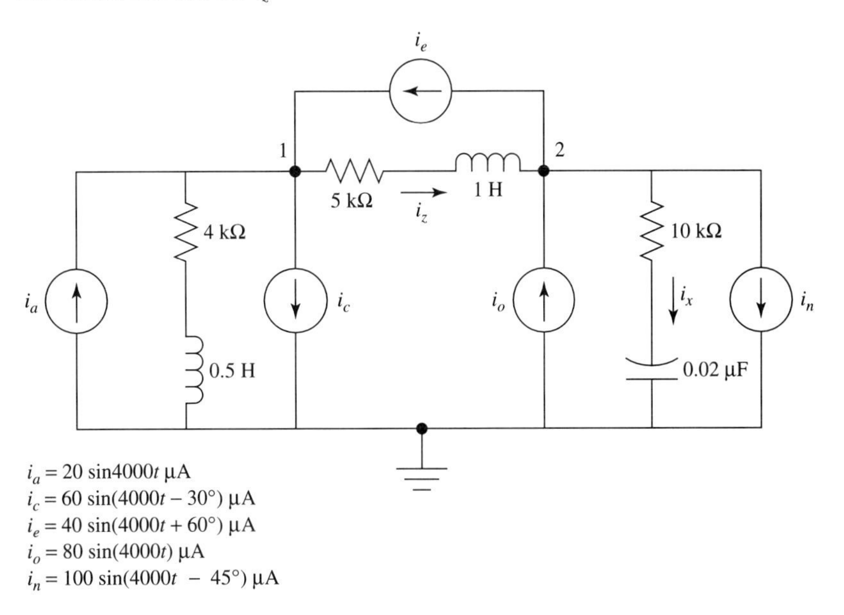

FIGURE 7.5

Solution We must first find the admittances of each branch in the circuit. Since each branch consists of the series combination of two elements, we will first find each impedance and then take the reciprocal of each impedance to find the admittance.

$$Z_1 = 4\text{K} + (j4000 \times 0.5) = 4\text{K} + j2\text{K} = 4.47\text{K}\underline{/26.6°}\ \Omega$$
$$Y_1 = 1/Z_1 = 1/(4.47\text{K}\underline{/26.6°}) = 0.22\underline{/-26.6°}\text{ mS}$$

$$Z_2 = 5\text{K} + (j4000 \times 1) = 5\text{K} + j4\text{K} = 6.4\text{K}\underline{/38.7°}\ \Omega$$
$$Y_2 = 1/Z_2 = 1/(6.4\text{K}\underline{/38.7°}) = 0.16\underline{/-38.7°}\text{ mS}$$

$$Z_3 = 10\text{K} - [j1/(4000 \times 0.02 \times 10^{-6})] = 10\text{K} - j12.5\text{K} = 16\text{K}\underline{/-51.3°}\ \Omega$$

$$Y_3 = 1/Z_3 = 1/(16\text{K}\underline{/-51.3°}) = 0.06\underline{/51.3°}\text{ mS}$$

We may now find the admittance coefficients using Equations 7.46 through 7.48.

$$Y_{11} = Y_1 + Y_2 = (0.22 \times 10^{-3}\underline{/-26.6^\circ}) + (0.16 \times 10^{-3}\underline{/-38.7^\circ})$$
$$= 0.38\underline{/-31.7^\circ}\text{ mS}$$
$$Y_{22} = Y_2 + Y_3 = (0.16 \times 10^{-3}\underline{/-38.7^\circ}) + (0.06 \times 10^{-3}\underline{/51.3^\circ})$$
$$= 0.17\underline{/-18.1^\circ}\text{ mS}$$
$$Y_{12} = Y_{21} = Y_2 = 0.16\underline{/-38.7^\circ}\text{ mS}$$

We now find the current source summations using Equations 7.49 and 7.50.

$$\Sigma I_{S_1} = I_a + I_e - I_c = (20\underline{/0^\circ}) + (40\underline{/60^\circ}) - (60\underline{/-30^\circ})\ \mu\text{A}$$
$$= 65.7\underline{/100.5^\circ}\ \mu\text{A}$$
$$\Sigma I_{S_2} = I_o - I_e - I_n = (80\underline{/0^\circ}) - (40\underline{/60^\circ}) - (100\underline{/-45^\circ})\ \mu\text{A}$$
$$= 37.6\underline{/106.5^\circ}\ \mu\text{A}$$

Placing these values for admittance coefficients and current source summations in Equations 7.44 and 7.45 gives us the following two simultaneous equations:

$$(0.38 \times 10^{-3}\underline{/-31.7^\circ})V_1 - (0.16 \times 10^{-3}\underline{/-38.7^\circ})V_2 = 65.7 \times 10^{-6}\underline{/100.5^\circ}$$
$$-(0.16 \times 10^{-3}\underline{/-38.7^\circ})V_1 + (0.17 \times 10^{-3}\underline{/-18.1^\circ})V_2 = 37.6 \times 10^{-6}\underline{/106.5^\circ}$$

To solve these two equations for the nodal voltages, we express each one as the ratio of two determinants as in previous examples.

$$V_1 = N_1/D \quad \text{and} \quad V_2 = N_2/D$$

where

$$D = \begin{vmatrix} (0.38 \times 10^{-3}\underline{/-31.7^\circ}) & (-0.16 \times 10^{-3}\underline{/-38.7^\circ}) \\ (-0.16 \times 10^{-3}\underline{/-38.7^\circ}) & (0.17 \times 10^{-3}\underline{/-18.1^\circ}) \end{vmatrix}$$
$$= (0.38 \times 10^{-3}\underline{/-31.7^\circ})(0.17 \times 10^{-3}\underline{/-18.1^\circ}) - (0.16 \times 10^{-3}\underline{/-38.7^\circ})(0.16 \times 10^{-3}\underline{/-38.7^\circ})$$
$$\mathbf{D = 4.36 \times 10^{-8}\underline{/-34.0^\circ}}$$

$$N_1 = \begin{vmatrix} (65.7 \times 10^{-6}\underline{/100.5^\circ}) & (-0.16 \times 10^{-3}\underline{/-38.7^\circ}) \\ (37.6 \times 10^{-6}\underline{/106.5^\circ}) & (0.17 \times 10^{-3}\underline{/-18.1^\circ}) \end{vmatrix}$$
$$= (65.7 \times 10^{-6}\underline{/100.5^\circ})(0.17 \times 10^{-3}\underline{/-18.1^\circ}) - (0.16 \times 10^{-3}\underline{/-38.7^\circ})(37.6 \times 10^{-6}\underline{/106.5^\circ})$$
$$\mathbf{N_1 = 5.59 \times 10^{-9}\underline{/9.8^\circ}}$$

$$N_2 = \begin{vmatrix} (0.38 \times 10^{-3}\underline{/-31.7^\circ}) & (65.7 \times 10^{-6}\underline{/100.5^\circ}) \\ (-0.16 \times 10^{-3}\underline{/-38.7^\circ}) & (37.6 \times 10^{-6}\underline{/106.5^\circ}) \end{vmatrix}$$
$$= (0.38 \times 10^{-3}\underline{/-31.7^\circ})(37.6 \times 10^{-6}\underline{/106.5^\circ}) - (65.7 \times 10^{-6}\underline{/100.5^\circ})(-0.16 \times 10^{-3}\underline{/-38.7^\circ})$$
$$\mathbf{N_2 = 2.46 \times 10^{-8}\underline{/69.3^\circ}}$$

(a) The nodal voltages are now found by substituting these values for D, N_1, and N_2 in the appropriate ratios.

$$V_1 = N_1/D = (5.59 \times 10^{-9}\underline{/9.8^\circ})/(4.36 \times 10^{-8}\underline{/-34.0^\circ}) = 0.128\underline{/43.8^\circ}\text{ V}$$
$$V_2 = N_2/D = (2.46 \times 10^{-8}\underline{/69.3^\circ})/(4.36 \times 10^{-8}\underline{/-34.0^\circ})$$
$$= 0.564\underline{/103.3^\circ}\text{ V}$$

Answer $V_1 = 128\underline{/43.8^\circ}$ mV, $V_2 = 564\underline{/103.3^\circ}$ mV.

(b) To find the sinusoidal branch currents, we use Ohm's Law to find the phasor currents and then convert them to sinusoidal form.

$$I_x = Y_3V_2 = (0.06 \times 10^{-3}\angle 51.3°)(564 \times 10^{-3}\angle 103.3°)$$
$$= 3.38 \times 10^{-5}\angle 154.6° \text{ A}$$

$$I_z = Y_2(V_1 - V_2) = (0.16 \times 10^{-3}\angle -38.7°)(0.128\angle 43.8° - 0.564\angle 103.3°)$$
$$= 8.18 \times 10^{-5}\angle -102.9° \text{ A}$$

Answer $i_x = 33.8 \sin(4000t + 154.6°)\ \mu\text{A}$, $i_z = 81.8 \sin(4000t - 102.9°)\ \mu\text{A}$.

7.3.3 Three-Node Circuit

The solution of three-node circuits is very similar to that of two-node circuits. The difference is the necessity to solve three simultaneous equations instead of two using similar admittance coefficients and current source summations.

The generalized nodal equations for a three-node circuit are:

$$Y_{11}V_1 - Y_{12}V_2 - Y_{13}V_3 = \Sigma I_{S_1} \qquad \textbf{(Eq. 7.51)}$$

$$-Y_{21}V_1 + Y_{22}V_2 - Y_{23}V_3 = \Sigma I_{S_2} \qquad \textbf{(Eq. 7.52)}$$

$$-Y_{31}V_1 - Y_{32}V_2 + Y_{33}V_3 = \Sigma I_{S_3} \qquad \textbf{(Eq. 7.53)}$$

V_1, V_2, and V_3 are the three nodal voltages. The coefficients of these nodal voltages are defined as follows:

Y_{11} = Sum of all admittances connected between node 1 and all other nodes **(Eq. 7.54)**

Y_{22} = Sum of all admittances connected between node 2 and all other nodes **(Eq. 7.55)**

Y_{33} = Sum of all admittances connected between node 3 and all other nodes **(Eq. 7.56)**

$Y_{12} = Y_{21}$ = Sum of all admittances connected between nodes 1 and 2 **(Eq. 7.57)**

$Y_{13} = Y_{31}$ = Sum of all admittances connected between nodes 1 and 3 **(Eq. 7.58)**

$Y_{23} = Y_{32}$ = Sum of all admittances connected between nodes 2 and 3 **(Eq. 7.59)**

Note again how, in the Equations 7.51 through 7.53, the sign of the admittance coefficient is ***positive*** if the subscripts are the same and ***negative*** if they are different.

The current source summations are defined as follows:

ΣI_{S_1} = Sum of all current sources connected between node 1 and all other nodes **(Eq. 7.60)**

ΣI_{S_2} = Sum of all current sources connected between node 2 and all other nodes **(Eq. 7.61)**

ΣI_{S_3} = Sum of all current sources connected between node 3 and all other nodes **(Eq. 7.62)**

The sign of each current source is determined in the same manner as in a two-node circuit. It is, in fact, determined in the same manner for a circuit with any number of nodes.

In finding ΣI_{S_1}, a current source is ***positive*** if its current is entering node 1 and ***negative*** if its current is leaving node 1.

In finding ΣI_{S_2}, a current source is ***positive*** if its current is entering node 2 and ***negative*** if its current is leaving node 2.

In finding ΣI_{S_3}, a current source is ***positive*** if its current is entering node 3 and ***negative*** if its current is leaving node 3.

EXAMPLE 7.10

(a) Find the three nodal voltages V_1, V_2, and V_3 in Fig. 7.6.
(b) Find the current I_x
given $Y_1 = 20\angle 50°$ S, $Y_2 = 40\angle -60°$ S, $Y_3 = 100\angle -30°$ S, $Y_4 = 60\angle 0°$ S, $Y_5 = 50\angle 45°$ S, $I_1 = 4\angle 0°$ A, $I_2 = 8\angle 40°$ A, $I_3 = 12\angle -70°$ A, and $I_4 = 10\angle -35°$ A.

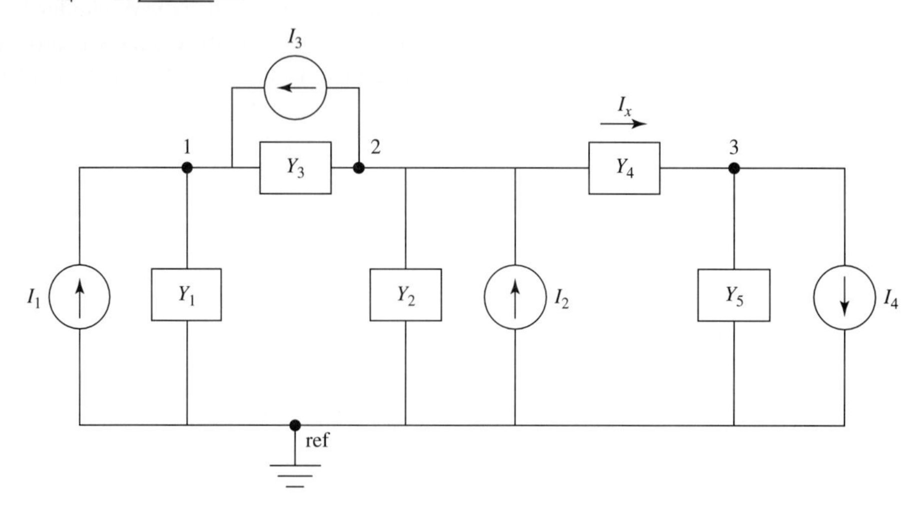

FIGURE 7.6

Solution The admittance coefficients are found using Equations 7.54 through 7.59.

$$Y_{11} = Y_1 + Y_3 = (20\angle 50°) + (100\angle -30°) = 105.3\angle -19.2° \text{ S}$$
$$Y_{22} = Y_2 + Y_3 + Y_4 = (40\angle -60°) + (100\angle -30°) + (60\angle 0°) = 186.9\angle -26.9° \text{ S}$$
$$Y_{33} = Y_4 + Y_5 = (60\angle 0°) + (50\angle 45°) = 101.7\angle 20.3° \text{ S}$$
$$Y_{12} = Y_{21} = Y_3 = 100\angle -30° \text{ S}$$
$$Y_{13} = Y_{31} = 0$$
$$Y_{23} = Y_{32} = Y_4 = 60\angle 0° \text{ S}$$

Using Equations 7.60 through 7.62 we will find the current source summations.

$$\Sigma I_{S_1} = I_1 + I_3 = (4\angle 0°) + (12\angle -70°) = 13.9\angle -54.3° \text{ A}$$
$$\Sigma I_{S_2} = I_2 - I_3 = (8\angle 40°) - (12\angle -70°) = 16.5\angle 83.0° \text{ A}$$
$$\Sigma I_{S_3} = -I_4 = -10\angle -35° \text{ A}$$

These values of admittance coefficients and current source summations will now be placed in the Equations 7.51 through 7.53:

$$(105.3\angle -19.2°)V_1 - (100\angle -30°)V_2 - (0)V_3 = 13.9\angle -54.3°$$
$$-(100\angle -30°)V_1 + (186.9\angle -26.9°)V_2 - (60\angle 0°)V_3 = 16.5\angle 83.0°$$
$$-(0)V_1 - (60\angle 0°)V_2 + (101.7\angle 20.3°)V_3 = -10\angle -35°$$

The nodal voltages may now be solved for by expressing each one as the ratio of two determinants in the same manner as in the previous examples of solving simultaneous equations:

$$V_1 = N_1/D \qquad V_2 = N_2/D \qquad V_3 = N_3/D$$

These determinants are set up in the usual manner by using the admittance coefficients and current source summations. Note how this procedure is the same regardless of the number of nodes.

$$D = \begin{vmatrix} (105.3\angle -19.2^\circ) & (-100\angle -30^\circ) & (0) \\ (-100\angle -30^\circ) & (186.9\angle -26.9^\circ) & (-60\angle 0^\circ) \\ (0) & (-60\angle 0^\circ) & (101.7\angle 20.3^\circ) \end{vmatrix}$$

$$= (105.3\angle -19.2^\circ)\begin{vmatrix} (186.9\angle -26.9^\circ) & (-60\angle 0^\circ) \\ (-60\angle 0^\circ) & (101.7\angle 20.3^\circ) \end{vmatrix}$$

$$- (-100\angle -30^\circ)\begin{vmatrix} (-100\angle -30^\circ) & (-60\angle 0^\circ) \\ (0) & (101.7\angle 20^\circ) \end{vmatrix} + (0)$$

$$= (105.3\angle -19.2^\circ)(1.54 \times 10^4\angle -8.1^\circ)$$
$$- (-100\angle -30^\circ)(1 \times 10^4\angle 170^\circ) \qquad \boldsymbol{D = 6.82 \times 10^5\angle -8.5^\circ}$$

$$N_1 = \begin{vmatrix} (13.9\angle -54.3^\circ) & (-100\angle -30^\circ) & (0) \\ (16.5\angle 83.0^\circ) & (186.9\angle -26.9^\circ) & (-60\angle 0^\circ) \\ (-10\angle -35^\circ) & (-60\angle 0^\circ) & (101.7\angle 20.3^\circ) \end{vmatrix}$$

$$= (13.9\angle -54.3^\circ)\begin{vmatrix} (186.9\angle -26.9^\circ) & (-60\angle 0^\circ) \\ (-60\angle 0^\circ) & (101.7\angle 20.3^\circ) \end{vmatrix}$$

$$- (-100\angle -30^\circ)\begin{vmatrix} (16.5\angle 83.0^\circ) & (-60\angle 0^\circ) \\ (-10\angle -35^\circ) & (101.7\angle 20.3^\circ) \end{vmatrix} + 0$$

$$= (13.9\angle -54.3^\circ)(1.54 \times 10^4\angle -8.1^\circ)$$
$$- (-100\angle -30^\circ)(2.16 \times 10^3\angle 113.9^\circ) \qquad \boldsymbol{N_1 = 1.25 \times 10^5\angle 11.6^\circ}$$

$$N_2 = \begin{vmatrix} (105.3\angle -19.2^\circ) & (13.9\angle -54.3^\circ) & (0) \\ (-100\angle -30^\circ) & (16.5\angle 83.0^\circ) & (-60\angle 0^\circ) \\ (0) & (-10\angle -35^\circ) & (101.7\angle 20.3^\circ) \end{vmatrix}$$

$$= (105.3\angle -19.2^\circ)\begin{vmatrix} (16.5\angle 83.0^\circ) & (-60\angle 0^\circ) \\ (-10\angle -35^\circ) & (101.7\angle 20.3^\circ) \end{vmatrix}$$

$$- (13.9\angle -54.3^\circ)\begin{vmatrix} (-100\angle -30^\circ) & (-60\angle 0^\circ) \\ (0) & (101.7\angle 20.3^\circ) \end{vmatrix} + 0$$

$$= (105.3\angle -19.2^\circ)(2.16 \times 10^3\angle 113.9^\circ)$$
$$- (13.9\angle -54.3^\circ)(1 \times 10^4\angle 170.3^\circ) \qquad \boldsymbol{N_2 = 1.1 \times 10^5\angle 67.4^\circ}$$

$$N_3 = \begin{vmatrix} (105.3\angle -19.2^\circ) & (-100\angle -30^\circ) & (13.9\angle -54.3^\circ) \\ (-100\angle -30^\circ) & (186.9\angle -26.9^\circ) & (16.5\angle 83.0^\circ) \\ (0) & (-60\angle 0^\circ) & (-10\angle -35^\circ) \end{vmatrix}$$

$$= (105.3\angle -19.2^\circ)\begin{vmatrix} (186.9\angle -26.9^\circ) & (16.5\angle 83.0^\circ) \\ (-60\angle 0^\circ) & (-10\angle -35^\circ) \end{vmatrix}$$

$$- (-100\angle -30^\circ)\begin{vmatrix} (-100\angle -30^\circ) & (13.9\angle -54.3^\circ) \\ (-60\angle 0^\circ) & (-10\angle -35^\circ) \end{vmatrix}$$

$$= (105.3\angle -19.2^\circ)(2.74 \times 10^3\angle 106.1^\circ)$$
$$= -(-100\angle -30^\circ)(1.82 \times 10^3\angle -60.1^\circ) \qquad \boldsymbol{N_3 = 1.07 \times 10^5\angle 81.8^\circ}$$

These values for N_1, N_2, N_3, and D are now substituted into the ratios to find the nodal voltages.

$$V_1 = N_1/D = (1.25 \times 10^5\angle 11.6^\circ)/(6.82 \times 10^5\angle -8.5^\circ) = 1.83 \times 10^{-1}\angle 20.1^\circ \text{ V}$$

$$V_2 = N_2/D = (1.1 \times 10^5\angle 67.4^\circ)/(6.82 \times 10^5\angle -8.5^\circ) = 1.61 \times 10^{-1}\angle 75.9^\circ \text{ V}$$

$$V_3 = N_3/D = (1.07 \times 10^5\angle 81.8^\circ)/(6.82 \times 10^5\angle -8.5^\circ) = 1.56 \times 10^{-1}\angle 90.3^\circ \text{ V}$$

Answer (*a*) $V_1 = 183\angle 20.1^\circ$ mV, $V_2 = 161\angle 75.9^\circ$ mV, $V_3 = 156\angle 90.3^\circ$ mV. To find the current I_x we use Ohm's Law and the results just obtained for the nodal voltages.

$$I_x = Y_4V_{23} = Y_4(V_2 - V_3) = (60\angle 0^\circ)[(0.16\angle 75.9^\circ) - (0.16\angle 90.3^\circ)]$$

Answer (*b*) $I_x = 2.4\angle -6.9^\circ$ A.

7.4 MESH ANALYSIS VS. NODAL ANALYSIS

In general, a circuit that can be solved using mesh analysis can also be solved using nodal analysis and vice versa. The method to be used in a given circuit depends primarily on personal preference.

It must be noted, however, that to use mesh analysis all power sources must be in the form of voltage sources. Any current source in the circuit must be converted to a voltage source before applying the mesh equations to the circuit.

Similarly, to use the method of nodal analysis all power sources must be in the form of current sources. Any voltage source in the circuit must be converted to a current source before applying the nodal equations to the circuit.

Source conversion was covered in Chapter 2. The reader is advised to review the section on source conversion before studying the next example.

EXAMPLE 7.11

The circuit of Fig. 7.7(a) contains one voltage source and one current source.

(a) Use mesh analysis to find V_a, the voltage across the resistor R_a.

(b) Use nodal analysis to find the voltage V_a. Compare the answers to parts (a) and (b) to confirm that both are the same.

Solution

(a) To use mesh analysis in the solution of this problem, we must first convert the current source, I_n, to a voltage source as follows:

$$E_n = I_nZ_n = (I_n)(R_n - jX_n) = (200 \times 10^{-3}\angle -30^\circ)(180 - j280) = 66.6\angle -87.26^\circ \text{ V}$$

The circuit of Fig. 7.7(a) is redrawn in Fig. 7.7(b) with the voltage source E_n replacing the current source. The mesh currents I_1 and I_2 have been added to the circuit.

From equations 7.27, 7.28, and 7.29:

$$Z_{11} = (R_a + R_c + jX_c) = (160 + 100 + j100) = (260 + j100)\ \Omega = 278.6\angle 21.0^\circ\ \Omega$$

$$Z_{22} = (R_a + R_e + R_n - jX_e - jX_n) = (160 + 200 + 180 - j320 - j280) = (540 - j600)\ \Omega = 807.2\angle -48.0^\circ\ \Omega$$

$$Z_{12} = Z_{21} = R_a = 160\angle 0^\circ\ \Omega$$

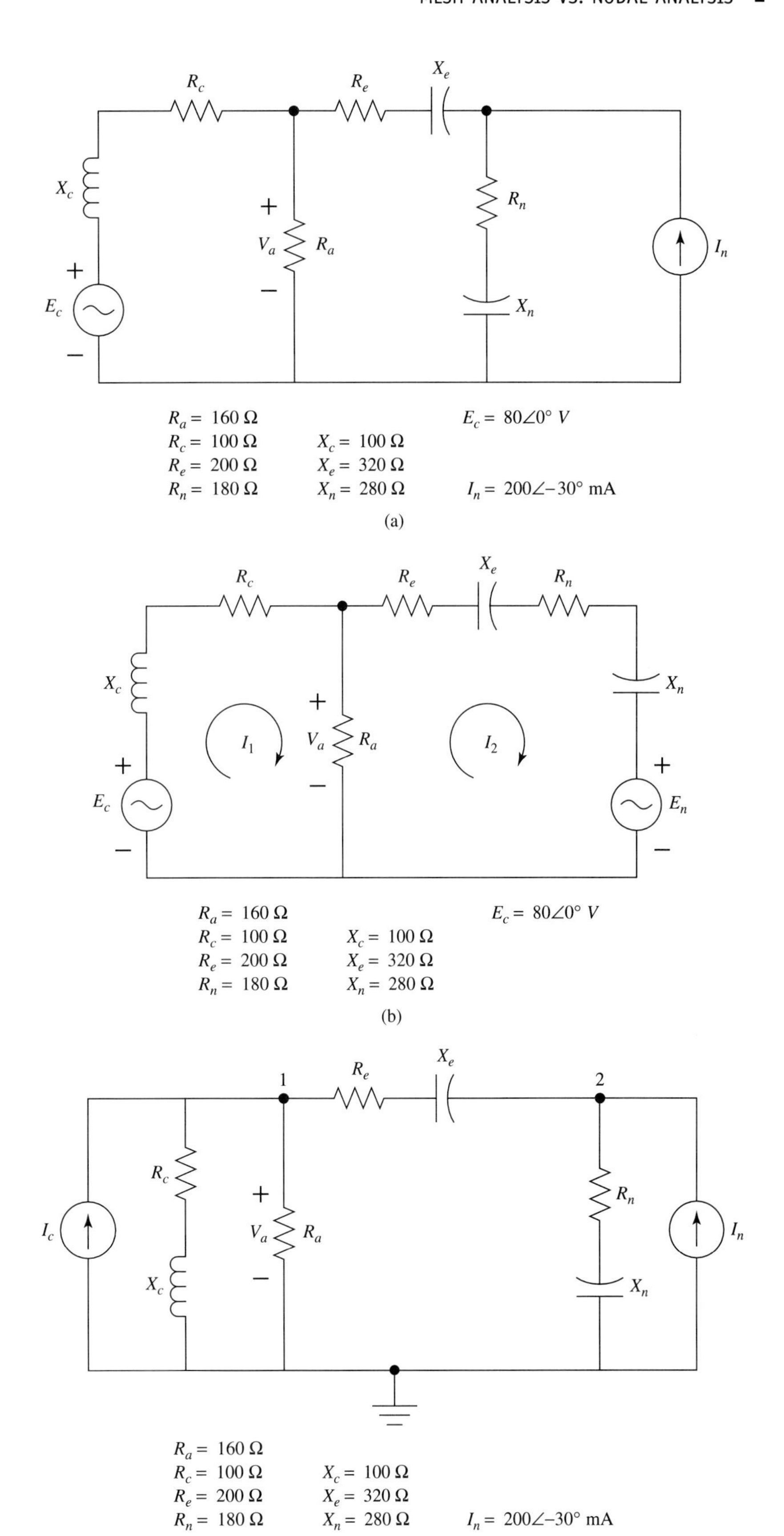
R_c
R_e
X_e
X_c
V_a
R_a
R_n
X_n
I_n
E_c
$R_a = 160\ \Omega$
$R_c = 100\ \Omega$
$R_e = 200\ \Omega$
$R_n = 180\ \Omega$
$X_c = 100\ \Omega$
$X_e = 320\ \Omega$
$X_n = 280\ \Omega$
$E_c = 80\angle 0°\ V$
$I_n = 200\angle -30°$ mA
(a)
I_1
I_2
E_n
$R_a = 160\ \Omega$
$R_c = 100\ \Omega$
$R_e = 200\ \Omega$
$R_n = 180\ \Omega$
$X_c = 100\ \Omega$
$X_e = 320\ \Omega$
$X_n = 280\ \Omega$
$E_c = 80\angle 0°\ V$
(b)
1
2
I_c
$R_a = 160\ \Omega$
$R_c = 100\ \Omega$
$R_e = 200\ \Omega$
$R_n = 180\ \Omega$
$X_c = 100\ \Omega$
$X_e = 320\ \Omega$
$X_n = 280\ \Omega$
$I_n = 200\angle -30°$ mA
(c)

FIGURE 7.7

From Equations 7.30 and 7.31:

$$\Sigma E_{S_1} = E_c = 80\angle 0^\circ \text{ V}$$
$$\Sigma E_{S_2} = -E_n = -66.6\angle -87.26^\circ \text{ V}$$

We now substitute these values in Equations 7.25 and 7.26.

$$(278.6\angle 21^\circ)I_1 - (160\angle 0^\circ)I_2 = 80\angle 0^\circ$$
$$-(160\angle 0^\circ) + (807.2\angle -48^\circ)I_2 = -66.6\angle -87.26^\circ$$

The mesh currents may now be expressed as the ratio of two determinants:

$$I_1 = N_1/D \quad \text{and} \quad I_2 = N_2/D$$

where

$$D = \begin{vmatrix} (278.6\angle 21^\circ) & (-160\angle 0^\circ) \\ (-160\angle 0^\circ) & (807.2\angle -48^\circ) \end{vmatrix}$$
$$= (278.6\angle 21^\circ)(807.2\angle -48^\circ) - (-160\angle 0^\circ)(-160\angle 0^\circ)$$
$$= \mathbf{2.02 \times 10^5\angle -30.3^\circ}$$

$$N_1 = \begin{vmatrix} (80\angle 0^\circ) & (-160\angle 0^\circ) \\ (-66.6\angle -87.26^\circ) & (807.2\angle -48^\circ) \end{vmatrix}$$
$$= (80\angle 0^\circ)(807.2\angle -48^\circ) - (-160\angle 0^\circ)(-66.6\angle -87.26^\circ)$$
$$= \mathbf{5.67 \times 10^4\angle -41.2^\circ}$$

$$N_2 = \begin{vmatrix} (278.6\angle 21^\circ) & (80\angle 0^\circ) \\ (-160\angle 0^\circ) & (-66.6\angle -87.26^\circ) \end{vmatrix}$$
$$= (278.6\angle 21^\circ)(-66.6\angle -87.26^\circ) - (80\angle 0^\circ)(-160\angle 0^\circ)$$
$$= \mathbf{1.78 \times 10^4\angle 72.6^\circ}$$

I_1 and I_2 are now found by substituting the values of N_1, N_2, and D into the preceding ratios.

$$I_1 = N_1/D = (5.67 \times 10^4\angle -41.2^\circ)/(2.02 \times 10^5\angle -30.3^\circ)$$
$$= 2.807 \times 10^{-1}\angle -10.9^\circ = 280.7\angle -10.9^\circ \text{ mA}$$
$$I_2 = N_2/D = (1.78 \times 10^4\angle 72.6^\circ)/(2.02 \times 10^5\angle -30.3^\circ)$$
$$= 8.812 \times 10^{-2}\angle 103.0^\circ = 88.1\angle 103.0^\circ \text{ mA}$$

The voltage, V_a, can now be found by applying Ohm's Law to the resistor R_a.

$$V_a = (I_1 - I_2)R_a$$
$$= [(2.807 \times 10^{-1}\angle -10.9^\circ) - (8.812 \times 10^{-2}\angle 103^\circ)](160\angle 0^\circ)$$
$$= 52.2\angle -25.2^\circ \text{ V}$$

Answer $V_a = 52.2\angle -25.2^\circ$ V.

(b) To use nodal analysis in the solution of this problem, we must first convert the voltage across E_c to a current source as follows:

$$I_c = E_c/Z_c = E_c/(R_c + jX_c) = (80\angle 0^\circ)/(100 + j100)$$
$$= 0.5657\angle -45^\circ = 565.7\angle -45^\circ \text{ mA}$$

The circuit of Fig. 7.7(a) is redrawn in Fig. 7.7(c) with the current source I_c replacing the voltage source E_c. In this circuit, the independent nodes 1 and 2 have been identified. Note that the voltage V_a is the same as the node voltage, V_1:

$$V_a = V_1$$

From Equations 7.46, 7.47, and 7.48:

$$Y_{11} = \frac{1}{R_e - jX_e} + \frac{1}{R_a} + \frac{1}{R_c + jX_c} = \frac{1}{200 - j320} + \frac{1}{160\angle 0°} + \frac{1}{100 + j100}$$

$$= \mathbf{12.95\angle -12.3° \text{ mS}}$$

$$Y_{22} = \frac{1}{R_e - jX_e} + \frac{1}{R_n - jX_n} = \frac{1}{200 - j320} + \frac{1}{180 - j280}$$

$$= \mathbf{5.65\angle -57.6° \text{ mS}}$$

$$Y_{12} = Y_{21} = \frac{1}{R_e - jX_e} = \frac{1}{200 - j320} = \mathbf{2.65\angle 58.0° \text{ mS}}$$

Using Equations 7.49 and 7.50:

$$\Sigma I_{S_1} = I_c = 565.7 \times 10^{-3}\angle -45°$$

$$\Sigma I_{S_2} = I_n = 200 \times 10^{-3}\angle -30°$$

The generalized nodal equations for this circuit are obtained by substituting these values for Y_{11}, Y_{22}, Y_{12}, Y_{21}, ΣI_{S_1}, and ΣI_{S_2} into Equations 7.44 and 7.45:

$$(12.95 \times 10^{-3}\angle -12.3°)V_1 - (2.65 \times 10^{-3}\angle 58.0°)V_2 = 565.7 \times 10^{-3}\angle -45°$$

$$-(2.65 \times 10^{-3}\angle 58.0°)V_1 + (5.65 \times 10^{-3}\angle 57.6°)V_2 = 200 \times 10^{-3}\angle -30°$$

The nodal equations may now be solved using the method of determinants:

$$V_1 = N_1/D \qquad V_2 = N_2/D$$

$$D = \begin{vmatrix} (12.95\angle -12.3°) & (-2.65\angle 58°) \\ (-2.65\angle 58°) & (5.65\angle 57.6°) \end{vmatrix}$$

$$= (12.95\angle -12.3°)(5.65\angle 57.6°) - (-2.65\angle 58°)(-2.65\angle 58°)$$

$$= 71.2\angle 40.0°$$

$$N_1 = \begin{vmatrix} (565.7\angle -45°) & (-2.65\angle 58°) \\ (200\angle -30°) & (5.65\angle 57.6°) \end{vmatrix}$$

$$= (565.7\angle -45°)(5.65\angle 57.6°) - (-2.65\angle 58°)(200\angle -30°)$$

$$= 3.71 \times 10^3\angle 14.8°$$

$$N_2 = \begin{vmatrix} (12.95\angle -12.3°) & (565.7\angle -45°) \\ (-2.65\angle 58°) & (200\angle -30°) \end{vmatrix}$$

$$= (12.95\angle -12.3°)(200\angle -30°) - (565.7\angle -45°)(-2.65\angle 58°)$$

$$= 3.66 \times 10^3\angle -22.6°$$

$$V_1 = N_1/D = (3.71 \times 10^3\angle 14.8°)/(71.2\angle 40.0°) = 52.1\angle -25.2° \text{ V}$$

$$V_2 = N_2/D = (3.66 \times 10^3\angle -22.6°)/(71.2\angle 40.0°) = 51.4\angle -62.6° \text{ V}$$

$$V_a = V_1$$

Answer $V_a = 52.1\angle -25.2°$ V. Note that this is the same result as that obtained using mesh analysis in part (a) of this example.

7.5 ANALYSIS WITH IDEAL SOURCES

Sometimes multimesh or multinode circuits include one or more voltage sources with zero series impedance and/or current sources with infinite parallel impedance. These voltage sources cannot be converted to equivalent current sources and vice versa.

In such cases, we must use basic circuit laws (KCL, KVL, Ohm's Law, voltage division, current division, etc.) to form the simultaneous equations we have been setting up using the "format" approach of the previous examples. These equations may then be solved in the normal manner using determinants or some other manner such as substitution.

EXAMPLE 7.12

The circuit of Fig. 7.8 is to be solved for the currents I_x, I_n, and I_a.

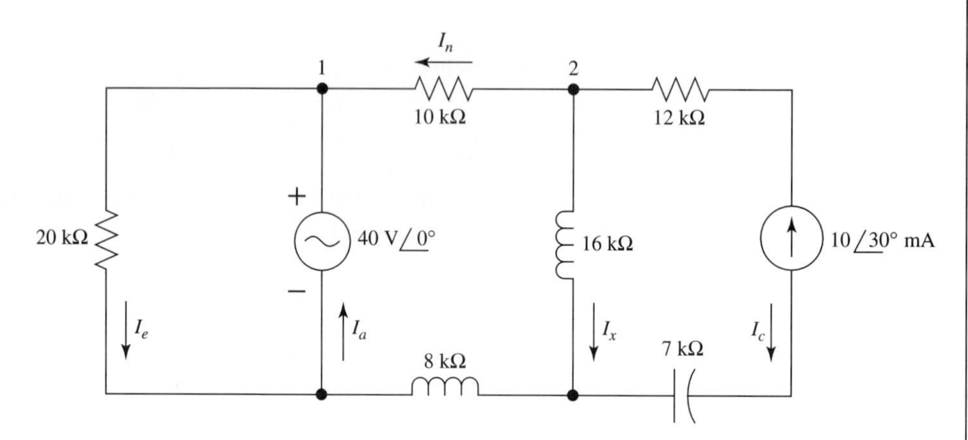

FIGURE 7.8

Solution Since the current source has no parallel impedance and the voltage source has no series impedance, neither one may be converted to the opposite form. We must therefore use basic circuit laws to form our simultaneous equations.

Using Ohm's Law: $I_e = (40\underline{/0°})/(20 \times 10^{-3}\underline{/0°}) = 2\underline{/0°}$ mA.

Using KCL at node 1: $I_n + I_a = I_e = 2 \times 10^{-3}\underline{/0°}$.

Using KCL at node 2: $I_n + I_x = 10 \times 10^{-3}\underline{/30°}$.

These two equations have three unknowns. We therefore need one more equation to solve for the currents.

Using KVL around the center loop:

$$(8 \times 10^3\underline{/90°})I_n - (16 \times 10^3\underline{/90°})I_x + (10^4\underline{/0°})I_n + (40\underline{/0°}) = 0$$

or

$$(8 \times 10^3\underline{/90°} + 10^4\underline{/0°})I_n - (16 \times 10^3\underline{/90°})I_x = -40\underline{/0°}$$
$$(1.28 \times 10^4\underline{/38.7°})I_n - (16 \times 10^3\underline{/90°})I_x = -40\underline{/0°}$$

The last equation may now be used to find an expression for I_n in terms of I_x as follows:

$$I_n = [(-40\underline{/0°}) + (16 \times 10^3\underline{/90°})I_x)]/(1.28 \times 10^4\underline{/38.7°})$$
$$= (3.1 \times 10^{-3}\underline{/141.3°}) + (1.25\underline{/51.3°})I_x$$

This expression for I_n is now substituted for I_n in the equation written above at node 2:

$$(3.1 \times 10^{-3}\underline{/141.3°}) + (1.25\underline{/51.3°})I_x + I_x = 10 \times 10^{-3}\underline{/30°}$$

Solving for I_x we get the following result:

$$I_x = [(10 \times 10^{-3}\underline{/30°}) - (3.1 \times 10^{-3}\underline{/141.3°})]/(1 + 1.25\underline{/51.3°})$$
$$= \mathbf{5.66 \times 10^{-3}\underline{/-13.2°}\ A}$$

Setting this value for I_x into the KCL expression at node 2: we may now solve for I_n:

$$I_n = (10 \times 10^{-3}\angle 30°) - I_x = (10 \times 10^{-3}\angle 30°) - (5.66 \times 10^{-3}\angle -13.2°)$$
$$= \mathbf{7.03 \times 10^{-3}\angle 63.4°\ A}$$

We may now solve for I_a by substituting this value for I_n into the above equation at node 1:

$$I_a = (2 \times 10^{-3}\angle 0°) - I_n = (2 \times 10^{-3}\angle 0°) - (7.03 \times 10^{-3}\angle 63.4°)$$
$$= \mathbf{6.39 \times 10^{-3}\angle -100.3°}$$

Answer $I_x = 5.66\angle -13.2°$ mA, $I_n = 7.03\angle 63.4°$ mA, $I_a = 6.39\angle -100.3°$ mA.

7.6 SUPERPOSITION THEOREM

Another method that may be used to analyze circuits that contains more than one source is the use of the Superposition Theorem. This theorem may be stated as follows:

> The response (i.e., voltage or current) due to several independent voltage and/or current sources acting together may be expressed as the sum of the responses due to each individual source acting alone.

In mathematical form, this theorem may be written as follows:

$$I_x = \Sigma(I_x)_n \quad \textbf{(Eq. 7.63)}$$

and

$$V_x = \Sigma(V_x)_n \quad \textbf{(Eq. 7.63a)}$$

where $(I_x)_n$, or $(V_x)_n$, is the response to source n acting alone.

When calculating the response due to any one of the sources acting alone, all the other sources must be set to zero. ***This means all other independent current sources are replaced by "opens" and all other independent voltage sources are replaced by "shorts."***

EXAMPLE 7.13

Find the current I_x in the circuit of Fig. 7.9(a) using the Superposition Theorem.

Solution The Superposition Theorem (Eq. 7.63) enables us to express the current, I_x, as the sum of two components as follows:

$$I_x = (I_x)_1 + (I_x)_2$$

where $(I_x)_1$ = The component of I_x due to source E_1 acting alone
$(I_x)_2$ = The component of I_x due to source E_2 acting alone

To find $(I_x)_1$ we redraw the circuit of Fig. 7.9(a) while setting the voltage source E_2 to zero by shorting it out. The resulting circuit appears in Fig. 7.9(b).

Note that the current $(I_x)_1$ in Fig. 7.9(b) is drawn in the branch in which I_x appeared in Fig. 7.9(a). To find $(I_x)_1$ we will first find the impedance, Z_{ab}. Using Ohm's Law, we will find the source current I_s. We will then use current division between the two parallel branches to find $(I_x)_1$.

$$Z_{ab} = (10 + j30) + [40\|(20 - j10)] = (10 + j30) + (14.7\angle -17.1°)$$
$$= 35.2\angle 46.9°\ \Omega$$
$$I_s = E_1/Z_{ab} = (200\angle 0°)/(35.2\angle 46.9°) = 5.68\angle -46.9°\ \text{A}$$
$$(I_x)_1 = (Y_x/Y_{mn})I_s$$
$$Y_x = 1/(20 - j10) = 44.7 \times 10^{-3}\angle 26.6°\ \text{S}$$

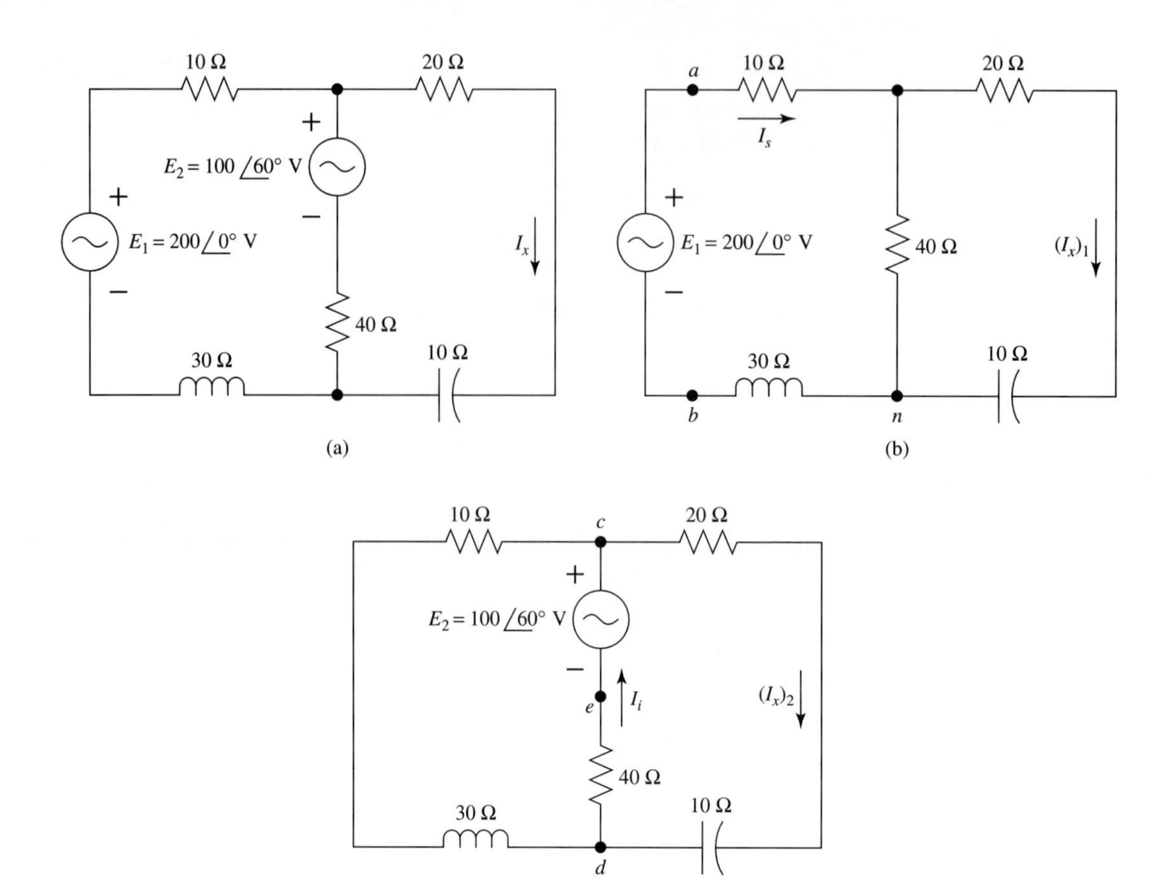

FIGURE 7.9

$$
\begin{aligned}
Y_{mn} &= Y_x + (1/40) = (44.7 \times 10^{-3}\angle 26.6^\circ) + (25 \times 10^{-3}\angle 0^\circ) \\
&= 68.0 \times 10^{-3}\angle 17.1^\circ \text{ S} \\
(I_x)_1 &= (Y_x/Y_{mn})(I_s) \\
&= [(44.7 \times 10^{-3}\angle 26.6^\circ)/(68.0 \times 10^{-3}\angle 17.1^\circ)](5.68\angle -46.9^\circ) \\
&= \mathbf{3.73\angle -37.4^\circ \text{ A}}
\end{aligned}
$$

To find $(I_x)_2$ we redraw the circuit of Fig. 7.9(a) while setting the voltage source E_1 to zero by shorting it out. The resulting circuit appears in Fig. 7.9(c).

Note that the current $(I_x)_2$ in Fig. 7.9(b) is drawn in the branch in which I_x appeared in Fig. 7.9(a). We will find $(I_x)_2$ in a manner similar to that used to find $(I_x)_1$. Namely, we will first find the impedance Z_{ce}. Using Ohm's Law we will find the input current I_i. Using current division, we will then find $(I_x)_2$.

$$
\begin{aligned}
Z_{ce} &= 40 + [(10 + j30)\|(20 - j10)] = 40 + (19.6\angle 11.3^\circ) = 59.3\angle 3.7^\circ \ \Omega \\
I_i &= E_2/Z_{ce} = (100\angle 60^\circ)/(59.3\angle 3.7^\circ) \text{ A} = 1.7\angle 56.3^\circ \text{ A} \\
Y_x &= 1/(20 - j10) = 44.7 \times 10^{-3}\angle 26.6^\circ \text{ S} \\
Z_{cd} &= (10 + j30)\|(20 - j10) = 19.6\angle 11.3^\circ \text{ S} \\
Y_{cd} &= 1/Z_{cd} = 1/(19.6\angle 11.3^\circ) = 51.0 \times 10^{-3}\angle -11.3^\circ \\
(I_x)_2 &= (Y_x/Y_{cd})(I_i) \\
&= [(44.7 \times 10^{-3}\angle 26.6^\circ)/(51.0 \times 10^{-3}\angle -11.3^\circ)](1.7\angle 56.3^\circ) \\
&= \mathbf{1.50\angle 94^\circ \text{ A}}
\end{aligned}
$$

We may now substitute the values for $(I_x)_1$ and $(I_x)_2$ into the superposition equation (7.64):

$$I_x = (I_x)_1 + (I_x)_2 = (3.7\angle -37.4^\circ) + (1.5\angle 94^\circ) = 2.94\angle -14.7^\circ$$

Answer $I_x = 2.94\angle -14.7^\circ$ A.

EXAMPLE 7.14

Find the sinusoidal voltage v_x in the circuit of Fig. 7.10(a) using the superposition theorem.

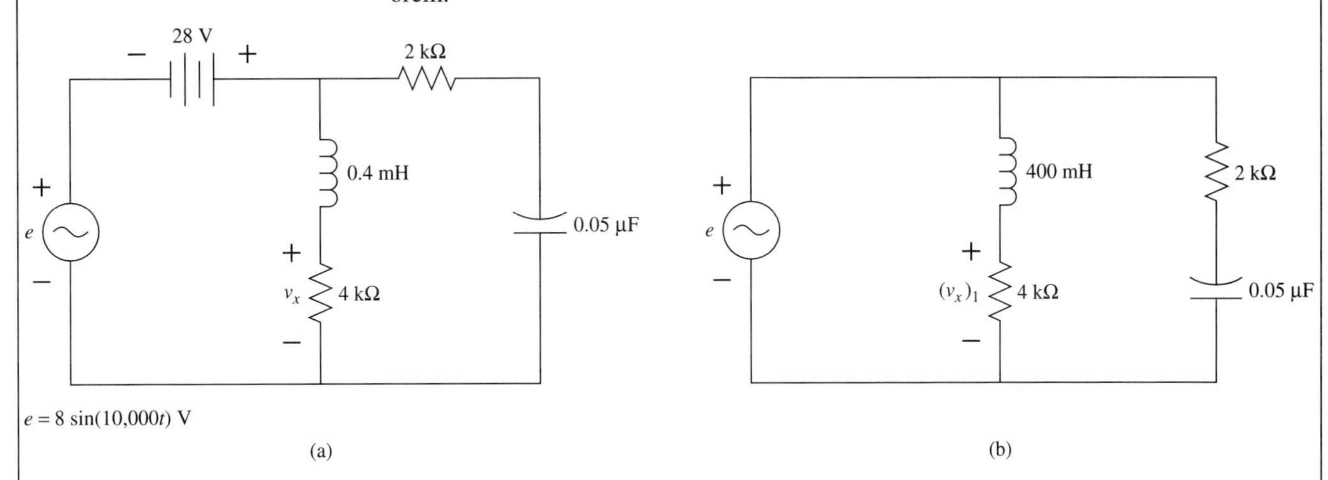

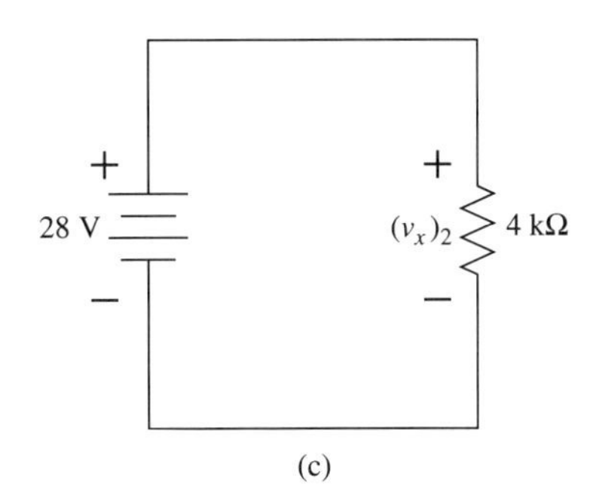

FIGURE 7.10

Solution We will designate the sources as follows:

AC voltage source = Source 1
DC voltage source = Source 2

From the Superposition Theorem, we may express the voltage v_x as the sum of two components as follows:

$(v_x)_1$ = Component of (v_x) due to source 1
$(v_x)_2$ = Component of (v_x) due to source 2

To find $(v_x)_1$, we must set source 2 to zero by shorting it out. This results in the circuit of Fig. 7.10(b).

In Fig. 7.10(b), the 4-kΩ resistor is in series with the inductor and the combination is in parallel with the source e. We may therefore find the voltage $(v_x)_1$ by voltage division between the resistor and inductor.

$$(v_x)_1 = [(4\text{K}\angle 0^\circ)(8\angle 0^\circ)]/[(4\text{K}\angle 0^\circ) + 10{,}000(400 \times 10^{-3}\angle 90^\circ)]$$
$$= 5.66\angle -45^\circ \text{ V}$$

$$(v_x)_1 = 5.66 \sin(10{,}000t - 45^\circ) \text{ V}$$

To find $(v_x)_2$, we must set source 1 to zero by shorting it out. This results in the circuit of Fig. 7.10(c). Note that because this is a dc circuit, the capacitor of the original circuit was replaced by an open circuit and the inductor was replaced by a short circuit.

In Fig. 7.10(c) the 4-kΩ resistor is directly across the dc supply. Therefore:

$$(v_x)_2 = 28 \text{ Vdc}$$

Using these values for $(v_x)_1$ and $(v_x)_2$ in the superposition equation:

$$(v_x) = (v_x)_1 + (v_x)_2 = 5.66 \sin(10{,}000t - 45°) + 28$$

Answer $v_x = 28 + 5.66 \sin(10{,}000t - 45°)$ V

Caution: A dc quantity cannot be added to a phasor. Note that in the previous problem, $(v_x)_1$ had to be converted to its sinusoidal form before it could be added to the dc voltage of $(v_x)_2$.

7.7 THEVENIN'S AND NORTON'S THEOREMS

In general, the term ***source*** can be applied not only to a simple voltage or current source but to an entire circuit as well. In many cases, a circuit acts as a source supplying power to a "load." Further, the load often can be different at different times. In such cases, it is convenient to be able to replace the source circuit by a much simpler one. This, as we are about to see, simplifies the calculation of load voltage and/or load current.

Two techniques are used to simplify source circuits that are ***linear*** and ***bilateral*** in nature. One uses a theorem known as Thevenin's Theorem, and the other uses a closely related theorem known as Norton's Theorem. (***Note:*** A ***linear*** circuit is one in which each element has the property of proportionality such that if any voltage or current in the circuit is multiplied by a constant, all other voltages and currents in the circuit are multiplied by the same constant. A ***bilateral*** circuit is one in which the properties of all the elements are independent of the direction of the current through them or the polarity of the voltage across them.)

7.7.1 Thevenin's Theorem

Thevenin's Theorem states that a linear, bilateral circuit may be replaced by "Thevenin's equivalent circuit," which is equivalent to the original circuit with respect to two, and only two, terminals. Thevenin's equivalent circuit consists of a voltage source, known as the ***Thevenin voltage*** in series with an impedance, known as the ***Thevenin impedance.***

Thevenin's Theorem is illustrated in Fig. 7.11. The circuit in Fig. 7.11(a) may be replaced by the Thevenin equivalent circuit of Fig. 7.11(b), which is equivalent to the circuit of Fig. 7.11(a) with respect to terminals a and b.

In the circuit of Fig. 7.11(b), E_{Th} = Thevenin voltage and Z_{Th} = Thevenin impedance. The ***Thevenin voltage,*** E_{th}, may be calculated using the following technique:

$E_{Th} = (E_{ab})_{oc}$ = open circuit voltage between terminals a and b in the original circuit.

This definition means that we must remove any load connected to terminals a and b and then calculate the voltage between these two terminals. This value is the Thevenin voltage, E_{Th}.

The ***Thevenin impedance,*** Z_{Th}, can be calculated using one of two methods.

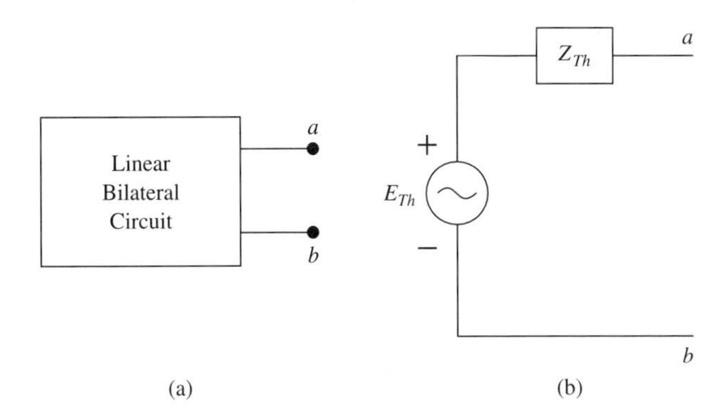

FIGURE 7.11

Method 1. Method 1 may be used if there are no dependent sources in the circuit. (See Section 7.8 for a discussion of dependent sources.)

Z_{Th} = impedance between terminals a and b of the original circuit with no load and all independent sources set to zero.

When method 1 is used, all independent voltage sources are set to zero and all independent current sources are opened. Any load connected to terminals a and b is removed, and the impedance between terminals a and b is calculated. This value is the Thevenin impedance, Z_{Th}.

Method 2. Method 2 ***may*** be used regardless of whether or not dependent sources are included in the circuit but ***must*** be used if dependent sources are included in the circuit.

$$Z_{\text{Th}} = (E_{ab})_{oc}/I_{sc}$$

where $(E_{ab})_{oc} = E_{\text{Th}}$ = open circuit voltage between terminals a and b

I_{sc} = short circuit current from terminal a to terminal b

EXAMPLE 7.15

The 10-kΩ resistor in series with the 8-kΩ capacitive reactance of Fig. 7.12(a) is considered the load on the rest of the circuit. Find:

(a) Thevenin's equivalent circuit with respect to terminals a and b

(b) I = the current through this load using the Thevenin circuit found in part (a)

Solution

(a) Thevenin's equivalent circuit is drawn in Fig. 7.12(b). To find E_{Th}, we remove the load as shown in Fig. 7.12(c). Since this results in an open circuit, no current can flow. There is therefore no voltage drop across any of the components in this circuit, and we may therefore calculate E_{Th} as follows:

$$E_{\text{Th}} = (E_{ab})_{oc} = 20\text{ V}\underline{/0°}\text{ V}$$

Since there are no dependent sources in this circuit, we may calculate the Thevenin impedance by using method 1. Setting the only voltage source in this circuit to zero, we calculate Z_{Th} as follows: Short the voltage source and redraw the circuit as shown in Fig. 7.12(d). In this circuit:

$$Z_{\text{Th}} = Z_{ab} = (5\text{K} + j30\text{K} - j2\text{K}) = (5\text{K} + j28\text{K})\ \Omega = \mathbf{28.4K\underline{/80.0°}\ \Omega}$$

(b) To find I, the original circuit is redrawn as Thevenin's equivalent circuit as in Fig. 7.12(e). Note that the current I in this circuit is the same as in the original circuit. Using Ohm's Law:

$$I = E_{\text{Th}}/(Z_{\text{Th}} + Z_L)$$

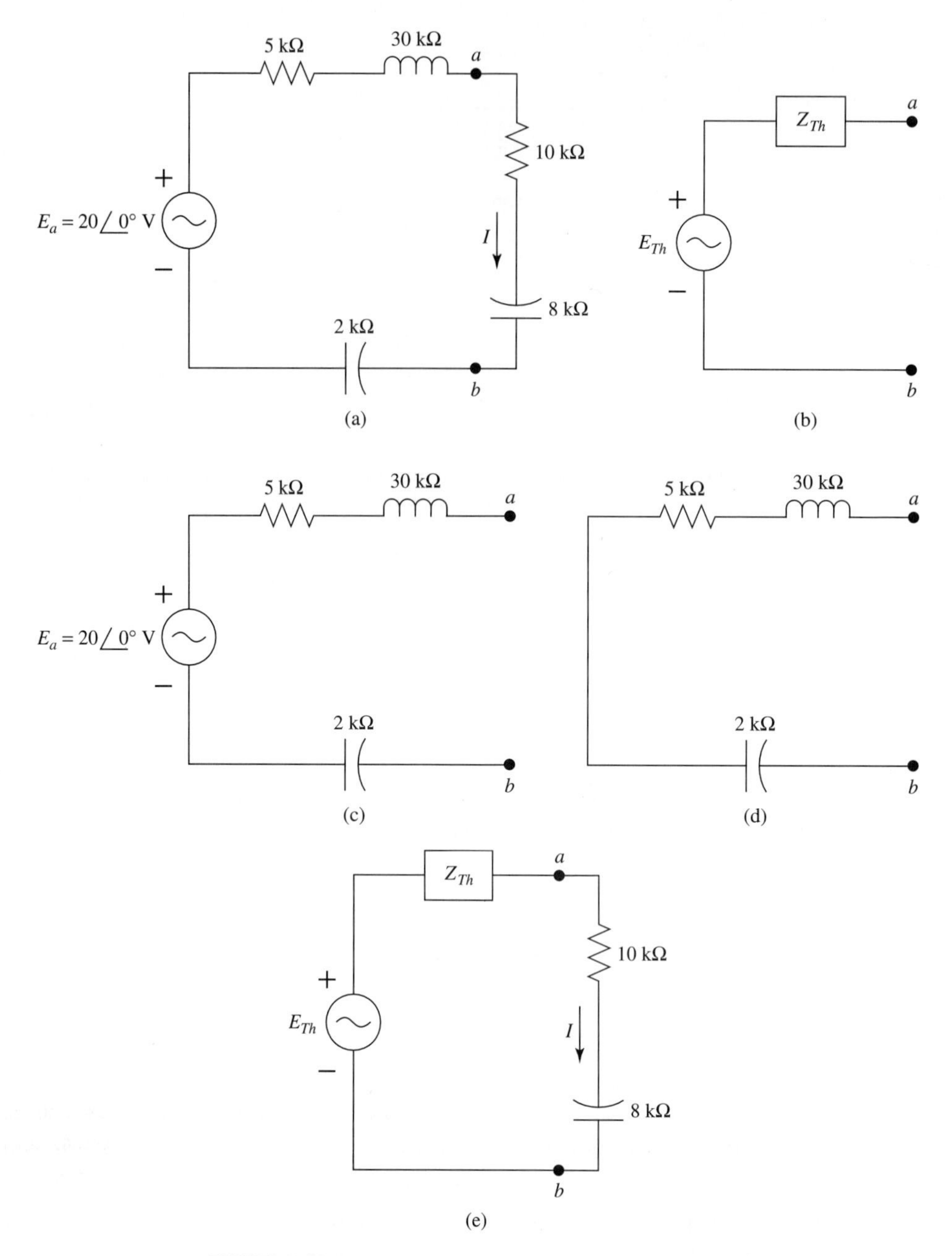

FIGURE 7.12

where Z_L is the load impedance.

$$I = (20\angle 0°)/(5\text{K} + j28\text{K} + 10\text{K} - j8\text{K}) = (20\angle 0°)/(15\text{K} + j20\text{K})$$
$$= 8 \times 10^{-4}\angle -53.1° \text{ A}$$

Answer $I = 0.8\angle -53.1°$ mA.

EXAMPLE 7.16

In Fig. 7.13(a), the 500-Ω resistor in series with the 200-Ω inductive reactance is considered as the load on the rest of the circuit connected to terminals 1 and 2.

(a) Find Thevenin's equivalent circuit for the source circuit between terminals 1 and 2.

(b) Find the load current, I_x, using Thevenin's equivalent circuit found in part (a).

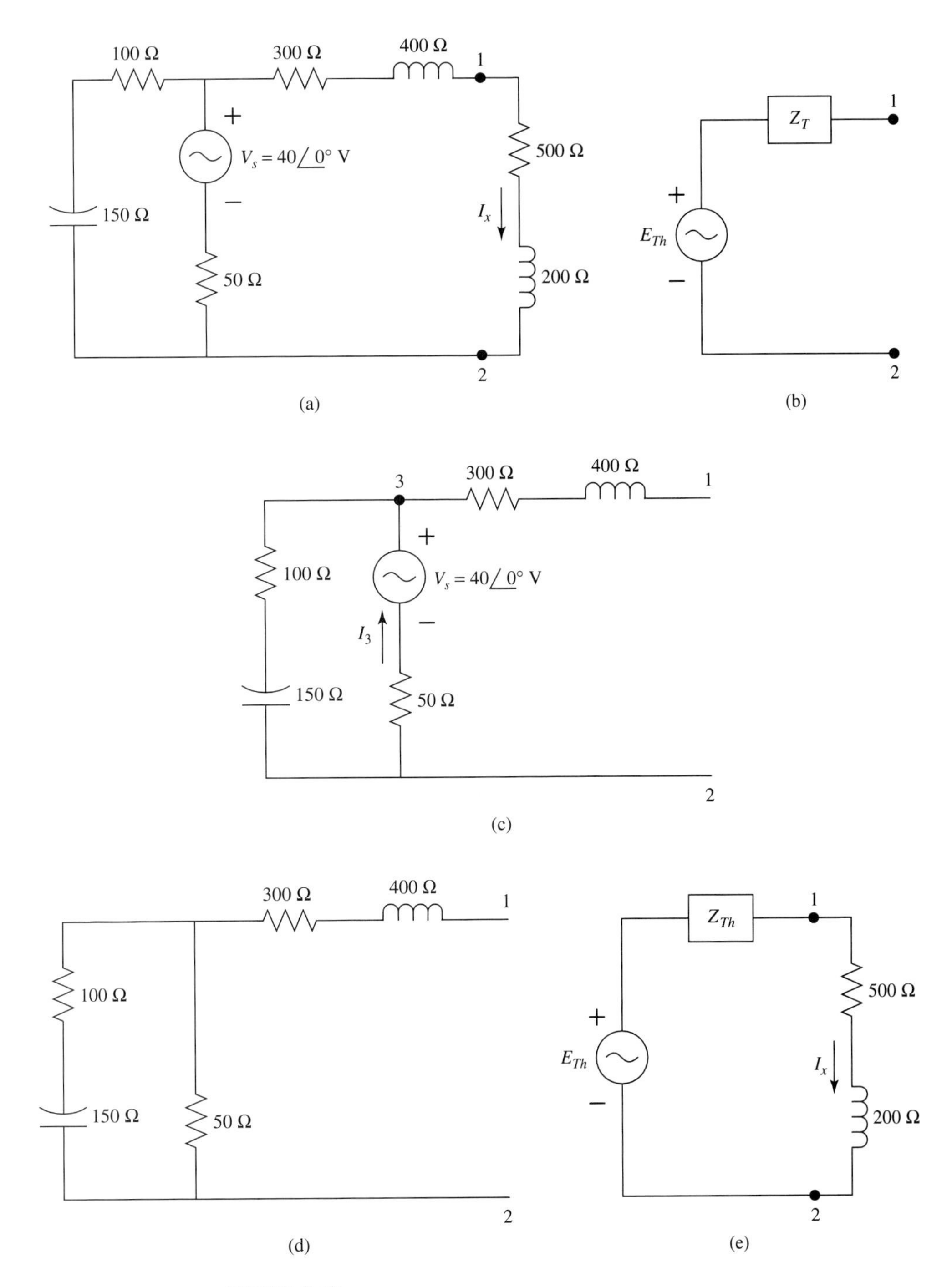

FIGURE 7.13

Solution

(a) Thevenin's equivalent circuit is drawn first in Fig. 7.13(b). To find E_{Th}, we redraw the original circuit with the load removed (open circuit). This is shown in Fig. 7.13(c). In this circuit, E_{Th} is the voltage between terminals 1 and 2, V_{12}.

Since no current flows through the 300-Ω resistor and 400-Ω inductive reactance in Fig. 7.13(c) there is no voltage drop across these components and

$$V_{12} = V_{32}$$

V_{32} is found by recognizing that the remainder of the circuit is a series circuit carrying the current called I_s. Using Ohm's Law and representing the impedance of the series circuit by Z_s:

$$I_s = V_s/Z_s = (40\angle 0^\circ)/(150 - j150) = 188.6\angle 45^\circ \text{ mA}$$

From KVL:

$$V_{32} = V_s - I_s(50) = (40\angle 0^\circ) - (188.6 \times 10^{-3}\angle 45^\circ)(50) = 34.0\angle -11.3^\circ$$

Answer $E_{\text{Th}} = 34\angle -11.3^\circ$ V.

Z_{Th} may be calculated by method 1. To do this, the single voltage source is set to zero and the original circuit is redrawn in Fig. 7.13(d) with the load removed. In this circuit, Z_{Th} is the voltage between terminals 1 and 2. This may be calculated as follows:

$$Z_{\text{Th}} = Z_{12} = (300\angle 0^\circ + (400\angle 90^\circ) + [(50\angle 0^\circ)(100 - j150)/(150 - j150)]$$
$$= 519.7\angle 48.9^\circ$$

Answer $Z_{\text{Th}} = 519.7\angle 48.9^\circ\ \Omega$.

(b) To find the load current, I_x, the load is placed across Thevenin's equivalent circuit in Fig. 7.13(e). In this circuit using Ohm's Law:

$$I_x = E_{\text{Th}}/(Z_{\text{Th}} + 500 + j200)$$
$$= (34\angle -11.3^\circ)/[(519.7\angle 48.9^\circ) + (500 + j200)] = 3.3 \times 10^{-2}\angle -46.4^\circ$$

Answer $I_x = 33.0\angle -46.4^\circ$ mA.

EXAMPLE 7.17

In the circuit of Fig. 7.14(a), the 4-kΩ resistor is the load on the source circuit between terminals a and b.

(a) Find Thevenin's equivalent circuit for the source between terminals a and b.
(b) Find the voltage across the load resistor, V_x.

Solution

(a) Thevenin's equivalent circuit is drawn in Fig. 7.14(b). Note that since this is a dc circuit, the Thevenin voltage will be dc and the Thevenin impedance will be a resistor.

E_{Th} is found as usual by removing the load and calculating the voltage across terminals a and b. The corresponding circuit is drawn in Fig. 7.14(c). Note that this is a series circuit with a current that we have labeled I_s. This current is calculated using Ohm's Law:

$$I_s = (16 + 12)/(3\text{K} + 2\text{K}) = 28/5\text{K} = 5.6 \text{ mA}$$
$$E_{\text{Th}} = V_{ab} = (3\text{K})I_s - 16 = (3 \times 10^3)(5.6 \times 10^{-3}) - 16 = 0.8\text{V}$$

Answer $E_{\text{Th}} = 0.8$V.

Thevenin's resistance is found by method 1. The original circuit is redrawn in Fig. 7.14(d) with the load removed and the voltage sources replaced by shorts.

In this circuit, looking into terminals a and b we see that the two resistors are in parallel. Therefore:

$$R_{\text{Th}} = 2\text{K}\|3\text{K} = (2 \times 10^3)(3 \times 10^3)/[(3 \times 10^3) + (2 \times 10^3)]$$
$$= 1.2 \times 10^3\ \Omega$$

Answer $R_{\text{Th}} = 1.2$ kΩ.

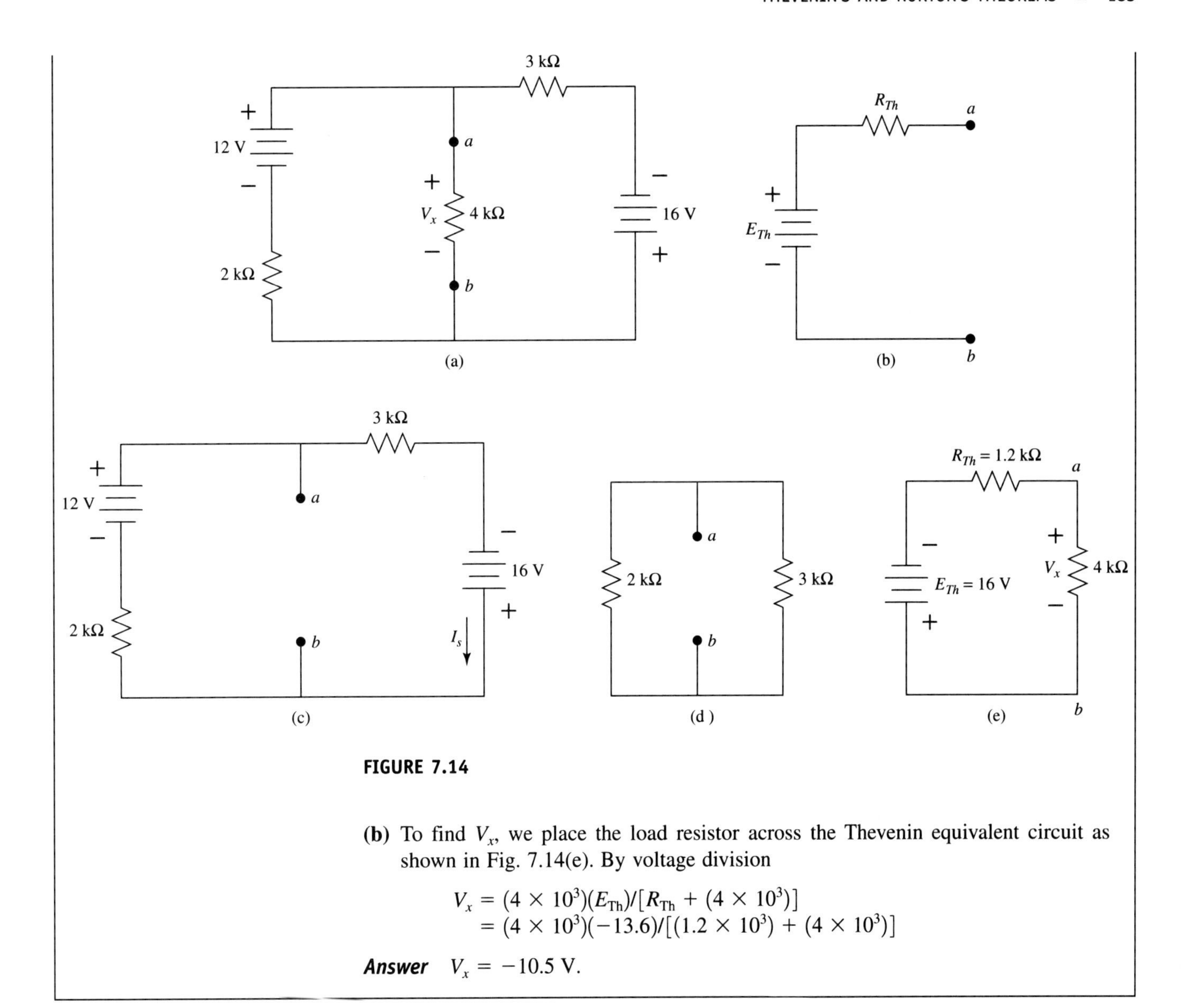

FIGURE 7.14

(b) To find V_x, we place the load resistor across the Thevenin equivalent circuit as shown in Fig. 7.14(e). By voltage division

$$V_x = (4 \times 10^3)(E_{\mathrm{Th}})/[R_{\mathrm{Th}} + (4 \times 10^3)]$$
$$= (4 \times 10^3)(-13.6)/[(1.2 \times 10^3) + (4 \times 10^3)]$$

Answer $V_x = -10.5$ V.

EXAMPLE 7.18

In the circuit of Fig. 7.15(a):

(a) Find Thevenin's equivalent circuit looking into terminals a and b, considering the 400-Ω resistor in series with the 1000-Ω inductive reactance as the load and the rest of the circuit as the source.

(b) Find the load current, I_x.

Solution

(a) Thevenin's equivalent circuit is drawn in Fig. 7.15(b). We find E_{Th} first by removing the load and redrawing the circuit as shown in Fig. 7.15(c). In this circuit, $E_{\mathrm{Th}} = V_{ab}$, the voltage between terminals a and b.

Since there are two sources in this circuit we will find V_{ab} using superposition. Therefore:

$$V_{ab} = (V_{ab})_1 + (V_{ab})_2$$

where $(V_{ab})_1$ = The component of V_{ab} that is due to V_1 acting alone
$(V_{ab})_2$ = The component of V_{ab} that is due to I_2 acting alone

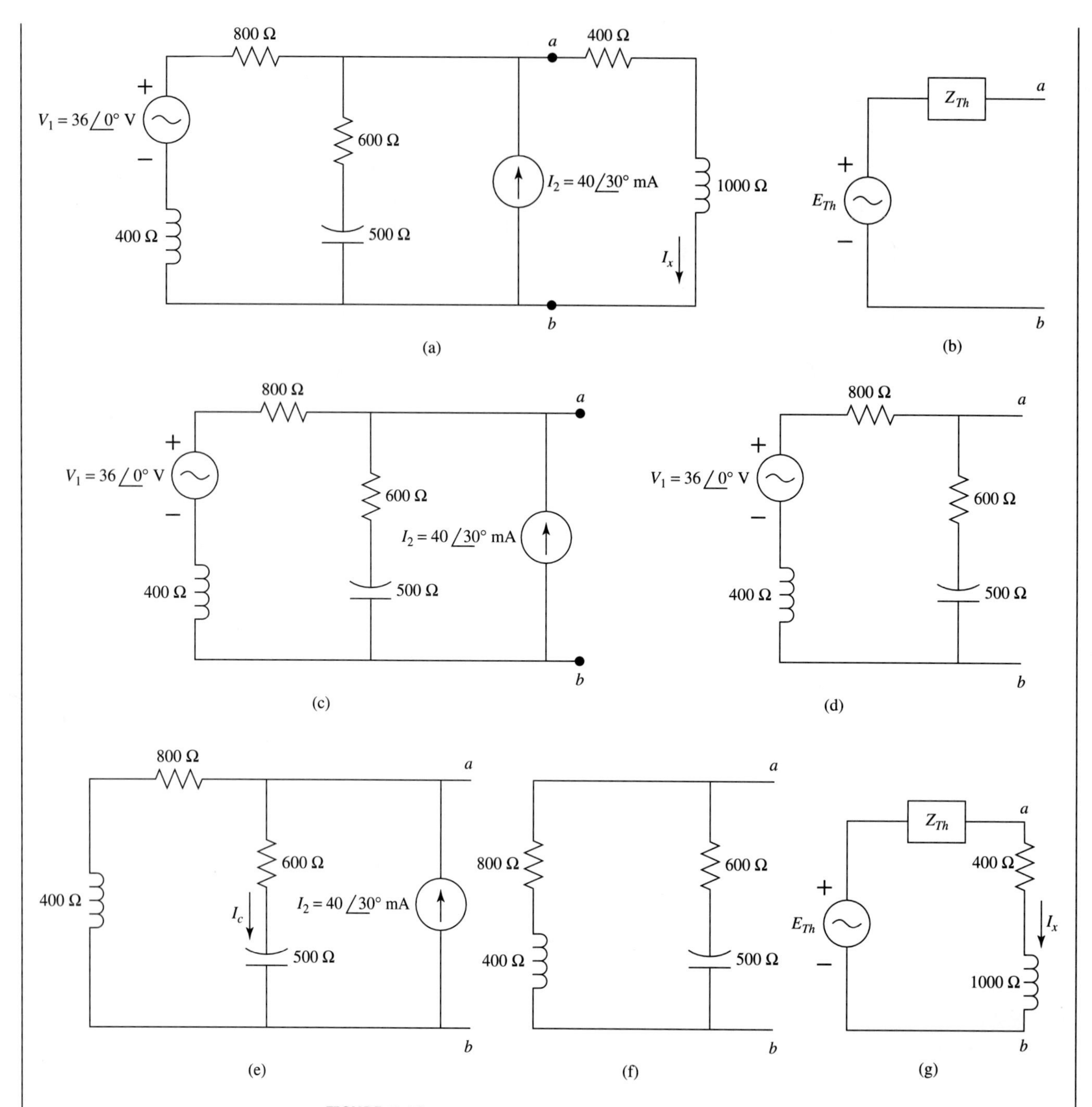

FIGURE 7.15

To find $(V_{ab})_1$, set I_2 to zero. The circuit of Fig. 7.15(c) is then redrawn as shown in Fig. 7.15(d). In this circuit, by voltage division:

$$(V_{ab})_1 = (600 - j500)(36\angle 0°)/(1400 - j100) = \mathbf{20.0\angle -35.7° \ V}$$

To find $(V_{ab})_2$, set V_1 to zero. The circuit of Fig. 7.15(c) is then redrawn as shown in Fig. 7.15(e). In this circuit, by current division:

$$I_c = (800 + j400)(I_2)/(1400 - j100) = 25.5 \times 10^{-3}\angle 60.7° \ \text{A}$$

From Ohm's Law:

$$(V_{ab})_2 = (25.5 \times 10^{-3}\angle 60.7^\circ)(600 - j500) = \mathbf{19.9\angle 20.9^\circ\ V}$$
$$V_{ab} = (V_{ab})_1 + (V_{ab})_2 = (20.0\angle -35.7^\circ) + (19.9\angle 20.9^\circ) = 35.1\angle -7.5^\circ\ \text{V}$$

Answer $E_{\text{Th}} = 35.1\angle -7.5^\circ$ V.

The Thevenin impedance, E_{Th}, will be found using method 1 because all sources are independent. Accordingly, the original circuit is redrawn in Fig. 7.15(f), with the load removed and both sources set to zero. In this circuit, Z_{Th} is the impedance looking into terminals a and b. This gives us two impedances in parallel as follows:

$$Z_{\text{Th}} = (800 + j400)(600 - j500)/(1400 - j100) = 497.7\angle -9.2^\circ\ \Omega$$

Answer $Z_{\text{Th}} = 497.7\angle -9.2^\circ\ \Omega$

(b) The load current is found by placing the load across the output terminals, a and b, of Thevenin's equivalent circuit as shown in Fig. 7.15(g). Using Ohm's Law:

$$I_x = (35.1\angle -7.5^\circ)/[(497.7\angle -9.2^\circ) + (400 + j1000)]$$
$$= 2.74 \times 10^{-2}\angle -53.4^\circ$$

Answer $I_x = 27.4\angle -53.4^\circ$ mA.

Note that in all the previous examples of Thevenin's equivalent circuit, Z_{Th} was found using method 1. This is because all these examples included only independent sources. Later in this chapter, examples with dependent sources will be used. These examples will demonstrate the use of method 2 to find Z_{Th}.

7.7.2 Norton's Theorem

Norton's Theorem states that a linear, bilateral circuit may be replaced by "Norton's equivalent circuit," which, just as in the case of Thevenin's Theorem, is equivalent to the original circuit with respect to two, and only two, terminals. Norton's equivalent circuit consists of a current source, known as the ***Norton current,*** in parallel with an impedance, known as the ***Norton impedance.***

Norton's Theorem is illustrated in Fig. 7.16. The circuit in Fig. 7.16(a) may be replaced by the Norton equivalent circuit of Fig. 7.16(b), which is equivalent to the circuit of Fig. 7.16(a) with respect to terminals a and b.

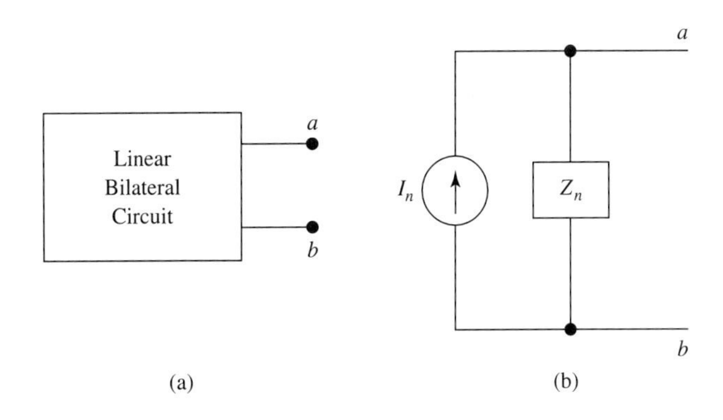

FIGURE 7.16

In the circuit of Fig. 7.16(b):

I_n = Norton current $\quad Z_n$ = Norton impedance

Note: When Fig. 7.16 is compared to Fig. 7.11 it is apparent that the two equivalent circuits are related by the following source conversion equations:

$$I_n = E_{Th}/Z_{Th} \qquad \textbf{(Eq. 7.64)}$$

and

$$Z_n = Z_{Th} \qquad \textbf{(Eq. 7.65)}$$

Therefore, once one equivalent circuit is known, the other can be found simply by using these conversion equations.

The ***Norton current,*** I_n, may be calculated using the following definition:

$I_n = (I_{ab})_{sc}$ = short circuit current from terminal a to b in the original circuit.

The technique used to find I_n therefore involves shorting the path between terminals a and b and calculating the current that flows from terminal a to terminal b in this shorted path. The resulting value is the Norton current, I_n.

The Norton impedance may be found in exactly the same manner as the Thevenin impedance because, by Equation 7.66, they are equal. Accordingly, the two methods discussed for finding the Thevenin impedance apply equally to the Norton impedance.

EXAMPLE 7.19

The 4-kΩ resistor in series with the 3-kΩ capacitive reactance is the ***load*** on the source circuit between terminals a and b of Fig. 7.17(a). Find:

(a) Norton's equivalent circuit for the source circuit between terminals a and b.

(b) The voltage across the load, V_o, in Fig. 7.17(a).

Solution

(a) The Norton equivalent circuit is drawn in Fig. 7.17(b).

We will find the Norton current I_n first. To do so, we place a short across terminals a and b as shown in Fig. 7.17(c). The current in this short from terminal a to terminal b is the Norton current, I_n. From Ohm's Law, and recognizing that the short provides us with a simple series circuit:

$$I_n = E/(1\text{K} + j7\text{K}) = (60\angle 0°)/(1\text{K} + j7\text{K}) = 8.5\angle -81.9° \text{ mA}$$

Answer $I_n = 8.5\angle -81.9°$ mA.

We now find Z_n using method 1 described in Section 7.7.1 for finding Z_{Th}. Accordingly, the load is removed, the voltage source is set to zero and the circuit is redrawn as in Fig. 7.17(d). In this figure, Z_n is the impedance looking into terminals a and b, which is the parallel combination of the two branches shown.

$$Z_n = (1\text{K} + j7\text{K})(7\text{K} - j1\text{K})/(8\text{K} + j6\text{K}) = 5\text{K}\angle 36.9° \ \Omega$$

Answer $Z_n = 5\text{K}\angle 36.9°\ \Omega$.

(b) To find the voltage across the load, V_o, we place the load across Norton's equivalent circuit as shown in Fig. 7.17(e). In this circuit, the parallel combination of Z_n and the load multiplied by I_n will give us the desired load voltage.

$$V_o = I_n \frac{Z_n(4\text{K} - j3\text{K})}{Z_n + 4\text{K} - j3\text{K}}$$

$$= (8.5 \times 10^{-3}\angle -81.9°)\frac{(5\text{K}\angle 36.9°)(4\text{K} - j3\text{K})}{(5\text{K}\angle 36.9°) + 4\text{K} - j3\text{K}} = 26.6\angle -81.9° \text{ V}$$

Answer $V_o = 26.6\angle -81.9°$ V.

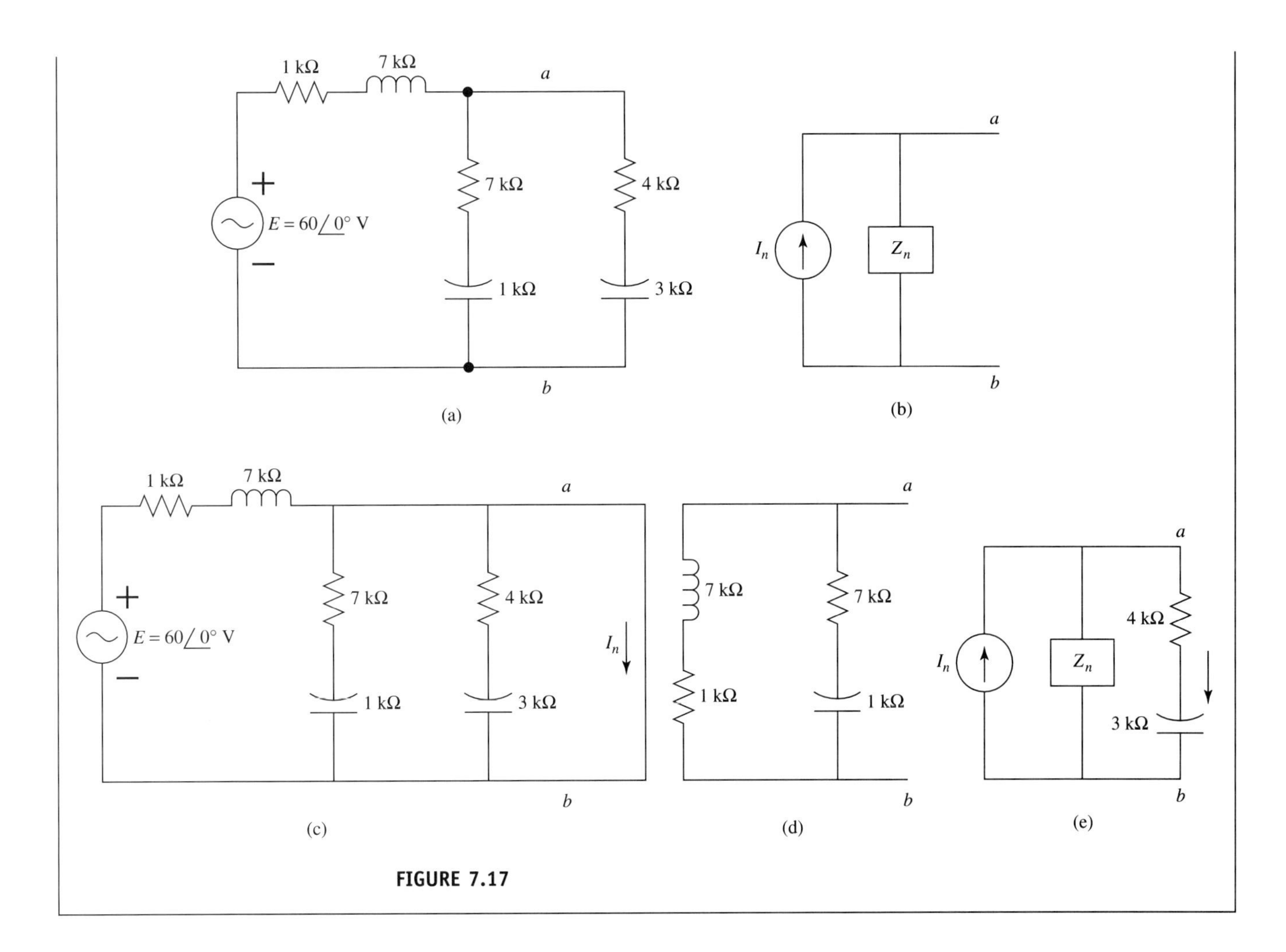

FIGURE 7.17

EXAMPLE 7.20

Find Thevenin's equivalent circuit for Example 7.19.

Solution Since we have already found Norton's equivalent circuit, finding Thevenin's equivalent circuit reduces simply to converting the Norton current source to its equivalent voltage source. Thevenin's equivalent circuit is drawn in Fig. 7.18 together with its Norton counterpart. Using Equations 7.64 and 7.65:

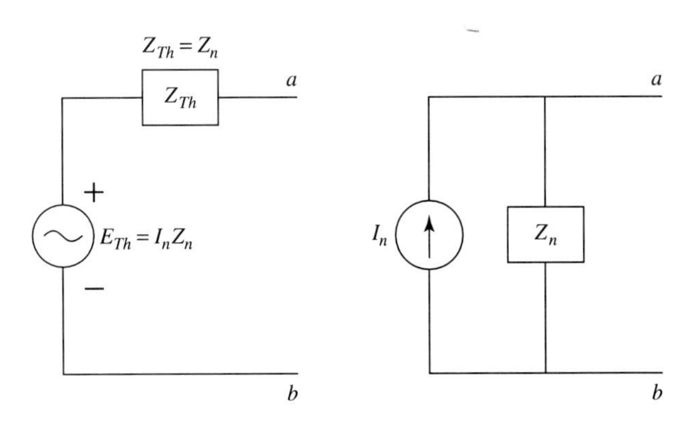

FIGURE 7.18

$$Z_{\text{Th}} = Z_n = 5\text{K}\angle 36.9^\circ\ \Omega$$

and

$$E_{\text{Th}} = I_n Z_n = (8.5 \times 10^{-3}\angle -81.9^\circ)(5\text{K}\angle 36.9^\circ) = 42.5\angle -45.9\ \text{V}$$

Answer $Z_{\text{Th}} = 5\text{K}\angle 36.9^\circ$, $E_{\text{Th}} = 42.5\angle -45^\circ$ V.

7.8 DEPENDENT SOURCES

Dependent voltage sources are ones in which the voltage output depends on another voltage or current, which may or may not be located in the same circuit as the dependent source. Similarly, ***dependent current sources*** are ones in which the current output depends on another voltage or current, which may or may not be located in the same circuit as the dependent source.

The accepted ***symbol for a dependent voltage or dependent current source is a diamond-shaped box.*** As we have seen, ***the symbol for an independent voltage or current source is a circle.***

When a dependent voltage or current source is dependent on a voltage or current located in a circuit ***different*** from the dependent source, all that is necessary is that the dependent source output be obtained before using it in the problem being considered. The analysis techniques are then virtually the same as for independent sources.

However, when the dependent source output is dependent on a voltage or current located in the ***same*** circuit as the dependent source, some modification of the techniques previously discussed is necessary.

7.8.1 Dependent Sources in Thevenin Equivalent Circuit Problems

EXAMPLE 7.21

Find the Thevenin equivalent circuit looking into terminals 1 and 2 of the source circuit of Fig. 7.19(a). The 1-MΩ resistor is considered the load.

Solution The first step in any Thevenin equivalent circuit problem is to draw the equivalent circuit. This is done in Fig. 7.19(b). Note that this circuit contains a dependent source. The output of this dependent source, however, depends on V_r, which is a voltage located in a different circuit from the dependent source. V_r must therefore be given, as it is in Fig. 7.19, in order to solve this problem.

Before finding E_{Th}, therefore, we simply calculate the value of the dependent source by multiplying V_r by $10\angle 0^\circ$ as indicated in the circuit diagram of Fig. 7.19(a). We then redraw the original circuit, removing the load in the normal manner as shown in Fig. 7.19(c), and find E_{Th} by calculating the voltage between terminals 1 and 2 of this circuit. By voltage division

$$E_{12} = (2 \times 10^6 - j3 \times 10^6)(120\angle 30^\circ)/(6 \times 10^6 - j3 \times 10^6) = 64.5\angle 0.2^\circ$$

Answer $E_{\text{Th}} = 64.5\angle 0.2^\circ$ V.

Z_{Th} may be found by method 1 because the voltage source in this circuit is dependent on a voltage in a ***different*** circuit. Accordingly, the original circuit is redrawn with the voltage source set to zero and the load removed as shown in Fig. 7.19(d). In this circuit:

$$\begin{aligned} Z_{\text{Th}} &= [(4 \times 10^6\angle 0^\circ)(2 \times 10^6 - j3 \times 10^6)/(6 \times 10^6 - j3 \times 10^6)] \\ &= 2.1\angle -29.7^\circ \end{aligned}$$

Answer $Z_{\text{Th}} = 2.1\angle -29.7^\circ\ \Omega$.

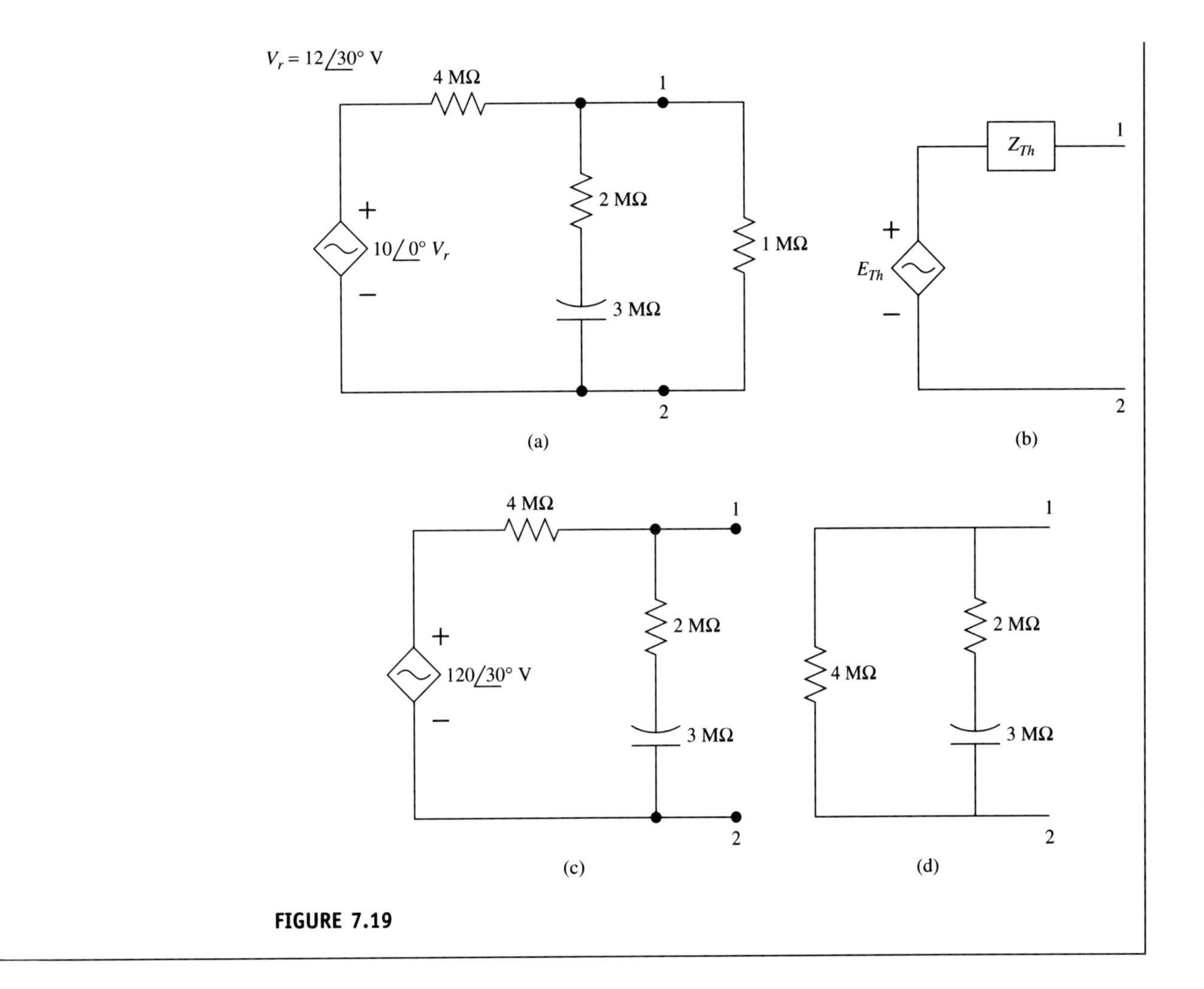

FIGURE 7.19

EXAMPLE 7.22

Find Thevenin's equivalent circuit for the source circuit between terminals a and b of Fig. 7.20(a) considering the R–L series circuit as the load.

Solution E_{Th} will be found in the usual manner (i.e., the open circuit voltage between terminals a and b). Since the output of the dependent voltage source depends on a current in the ***same*** circuit, however, Z_{Th} must be found by using method 2.

Thevenin's equivalent circuit is drawn in Fig. 7.20(b). We will find E_{Th} using the circuit of Fig. 7.20(c), which is the original circuit with the load removed.

$$E_{\text{Th}} = (V_{ab})_{oc} = V_o + (160\angle 0°)I_o = V_o + (160\angle 0°)(5\angle 30°)$$
$$= V_o + (800\angle 30°)$$

Using KVL:

$$(20I_o) - V_o - (240\angle 0°)I_o = 0$$

or

$$V_o = (20)(5\angle 30°) - (240\angle 0°)(5\angle 30°)$$
$$V_o = -1100\angle 30° \text{ V}$$

Substituting this value of V_o into the expression for E_{Th}:

$$E_{\text{Th}} = (-1100\angle 30°) + (800\angle 30°) = -300\angle 30°$$

Answer $E_{\text{Th}} = 300\angle -150°$ V.

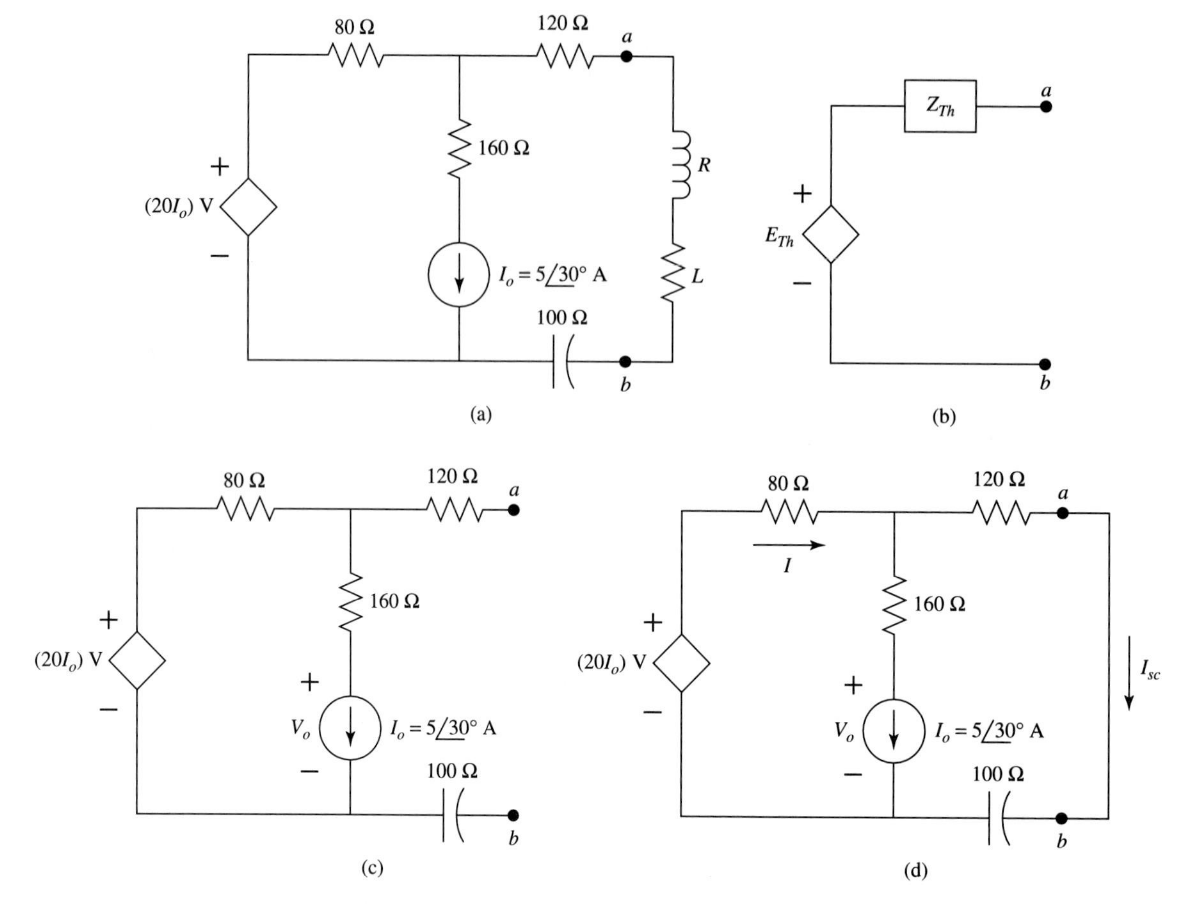

FIGURE 7.20

To use method 2 of finding Z_{Th}, the original circuit is drawn in Fig. 7.20(d) with a short placed across terminals a and b. This circuit may be solved using any of the procedures discussed previously (e.g., mesh analysis, nodal analysis, circuit laws, etc.). We will use circuit laws:

From KVL around the left loop:

$$(20I_o) - V_o - (160\angle 0^\circ\, I_o) - (80\angle 0^\circ)I = 0$$

or

$$(20)(5\angle 30^\circ) - V_o - (160\angle 0^\circ)(5\angle 30^\circ) - (80\angle 0^\circ\, I) = 0$$

$$V_o = (-700\angle 30^\circ) - 80\angle 0^\circ\, I \qquad \textbf{(Eq. 7.66)}$$

From KCL: $I = I_o + I_{sc} = (5\angle 30^\circ) + I_{sc}$.

Substituting this value of I into Equation 7.66 for V_o:

$$V_o = (-700\angle 30^\circ) - (80\angle 0^\circ)[(5\angle 30^\circ) + I_{sc}]$$

$$V_o = (-1100\angle 30^\circ) - (80\angle 0^\circ)\, I_{sc} \qquad \textbf{(Eq. 7.67)}$$

From KVL around the right loop:

$$V_o - (100\angle -90^\circ)I_{sc} - (120\angle 0^\circ)I_{sc} + (160\angle 0^\circ)(5\angle 30^\circ) = 0$$

$$V_o = (156.2\angle -39.8^\circ)I_{sc} - (800\angle 30^\circ)$$

Substituting this expression of V_o into Equation 7.68:

$$(156.2\angle -39.8^\circ)I_{sc} - (800\angle 30^\circ) = (-1100\angle 30^\circ) - (80\angle 0^\circ)I_{sc}$$

Solving for I_{sc}:

$$I_{sc} = 1.34\angle -123.4^\circ \text{ A}$$

From method 2:

$$Z_{Th} = E_{oc}/I_{sc} = E_{Th}/I_{sc} = (300\angle -150^\circ)/(1.34\angle -123.4^\circ) = 223.9\angle -26.6^\circ$$

Answer $Z_{Th} = 223.9\angle -26.6^\circ\ \Omega$.

7.8.2 Dependent Sources in Norton's Equivalent Circuit

Circuit problems involving Norton's equivalent circuit are handled in the same way as those involving Thevenin's equivalent circuit. Since $Z_n = Z_{Th}$, the Norton impedance is found using either method 1 or method 2 depending upon whether or note the dependent sources are functions of parameters in the same circuit or in different circuits.

The Norton current, I_n, is of course equal to the short circuit current between the two terminals under consideration (I_{sc}). It must be found as part of the Norton equivalent circuit.

7.8.3 Dependent Sources in Mesh and Modal Analysis

When dependent sources appear in circuits that are to be solved using mesh or nodal analysis, the techniques employed are very similar to those used when only independent sources are included in the circuit.

A. If the dependent sources are functions of parameters located in ***different*** circuits, then the values of the dependent sources are first determined and then the mesh or nodal analysis is performed exactly as in the case of independent sources.

B. If the dependent sources are functions of parameters located in the ***same*** circuit, then:

1. The mesh or nodal equations are first set up in the same manner as in the case of independent sources.
2. The dependent quantities are then expressed as functions of the mesh currents (or nodal voltages). The equations can then be rearranged so that they appear as usual with the mesh currents (or nodal voltages) as the unknowns. These equations are then solved in the usual manner using determinants.

■ SUMMARY

- Mesh (or nodal) analysis may be used to solve a circuit that consists of two or more meshes (or nodes). To use mesh analysis, all sources in the circuit must be voltages. To use nodal analysis, all sources in the circuit must be currents.
- The method of determinants is the preferred method of solving the simultaneous equations that result from performing mesh or nodal analysis of a circuit, especially when three or more mesh currents (or nodal voltages) are involved.
- The Superposition Theorem allows us to solve a circuit containing more than one voltage and/or current source by finding the response to each source individually (acting alone) and then adding the individual responses to determine the final result. In finding the response due to one source, all the other sources in the circuit are set to zero. (Voltage sources are replaced by shorts and current sources are replaced by opens.)
- Thevenin's Theorem allows us to replace a relatively complicated circuit by a simple series circuit consisting of a voltage source (known as the Thevenin voltage) and a single impedance (known as the Thevenin impedance).
- The advantage of using Thevenin's equivalent circuit is that load voltage and load current may be much more easily calculated using the equivalent circuit rather than using the more complicated original circuit. This is especially true when the load changes and its new voltage and/or current must be recalculated several times.
- Norton's equivalent circuit is the current source equivalent of Thevenin's equivalent circuit. It consists of a current source in parallel with an impedance. The Norton impedance is equal to the Thevenin impedance and the Norton current is equal to the Thevenin voltage divided by the Thevenin impedance. Either equivalent circuit may be used in any given problem, and it is generally personal preference that determines which one is actually used.
- Dependent sources are voltage and current sources whose values depend on other voltages or currents. Modifications to the analysis techniques used when independent sources only are included in the circuit are necessary. The modifications depend on whether the dependent sources are functions of other voltages or currents located in the same circuit as the dependent sources or in different circuits.

■ PROBLEMS

7.1 (a) Write the mesh equations for the circuit of Fig. P7.1.
(b) Solve for the two mesh currents.

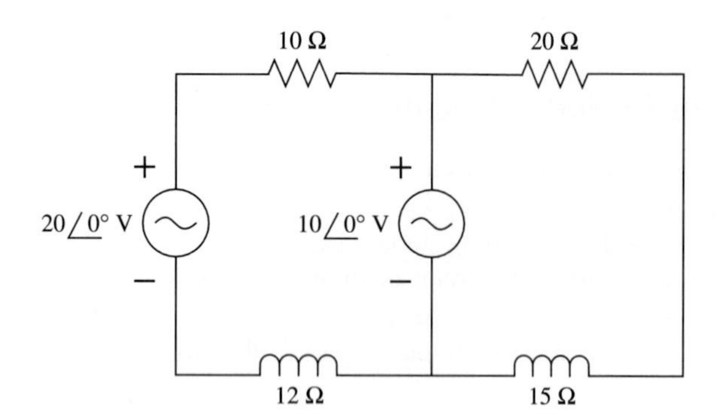

FIGURE P7.1

7.2 (a) Write the mesh equations for the circuit of Fig. P7.2.
(b) Solve for the two mesh currents.

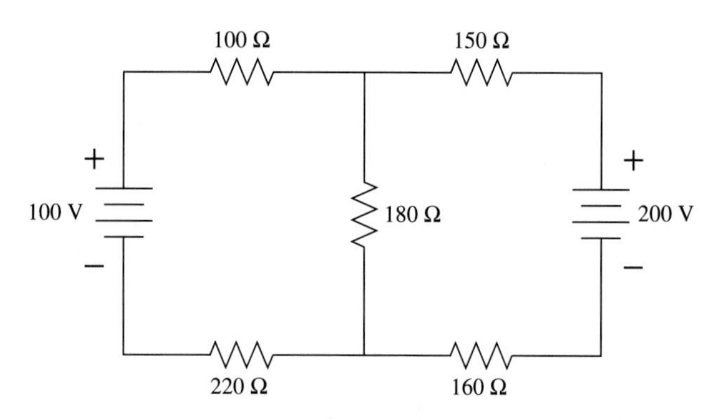

FIGURE P7.2

7.3 (a) Write the mesh equations for the circuit of Fig. P7.3.
(b) Solve for the two mesh currents.
(c) Find the current I_a.
(d) Find the current I_c.
(e) Find the current I_x.

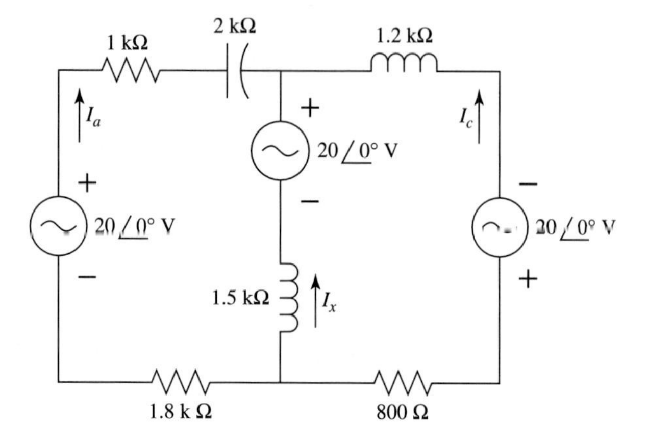

FIGURE P7.3

7.4 (a) Write the mesh equations for the circuit of Fig. P7.4.
(b) Solve for the two mesh currents.
(c) Find the voltage V_x.

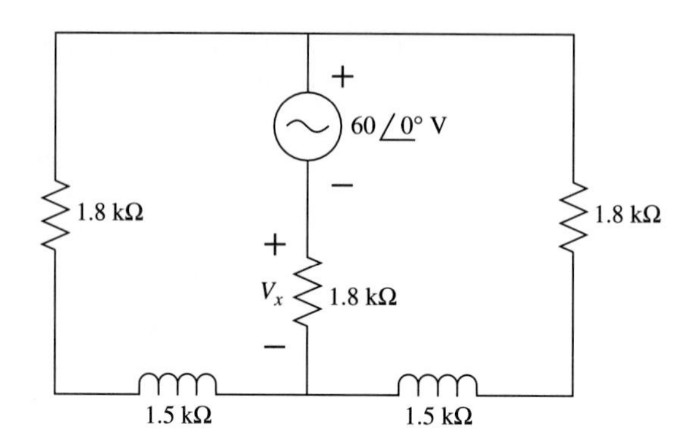

FIGURE P7.4

7.5 (a) Write the mesh equations for the circuit of Fig. P7.5.
(b) Solve for the two mesh currents.
(c) Find the current I_x.

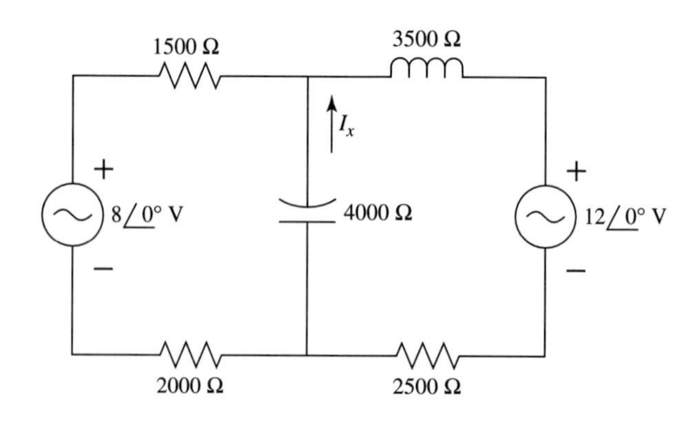

FIGURE P7.5

7.6 (a) Write the mesh equations for the circuit of Fig. P7.6.
(b) Find the three mesh currents.
(c) Find the current I_x.

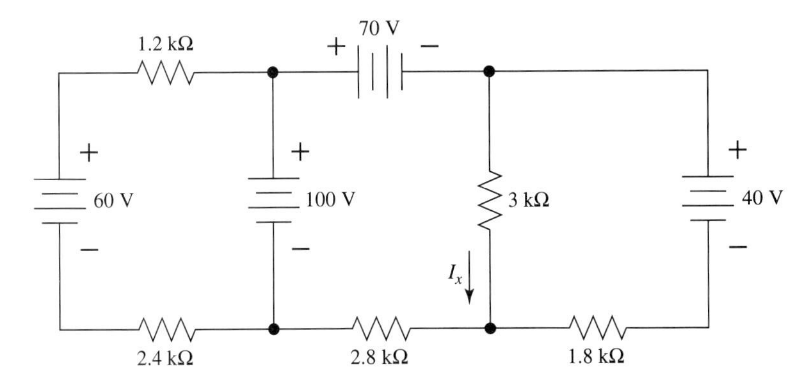

FIGURE P7.6

7.7 **(a)** Write the mesh equations for the circuit of Fig. P7.7.
(b) Find the three mesh currents.
(c) Find the voltage V_x.

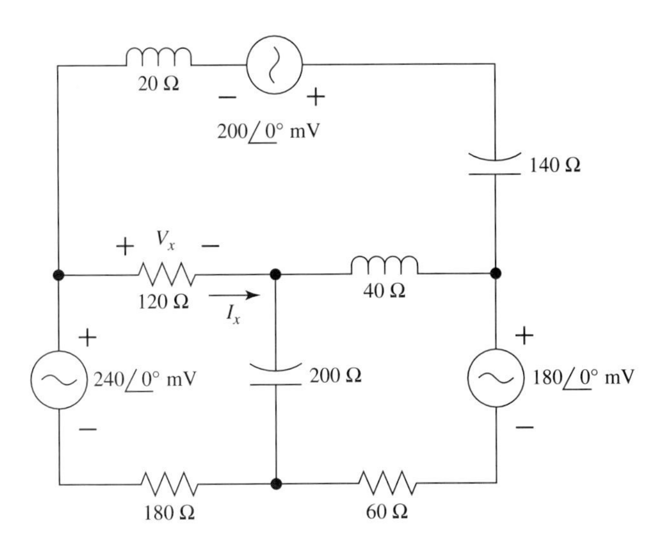

FIGURE P7.7

7.8 **(a)** Write the nodal equations for the circuit of Fig. P7.8.
(b) Solve for the two nodal voltages.

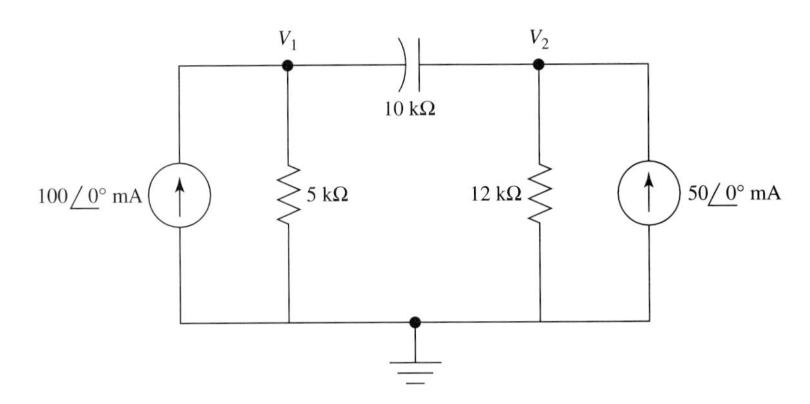

FIGURE P7.8

7.9 **(a)** Write the nodal equations for the circuit of Fig. P7.9.
(b) Solve for the two nodal voltages.
(c) Find the current I_a.
(d) Find the current I_c.

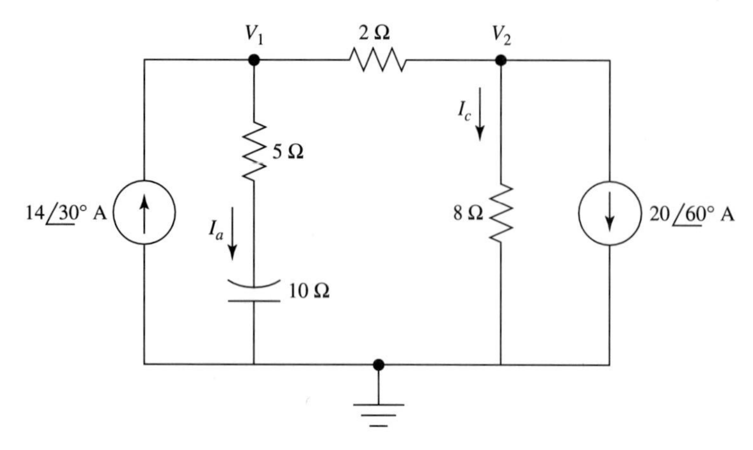

FIGURE P7.9

7.10 **(a)** Write the nodal equations for the circuit of Fig. P7.10.
(b) Solve for the two nodal voltages.
(c) Find the current I_a.

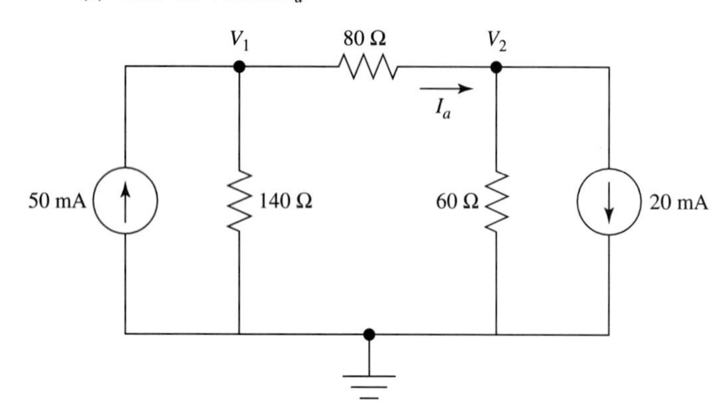

FIGURE P7.10

7.11 **(a)** Write the nodal equations for the circuit of Fig. P7.11.
(b) Solve for the two nodal voltages.
(c) Find the voltage V_x.

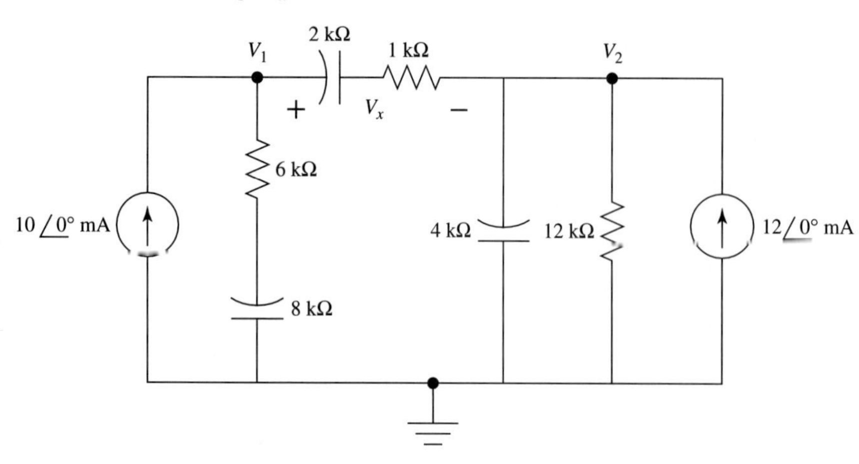

FIGURE P7.11

7.12 **(a)** Write the nodal equations for the circuit of Fig. P7.12.
(b) Solve for the two nodal voltages.
(c) Find the current I_x.
(d) Find the voltage V_a.

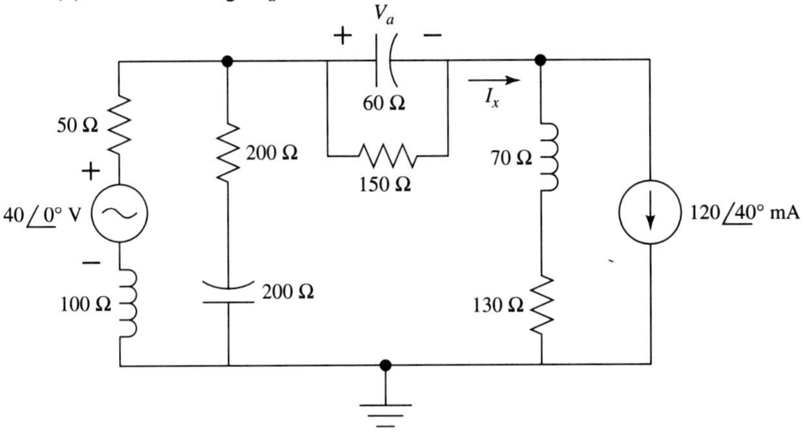

FIGURE P7.12

7.13 **(a)** Write the nodal equations for the circuit of Fig. P7.13.
(b) Solve for the three nodal voltages.
(c) Find the current I_x.

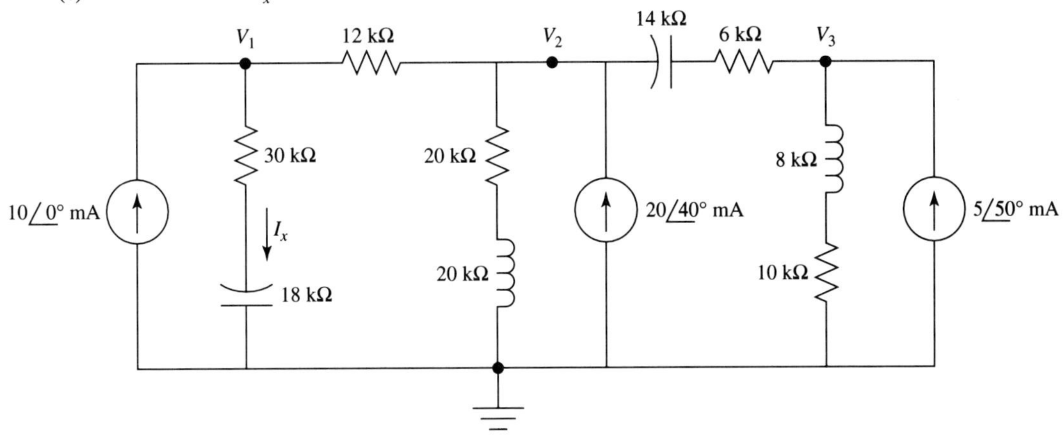

FIGURE P7.13

7.14 **(a)** Write the nodal equations for the circuit of Fig. P7.14.
(b) Solve for the three nodal voltages.
(c) Find the voltage V_x.

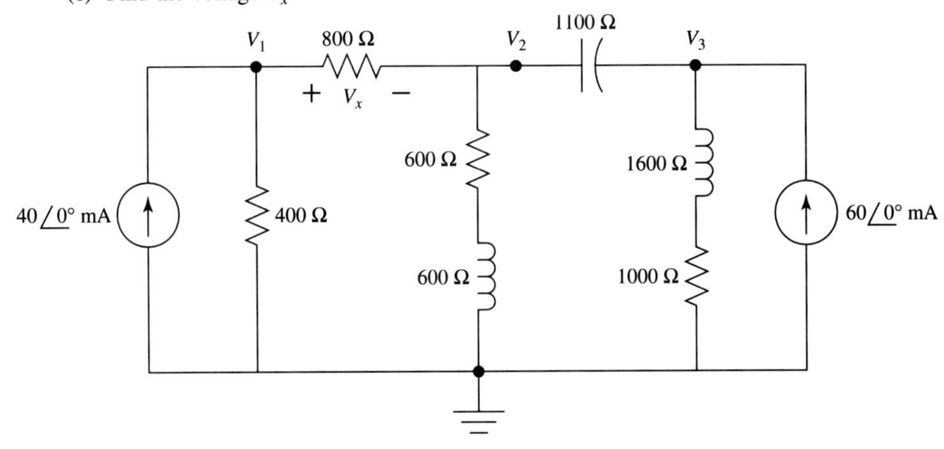

FIGURE P7.14

7.15 **(a)** Write the nodal equations for the circuit of Fig. P7.15.
(b) Solve for the three nodal voltages.
(c) Find the current I_x.

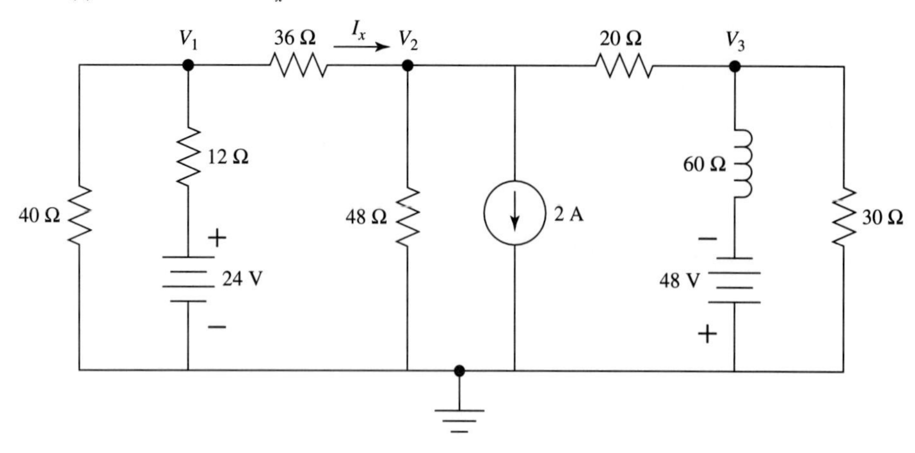

FIGURE P7.15

7.16 **(a)** Write the nodal equations for the circuit of Fig. P7.3.
(b) Find the currents I_a, I_c, and I_x, and compare these values with the results of problem 7.3(c), (d), and (e).

7.17 Find I_x in the circuit of Fig. P7.5 using nodal analysis and compare the result with that obtained in problem 7.5.

7.18 Find the phasor current I_x in the circuit of Fig. P7.18 using the principle of superposition.

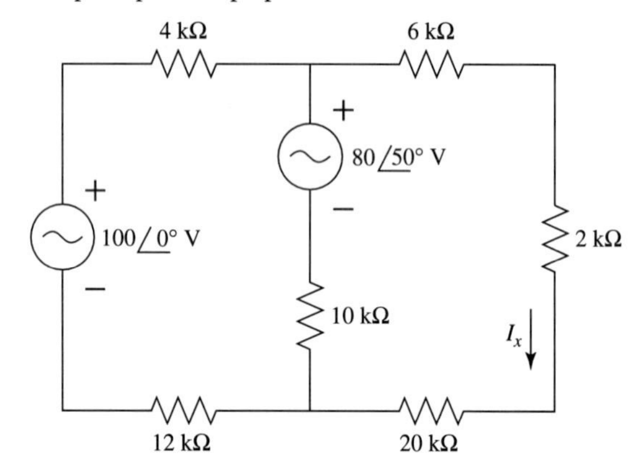

FIGURE P7.18

7.19 Find the phasor voltage V_x in the circuit of Fig. P7.19 using the principle of superposition.

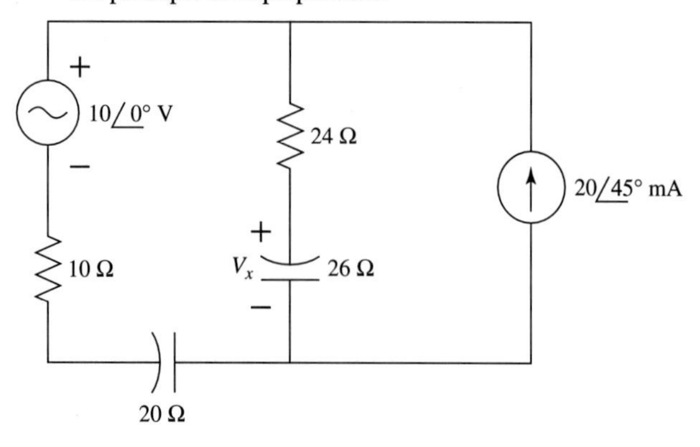

FIGURE P7.19

7.20 Use the principle of superposition to find the phasor current I_c in the circuit of Fig. P7.20.

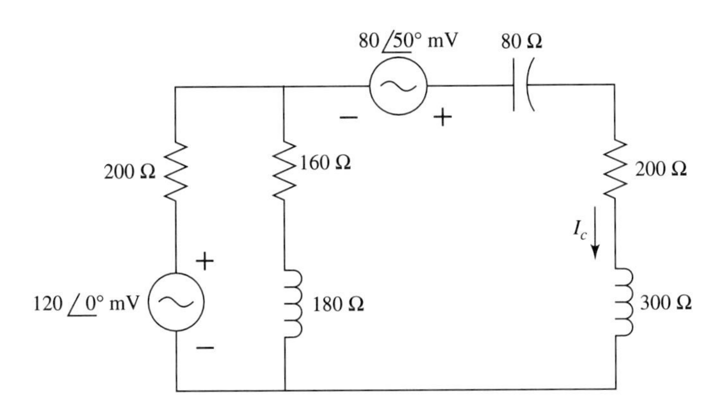

FIGURE P7.20

7.21 Use the principle of superposition to find the current I_x in the circuit of Fig. P7.21.

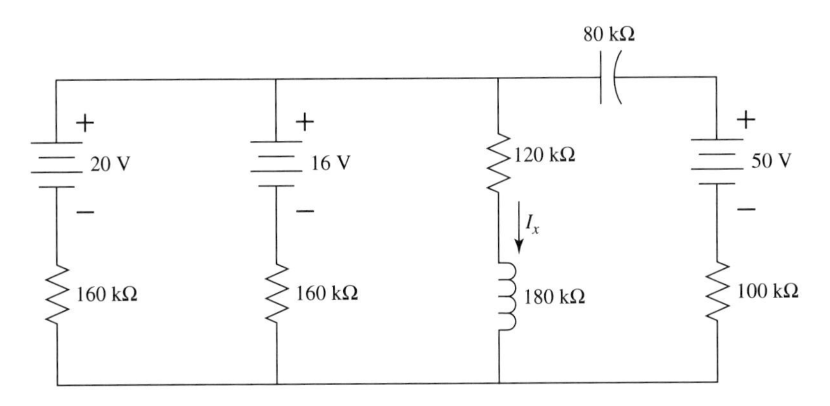

FIGURE P7.21

7.22 Use the principle of superposition to find the sinusoidal voltage across the .04-μF capacitor of Fig. P7.22.

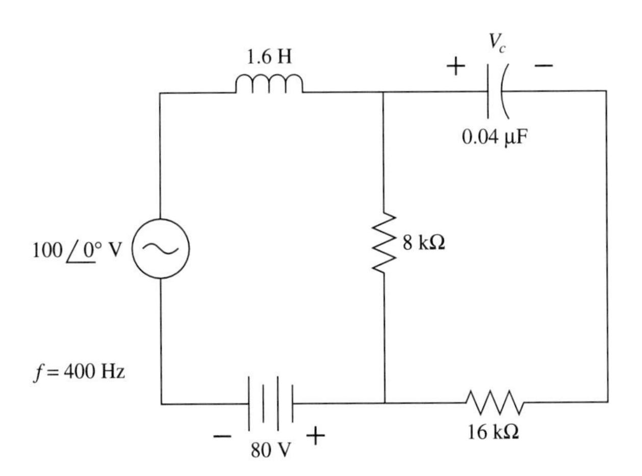

FIGURE P7.22

7.23 Find the sinusoidal current through the 100-mH inductor of Fig. P7.23 using the principle of superposition.

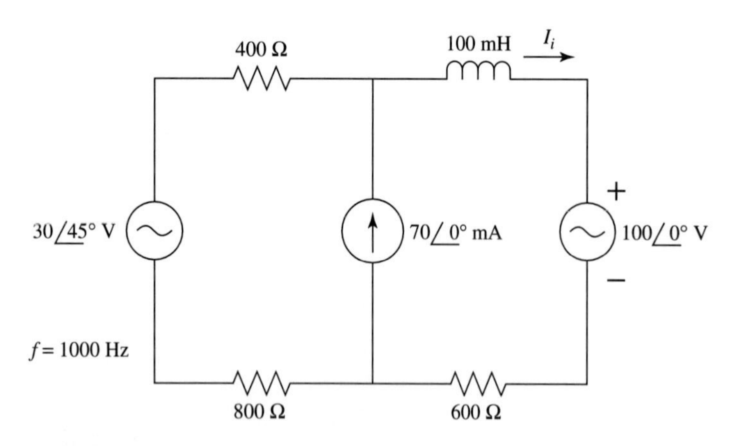

FIGURE P7.23

7.24 Use the principle of superposition to find the sinusoidal voltage V_x in the circuit of Fig. P7.24.

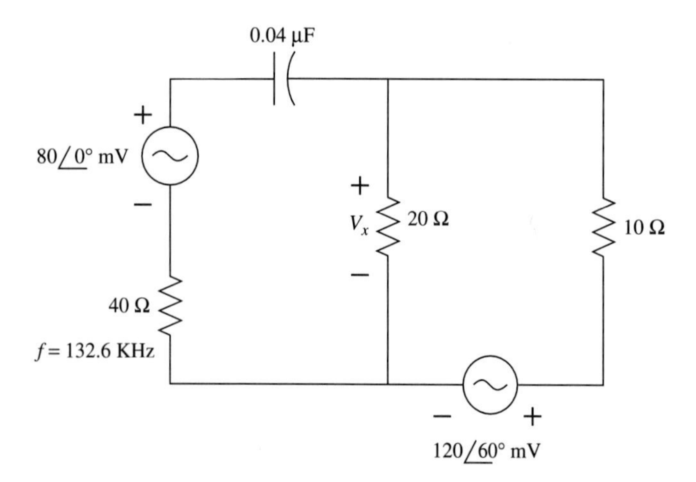

FIGURE P7.24

7.25 Use the principle of superposition to find the sinusoidal current I_x in the circuit of Fig. P7.25.

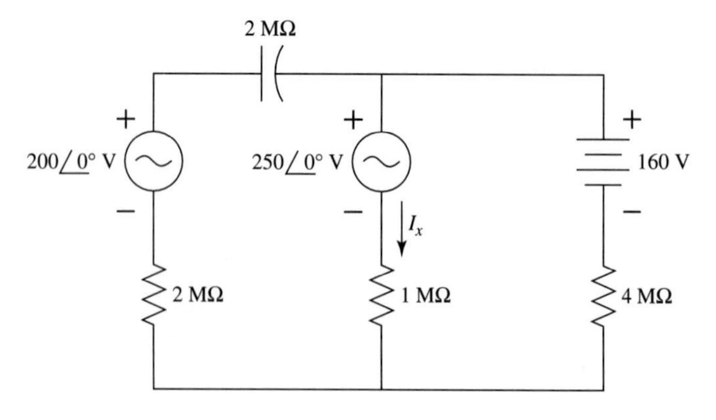

FIGURE P7.25

7.26 Find the voltage V_a in the circuit of Fig. P7.26 using the principle of superposition.

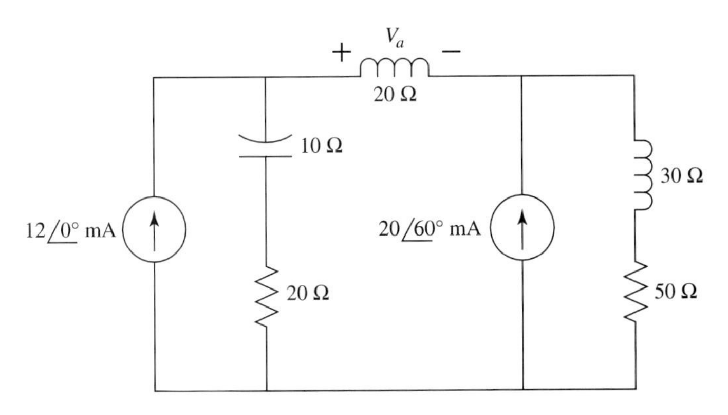

FIGURE P7.26

7.27 Find Thevenin's equivalent circuit for the source circuit of Fig. P7.27 between terminals *a* and *b* considering the 200-Ω resistor as the load.

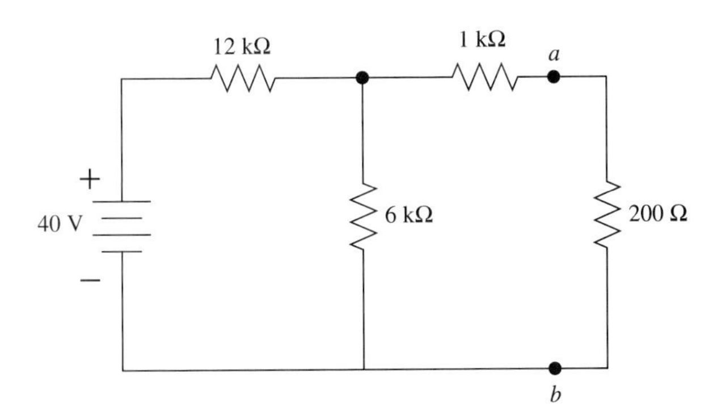

FIGURE P7.27

7.28 Find Thevenin's equivalent circuit for the source circuit of Fig. P7.28 between terminals *a* and *b* considering the unknown capacitor as the load.

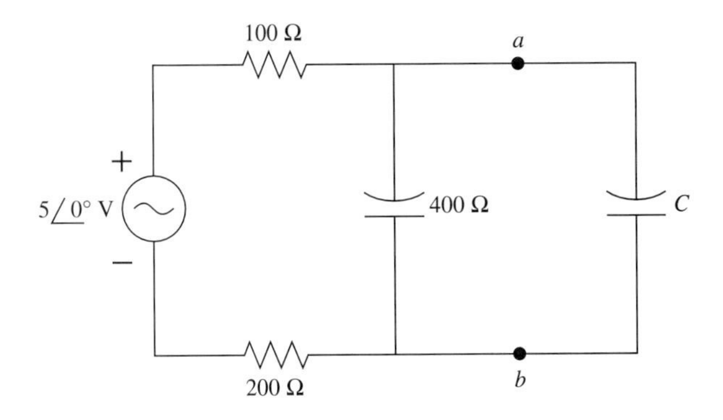

FIGURE P7.28

7.29 **(a)** Find Thevenin's equivalent circuit for the source circuit of Fig. P7.29 between terminals 1 and 2 considering the series R–L circuit as the load.

(b) Find the phasor current through the load, I_a, using the equivalent circuit found in part (a).

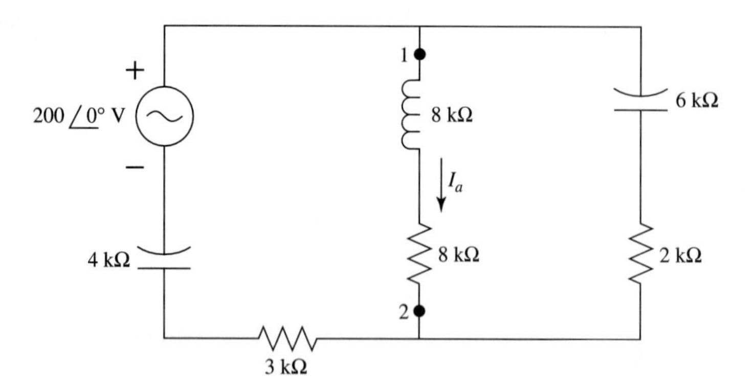

FIGURE P7.29

7.30 Use Thevenin's equivalent circuit to find the phasor voltage across the 0.02-μF load capacitor of Fig. P7.30.

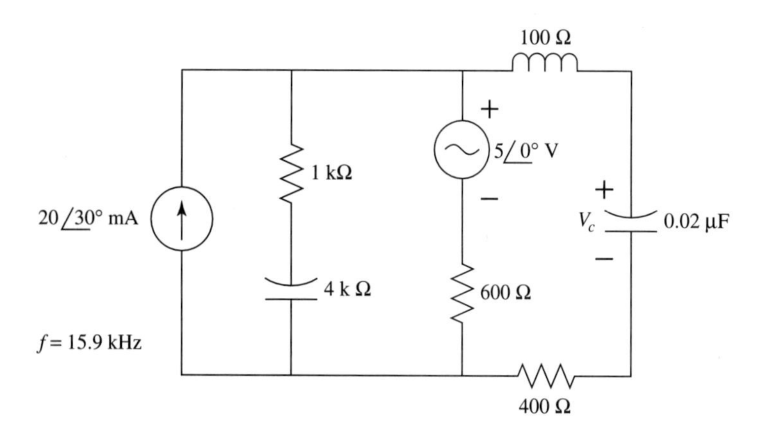

FIGURE P7.30

7.31 Use Thevenin's equivalent circuit to find the current through the 10-kΩ load resistor of Fig. P7.31.

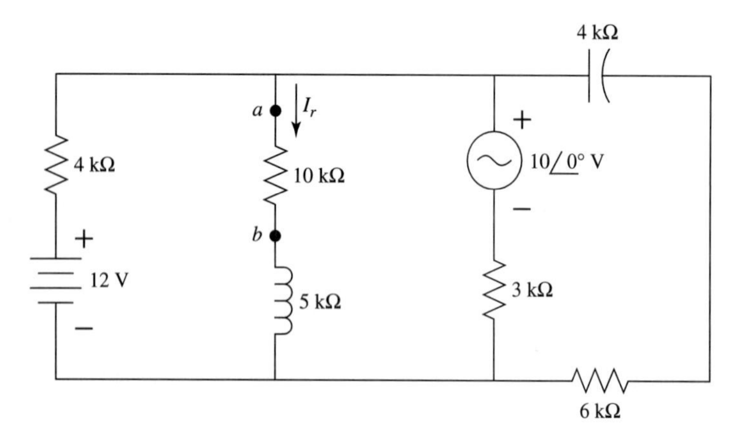

FIGURE P7.31

7.32 Use Thevenin's equivalent circuit to find the sinusoidal voltage across the load in Fig. P7.32. The series combination of the 4-kΩ resistor and the 0.6 μF capacitor is the load in this circuit.

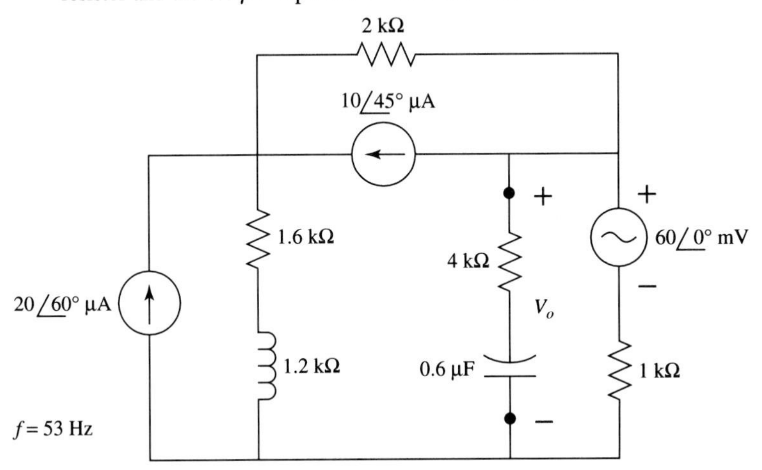

FIGURE P7.32

7.33 Find the voltage across the 400-Ω capacitive reactance load in the circuit of Fig. P7.33 considering the rest of the circuit between terminals a and b as the source circuit. Use Thevenin's equivalent circuit in your solution.

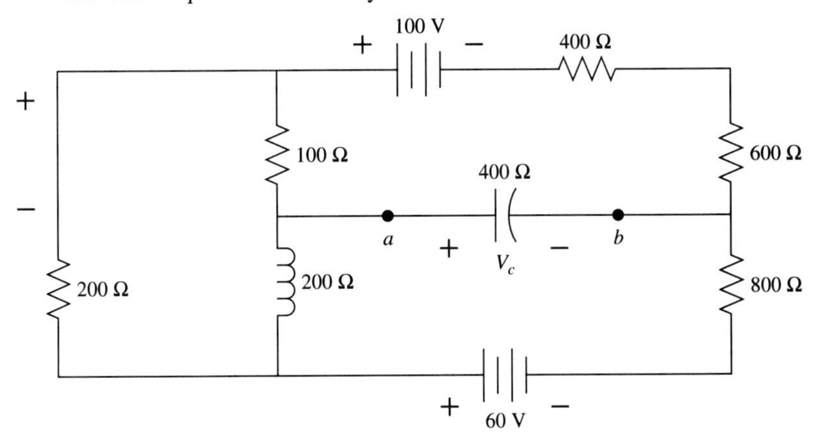

FIGURE P7.33

7.34 Use Norton's equivalent circuit with respect to terminals 1 and 2 of Fig. P7.34 to find the current through the 400-Ω load resistor.

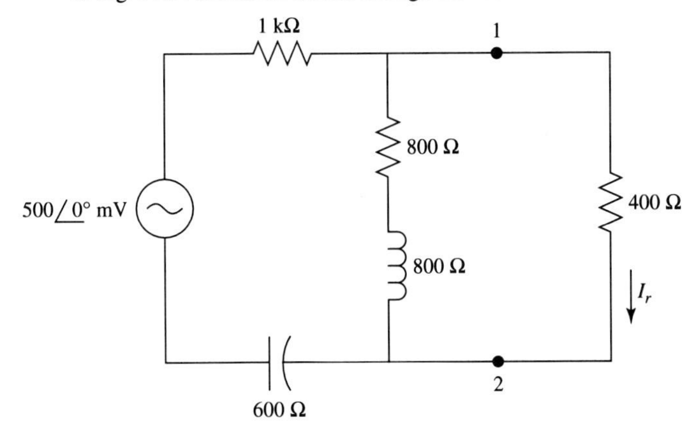

FIGURE P7.34

7.35 Find the voltage across the 4-kΩ capacitive reactance load in the circuit of Fig. P7.35 using Norton's equivalent circuit.

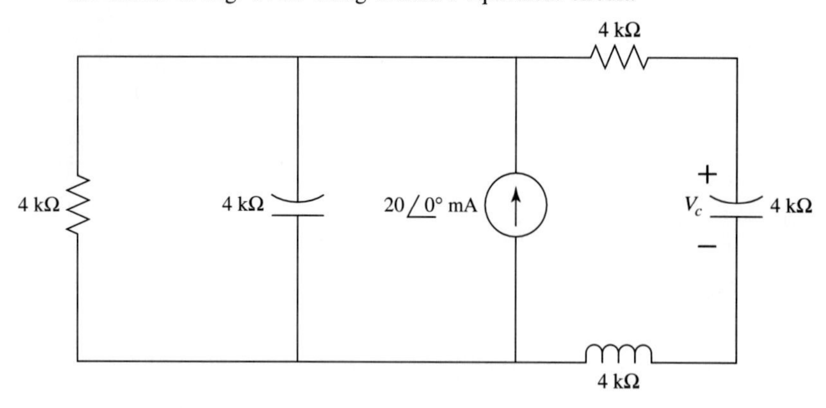

FIGURE P7.35

7.36 Repeat problem 7.27 using Norton's equivalent circuit. Compare your answer with that obtained in problem 7.27.

7.37 Find Norton's equivalent circuit for the circuit of Fig. P7.37 between terminals *a* and *b*.

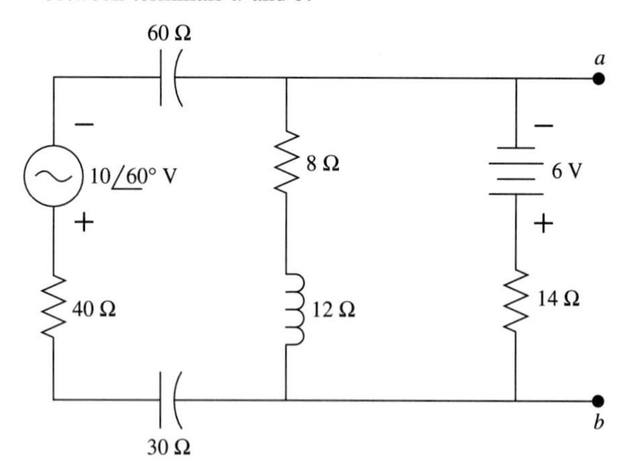

FIGURE P7.37

7.38 Find the voltage across the load in Fig. P7.32 using Norton's equivalent circuit. Find Norton's equivalent circuit using the results of problem 7.32.

7.39 Use Norton's equivalent circuit to find the current through the series branch load consisting of the 600-Ω resistor and 500-Ω inductive reactance in the circuit of Fig. P7.39.

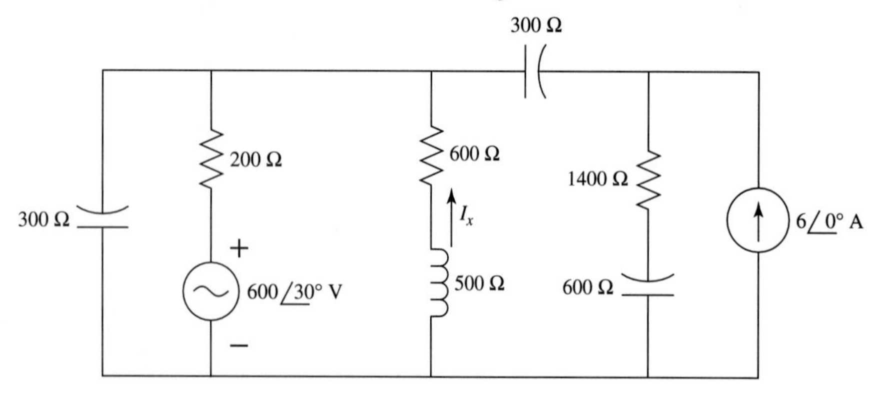

FIGURE P7.39

7.40 State the primary advantage of Thevenin's and Norton's equivalent circuits.

7.41 Write the mesh equations for the two-mesh circuit of Fig. P7.41 such that the mesh currents are the only unknowns that appear in these equations.

$I_x = 20\angle 0°$ mA
10 Ω
30 Ω
20 Ω
+
$10I_x$ V
−
100∠30° mV
60 Ω
40 Ω
20 Ω

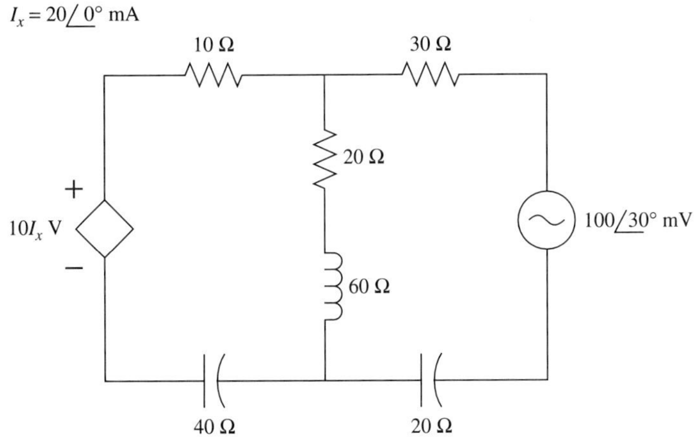

FIGURE P7.41

7.42 Write the mesh equations for the two-mesh circuit of Fig. P7.42 such that the mesh currents are the only unknowns that appear in these equations.

10 Ω
30 Ω
20 Ω
+
I_x
$10I_x$ V
−
100∠30° mV
60 Ω
40 Ω
20 Ω

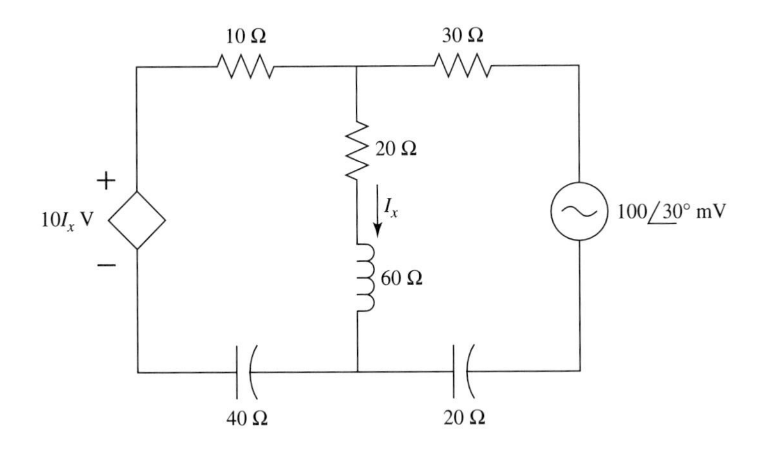

FIGURE P7.42

7.43 Find Thevenin's equivalent circuit for the source circuit of Fig. P7.43 between terminals *a* and *b* considering the 10-kΩ resistor as the load.

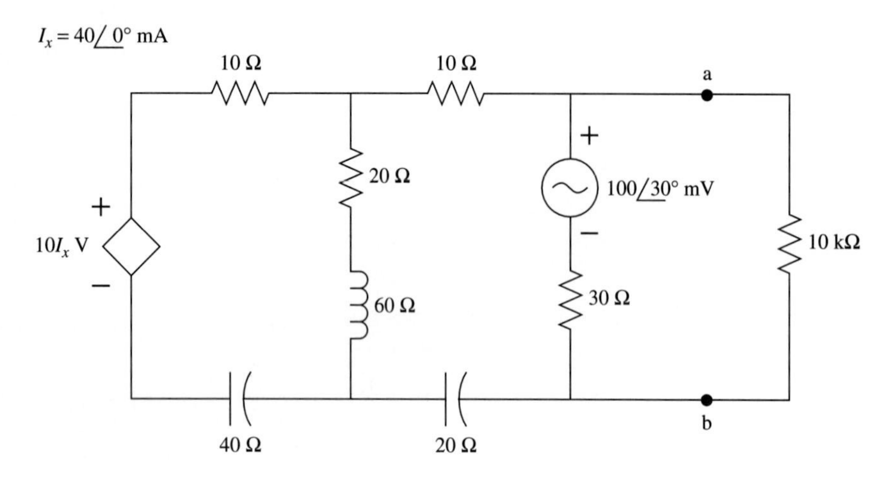

FIGURE P7.43

7.44 Find Thevenin's equivalent circuit for the source circuit of Fig. P7.44 between terminals 1 and 2 considering the 1 μF capacitor as the load.

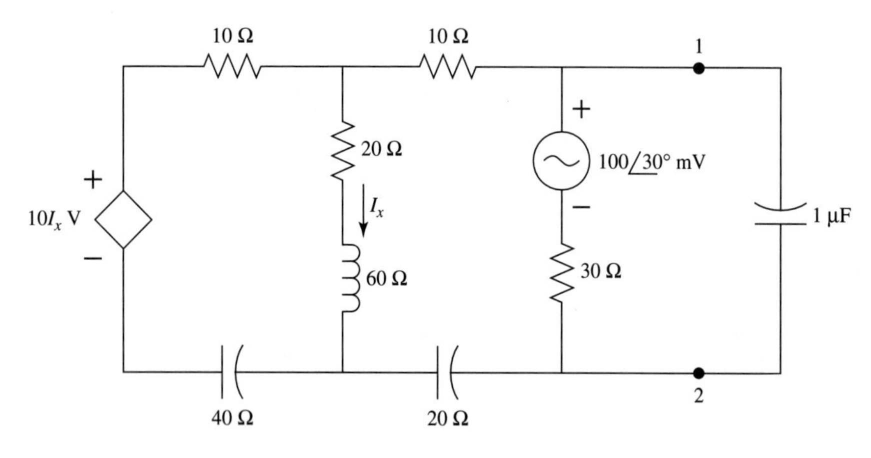

FIGURE P7.44

CHAPTER 8

FREQUENCY RESPONSE OF COMMON CIRCUITS

INTRODUCTION

In Chapter 1 we discussed the use of exponents. Exponents are a shorthand way that mathematicians use for writing repeated factors in a multiplication problem. Through the use of exponents and scientific notation we are able to simplify the work needed to solve a multiplication problem. In scientific notation we limit ourselves to integer values (both positive and negative) for exponents of the number 10. We will now generalize the concept of exponents to include the concept of logarithms.

8.1 PROPERTIES OF EXPONENTS

8.1.1 Logarithms

Consider the function $Y(x) = 10^x$, where x is a real number that can have any value: negative, positive, or zero. This mathematical equation, called an exponential function, has the property that ***for any given value of x the corresponding value of Y is always real, positive, and greater than zero.*** It is a single-valued function of x. This means that ***every value of x produces a unique value of Y. Different values of x produce different values of Y.*** This equation is plotted in Fig. 8.1 for values of x between 0 and 1.

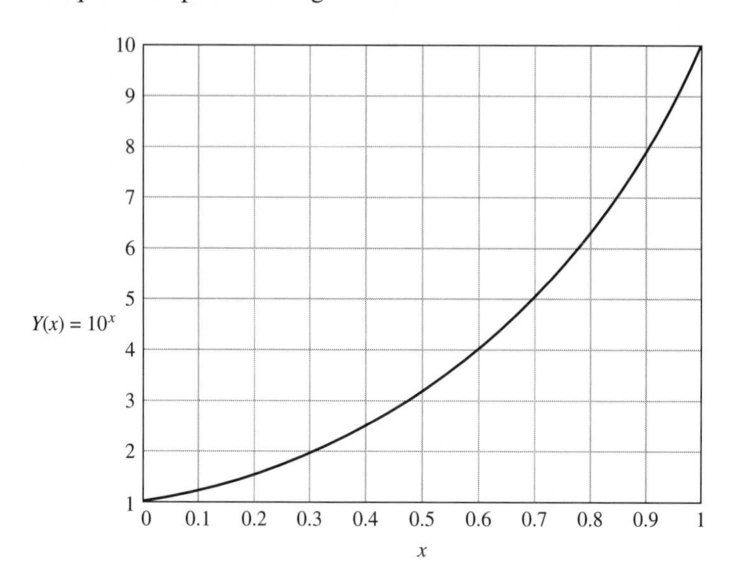

FIGURE 8.1
$Y(x) = 10^x$

Examination of this plot reveals that as x goes from 0 to 1, the value of 10^x takes on all values between 1 and 10. If we consider this plot together with the concept of scientific notation, in which any number is expressible as a number between 1 and 10 times 10 to some power, we observe that ***every positive number can be expressed as being equal to 10 to some power.*** Thus, for ***every*** positive value of ***Y*** it follows that:

$$Y = 10^x \qquad \textbf{(Eq. 8.1)}$$

expresses any number in exponential form. The exponent x is called the logarithm of the number Y to the base 10. This is written mathematically as:

$$x = \log_{10}(Y) \qquad \textbf{(Eq. 8.2)}$$

The subscript 10 reminds us that 10 is the base of the logarithm. That is, 10 is the number that when raised to the power x equals the number Y. Logarithms with the base 10 are called common logarithms.

Logarithms are not limited to the base 10. Any number greater than 1 can be used for the base of a logarithm system, because any number raised to a sufficiently high power can be used to represent another number. One value commonly used for the base of a logarithmic system is e, which is a transcendental number equal to 2.71828. . . . Very often in circuit analysis, powers of e are used to describe the response of a circuit to a dc input voltage. These logarithms, which obey all of the same rules that base 10 logarithms obey, are very important in circuit analysis. They are called natural logarithms.

The concept of a logarithm (which is the same as an exponent) allows us to convert multiplication and division of numbers into a corresponding addition and subtraction of logarithms representing those numbers. The only requirement for using these logarithms is to have a table of logarithms available. As was in the case of the trigonometric functions, which were defined and introduced in Chapter 3, modern calculators have eliminated the need for trigonometric tables. They likewise have replaced old-fashioned logarithm tables with buttons on a calculator. The operation of finding a logarithm has been reduced to pushing a button marked ***log*** on a calculator for base 10 logarithms and a button marked ***ln*** for natural logarithms.

The following rules summarize some important properties of logarithms. The letter β in the equations can be any positive number greater than 1 that is used as the base of the logarithm. It may represent 10 or e or any number greater than 1.

Rules for Logarithms

$$\log_\beta(1) = 0 \quad \textbf{(Eq. 8.3)}$$

$$\log_\beta(\beta) = 1 \quad \textbf{(Eq. 8.4)}$$

$$\log_{10}(10) = 1 \quad \textbf{(Eq. 8.4a)}$$

$$\log_e(e) = 1 \quad \textbf{(Eq. 8.4b)}$$

$$\log_\beta(\beta^x) = x \quad \textbf{(Eq. 8.5)}$$

$$\log(a \cdot b) = \log a + \log b \quad \textbf{(Eq. 8.6)}$$

$$\log(a/b) = \log(a) - \log(b) \quad \textbf{(Eq. 8.7)}$$

$$\log a^x = x \log a \quad \textbf{(Eq. 8.8)}$$

We see from Equation 8.3 and Fig. 8.1 that the logarithm of 1 is 0. Equation 8.3 is true regardless of the base value we choose. For most of our work we will use *10* as the base value. With this choice we observe that $10^0 = 1$ and 10^1 equals 10. From this it follows that any number greater than 1 and less than 10 must have a logarithm greater than 0 and less than 1 (i.e., a decimal fraction). If we apply Equation 8.6 to a number represented in scientific notation, it shows that the logarithm of any number has an integer part (called a characteristic) and a decimal part (called a mantissa). The mantissa represents the value of the number and the integer part represents the factor of 10 that multiplies it when the number is expressed in scientific notation.

EXAMPLE 8.1

Find the logarithm to the base 10 of 3, 30, 300, and 3000.

Solution These numbers can be written as $3 = 3 \times 10^0$, 3×10^1, 3×10^2, and 3×10^3 respectively. We can use the calculator to find the logarithm of each of these numbers or just the mantissa. The calculator gives us the value of the mantissa as:

$$\log 3 = 0.4771212$$

and using Equations 8.5, 8.6 and 8.8 the log of 30 becomes:

$$\log 30 = \log(3 \cdot 10) = \log 3 + \log 10 = 0.4771212 + 1 = 1.4771212$$

Similarly,

$$\log 300 = \log(3 \cdot 10^2) = \log 3 + 2 \log 10 = 0.4771212 + 2 = 2.4771212$$

and

$$\log 3000 = \log(3 \cdot 10^3) = \log 3 + 3 \log 10 = 0.4771212 + 3 = 3.4771212$$

Verify these results using your own calculator. Note that the calculator displays both the characteristic and the mantissa at the same time.

EXAMPLE 8.2

Find the logarithm to the base 10 of 5, 0.5, 0.05, and 0.0005.

Solution Each of these numbers can be written in scientific notation. Thus:

$$5 = 5 \times 10^{0} \qquad 0.5 = 5 \times 10^{-1} \qquad 0.05 = 5 \times 10^{-2} \qquad 0.0005 = 5 \times 10^{-4}$$

From our calculator we find:

$$\log 5 = 0.69897$$

thus we can find the others as:

$$\log 0.5 = \log(5 \cdot 10^{-1}) = \log 5 - \log 10 = 0.69897 - 1 = -0.30103$$
$$\log 0.05 = \log(5 \cdot 10^{-2}) = \log 5 - 2 \log 10 = 0.69897 - 2 = -1.30103$$
$$\log 0.0005 = \log(5 \cdot 10^{-4}) = \log 5 - 4 \log 10 = 0.69897 - 4 = -3.30103$$

Once again verify the results using your calculator to observe that it does again yield the quoted results.

Although we can find the value of each of these numbers directly from the calculator, Examples 8.1 and 8.2 demonstrate the use of Equations 8.5, 8.6, and 8.8. From these examples we note that for common logarithms (logarithms to the base 10):

1. Numbers between 1 and 10 have logarithms between 0 and 1.
2. Numbers greater than 10 have logarithms with positive whole number (characteristic) and decimal parts (mantissa).
3. Numbers less than 1 and greater than 0 have negative logarithms.

EXAMPLE 8.3

Using logarithms find the value of 342·267

Solution Equation 8.6 can be applied to this problem. From Equation 8.6 we find that

$$\begin{aligned} \log(342 \cdot 267) &= \log(342) + \log(267) \\ \log(342) &= 2.534026106 \\ \log(267) &= 2.426511261 \\ \log(342) + \log(267) &= 4.960537367 \end{aligned}$$

Using Equation 8.1 to find the value of Y we see that:

$$Y = 10^{4.960537367} = 91314$$

EXAMPLE 8.4

Using logarithms find the value for 125/50

Solution

$$\begin{aligned} \log(125) &= 2.096910 \\ \log(50) &= 1.698970 \end{aligned}$$

Using Equation 8.7 we find:

$$\begin{aligned} \log(125/50) &= \log(125) - \log(50) \\ \log(125/50) &= 2.096910 - 1.698970 \\ &= 0.3980 \end{aligned}$$

From Equation 8.1 we find:

$$Y = 10^{0.3980} = 2.5$$

8.2 USE OF LOGARITHMS IN POWER MEASUREMENT: DECIBELS

Logarithms find many uses in the calculation of circuit properties as a result of their ability to represent a large range of values in a simple way. One such use is in the calculation of the power gain of a circuit or electronic system. We define the power gain of a circuit or system as:

$$\text{Power Gain} = \frac{\text{Power output}}{\text{Power input}} \qquad \textbf{(Eq. 8.9)}$$

If the output power is smaller than the input power then the circuit is said to have a power loss or a power gain of less than 1. Consider a radio transmitter and receiver as an example of an electronic system. The transmitter broadcasts many watts of power, but any individual receiver gets at its input terminals only microwatts of that power. Clearly, in this case, the power of the transmitter delivered to an individual receiver is a very small part of the total power it sends out. The power loss for this system is significant, and the power gain is therefore much smaller than 1. On the other hand the power taken from the transmitter microphone that the broadcaster uses is many times smaller than the output power of the transmitter. In this case the ratio may be tens of thousands times the input power. The wide range of possible ratios in describing such systems suggests that we use a logarithmic scale to express results of power measurement gains.

We define power gain, expressed in units called bels, as the logarithm of the output power divided by the input power. Thus, with this definition:

$$\text{Power Gain} = \log_{10} \frac{\text{Power output}}{\text{Power input}} \qquad \textbf{(Eq. 8.10)}$$

A power ratio of 10 : 1 results in a power gain of 1 bel, because the log of 10 is 1. A power ratio of 100 : 1 is 2 bels. For most circuits a bel (named in honor of Alexander Graham Bell) is too large a unit to deal with. We therefore define a smaller unit called the decibel. By definition we assign:

$$\text{10 decibels (abbreviated 10 dB)} = \text{1 bel} \qquad \textbf{(Eq. 8.11)}$$

from which it follows:

$$\text{Power Gain (dB)} = 10 \log_{10} \frac{\text{Power output}}{\text{Power input}} \qquad \textbf{(Eq. 8.12)}$$

With this definition we see that a power ratio of 10 : 1 is described as a power gain of 10 decibels and a power ratio of 100 : 1 is described as a power gain of 20 decibels.

EXAMPLE 8.5

A 50-kW transmitter delivers 100 μW of power to a receiver. How many dB loss does this represent?

Solution

$$\begin{aligned} \text{Power Gain} &= 10 \log(100\ \mu\text{W}/50\ \text{kW}) \\ &= 10 \log(2 \times 10^{-9}) = -87\ \text{dB} \end{aligned}$$

This represents a power loss of 87 dB.

EXAMPLE 8.6

How many dB does a power ratio of 2 : 1 represent? How about a power ratio of 200 : 1?

Solution A power ratio of 2 : 1 represents $10 \log_{10}(2) = 3$ dB gain. A power ratio of 200 : 1 represents $10 \log_{10}(200) = 23$ dB gain.

8.2.1 Equivalent Calculations of Decibels

Power in an electrical circuit can be expressed as a function of the voltage in the circuit or the current through the circuit. We can calculate decibels from a knowledge of the voltage or the current in an electrical circuit. For example we can express power as:

$$\text{Power} = V^2/R \qquad \textbf{(Eq. 8.13)}$$

$$\text{Power} = I^2R \qquad \textbf{(Eq. 8.14)}$$

Using the definition of Equation 8.13 in the definition for the power gain equation (8.12), we find that:

$$\begin{aligned}\text{Power Gain} &= 10 \log_{10} \frac{(V_{\text{output}})^2/R_{\text{output}}}{(V_{\text{input}})^2/R_{\text{input}}} \\ &= 20 \log_{10} \frac{V_{\text{output}}}{V_{\text{input}}} + 10 \log_{10} \frac{R_{\text{input}}}{R_{\text{output}}}\end{aligned} \qquad \textbf{(Eq. 8.15)}$$

Similarly, using Equation 8.14 we find:

$$\begin{aligned}\text{Power Gain} &= 10 \log_{10} \frac{(I_{\text{output}})^2/R_{\text{output}}}{(I_{\text{input}})^2/R_{\text{input}}} \\ &= 20 \log_{10} \frac{I_{\text{output}}}{I_{\text{input}}} + 10 \log_{10} \frac{R_{\text{output}}}{R_{\text{input}}}\end{aligned} \qquad \textbf{(Eq. 8.16)}$$

EXAMPLE 8.7

Consider a circuit whose properties are as given here. Calculate the power gain in dB:

Input voltage = 10 Vrms	Output voltage = 5 Vrms
Input current = 100 mArms	Output current = 100 mArms
Input resistance = 100 Ω	Output resistance = 50 Ω
Input power = 1 W	Output power = 0.5 W

Solution Using Equation 8.12 we find:

$$\text{Power gain} = 10 \log(P_{\text{out}}/P_{\text{in}}) = 10 \log(0.5/1) = -3 \text{ dB}$$

Using Equation 8.15:

$$\begin{aligned}\text{Power gain} &= 20 \log_{10}(V_{\text{output}}/V_{\text{input}}) + 10 \log_{10}(R_{\text{input}}/R_{\text{output}}) \\ &= 20 \log(5/10) + 10 \log(100/50) \\ &= -6 \text{ dB} + 3 \text{ dB} = -3 \text{ dB}\end{aligned}$$

Using Equation 8.16:

$$\begin{aligned}\text{Power gain} &= 20 \log_{10}(I_{\text{output}}/I_{\text{input}}) + 10 \log_{10}(R_{\text{output}}/R_{\text{input}}) \\ &= 20 \log(100 \text{ mA}/100 \text{ mA}) + 10 \log(50/100) \\ &= 0 \text{ dB} - 3 \text{ dB} = -3 \text{ dB}\end{aligned}$$

Note that no matter which formula we use to compute the performance of the circuit we get the same answer for the power ratio: −3 dB. That is because no matter how it is calculated, the decibel value is only a function of the power ratio, not the voltage or current ratio.

8.2.2 Decibel Scales

Decibels are useful even if there are no direct output/input power ratio properties to compare. We can instead use standard reference levels to compare signals in a circuit to standard signal levels we establish. For example, the power of a signal into a circuit can be referenced to a 1-mW signal level, and the result is a power measurement called dbm, which compares the power in a circuit to a 1-mW reference level. Equation 8.12 thus becomes:

$$\text{dbm} = 10 \log_{10}(P_{\text{output}}/0.001 \text{ W}) \qquad \textbf{(Eq. 8.17)}$$

Alternatively, we can use Equation 8.15 as a starting equation.

$$\text{Power gain} = \text{dbm} = 10 \log_{10} \frac{(V_{\text{output}})^2/R_{\text{output}}}{(V_{\text{input}})^2/R_{\text{input}}}$$

$$\text{dbm} = 20 \log_{10} V_{\text{output}} - 10 \log_{10} R_{\text{output}} - 10 \log_{10}(0.001) \qquad \textbf{(Eq. 8.18)}$$

$$\text{dbm} = 20 \log_{10} V_{\text{output}} - 10 \log_{10} R_{\text{output}} + 30$$

This equation, commonly used by telephone and audio technicians, relates dbm to a voltage reading in a circuit. For their work the resistance level is assumed to be 600 Ω. This converts the equation to:

$$\text{dbm} = 20 \log_{10} V_{\text{output}} + 2.218 \qquad \textbf{(Eq. 8.19)}$$

Many audio voltmeters are calibrated using this equation, and thereby relative power can be read directly in dbm on an audio voltmeter. Other standard values exist that produce logarithmic scales. These are generally discussed in more advanced texts. One common value is a 1-W power level. The corresponding unit is labeled as dbw, which stands for dB relative to one watt.

EXAMPLE 8.8

What voltage level corresponds to 0 dbm? What dbm level corresponds to 5 volts?

Solution Using Equation 8.19 we find that $0 \text{ dbm} = 20 \log_{10} V_{\text{output}} + 2.218$, from which we find that:

$$-2.218 = 20 \log_{10} V_{\text{output}}$$

$$V_{\text{output}} = 10^{-0.1109} = \mathbf{0.775\ V}$$

Starting with Equation 8.19 we find that 5 Vrms equals:

$$? \text{ dbm} = 20 \log_{10} V_{\text{output}} + 2.218$$

$$= 20 \log_{10} 5 + 2.218$$

$$5 \text{ V} = \mathbf{16.2\ dbm}$$

8.2.3 Transfer Function Properties

All electrical circuits constructed using capacitors and/or inductors have properties that change as the frequency of operation changes. Even for a purely resistive circuit, performance varies with frequency because of the stray capacitance and inductance that physical components introduce into a theoretical circuit. Thus, every electrical circuit varies as the frequency of operation changes. One way of expressing the changes in behavior is to construct ratios of circuit parameters and express these ratios as a function of frequency. ***These ratios are called transfer functions.***

One of the most commonly used transfer functions is the ***ratio of output voltage to input voltage.*** This ***voltage transfer ratio*** describes how the output changes with respect to the input as the frequency of operation varies. Such a ratio removes any amplitude dependence because it expresses a ratio of output signal to input signal. Thus, because the circuits we study are linear, if the input signal doubles then the output signal doubles, thereby not affecting the transfer ratio, which by construction is amplitude independent. In addition to ***voltage*** transfer ratios we also use a ***current transfer ratio*** to relate the output current to the input current in a circuit. All of these ratios vary with frequency as a result of using capacitors, inductors, and resistors. Although the impedance of each of these elements is only a simple function of frequency, their combined effect on each other can produce some very complex circuit properties.

Resistors are frequency independent. Their impedance is not a function of frequency. When a resistor is placed in series with an inductor, the circuit properties are such that at low frequencies the combination of elements behaves like a resistor but at high frequencies it behaves like an inductor. The opposite is true for a resistor in parallel with an inductor. At low frequencies the circuit behaves like an inductor, but at high frequencies the combined element behaves like a resistor. ***Notice that no new elements are created by combining a resistor and an inductor in series or in parallel. When we repeat the high- and low-frequency argument for a resistor and capacitor in series and parallel, the same general behavior is exhibited. That is, the circuit behaves like either a resistor or a capacitor—no new element is created. The result of this behavior is that any circuit we consider is directly related to only the properties of resistors, capacitors, and inductors—no new elements are ever created.***

Whether we use mesh equations or node equations to help us solve for the transfer function, we ultimately resort to the use of determinants to solve the linear equations that result. The equations we solve are always linear equations whose coefficients are complex numbers. Determinants we form are made up of combinations of complex numbers. The solution of these determinants produces the transfer function we seek. Each determinant has a value that is made up of the sum and differences of the product of determinant elements. As a result, the form that transfer functions assume is a ratio of two polynomials constructed of integer powers of $j\omega$. If we substitute the letter s for $j\omega$ we observe that the resulting transfer function is a function of two polynomials of integer powers of s with real coefficients.

Consider a polynomial: $a_n s^n + a_{n-1}s^{n-1} + \cdots + a_0$. If we set this polynomial of order n equal to zero, there are exactly n values of s that will satisfy this equation. These values are unique for the polynomial and are called roots of the polynomial. The original polynomial can be factored into the product of n terms, each of which is of the form $(s + r_i)$. Any transfer function is constructed of two such polynomials. The roots of the numerator polynomial are called ***zeros*** of the transfer function while the roots of the denominator are called ***poles*** of the transfer function. The polynomial nature of the transfer function means that ***any transfer function can be written as:***

$$T(s) = \frac{(s + z_1)(s + z_2)\cdots(s + z_n)}{(s + p_1)(s + p_2)\cdots(s + p_k)} \qquad \textbf{(Eq. 8.20)}$$

If we factor out the product of the z_i and the p_i from the top and bottom polynomials we can rewrite Equation 8.20 as:

$$T(s) = \mathrm{K}\frac{\left(\frac{s}{z_1} + 1\right)\left(\frac{s}{z_2} + 1\right)\cdots\left(\frac{s}{z_n} + 1\right)}{\left(\frac{s}{p_1} + 1\right)\left(\frac{s}{p_2} + 1\right)\cdots\left(\frac{s}{p_k} + 1\right)} \qquad \textbf{(Eq. 8.21)}$$

where $\mathrm{K} = \dfrac{z_1 \cdot z_2 \cdot \cdots z_n}{p_1 \cdot p_2 \cdot \cdots p_k}$

8.3 TRANSFER FUNCTIONS VERSUS FREQUENCY: LOGARITHMS

A plot of the transfer function of a circuit as the frequency varies enables one to observe the frequency response of a network. If the transfer function amplitude is plotted in decibels, the resultant plot displays the behavior of the transfer function over a range of frequencies. Such plots, because of the mathematical structure of the transfer function, are simple curves that allow rapid and simple evaluation of a circuit's performance. In addition, we shall see that they are easy to draw after a few underlying techniques are mastered.

Using Equation 8.21, the dB gain of a voltage transfer function can be written as:

$$\text{dB gain} = 20\log_{10}|T(s)|$$

$$= 20\log_{10} \mathrm{K}\frac{\left(\frac{s}{z_1}+1\right)\left(\frac{s}{z_2}+1\right)\cdots\left(\frac{s}{z_n}+1\right)}{\left(\frac{s}{p_1}+1\right)\left(\frac{s}{p_2}+1\right)\cdots\left(\frac{s}{p_k}+1\right)}$$

This can be simplified because the logarithm of product of terms equals the sum of the logarithms of the product. Using Equation 8.5 this becomes:

$$= 20\log_{10}\mathrm{K} + 20\log_{10}\left(\frac{s}{z_1}+1\right) + 20\log_{10}\left(\frac{s}{z_2}+1\right) + \cdots +$$
$$20\log_{10}\left(\frac{s}{z_n}+1\right) - 20\log_{10}\left(\frac{s}{p_1}+1\right) - 20\log_{10}\left(\frac{s}{p_2}+1\right) - \cdots -$$
$$20\log_{10}\left(\frac{s}{p_k}+1\right) \qquad \textbf{(Eq. 8.22)}$$

Equation 8.22 states that the log of the transfer function is composed of a sum of individual single terms contained in the original transfer function (Eq. 8.21). To plot the transfer function we can add graphically the results from each pole and zero term. This is much easier than trying to evaluate the entire transfer function at once.

8.3.1 Logarithms as a Graphic Aid

We have seen the use of logarithms to define power ratios and for power-level comparisons. This use arose from a need to express large differences in a simple way. Electrical circuits, if they are to be useful, must often operate over a large range of frequencies. We often plot the behavior of a circuit as a function of frequency as a way to represent graphically how the circuit properties change. These graphs provide great insight into performance of the circuit.

Suppose that we wish to plot the properties of a circuit as the radian frequency varies between 1 rad/s and 10,000 rad/s (2π Hz to $20{,}000\pi$ Hz). We would first calculate the equation to be plotted or measure the circuit property as a function of ω. Customarily, we plot the frequency as the x axis of the graph and the circuit property [in this case $Y(\omega)$] as the y axis.

Consider this circuit transfer function:

$$Y(\omega) = \frac{1}{1+\frac{(\omega-5000)^2}{5000^2}}$$

This transfer function represents the ratio of the output signal level divided by input signal level as the frequency varies. It is plotted on the y axis of Fig. 8.2 versus ω, (which is plotted on the x axis), for values of ω between 1 and 10,000 rad/s. If we wished to plot the same transfer function equation over a wider frequency range (of, say, 1 rad/s to

1,000,000 rad/s), the frequency axis would have to be one hundred times larger to keep the same frequency resolution shown in Fig. 8.2. This is impossible on a normal piece of paper. Figure 8.3 shows the effect of enlarging the frequency span. Note that the detailed behavior of the curve in the frequency range of 1 rad/s to 10,000 rad/s is lost.

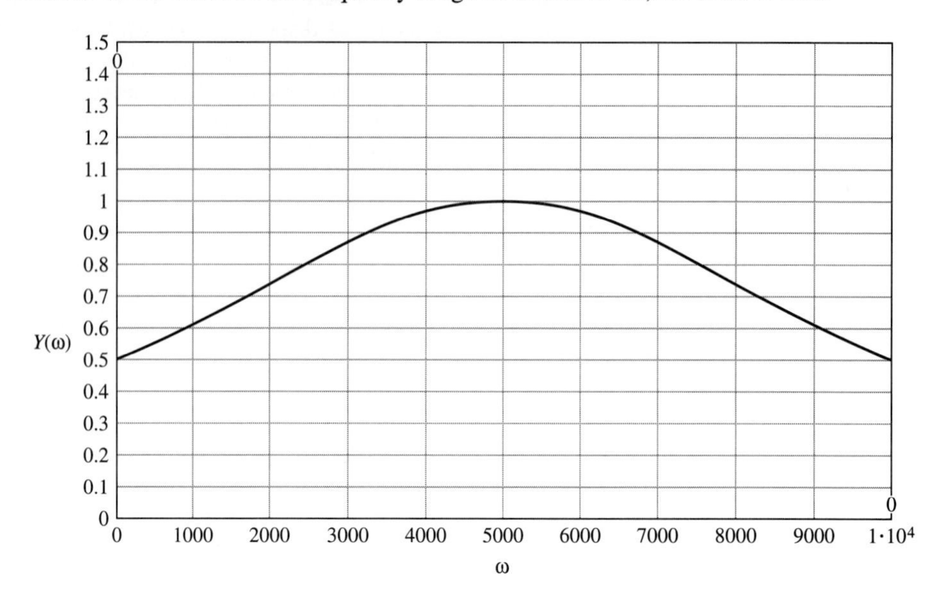

FIGURE 8.2
Y(ω) versus ω

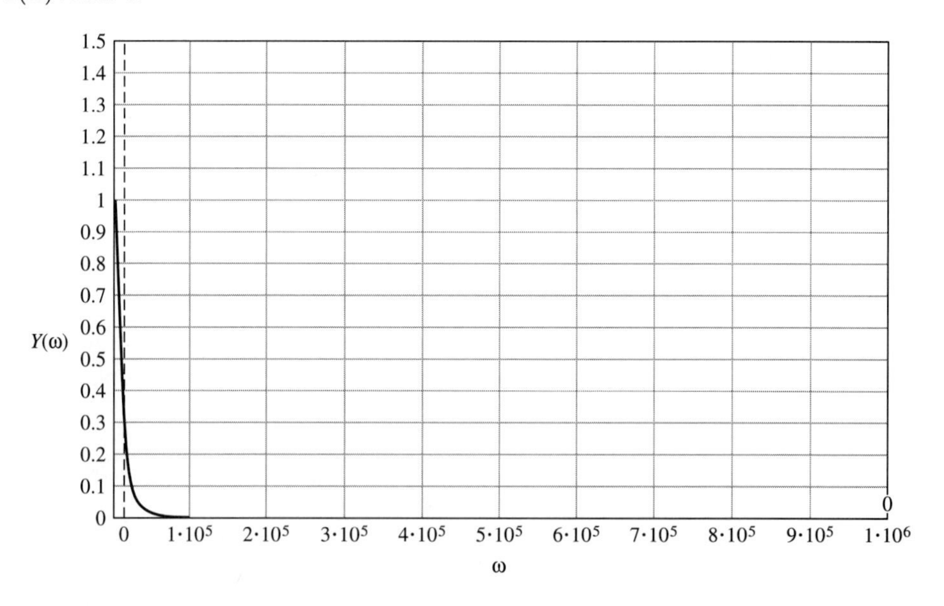

FIGURE 8.3
Y(ω) versus ω

When we compare the two graphs we see that Fig. 8.3 allots very little space to the graph in the 1 to 10,000 rad/s region. Most of the space in the graph is for higher frequencies (100,000 to 1,000,000 rad/s). If we plot the logarithm of ω rather than ω on the x axis, the character of the curve changes. Figure 8.4 shows a plot of $Y(\omega)$ versus $\log(\omega)$. Since $\log(\omega)$ is directly related to ω, we can label the curve with values of ω rather than its logarithm.

We have plotted the common logarithm of the frequency rather than the frequency itself in Fig. 8.4. Note that the curve's variation is clearly shown for low-frequency as well as for high-frequency values. Even Fig. 8.2 does not show the detail that appears in Fig. 8.4. This great display of detail is a characteristic of logarithmic plots. Whereas the frequency ratio is 1,000,000 : 1, the logarithm of the frequency ratio is 6. Notice that the

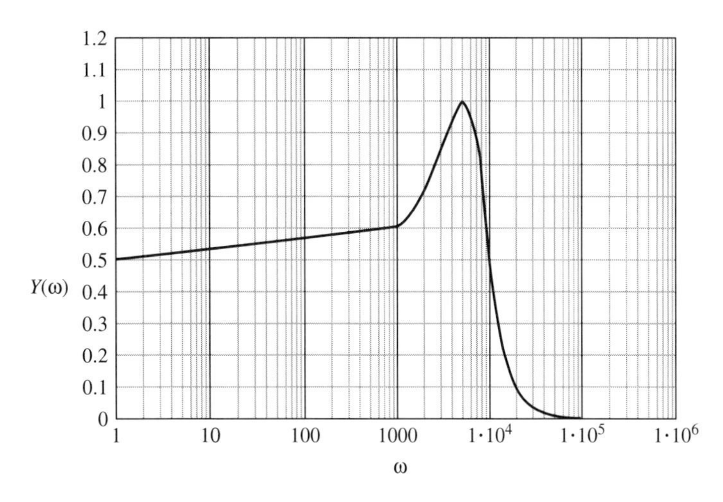

FIGURE 8.4
Y*(ω) *versus* ω *on a Logarithmic Scale

x axis has the same amount of space between $\omega = 1$ and $\omega = 10$ as it does between $\omega = 100{,}000$ and $\omega = 1{,}000{,}000$ rad/s. This is a direct result of the logarithmic plot using the base 10. The frequency range is divided into 6 frequency bands. Each of these bands represents a frequency ratio of 10 : 1. This is called a ***decade frequency ratio,*** and each decade occupies the same space on a logarithmic plot. That is, any decade of frequency (ratio of 10 : 1 in frequency) occupies the same amount of space as any other decade of frequency. This is a characteristic of logarithmic plots using the base 10, namely, that every decade change of value occupies as much space as any other decade change of value. Figure 8.4 has six equal regions in the graph on the x axis corresponding to the six decades of frequency variation.

A 2 : 1 frequency ratio is called an octave. This represents a smaller change in frequency than a decade, which is 10 : 1. On a logarithmic plot of frequency, an octave represents approximately 30 percent of the distance between decades of frequency on a logarithmic plot. This smaller ratio of frequencies is useful for describing in greater detail small regions of the circuit behavior plot where rapid change in circuit properties is taking place. Analysis of low-frequency circuits is usually calculated with octave-size changes in frequency. The curves for octaves and decades are the same, except the curves for decades generally are over a wide frequency range.

EXAMPLE 8.9

A piano spans seven octaves of frequency. What is the frequency ratio of the highest frequency to the lowest frequency on a piano? How many decades is this?

Solution

7 octaves = 2^7 = **128**

The highest note on a piano is 128 times higher in frequency than the lowest note.

Log(128) = 2.10, therefore 128 : 1 = **2.1 decades**

1 octave = 0.30 decades **(Eq. 8.23)**

In addition to plotting the frequency logarithmically, we plot transfer functions and the impedance of the circuit, which vary widely over the frequency range being considered, logarithmically. In this case, both the x axis and the y axis are being plotted logarithmically. When a property varies over several orders of magnitude, it is convenient to plot the logarithm of this property rather than the property itself. In this way each decade of change in circuit property occupies the same amount of space as any other decade of change. This allows all values to be equally represented in the plot of properties vs. frequency. This provides for a better understanding of the operation of the circuit.

EXAMPLE 8.10

Plot the magnitude of the reactance of a 0.1-μF capacitor as the radian frequency varies from 1000 to 1,000,000 rad/s.

Solution The reactance of a capacitor as a function of frequency is given as:

$$X_C = 1/\omega C$$

At 1000 rad/s the reactance equals:

$$\begin{aligned} X_C &= 1/(1000)(1 \times 10^{-7}) \\ &= 10{,}000\ \Omega \end{aligned}$$

whereas at 1,000,000 rad/s the reactance equals 10 Ω. The result is plotted in Fig. 8.5, which is a linear plot of reactance versus radian frequency. Observe that most of the curve cannot be seen as it is located along the x axis because its magnitude is so small.

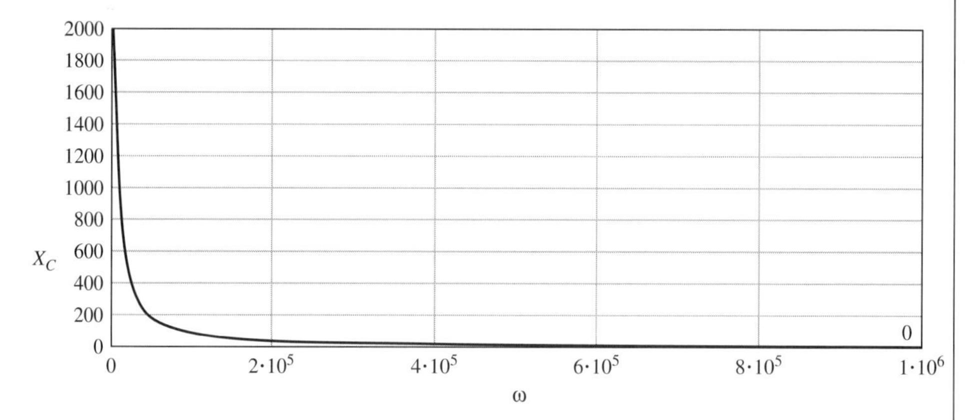

FIGURE 8.5
Reactance of a capacitor: Linear Scale

If we take the logarithm of the expression for the reactance, then the equation becomes:

$$\begin{aligned} \log X_C &= \log(1/\omega C) \\ \log X_C &= \log(1) - \log(\omega) - \log(C) \\ &= 7 - \log(\omega) \end{aligned}$$

This is the equation of a straight line on a log–log plot, as Fig. 8.6 aptly demonstrates.

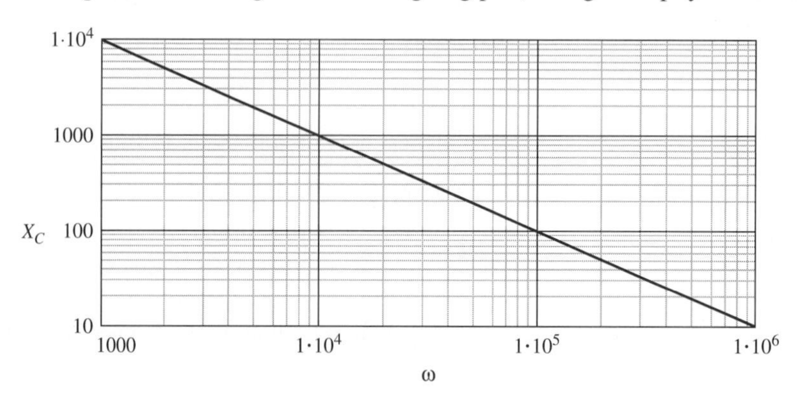

FIGURE 8.6
Reactance of a capacitor: Logarithmic Scale

Note: We always use logarithmic scales when we wish to present circuit properties over several decades of frequency. In that way we can examine circuit behavior over a wide range of frequencies with equal emphasis on each decade of the range.

As frequency varies, the properties of circuits—such as impedance and phase angle, for example—also vary. If the change in properties is large (several orders of magnitude) over the frequency range, we use a logarithmic scale to display the property variations over that range. Under these conditions we use a logarithmic graphing axis for the property and a logarithmic graphing axis for the frequency. Such a plot is called a log–log plot and is often used in circuit analysis. When a property does not change over a wide range as the frequency varies, we may replace the logarithmic axis for the property with a linear one. In this instance the plot is called a semilogarithmic (or semilog) plot.

EXAMPLE 8.11

A 100-kΩ resistor is placed in parallel with a capacitor of 10,000 pF. Plot the impedance of the combination and the phase angle of the impedance as ω varies from 1 to 1,000,000 rad/s.

Solution The impedance of a capacitor in parallel with a resistor is found by the parallel impedance formula to be:

$$Z = R/(1 + j\omega RC). \qquad \textbf{(Eq. 8.24)}$$

The magnitude of this impedance can be shown to be:

$$|Z| = R/\sqrt{[1 + (\omega RC)]^2} \qquad \textbf{(Eq. 8.25)}$$

while the phase angle is given as:

$$\theta = -\arctan(\omega RC). \qquad \textbf{(Eq. 8.26)}$$

For the set of values given this becomes:

$$|Z| = 10^5/\sqrt{[1 + (\omega \cdot 10^{-3})]^2} \quad \text{and} \quad \theta = -\arctan(\omega \cdot 10^{-3})$$

If we evaluate the impedance we find that the impedance at $\omega = 1$ is 100 kΩ and the impedance at $\omega = 1{,}000{,}000$ is only 100 Ω. This represents a ratio of 1000 : 1 in impedance, which is too large a range to be plotted on a linear axis. Thus, the impedance magnitude must use a logarithmic scale to plot the impedance magnitude variation.

When we examine the phase angle of the impedance as ω varies from 1 to 1,000,000, we see that the phase angle goes from 0° for small values of ω to −90° for

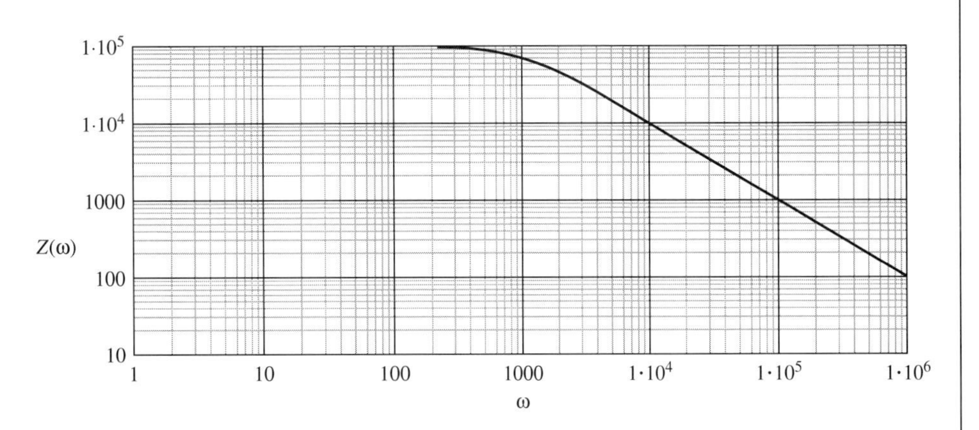

FIGURE 8.7

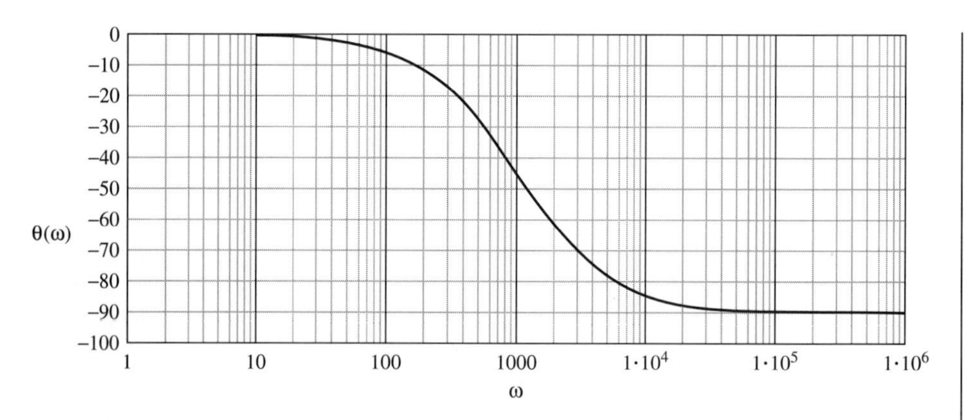

FIGURE 8.8

large values. This variation of phase angle can easily be plotted on a linear axis. Thus the phase angle versus frequency plot can use a semilog plot. The results of the two different plots are shown in Figs. 8.7 and 8.8.

8.3a PLOTTING A CURVE: THE ASYMPTOTIC APPROACH

To plot the curves for Example 8.11 required us to evaluate Equations 8.25 and 8.26 for many values of ω. This approach to curve plotting is tedious, and the chance for errors is large. Let us examine the behavior of the curves to look for clues that can help us plot the curves more easily. In particular, let us see how the curves behave versus frequency.

Consider Equations 8.24 and 8.25:

$$Z = R/(1 + j\omega RC). \qquad \textbf{(Eq. 8.24)}$$

$$|Z| = R/\sqrt{1 + (\omega RC)^2} \qquad \textbf{(Eq. 8.25)}$$

Examination of Equation 8.24 reveals that the variation of impedance with frequency is caused by the term in the denominator of Equation 8.24, namely, $1 + j\omega RC$. The magnitude of this complex number depends on the relationship between the real and the imaginary parts of the complex number. As shown in Equation 8.25, the magnitude of the complex number is the square root of the sum of the squares of the real and the imaginary parts. ***If the real part of a complex number is 10 times the magnitude of its imaginary part, then the imaginary part affects the magnitude of the impedance by less than 0.5%.*** The impedance under these conditions, therefore, is approximately constant and equal to R (100 kΩ). Figure 8.7 demonstrates this property very well.

When the imaginary part of the denominator is 10 times the real part of the denominator, the real part has a very small effect on the impedance. In fact, as was the case for $\omega RC = 0.1$ earlier, the error caused by neglecting the real part of the impedance is less than 0.5 percent. Examination of Fig. 8.7 reveals that the plot of the magnitude of the impedance is a straight line for values of $\omega RC = 10$ or more. If we extend this straight line backward, we see that the exact curve does not differ from the straight line by very much even at $\omega RC = 1$.

On the log–log plot, values of ωRC less than 0.1 produce a curve that is a straight line parallel to the (frequency) x axis. Values of ωRC greater than 10 produce a curve that is a straight line at some constant slope. These curves are shown in Fig. 8.9 along with the original curve. ***Note how similar the curve made up of the two straight lines is to the original curve.*** The lines obtained by considering the behavior for large and small values of ωRC are called asymptotes of the actual curve drawn. ***An asymptote is a line that a curve approaches but never quite reaches.*** If we extend the right side of the line for small values of ω and the left side of the line for large values of ω we note that the

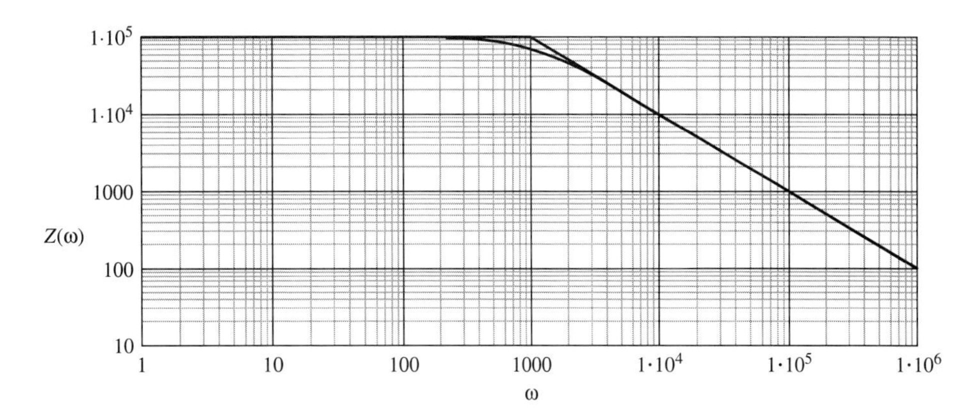

FIGURE 8.9

intersection of the two straight lines occurs at $\omega RC = 1$. At this point the real and the imaginary values of the denominator of Equation 8.24 are equal. It is also the point at which the actual curve is farthest from its asymptotes. ***For the plot of Equation 8.7, R = 100,000 Ω and C = 10,000 pF, from which RC = 0.001. The value of ωRC = 1 then is at ω = 1000, as shown on the curve.***

Consider Equation 8.26:

$$\theta = -\arctan(\omega RC) \quad \textbf{(Eq. 8.26)}$$

This equation expresses the phase angle behavior of the RC network as a function of frequency. Note that the angle approaches 0° as ωRC approaches zero. Since we are plotting the phase angle as a function of frequency on a logarithmic frequency axis, the phase angle never quite gets to 0°. For values of ωRC greater than 10 the phase angle approaches $-90°$. However, for any finite value of ω the phase angle never quite gets to 90°. Thus, the phase angle has asymptotes of 0° and 90°, both parallel to the x axis. These two asymptotes can be joined by a straight line drawn between the values of $\omega RC = 0.1$, $\theta = 0$, and $\omega RC = 10$, $\theta = -90°$. Finally, note that at $\omega RC = 1$ the phase angle is $-45°$. The curve of phase angle versus ω is shown in Fig. 8.8. The original curve, the straight line segments, and the asymptotes that approximate it are shown in Fig. 8.10 along with the original curve. While the fit is not as good for the phase angle versus frequency curve, it provides sufficient resolution for most applications.

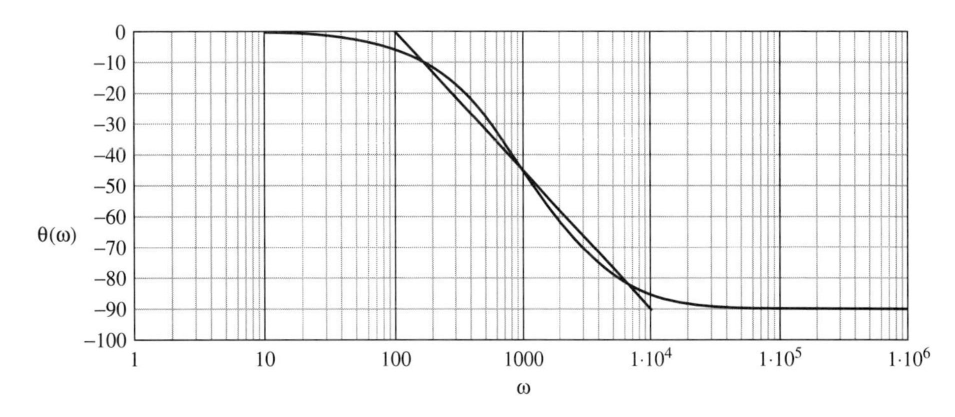

FIGURE 8.10

8.4 FILTERS

A filter is an electrical circuit that allows signals contained in a desired band of frequencies to pass through with no loss and prevents all signals whose frequencies are not in the selected band from appearing in the output of the filter. Such a device allows for the selection of only desired signals in the presence of undesired signals. One simple example of the use of filters is in the input of a radio receiver. All of the possible stations are present at the input to the radio, but only one station at a time is allowed through the filter (input stage of the radio). This is accomplished by adjusting the properties of the filter so that only the frequencies representing the station are within the filter passband region. Filters of several different types are discussed here. They are only a small sample of the filters that are available to meet the needs of electronic designers.

Each of the filter types introduced here is explained in two different settings. First the ideal filter is described in terms of its properties and definitions. Following the definition of an ideal filter, we describe the properties of a practical filter that has similar properties. It is impossible to achieve the properties of an ideal filter in the real world, but by increasing the number of circuit elements, electronics designers can come very close to the properties of an ideal filter. Several simple types of filters are described here, but the number of implementations of ideal filters is far greater than the small number demonstrated.

8.4.1 Low-Pass Filters

Figure 8.11 shows the transfer function for an ideal low-pass filter. This filter has a transfer function equal to 1 for all frequencies less than ω_c and a transfer function equal to 0 for all frequencies greater than ω_c. Such a device can be used to separate lower-frequency signals from signals of higher frequency. We will define ω_c as the cutoff frequency of the low-pass filter. This means that if the input signal frequency is less than ω_c, the signal will pass through the network with the same amplitude as it entered the network. Signals greater than ω_c will be eliminated—that is, they will not appear in the output signal. The filter is called a low-pass filter because signals of all frequencies lower than the cutoff frequency will pass through the filter and those signals whose frequency is above the cutoff frequency will not pass through the filter.

FIGURE 8.11

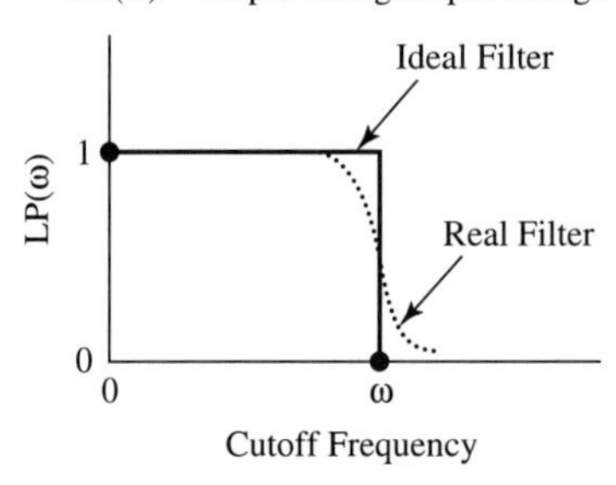

Notice that the cutoff edge of the ideal filter is vertical. A frequency of only slightly more than the cutoff frequency is completely eliminated. In any real design of a low-pass filter, it is impossible to achieve this result. Frequencies above the cutoff frequency will pass through the filter with some finite loss. The loss may be large, but it is not 100%. In the region of the cutoff frequency there will be a band of frequencies that will be only slightly attenuated. As the difference between the cutoff frequency and the actual frequency becomes larger, the loss will increase. This results in a finite slope for the cutoff region of the low-pass filter. This effect is shown as the dotted line in Fig. 8.11.

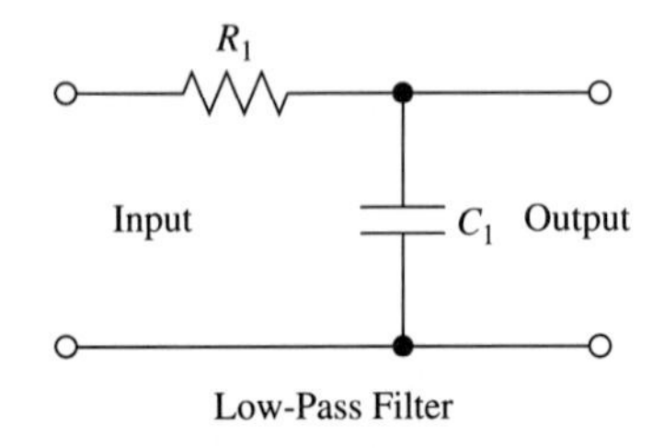

FIGURE 8.12

For our example of a simple low-pass filter, we will choose a series combination of a resistor and capacitor as shown in Fig. 8.12. The ratio of output voltage to input voltage (the circuit transfer ratio) can be found by applying the Voltage Divider Theorem to the circuit given. We write the transfer function of this low-pass filter as:

$$\text{LP}(\omega) = \frac{1/j\omega C_1}{R_1 + (1/j\omega C_1)} = \frac{1}{1 + j\omega R_1 C_1} \qquad \textbf{(Eq. 8.27)}$$

Note that for $\omega = 0$ the value of the transfer function is equal to 1. The denominator of the transfer function causes the transfer function variation with frequency that is a complex function of ω. As we have seen before, the magnitude of the denominator is 1 if the real part of the complex number is 10 times the magnitude of the imaginary part. Similarly the magnitude of the denominator is equal to the magnitude of the imaginary part if the magnitude of the imaginary part is 10 or greater times the real part, which in this case is a constant 1.

The critical value of ω for this denominator expression is that value of ω for which the real part and the imaginary part are equal. This value of ω is called the cutoff frequency of the low-pass filter. For values of ω greater than the cutoff frequency, the denominator has a larger imaginary part than real part, and below this value for ω the real part of the filter transfer function is larger than the imaginary part. From the expression for the transfer function (Eq. 8.27) we find that the cutoff frequency ω_c is given as

$$\omega_c = \frac{1}{R_1 C_1} \text{ rad/s} \qquad \textbf{(Eq. 8.28)}$$

We can convert this radian frequency to hertz by dividing ω_c by 2π. Thus the cutoff frequency of the low-pass filter shown is:

$$f_c = \frac{1}{2\pi R_1 C_1} \text{ Hz} \qquad \textbf{(Eq. 8.29)}$$

Equation 8.30 expresses the transfer function in terms of decibels. In displaying properties of filters, we normally plot the decibel value of the transfer function on the y axis versus the logarithm of the frequency on the x axis. Such a plot exhibits asymptotes and is easy to evaluate. ***It is called a Bode Plot.*** For the low-pass filter described here, the transfer function written in decibels is given as:

$$\begin{aligned} LP(\omega) &= [20 \log(1) - 20 \log[1 + (\omega R_1 C_1)^2]^{1/2} \text{ dB} \\ &= [20 \log(1) - 20 \log[1 + (\omega/\omega_c)^2]^{1/2} \text{ dB} \end{aligned} \qquad \textbf{(Eq. 8.30)}$$

where $\quad \omega_c = \dfrac{1}{R_1 C_1}$

If we solve for the value of the transfer function at the cutoff frequency, we find that LP(ω) equals -3 dB. That is, the output power is smaller than the input power into the filter by a factor of two. The relationship between the -3 dB point and the cutoff frequency is unique. All filters are defined as having a cutoff frequency at the -3 dB point(s) of their curve. No matter how complex the filter is, the cutoff frequency is always defined by the frequency at which the transfer ratio is 3 dB below the peak output of the filter. This point is sometimes called the half-power point. Cutoff frequency, half-power point, and -3 dB point are one in the same regardless of the filter being considered.

Using the rule that the value of a complex number is equal to the real component of the complex number for values of the imaginary part equal to less than 0.1, we find the value of the transfer function to be 0 dB for $\omega R_1 C_1$ less than 0.1. Similarly, for $\omega R_1 C_1$ greater than 10 we find that $T(\omega) = -20 \log(\omega R_1 C_1)$ dB. These expressions for small and large values of $\omega R_1 C_1$ represent the asymptotes of the Bode Plot for the low-pass filter. We find (see Fig. 8.13) that these two asymptotes intersect at $\omega R_1 C_1 = 1$ (the cutoff frequency). For frequencies above the cutoff frequency, the asymptote falls at the rate of 20 dB/decade or 6 dB/octave for the straight-line portion. This is a characteristic of simple R–C filters. It means that for every decade increase in frequency the attenuation of the output signal power is reduced by a factor of 100, whereas the voltage ratio is reduced by 10.

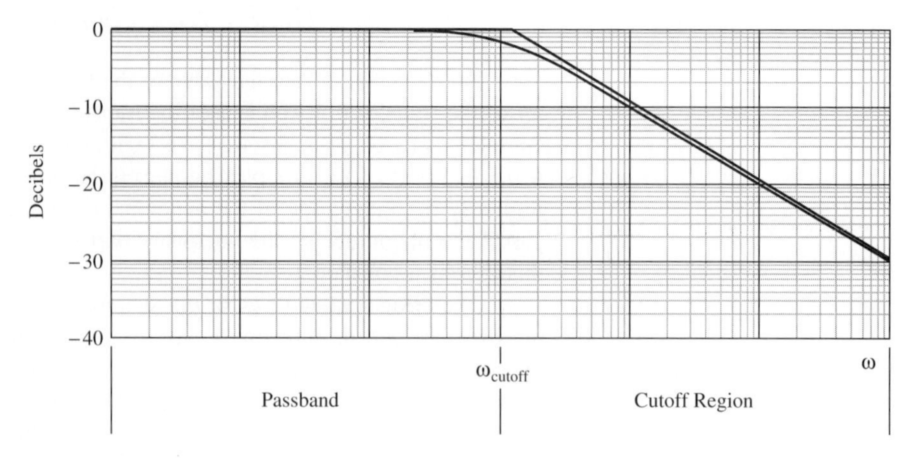

FIGURE 8.13

Consider two harmonically related signals (one frequency is an integer multiple of the other) that pass through a filter. Adding these two signals for all instants of time produces a waveform that is dependent not only on the amplitude of each but the phase of each as well. Since the two signals are harmonically related, the result of this addition repeats with the period of the lower-frequency signal. If we wish to keep the waveform the same as it passes through the filter so that the output looks the same as the input, then the phase angle should be proportional to the frequency of the waveform. That is, the phase angle versus frequency curve should ideally be a straight-line relationship. Expressing this mathematically we can say that:

$$\theta = k\omega \qquad \textbf{(Eq. 8.31)}$$

where k is some constant that produces no distortion in the waveform due to phase errors. This phase angle behavior will not distort the relationship between the frequencies passing through and therefore will not affect the waveform. For the actual low-pass filter we find that the relationship between the phase angle and the frequency is given by:

$$\theta = -\arctan(\omega R_1 C_1) = -\arctan(\omega/\omega_c) \qquad \textbf{(Eq. 8.32)}$$

The curve of the phase angle versus the frequency is shown in Fig. 8.14. Values of θ less than 45° represent the pass band of the filter. Values of θ more than 45° are in the cutoff region of the filter, where signals are attenuated.

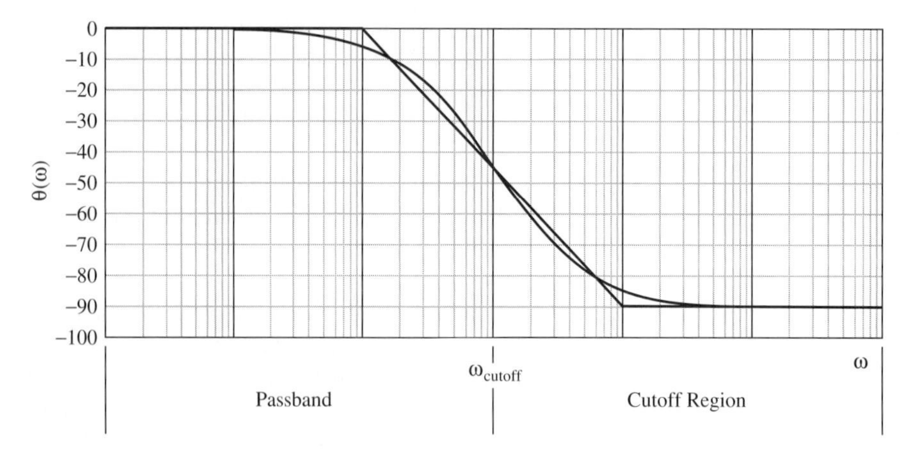

FIGURE 8.14

8.4.2 High-Pass Filters

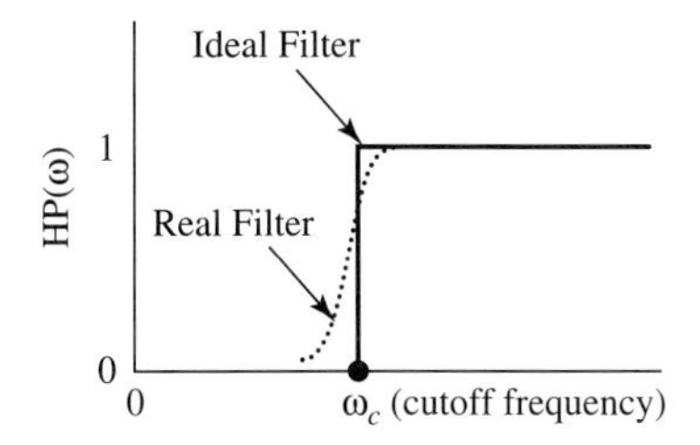

FIGURE 8.15

A high-pass filter is the complement of a low-pass filter. That is, it will allow signals of frequency higher than the cutoff frequency to pass through unaffected but will reduce the output amplitude of all signals whose frequency is lower than the cutoff frequency to zero. The transfer function of an ideal high-pass filter is shown in Fig. 8.15. Like a low-pass filter it allows signals in a band of frequencies to pass through the circuit but blocks signals whose frequencies lie outside the range of its bandpass characteristics. As was true for the low-pass filter, it is impossible to build a real filter that has the properties shown in Fig. 8.15. The cutoff sharpness, as shown by the vertical edge of the filter at ω_c, cannot be duplicated in building a real high-pass filter. Signals just below cutoff frequency will have small losses. As the frequency decreases, the losses will increase.

We will demonstrate the properties of a high-pass filter with the circuit of Fig. 8.16. The transfer function of this circuit is obtained by applying the Voltage Divider Theorem to the circuit and then simplifying to form the transfer function, HP:

$$\begin{aligned}\mathrm{HP}(\omega) &= \frac{1}{1 + (R_1/j\omega L_1)} \\ &= \begin{cases} 1 & \text{for } R_1/\omega L_1 < 0.1 \\ j\omega L_1/R_1 = j\omega/\omega_c & \text{for } \omega L_1/R_1 > 10 \end{cases}\end{aligned} \qquad \textbf{(Eq. 8.33)}$$

FIGURE 8.16

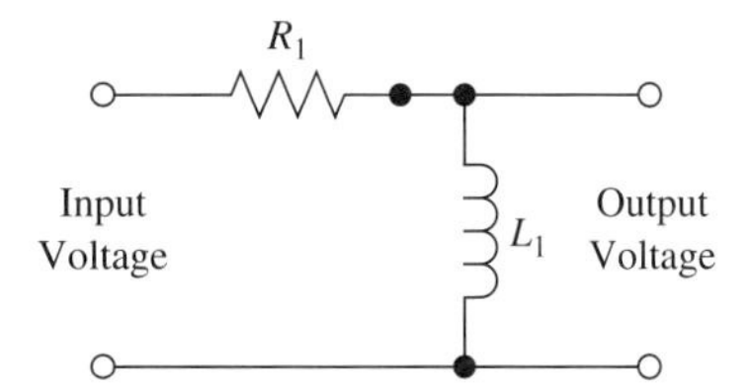

Examination of Equation 8.33 shows that there is attenuation for frequencies that are small compared to the cutoff frequency. By comparison with the results obtained earlier for the case of the low-pass filter, we define the cutoff frequency as that frequency at which the real part and the imaginary part of the transfer function denominator are equal. Thus the cutoff frequency for the high-pass filter is the same as the cutoff frequency of the low-pass filter, namely:

$$\omega_c = \frac{R_1}{L_1}\ \text{rad/s} \qquad f_c = \frac{R_1}{2\pi L_1}\ \text{Hz} \qquad \textbf{(Eq. 8.28)}$$

As we did for the low-pass filter, we write the transfer function expressed in decibels as:

$$\begin{aligned}\mathrm{HP}(\omega) &= 20\log(1) - 20\log[1 + (R_1/\omega L_1)^2]^{1/2} \\ &= 20\log(1) - 20\log[1 + (\omega_c/\omega)^2]^{1/2}\end{aligned} \qquad \textbf{(Eq. 8.34)}$$

The phase angle behavior with ω becomes:

$$\theta = \arctan(R_1/\omega L_1) = \arctan(\omega_c/\omega) \qquad \textbf{(Eq. 8.35)}$$

Equations 8.34 and 8.35 are plotted in Figs. 8.17 and 8.18. Notice the asymptotic behavior of the two curves and compare the results with the result for the low-pass filter. There are many similarities between the two filter types. Below the cutoff frequency, the loss increases by 20 dB for every decade reduction in frequency. This corresponds to the 20 dB loss per decade increase above the cutoff frequency in the case of the low-pass filter. In the high-pass filter passband, the phase angle is less than 45° but it is positive. For the low-pass filter the angle is less than 45° but the phase angle is negative.

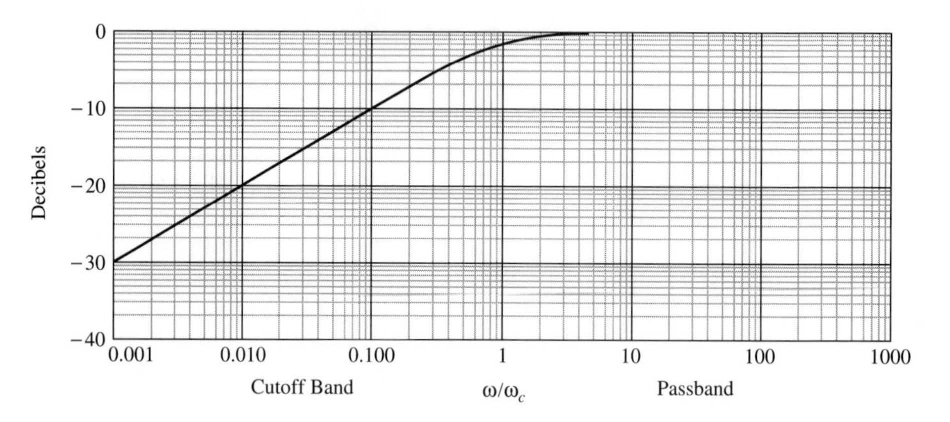

FIGURE 8.17

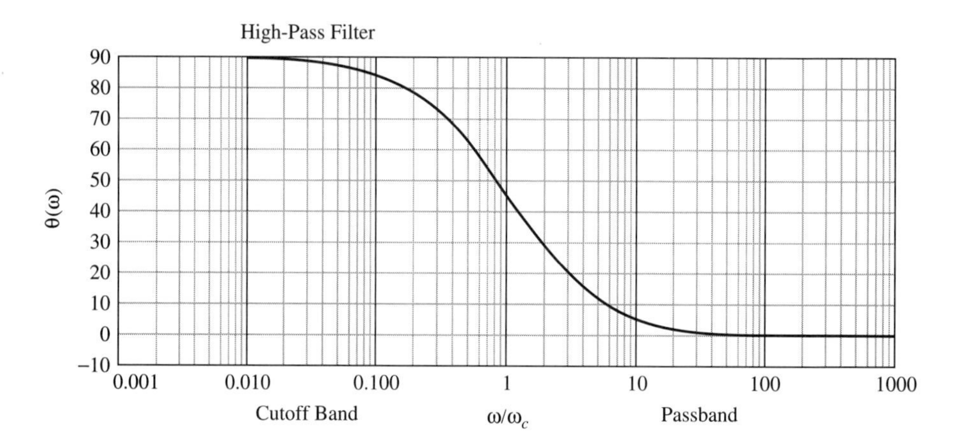

FIGURE 8.18

8.4.3 Bandpass Filters

In addition to the low-pass filter and the high-pass filter, other types of filters are used to select signals in a band of frequencies from the complete spectrum. One such filter is called a bandpass filter. This filter, whose transfer function behavior is shown in Fig. 8.19, has two cutoff frequencies. The lower cutoff frequency is called ω_1, and the upper cutoff frequency is called ω_2. This filter selects a band of frequencies from the spectrum and rejects all others. One can think of a bandpass filter as a set of two filters in tandem. These may be a low-pass filter with a cutoff frequency of ω_2 and a high-pass filter with a cutoff frequency of ω_1. The filters may be arranged in any order with either one first. They are arranged in series, with the output of one filter acting as the input of the following filter.

A filter with vertical cutoff frequency edges cannot be built for the bandpass filter, just as it could not be built for the low-pass or the high-pass filters previously discussed.

FIGURE 8.19

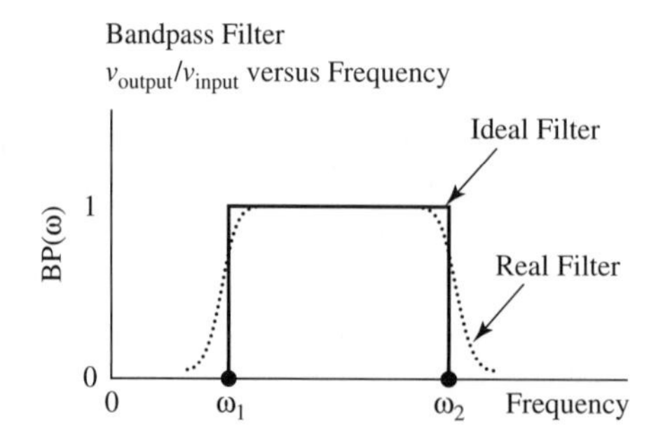

We can use the results obtained earlier for the low-pass filter and the high-pass filter and define a transfer function that is the product of the two transfer functions. In this way we can plot the frequency response for the amplitude and phase-angle performance of the bandpass filter. We can use the transfer functions for the low-pass filter and the high-pass filter to obtain a transfer function for a bandpass filter. The filter thus generated is not an ideal filter and has no sharp sides, but it is useful in explaining the properties of a bandpass filter. The transfer function for a bandpass filter made from cascaded high- and low-pass filters becomes:

$$\mathrm{BP}(\omega) = \left(\frac{1}{1 + j\omega R_2 C_2}\right)\cdot\left(\frac{1}{1 + \dfrac{1}{j\omega R_1 C_1}}\right) = \mathrm{BP}(\omega) \qquad \textbf{(Eq. 8.37)}$$

where $\omega_1 = 1/R_1C_1$ and $\omega_2 = 1/R_2C_2$ are the lower and upper cutoff frequencies respectively for the bandpass filter. $\mathrm{BP}(\omega)$ can be rewritten in terms of its magnitude and phase as:

$$\begin{aligned} |\mathrm{BP}(\omega)| &= \frac{1}{\left[\left(1 + \dfrac{R_2C_2}{R_1C_1}\right)^2 + \left(\omega R_2 C_2 - \dfrac{1}{\omega R_1 C_1}\right)^2\right]^{1/2}} \\ &= \frac{1}{\left[\left(1 + \dfrac{\omega_1}{\omega_2}\right)^2 + \left(\dfrac{\omega}{\omega_2} - \dfrac{\omega_1}{\omega}\right)^2\right]^{1/2}} \end{aligned} \qquad \textbf{(Eq. 8.38)}$$

$$\begin{aligned} \theta\,(\text{phase}) &= \arctan\left(\frac{\omega_1}{\omega}\right) - \arctan\frac{\omega}{\omega_2} \\ &= \arctan\left(\frac{1}{\omega R_1 C_1}\right) - \arctan \omega R_2 C_2 \end{aligned} \qquad \textbf{(Eq. 8.39)}$$

Equation 8.27 can also be rewritten expressed as a decibel quantity. Remember that the product of the logarithm of two terms is the sum of the logarithms of each term. Similarly the operation of division of terms can be expressed as the difference of two logarithms. With this in mind we can write:

$$\begin{aligned} \mathrm{BP}(\omega) &= 20\log(1) - 20\log[1 + (\omega R_2 C_2)^2]^{1/2} \\ &\quad - 20\log[1 + (1/\omega R_1 C_1)^2]^{1/2}\ \mathrm{dB} \\ &= 20\log(1) - 20\log[1 + (\omega/\omega_2)^2]^{1/2} - 20\log[1 + (\omega_1/\omega)^2]^{1/2}\ \mathrm{dB} \end{aligned} \qquad \textbf{(Eq. 8.40)}$$

A study of Equation 8.40 shows that the transfer function is equal to:

$$\mathrm{BP}(\omega) = -20\log[1 + (\omega R_1 C_1)^2]^{1/2}\ \mathrm{dB}$$

for low frequencies and that the asymptotic behavior for this region is the same as for a high-filter, namely -20 dB/decade as the frequency drops far from the cutoff frequency (ω_1). For frequencies greater than ω_1 but less than ω_2, the asymptote is parallel to the 0 dB line. Finally, for frequencies greater than ω_2:

$$\mathrm{BP}(\omega) = -20\log[1 + (\omega R_2 C_2)^2]^{1/2}\ \mathrm{dB}$$

which has the same asymptotic behavior as a low-pass filter and falls at 20 dB per decade for frequencies above ω_2.

Figure 8.20 shows a possible implementation of a bandpass filter. Notice that it is made up of a high-pass filter section followed by a low-pass filter section. The Bode Plot (Fig. 8.21) reveals two attenuation curves (above and below the bandpass region). As was the case in the high-pass and low-pass filters, the slope of the curves far from the bandpass region is the same and is equal to -20 dB per decade. However, the transfer function

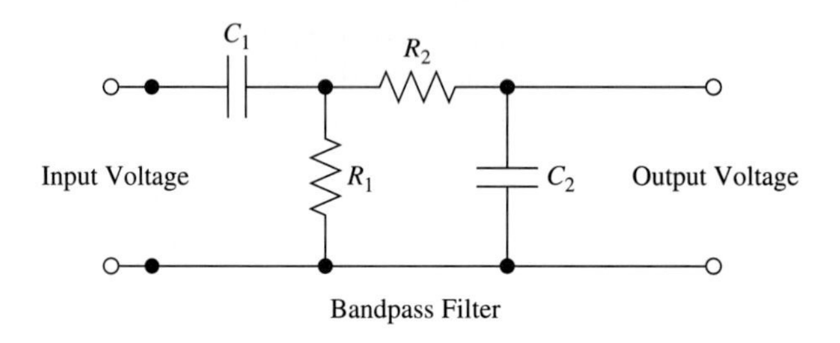

FIGURE 8.20

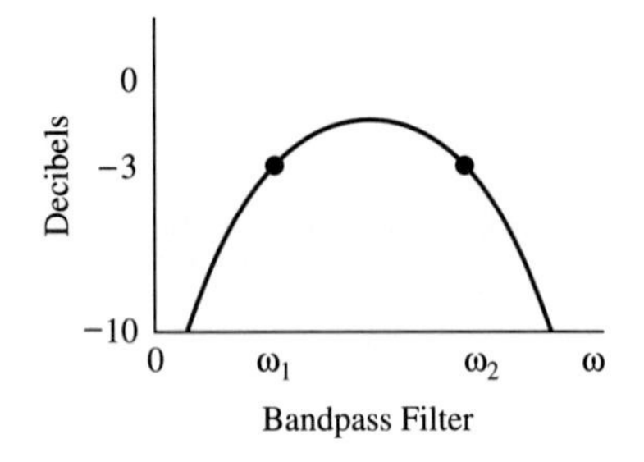

FIGURE 8.21

for the circuit shown is not the same as Equation 8.37. It can be shown with a simple mesh analysis that the transfer function for the circuit shown becomes:

$$|\mathrm{BP}(\omega)| = \frac{1}{\left[\left(1 + \frac{(R_1 + R_2)C_2}{R_1 C_1}\right)^2 + \left(\omega R_2 C_2 - \frac{1}{\omega R_1 C_1}\right)^2\right]^{1/2}}$$

$$= \frac{1}{\left[1 + \frac{(\omega_1/\omega_2)(R_1 + R_2)}{R_2}\right]^2 + \{[(\omega/\omega_2) - (\omega_1/\omega)]^2\}^{1/2}} \qquad \textbf{(Eq. 8.41)}$$

which reduces to Equation 8.37 if R_1 is much smaller than R_2. One interesting difference between the bandpass filter and the high- and low-pass filters is that the bandpass filter has some attenuation in the passband region unless $(\omega_1/\omega_2)(R_1 + R_2)/R_2$ is small compared to 1. This is a result of the interaction between the high-pass and the low-pass sections of the filter as implemented.

8.4.4 Band Rejection Filters

Figure 8.22 shows the transfer function for a band rejection filter. This type of filter cannot be constructed from a simple combination of low-pass and high-pass filters. As its name implies, it allows all but a select band of frequencies to pass through the filter unattenuated. It is the complement of the bandpass filter. One implementation of this filter

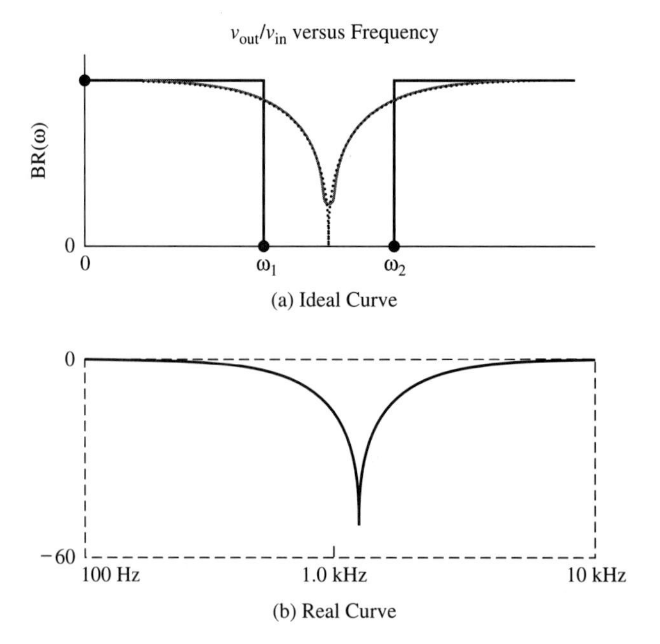

FIGURE 8.22
Band rejection filter

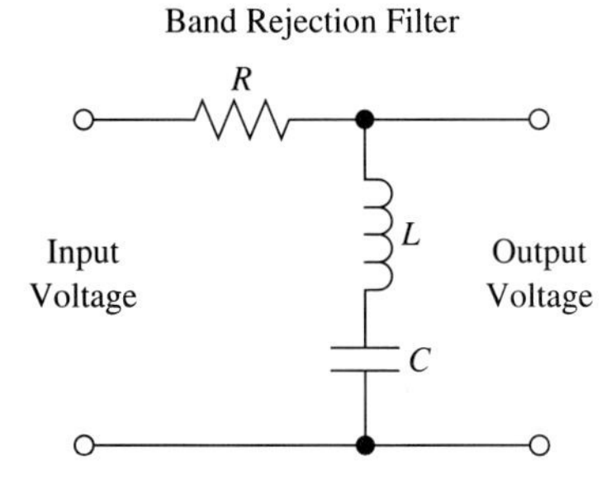

FIGURE 8.23

is shown in Fig. 8.23. The transfer function for this filter can be obtained from an application of mesh analysis as:

$$\mathrm{BR}(\omega) = \frac{1}{1 + j\dfrac{\omega RC}{(1 - \omega^2 LC)}} = \frac{1}{\left(1 + \dfrac{\omega RC}{1 - \omega^2 LC}\right)^{1/2}} \quad \textbf{(Eq. 8.42)}$$

The phase behavior of this filter is derived from the transfer function as:

$$\theta\,(\text{phase angle}) = -\arctan \frac{\omega RC}{(1 - \omega^2 LC)} \quad \textbf{(Eq. 8.43)}$$

Note that for small as well as large values of ω, the filter has a phase angle of 0°. In the rejection region the phase swings rapidly between −90° and +90°.

The peak attenuation of this filter occurs for the frequency at which $1 - \omega^2 LC = 0$. This occurs at:

$$\omega_0 = \frac{1}{\sqrt{L_1 C_1}}\ \text{rad/s} \quad \textbf{(Eq. 8.44)}$$

or

$$f_0 = \frac{1}{2\pi\sqrt{L_1 C_1}}\ \text{Hz} \quad \textbf{(Eq. 8.45)}$$

Note that f_0 is called the resonant frequency of the band rejection filter. The behavior of this filter is unlike that of the other filters that we have studied. It relies on a condition called resonance that will be discussed further in Chapter 9, which deals with tuned circuits.

■ SUMMARY

- A logarithm is an exponent of a number. The number is called the base of the logarithm. Logarithms have an integer part called a characteristic and a decimal part called a mantissa.
- Common logarithms use the base 10. These logarithms obey the rules of exponents and can be used in electrical calculations. They are written as $\log_{10}$.
- Logarithms that use base e are called natural logarithms and are written as ln, which stands for natural logarithm; e is a number that is equal to approximately 2.71828. Like π it is a transcendental number and goes on forever without repeating its decimal value.
- Logarithms allow a multiplication to be rewritten as an addition or a subtraction.
- A measurement quantity relating power levels in a circuit is called a decibel. A decibel is defined as 10 times the $\log_{10}$ of a power ratio (P_2/P_1).
- Decibel measurements allow wide variations in size to be expressed using much smaller numbers.
- An expression that relates output signal power to input signal power as a function of frequency is called a transfer function. The use of logarithms allows the behavior of electrical circuits to be easily displayed over several decades of frequency variation. This variation enables analysis of how circuits change as the applied frequency of the signal varies.
- Circuits that change their behavior as a function of the applied signal frequency are called filters.
- A filter that allows only low frequencies to pass through the circuit is called a low-pass filter.
- A circuit that allows only high frequencies to pass through a circuit is called a high-pass filter.
- A circuit that allows only a band of frequencies to pass through a circuit is called a bandpass filter.
- A circuit that rejects only a band of frequencies but passes all others is called a band rejection filter.

■ PROBLEMS

8.1 Find the values of the following logarithms. List their characteristic and mantissa separately:
(a) 3.756 **(b)** 23.2 **(c)** 1345
(d) 0.0123 **(e)** 0.987

8.2 Use logarithms to solve the following problems:
(a) 3.14·6.79/2.82
(b) 163·456/(22·13)
(c) 237·15/23
(d) 345·297/15.9
(e) 2.34/(3.67·5.73·239)

8.3 Repeat problem 1 using natural logarithms rather than common logarithms.

8.4 For the following values of x find the value of 10^x and e^x:
(a) 3.76 **(b)** -2.9 **(c)** 0.235 **(d)** -1.346
(e) -0.123

8.5 Show that $\log_{10} x = 0.434 \ln x$.

8.6 Show that $2.30 \log_{10} x = \ln x$.

8.7 Three amplifiers of gain 40, 60, and 120 are placed in series with other to form one large amplifier. Find the dB gain of each amplifier and the gain of the total amplifier in dB. A 0 dBm signal is fed into the amplifier. Calculate the dbm rating of the output signal. How many volts does this represent?

8.8 An amplifier with an input impedance of 200 Ω and a voltage gain of 50 delivers its power into an 8-Ω speaker. If the input signal to the amplifier is 1 V:
(a) What is the dB gain of the amplifier?
(b) What is the output power?
(c) What is the output voltage of the amplifier?

8.9 A 3-μV signal is picked up by a 50-Ω cable. How many dbm does this signal represent?

8.10 A frequency of 2 kHz is to be increased by 3.5 octaves. What is the new frequency? How many decades does this represent?

8.11 Plot the reactance of an inductor of 2 H on a log–log scale as the frequency goes from 10 Hz to 10 kHz. On the same set of axes plot the reactance of a 1-μF capacitor.

8.12 On a log–log graph plot the impedance of a 2000-Ω resistor in parallel with a 1-H inductor as the frequency varies from 10 Hz to 100 kHz. Draw the asymptotes and describe the behavior of the circuit. Plot the phase angle of the impedance versus frequency over the same range.

8.13 A low-pass filter is needed to filter a signal. If the cutoff frequency is to be 2500 Hz and the resistance used is to be 10,000 Ω:
(a) What size should the capacitor be?
(b) At what frequency is the phase shift $-30°$?
(c) At what frequency is the attenuation 5 dB?
(d) What is the phase angle at this frequency?

8.14 Draw the amplitude and phase angle plots for the filter of problem 8.13 for a frequency range of 100 Hz to 20 kHz.

8.15 The filter of problem 8.13 is to be built using an inductor in place of the capacitor.
(a) How does this change the circuit? Draw the new circuit.
(b) What size inductor should be used?
(c) How does this change affect the plots of amplitude and phase for the filter?

8.16 Design a high-pass filter with a cutoff frequency of 3000 Hz using a 4.7-kΩ resistor.
(a) What size should the capacitor be?
(b) At what frequency is the phase shift 30°?
(c) At what frequency is the attenuation 8 dB?
(d) What is the phase angle at this frequency?

8.17 Draw the transfer function amplitude and phase for problem 8.16. Include the asymptotes.

8.18 At what frequency does the filter of problem 8.16 have 60 dB of attenuation?

8.19 For the filter of problem 8.16 at what frequency is the phase angle?
(a) 5° **(b)** 10° **(c)** 15°

8.20 For the band rejection filter of Figure 8.23 find the transfer function if $R = 2.2$ kΩ, $C = 0.1$ μF, and the resonant frequency is 3000 Hz. Plot the curves for amplitude and phase of the resultant filter. Assume that the inductor has a 10-Ω resistance.

CHAPTER 9

R–L–C CIRCUITS

INTRODUCTION

A circuit consisting of a series or parallel combination of resistance, inductance, and capacitance is called an *R–L–C* circuit. These circuits have properties that vary as the frequency of a signal applied to the circuit varies. The properties can be adjusted by changing the values of the resistance, inductance, and capacitance to create all kinds of filters. In this chapter we will study the properties of these circuits to understand why they behave as they do. In addition we will study how the values of the elements making up the circuit can be selected to create desired circuit properties.

9.1 THE SERIES R–L–C CIRCUIT

Figure 9.1 shows a simple series circuit consisting of a resistor R, an inductor L, and a capacitor C. In the circuit shown all of the elements are in series. This series circuit is known as a series $R–L–C$ circuit because of the letters customarily used to identify resistors, inductors, and capacitors respectively.

The resistance in any circuit is an inescapable consequence of using real components rather than ideal components to fabricate a circuit. We cannot construct a coil of wire (an inductor) without considering the resistance of the wire used in its construction. A capacitor cannot be constructed that does not have some series resistance associated with the foil and the leads of the capacitor. ***The total resistance in this series circuit is the sum of the resistance associated with each element plus the resistance added to the circuit on purpose. We consider the total resistance to be one element and have labeled the element R.*** We will find that the value of the total resistance plays a major role in the performance of the circuit.

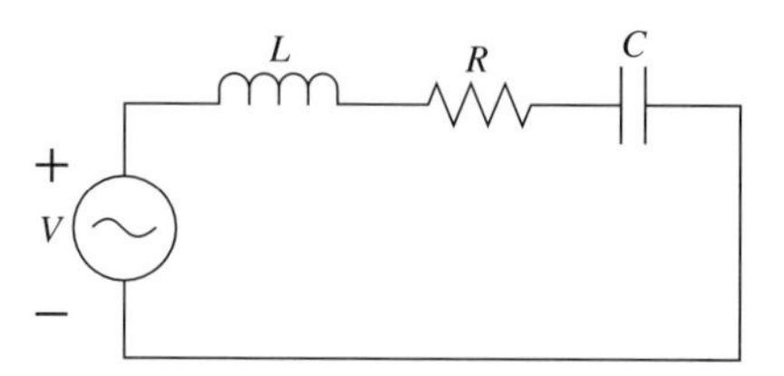

FIGURE 9.1
Series R–L–C circuit

Understanding of the behavior of this circuit starts with the study of its impedance as a function of frequency. The impedance of this circuit may be written as

$$Z(\omega) = R + jX = R + j\omega L - j/\omega C = R + j(\omega L - 1/\omega C) \qquad \textbf{(Eq. 9.1)}$$

Note that the impedance has a real part R that is independent of frequency and an imaginary part jX that is a function of frequency. The minimum impedance of the circuit will occur at a frequency ω_0 for which the imaginary part of the impedance is zero. At this frequency, the circuit is said to be at resonance and its impedance will be purely resistive and equal to R. If we apply a voltage source of this frequency, the circuit will behave as if it is purely resistive. Setting the imaginary part equal to zero we find the resonant frequency. Thus:

$$\omega_0 L - (1/\omega_0 C) = 0$$

from which we find:

$$\omega_0 L = 1/\omega_0 C$$
$$\omega_0^2 = 1/LC \qquad \textbf{(Eq. 9.2)}$$
$$\omega_0 = 1/\sqrt{LC}\ \text{rad/s}$$

but

$$\omega_0 = 2\pi f_0$$

Therefore:

$$f_0 = \frac{1}{2\pi\sqrt{LC}}\ \text{Hz} \qquad \textbf{(Eq. 9.3)}$$

Equation 9.3 relates the signal frequency in hertz for which the tuned circuit has the minimum impedance to the value of the inductance and the value of the capacitance of the circuit. This frequency, denoted f_0, is called the ***series resonant frequency*** of the circuit of Fig. 9.1. The value of L (inductance) is expressed in henries, and the value of C (capacitance) is expressed in farads. At the resonant frequency the circuit impedance equals R, the total series resistance. Since the impedance is ***real*** at the resonant frequency, the current through the circuit is related to the voltage by Ohm's Law and is in phase with the voltage

$$I_{max} = V/R \qquad \textbf{(Eq. 9.4)}$$

EXAMPLE 9.1

Find the resonant frequency of an inductor of 0.1 H in series with a capacitor of 2 μF.

Solution Using Equation 9.3 we find:

$$f_0 = 1/2\pi\sqrt{LC} = 1/2\pi\sqrt{0.1\cdot 2\cdot 10^{-6}} = 356 \text{ Hz}$$

EXAMPLE 9.2

A circuit is desired that will be resonant at 2 MHz. If a 10-μH inductor is available, what should be the value of the capacitor used to form the circuit?

Solution Using Equation 9.2 we find:

$$\begin{aligned}\omega_0^2 &= 1/LC \\ C &= 1/(2\ \pi)^2(f_0)^2L = 1/(2\ \pi)^2\ (2\cdot 10^6)^2(10\cdot 10^{-6}) \\ &= 633\cdot 10^{-12}\text{ F} = 633\text{ pF} \\ &= 633\text{ pF}\end{aligned}$$

At resonance the voltage source is the same as the voltage across the resistor because the reactive components of the circuit sum to zero. The inductor has an impedance equal to $Z_L = j\omega_0 L$, and the capacitor has an impedance equal to $Z = -j/\omega_0 C$. These two impedances are equal in magnitude but differ in sign. The inductor voltage leads the current by 90° while the capacitor voltage lags the current by 90°. Since the magnitude of the impedance for both the capacitor and inductor is the same, the phase angle of 180° between their voltages causes the voltage across the inductor to negate the voltage across the capacitor. ***This does not mean that there is no voltage across the capacitor and inductor but rather that the sum of the voltages across the capacitor and inductor equals zero.***

Let us plot a phasor diagram for the circuit to graphically show the relationship between the current in the circuit and the voltage across each element. A phasor diagram of the complete circuit can be constructed from the sum of the phasor diagrams for each element. Since the current I is common to all of the circuit elements, we will use it as a reference and plot it on the x axis for each element's phasor diagram. The voltage across the resistor is in phase with the current and therefore is plotted along the x axis. Since the voltage across the inductor leads the current by 90°, the voltage is drawn in the positive y direction on the y axis. Finally, the voltage across the capacitor lags the current by 90°, so it is drawn on the negative y axis. The phasor diagrams are shown in Fig. 9.2. ***Since the input voltage is the sum of the phasor voltages in the circuit, the phasor representation of the circuit is the sum of the phasor voltages for each element.*** It is equal in magnitude to the resistor voltage phasor and, like the resistor phasor voltage, is in phase with the current.

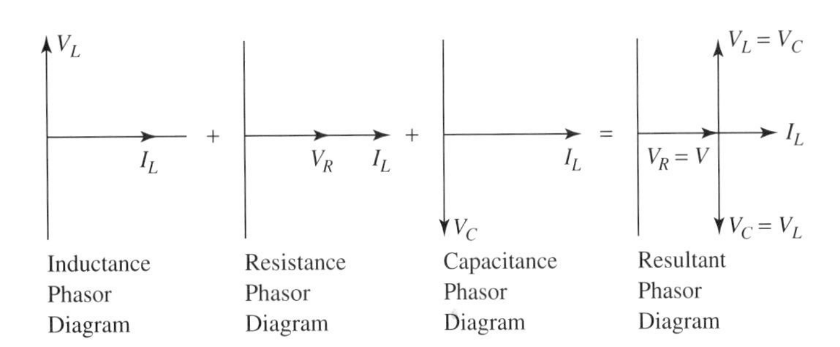

FIGURE 9.2
Phasor diagrams at resonance

At frequencies below resonance the reactance of the capacitor is larger than the reactance of the inductor. This means that the voltage across the capacitor is larger than the voltage across the inductor. The capacitor voltage is therefore larger than the inductor voltage, and the net reactive component of voltage is capacitive in nature. Thus, below resonance the circuit behaves like a capacitor in series with a resistor and the input voltage lags the input current. This is shown graphically in Fig. 9.3.

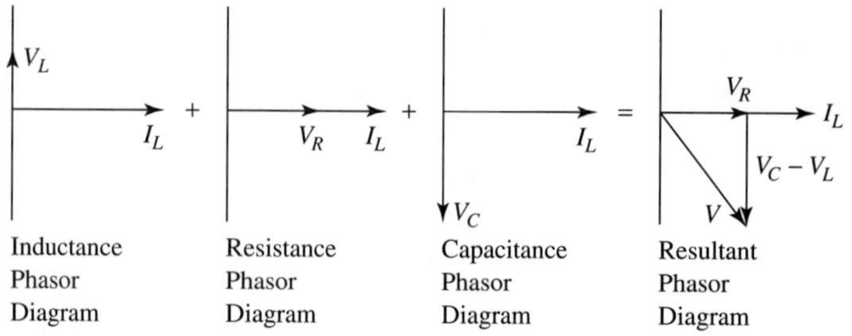

V_L I_L + V_R I_L + V_C I_L = V_R I_L $V_C - V_L$ V

Inductance Phasor Diagram | Resistance Phasor Diagram | Capacitance Phasor Diagram | Resultant Phasor Diagram

FIGURE 9.3
Phasor diagrams below resonance

Above the resonant frequency the reactance of the inductor exceeds the reactance of the capacitor. The voltage across the inductor is larger than the voltage across the capacitor, and the circuit behaves like a series R–L circuit. The input voltage leads the input current. This is shown in Fig. 9.4.

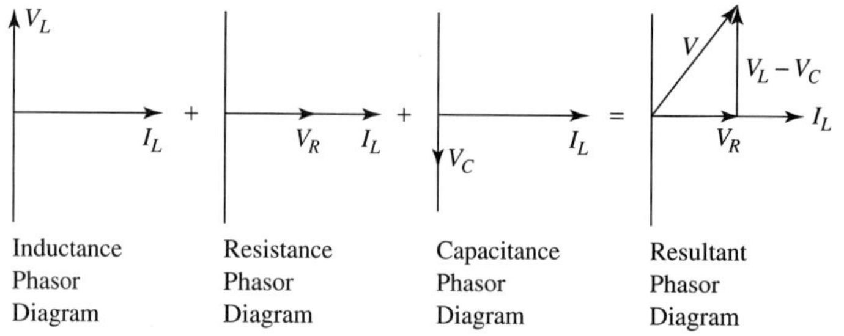

V_L I_L + V_R I_L + V_C I_L = V $V_L - V_C$ I_L V_R

Inductance Phasor Diagram | Resistance Phasor Diagram | Capacitance Phasor Diagram | Resultant Phasor Diagram

FIGURE 9.4
Phasor diagrams above resonance

Equation 9.1, which relates the impedance of the circuit to the frequency of the applied signal, may be normalized to express the frequency behavior of the circuit as a function of the dimensionless variable ω/ω_0. This allows us to plot curves that arc independent of the actual resonant frequency. They are called universal curves because they show the behavior of all series R–L–C circuits independent of the circuit's actual resonant frequency.

$$\begin{aligned} Z(\omega) &= R + j[\omega L - (1/\omega C)] \\ &= R + j\omega L[1 - 1/(\omega^2 LC)] \\ &= R + j\omega L[1 - (\omega_0^2/\omega^2)] \end{aligned} \qquad \textbf{(Eq. 9.1)}$$

Multiplying L by (ω_0/ω_0) we can rewrite the equation as:

$$Z(\omega) = R + j\omega_0 L \frac{\omega^2 - \omega_0^2}{\omega_0/\omega}$$

But

$$\frac{\omega^2 - \omega_0^2}{\omega_0/\omega} = \frac{(\omega^2/\omega_0^2) - 1}{\omega/\omega_0}$$

Therefore:

$$Z(\omega) = R + j\omega_0 L \frac{(\omega^2/\omega_0^2) - 1}{\omega/\omega_0} \quad \textbf{(Eq. 9.5)}$$

Equation 9.5 is the normalized form Equation of 9.1 and contains the factor ω/ω_0. The ratio ω/ω_0, called a normalized frequency variable, expresses all frequency values as a percentage of the resonant central frequency $\omega/\omega_0 = 1$. To convert the curve to a given central frequency, we simply multiply the curve's normalized frequency value by the frequency ω_0. Note that $\omega/\omega_0 = 1$ is the same as saying that $\omega = \omega_0$. The advantage of using ω/ω_0 rather than ω as a variable is that when plots are universal and independent of the actual resonant frequency, a single plot can be used for any resonant frequency because the curves are functions of only the relative frequency.

We note from Equation 9.5 that the impedance at resonance is purely resistive and equal to R because at $\omega/\omega_0 = 1$ the imaginary part of the impedance equals zero. For frequencies lower than resonance $(\omega^2/\omega_0^2)/ - 1$ is negative. This means that the reactance is negative, which means that the capacitive component of the reactance dominates the imaginary part of the impedance. Above resonance, where ω/ω_0 is greater than 1, the reactance is a positive value, which means it is inductive in nature.

If the impedance is converted from a rectangular representation to a polar representation, graphs of circuit behavior versus frequency may be plotted for magnitude and phase angle properties separately, as was done in Chapter 8. Equation 9.5 is converted to a magnitude by taking the square root of the sum of the squares of the real and imaginary parts of the impedance. The corresponding phase behavior is calculated by taking the ratio of the imaginary part of the impedance and dividing it by the real part of the impedance. This ratio is the arctangent of the phase angle. Making this conversion we find that:

$$|Z(\omega)| = R\sqrt{1 + \frac{(\omega_0 L)^2}{R^2} \frac{[(\omega/\omega_0)^2 - 1]^2}{(\omega/\omega_0)^2}} \quad \textbf{(Eq. 9.6)}$$

$$\theta(\omega) = \text{Arctan}\left(\frac{\omega_0 L}{R} \cdot \frac{(\omega/\omega_0)^2 - 1}{\omega/\omega_0}\right) \quad \textbf{(Eq. 9.7)}$$

Examination of Equations 9.6 and 9.7 reveals that both equations rely on the factor $(\omega_0 L)/R$. This factor is called the ***quality factor of the series*** **R–L–C** ***circuit*** and is denoted by the symbol $\boldsymbol{Q_s}$. Thus:

$$Q_s = (\omega_0 L)/R = X_L/R \quad \textbf{(Eq. 9.8)}$$

The factor $\boldsymbol{Q_s}$ is the ratio of the reactance of the inductor at resonance to the resistive component of the circuit. By definition, at resonance the magnitude of the inductive reactance is equal to the magnitude of the capacitive reactance. This means that we may also write an expression for the value of $\boldsymbol{Q_s}$ using the formula for the reactance of the capacitor at resonance. Thus we find that

$$Q_s = 1/\omega_0 RC = X_C/R \quad \textbf{(Eq. 9.9)}$$

Several different interpretations can be made of this quality factor $\boldsymbol{Q_s}$. For example, if we multiply the top and bottom of Equation 9.8 by the magnitude of the current in the circuit I we find:

$$Q_s = I(\omega_0 L)/IR = IX_L/IR = |V_L/V_R| \quad \textbf{(Eq. 9.10)}$$

This is the voltage magnitude across the inductor (or the capacitor if we use Eq. 9.9) divided by the voltage magnitude across the resistor, both evaluated at the resonant frequency. By proper choice of component values it is quite possible to get values of Q_s ranging from less than 1 to over 100. As a result, at resonance the voltage across the inductor may be many times larger than the voltage across the resistor. The voltage across the resistor, however, is the same as the voltage applied to the circuit. ***It is amazing but true that the voltage across the inductor (or capacitor) can be over 100 times as large as the voltage applied to the circuit.*** This phenomenon, called ***resonant boost,*** is easily demonstrated in the laboratory.

If we multiply Equation 9.8 by the square of the current magnitude we get:

$$Q_s = I^2(\omega_0 L)/I^2R = I^2X_L/I^2R = P_L/P_R \qquad \textbf{(Eq. 9.11)}$$

This ratio is the power stored (reactive power) in the inductor divided by the power dissipated in the series resistance. The ratio of impedances, the ratio of voltages, and the ratio of power stored to power dissipated are three different interpretations for the same quantity Q_s. Any of the three ratios can be used for the definition of the value of Q_s of a series circuit.

One more representation can be used for the value of Q_s. We can eliminate the resonant frequency dependence from the expression for Q_s through the use of Equation 9.2. Substitution of Equation 9.2 into Equation 9.8 yields:

$$Q_s = \frac{1}{R}\sqrt{L/C} \qquad \textbf{(Eq. 9.12)}$$

This final expression for Q_s demonstrates that Q_s is a function of only the physical parameters of the circuit. The value of Q_s is strongly dependent on the total resistance in the circuit. This resistance is the sum of the resistance associated with the voltage source, the resistance of the series resistor in the circuit, the resistance associated with the capacitor, and the resistance of the inductor itself.

If the inductor is considered by itself, one can use Equation 9.8 to calculate its individual value of Q_s. This value is usually specified for inductors and designated simply Q_L, since it relates only to the inductor and not the complete circuit. In creating circuits with high values of Q_s, it is the inductor that generally limits the maximum circuit quality factor. Desired values for Q_s can only be achieved with coils whose Q is greater than the desired circuit value for Q_s. This is a consequence of the observation that the total resistance can never be less than the resistance of the coil itself. Therefore the Q_s of the circuit must be lower than the value for the coil itself because of the added resistance not due to the coil.

Using this definition of Q_s, Equations 9.6 and 9.7 can be rewritten as:

$$|Z(\omega)| = R\sqrt{1 + Q_s^2\frac{[(\omega/\omega_0)^2 - 1]^2}{(\omega/\omega_0)^2}} \qquad \textbf{(Eq. 9.13)}$$

$$\theta(\omega) = \text{Arctan}\, Q_s\frac{(\omega/\omega_0)^2 - 1}{\omega/\omega_0} \qquad \textbf{(Eq. 9.14)}$$

Figure 9.5 shows that a plot of $|Z(\omega)|$ versus ω/ω_0 for $R = 100\ \Omega$ and several different values of Q_s. Notice that the minimum value of resistance is 100 Ω for all plots but that the shape of the curve changes as the value of Q_s changes. If we consider the frequency for which the impedance has equal real and imaginary parts, $\{Q_s^2\,[(\omega/\omega_0)^2 - 1]^2/(\omega/\omega_0)^2 = 1\}$, for example, we find that as the value of Q_s increases, the frequency for equal real and imaginary parts is closer to the resonant frequency. The difference in frequency between the upper and lower points corresponding to equal impedance levels becomes smaller. Correspondingly, larger values of Q_s result in larger values of $Z(\omega)$ for a given percentage frequency deviation from resonance. At resonance the impedance for each curve is the same value, namely, $Z = R = 100\ \Omega$.

Figure 9.6 is a plot of the phase angle of the impedance versus ω/ω_0 corresponding to the values of Q_s of Figure 9.5. The phase angle approaches $-90°$ for very low frequencies, passes through zero degrees at resonance and approaches $+90°$ for frequencies

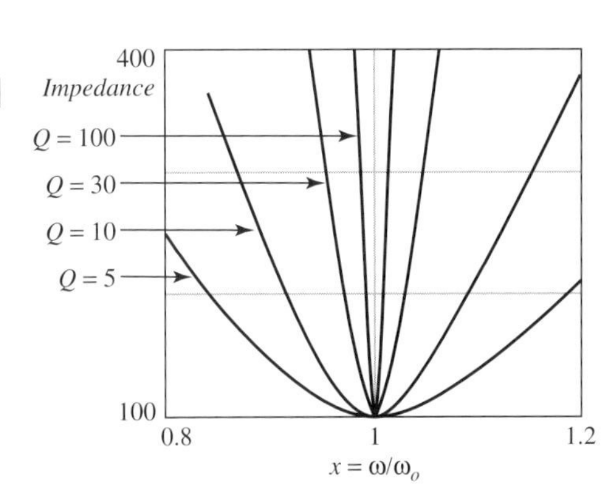

FIGURE 9.5
$|Z(\omega)|$ *versus* $x[x = (\omega/\omega_0)]$

far above resonance. Observation of Equation 9.14 reveals a factor of Q_s as a multiplier of the phase-angle expression. This means that for all values of ω/ω_0 away from resonance, larger values of Q_s produce larger magnitude phase angles for a given frequency. At low frequencies the current leads the voltage through the circuit; at high frequencies the current lags the voltage.

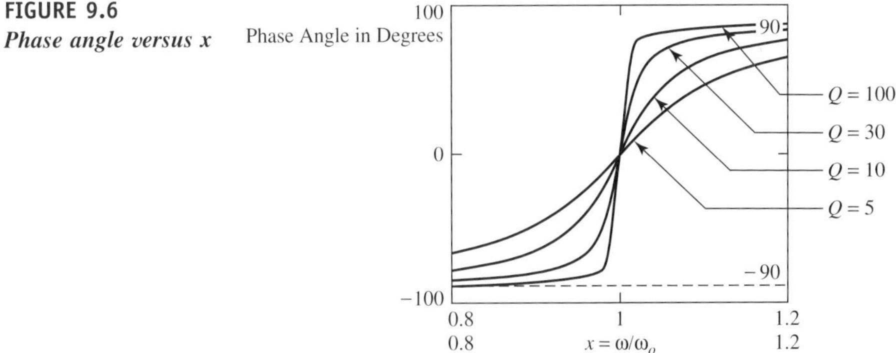

FIGURE 9.6
Phase angle versus x

9.1.1 Selectivity of Series *R–L–C* Circuits

Series R–L–C circuits can be used as filters to select only those signals within a specified band of frequencies from the entire frequency spectrum. Consider Fig. 9.7. In this circuit the output is taken across the resistor element of the series circuit. Using the Voltage Divider Theorem, the value of that output signal can be shown to be:

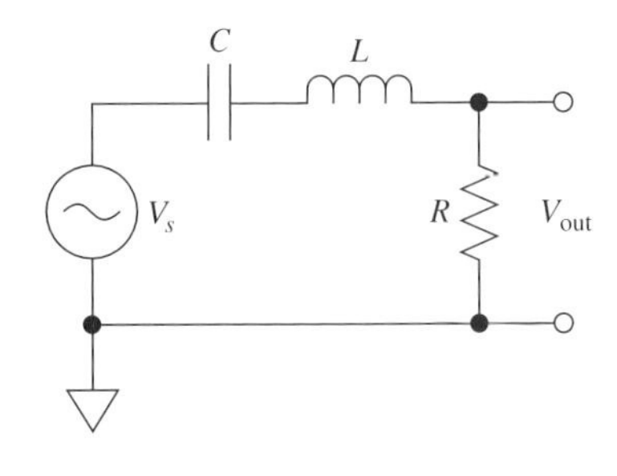

FIGURE 9.7

$$V_{out}(\omega) = \frac{R}{Z(\omega)} V_s \qquad \textbf{(Eq. 9.15)}$$

Using Equation 9.13 for the value of $Z(\omega)$, the voltage transfer function of the circuit may be written as:

$$\frac{|V_{out}(\omega)|}{|V_s(\omega)|} = \frac{1}{\sqrt{1 + Q_s^2 \dfrac{[(\omega/\omega_0)^2 - 1]^2}{(\omega/\omega_0)^2}}} \qquad \textbf{(Eq. 9.16)}$$

The corresponding phase angle behavior can be expressed as:

$$\theta(\omega) = -\arctan Q_s \frac{(\omega/\omega_0)^2 - 1}{\omega/\omega_0} \qquad \textbf{(Eq. 9.17)}$$

Figure 9.8 shows the frequency behavior of the transfer function magnitude. As we did in Chapter 8, ***we define the bandwidth of the circuit to equal the frequency interval between the upper- and the lower-frequency half-power points (same as 3 dB***

points) of the curve. Note that as the value of Q_s increases, the bandwidth of the circuit decreases.

FIGURE 9.8

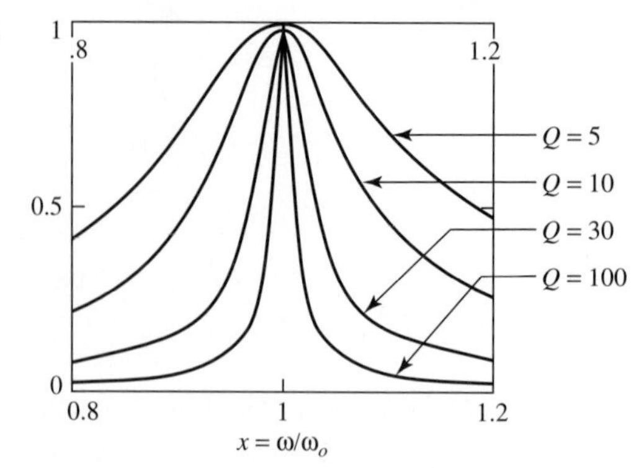

The phase angle behavior of the transfer function is shown in Fig. 9.9. It is related to the curves of Fig. 9.7, which plot the phase angle of the impedance of the circuit. Equation 9.15 has $Z(\omega)$ in the denominator. This causes the phase-angle behavior to differ in sign (see Eqs. 9.14 and 9.17).

FIGURE 9.9

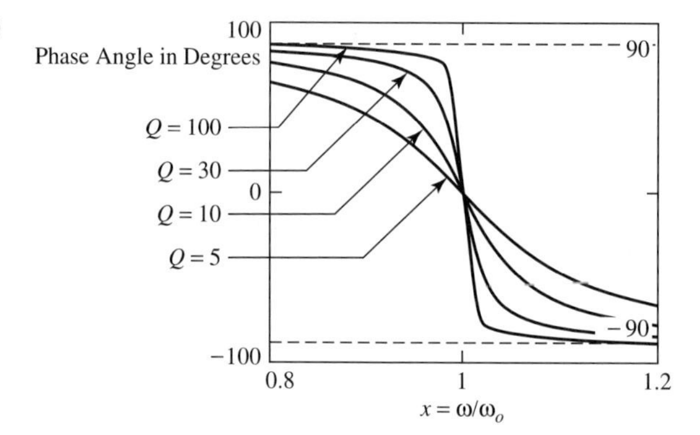

At the half-power points on the transfer curve the real and imaginary parts of the transfer function are equal and therefore the arctangent of the phase angle equals ± 1. This corresponds to a phase angle of $\pm 45°$. Using this fact we can solve for the upper and lower half-power frequencies.

Substituting into Equation 9.17, where $\theta = 45°$, we find that:

$$\pm 1 = \boldsymbol{Q_s}\frac{(\omega/\omega_0)^2 - 1}{\omega/\omega_0}$$

because the tangent of $\pm 45°$ equals ± 1. Setting $x = \omega/\omega_0$ we can write this equation as:

$$\pm 1 = \boldsymbol{Q_s}\frac{(x^2 - 1)}{x}$$

which converts to the quadratic equation set:

$$x^2 \mp \frac{x}{Q_s} - 1 = 0$$

This set is actually two different quadratic equations—one with a positive sign for the linear coefficient of x and one with a negative sign. The two equations have four roots between them: two for each of the equations. Solving for x by use of the quadratic formula we find that:

$$x = \pm\frac{1}{2\boldsymbol{Q_s}} \pm \sqrt{\left(\frac{1}{2\boldsymbol{Q_s}}\right)^2 + 1}$$

There are two solutions for each of the equations. Since the value of x must be positive, so that the solution corresponds to a real frequency greater than zero, and the square root is larger than the factor $-1/(2Q_s)$, it follows that the two values for x are given by:

$$x_1 = -\frac{1}{2Q_s} + \sqrt{\left(\frac{1}{2Q_s}\right)^2 + 1} \qquad \textbf{(Eq. 9.18)}$$

$$x_2 = +\frac{1}{2Q_s} + \sqrt{\left(\frac{1}{2Q_s}\right)^2 + 1} \qquad \textbf{(Eq. 9.19)}$$

where $x_1 = \omega_1/\omega_0$ and $x_2 = \omega_2/\omega_0$.

Equation 9.18 yields the formula for the lower half-power-point frequency; Equation 9.19 yields the formula for the upper half-power-point frequency. Substituting $x_1 = \omega_1/\omega_0$ and $x_2 = \omega_2/\omega_0$, we find the results we seek:

$$\omega_1 = \omega_0\left(-\frac{1}{2Q_s} + \sqrt{\left(\frac{1}{2Q_s}\right)^2 + 1}\right) \qquad \textbf{(Eq. 9.20)}$$

$$\omega_2 = \omega_0\left(+\frac{1}{2Q_s} + \sqrt{\left(\frac{1}{2Q_s}\right)^2 + 1}\right) \qquad \textbf{(Eq. 9.21)}$$

We note that Equations 9.18 and 9.19 can also be written in terms of frequency in hertz rather than in radian frequency measure. Using this approach the equations become:

$$f_1 = f_0\left(-\frac{1}{2Q_s} + \sqrt{\left(\frac{1}{2Q_s}\right)^2 + 1}\right) \qquad \textbf{(Eq. 9.20a)}$$

$$f_2 = f_0\left(\frac{1}{2Q_s} + \sqrt{\left(\frac{1}{2Q_s}\right)^2 + 1}\right) \qquad \textbf{(Eq. 9.21a)}$$

Careful analysis of Equations 9.20 and 9.21 reveals that the transfer function curve's half-power-point frequencies are not symmetric with respect to the frequency ω_0 but rather are symmetric with respect to a somewhat higher frequency given by the expression $\omega_0\sqrt{[1/(2Q_s)]^2 + 1}$, which is dependent on Q_s as well as on the resonant center frequency ω_0. For values of Q_s greater than 5, however, the error in symmetry of the half-power points is less than 0.5 percent. From this it follows that for all Q_s greater than 5, the half-power frequencies are symmetric around ω_0.

EXAMPLE 9.3

A circuit is adjusted to have a resonant frequency of 10 MHz. If the value of Q_s for the circuit is 10, find the lower and upper cutoff frequencies for the circuit.

Solution Substituting into Equation 9.18a we obtain the lower cutoff frequency as:

$$f_1 = f_0\left(-\frac{1}{2Q_s} - \sqrt{\left(\frac{1}{2Q_s}\right)^2 + 1}\right) = 10^7\left(-\frac{1}{20} + \sqrt{\left(\frac{1}{20}\right)^2 + 1}\right)\text{hz}$$

$$= 10^7(0.9512) = 9.512 \text{ MHz}$$

Similarly, using Equation 9.19a we obtain the upper cutoff frequency as:

$$f_2 = f_0\left(+\frac{1}{2Q_s} + \sqrt{\left(\frac{1}{2Q_s}\right)^2 + 1}\right) = 10^7\left(\frac{1}{20} + \sqrt{\left(\frac{1}{20}\right)^2 + 1}\right)$$

$$= 10^7(1.0512) = 10.512 \text{ MHz}$$

Looking at the results we find that the lower cutoff frequency is 488 kHz below the resonant frequency and the upper cutoff frequency is 512 kHz above the resonant frequency. This is a difference of 24 kHz or 2.4 percent of the bandwidth.

The quantity $\omega_2 - \omega_1$ represents the bandwidth of the circuit. Using Equations 9.20 and 9.21 to form this difference produces the equation:

$$\omega_2 - \omega_1 = \frac{\omega_0}{Q_s} \quad \textbf{(Eq. 9.22)}$$

from which one more definition for Q_s appears, namely:

$$Q_s = \frac{\omega_0}{\omega_2 - \omega_1} \quad \textbf{(Eq. 9.23)}$$

This formula, which relates the ratio of resonant frequency to bandwidth, can also be written as:

$$Q_s = \frac{f_0}{f_2 - f_1}\ \text{Hz} \quad \textbf{(Eq. 9.23a)}$$

where frequencies are expressed in hertz instead of radians per second. Equations 9.23 and 9.23a can be used to experimentally measure the actual value of Q_s for a circuit and are therefore very valuable in circuit design. Another interesting relationship between the cutoff frequencies and the resonant frequency ω_0 can be obtained by forming the product of ω_1 and ω_2 as given by Equations 9.20 and 9.21. We discover that for all values of Q_s:

$$\omega_1 \cdot \omega_2 = (\omega_0)^2 \quad \textbf{(Eq. 9.24)}$$

$$f_1 \cdot f_2 = (f_0)^2 \quad \textbf{(Eq. 9.24a)}$$

EXAMPLE 9.4

A radio station in the AM band has a center frequency of 1010 kHz and a bandwidth of 15 kHz. What is the value of Q_s that they can use for their radio transmitter circuits?

Using Equation 9.23a we can solve for the value of Q_s:

$$Q_s = \frac{f_0}{f_2 - f_1}\ \text{Hz}$$

$$= \frac{1010}{15} = 67.3$$

EXAMPLE 9.5

A circuit is needed to operate at 100 kHz with a bandwidth of 8 kHz. If the voltage source for this circuit has a magnitude of 1 V, has an internal resistance of 25 Ω, and the coils available have a value of a resistance of 20 Ω and a $Q = 50$ find the values of R, L, and C, which make up the circuit. Find the magnitude of the current through the circuit and the voltage across the capacitor, the inductor and the resistor.

Solution Using Equation 9.23a yields a value of $Q_s = 100/8 = 12.5$ for the circuit. If the value of series resistance is R, then by Equation 9.8 we find that:

$$12.5 = \omega_0 L/(25 + R + 20)$$

and for the coil itself we find:

$$50 = \omega_0 L/20$$

from which we find that

$$\omega_0 L = 1000 \text{ and } R = 35\ \Omega$$

From the impedance of the coil we find the value of the inductance as

$$L = 1000/(2\pi \cdot 10^5) = 1.59 \text{ mH}$$

But the impedance at resonance of the capacitor is also 1000 Ω, from which the value of the capacitor to be used is:

$$C = 1/(2\pi \cdot 10^5 \cdot 1000) = 0.00159\ \mu\text{F}$$

At resonance the circuit is purely resistive and equal to 80 Ω. From Ohm's Law this means that the current in the circuit is given by:

$$I = V/R_{\text{total}} = 1/80 = 12.5 \text{ mA}$$

The voltage across the series resistance is:

$$V_R = 12.5 \cdot 10^{-3} \cdot 35 = 0.4375 \text{ V}$$

The voltage across the capacitor is:

$$V_C = 12.5 \cdot 10^{-3} \cdot 10^3 = 12.5 \text{ V}$$

The voltage across the inductor is the algebraic sum of the drop across its resistance and its reactance. The drop across its reactance is 12.5 V (same as the capacitance drop), and the drop across its resistance is given as:

$$V_R = 12.5 \cdot 10^{-3} \cdot 20 = 0.250 \text{ V}$$

The total drop is the vector sum of two voltages. The resistor voltage is in phase with the current, and the inductance drop leads the current by 90°. We add these by taking the square root of the sum of the squares. Thus the voltage across the inductor becomes:

$$V_L = \sqrt{(12.5)^2 + (0.25)^2} = 12.502 \text{ V}$$

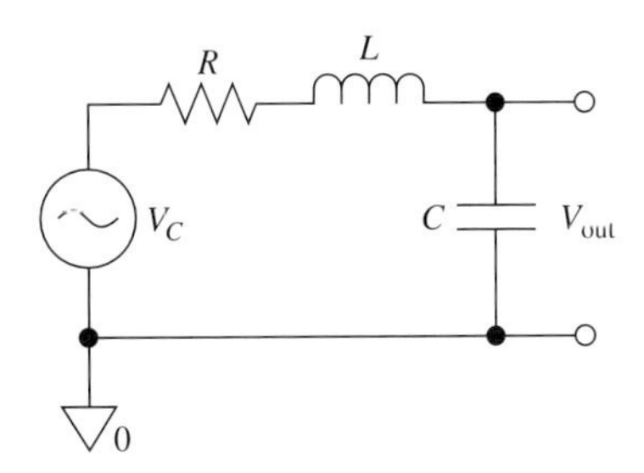

FIGURE 9.10

Before we leave the topic of series resonant circuits, let us consider the output of a related circuit. Figure 9.10 shows a circuit in which the output is taken across the capacitor rather than the resistor. It can be shown that the frequency for which the output amplitude across the capacitor is a maximum is given by:

$$\omega_c = \omega_0 \sqrt{1 - \frac{1}{2(Q_s)^2}} \qquad \textbf{(Eq. 9.25)}$$

Repeating the same analysis for the condition where the output signal is taken across the inductor results in the equation:

$$\omega_L = \omega_0 / \left(\sqrt{1 - \frac{1}{2(Q_s)^2}}\right) \qquad \textbf{(Eq. 9.26)}$$

If we consider the form of Equations 9.25 and 9.26, it appears that values of Q_s less than $(\sqrt{2})/2$ cause the equations to yield imaginary values for ω. ***Therefore, for the series circuit equations to be valid it follows that Q_s must be greater than $(\sqrt{2})/2$.*** If the values of Q_s are not greater than $(\sqrt{2})/2$, then the series circuit has no resonant peak and resonance does not occur. This result is usually only of theoretical interest because circuits employing *R–L–C* networks have values of Q_s much greater than this value.

9.2 PARALLEL *R–L–C* CIRCUITS

A parallel R–L–C circuit is shown in Fig. 9.11. In this circuit the voltage source of Fig. 9.1 has been replaced by a current source and the circuit elements (R_p, L_p, and C_p) are all in parallel with the source. At any frequency, the inductor and capacitor currents are 180° out of phase with each other and therefore the reactive current taken from the source is the difference between the capacitive and the inductive currents. At resonance the impedance of the capacitor and the inductor are the same, so their current magnitudes are the same. This means that the total reactive current sum from the source totals zero. The source current passes only through the resistor. This implies that the maximum impedance of the circuit occurs at the resonant frequency and is purely resistive in nature because the reactive elements in parallel with the resistor draw no net current from the source.

Calculation of the impedance of the circuit of Fig. 9.11 is done in two steps. First calculate the admittance and then take its reciprocal to find the impedance. The admittance of the parallel R–L–C circuit shown is given as:

$$Y(\omega) = \frac{1}{R_p} + j\omega C_p + \frac{1}{j\omega L_p} \qquad \textbf{(Eq. 9.27)}$$

FIGURE 9.11

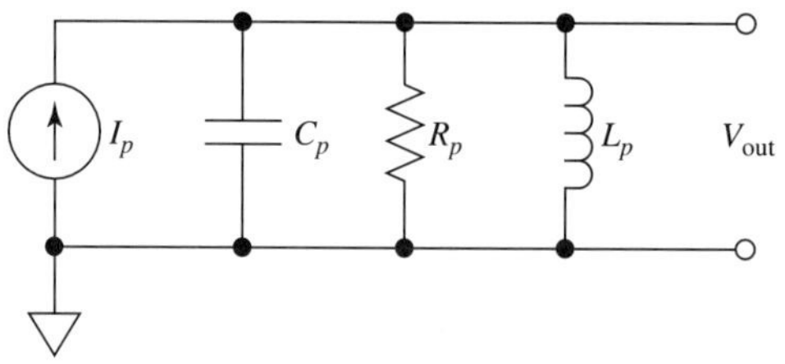

I_p C_p R_p L_p V_{out}

At resonance the imaginary part of the admittance disappears. Setting the imaginary part of the admittance to zero we find that the resonant frequency is given as:

$$\omega_0 = 1/\sqrt{L_p C_p}\ \text{rad/s} \qquad \textbf{(Eq. 9.28)}$$

$$f_0 = 1/2\pi\sqrt{L_p C_p}\ \text{Hz} \qquad \textbf{(Eq. 9.28a)}$$

This is the same expression we saw for the series resonant circuit (see Eq. 9.2). Using Equation 9.28 we can simplify the expression for the circuit admittance. Thus:

$$\begin{aligned}
Y(\omega) &= \frac{1}{R_p} + j\omega C_p\left(1 - \frac{1}{\omega^2 L_p C_p}\right) \\
&= \frac{1}{R_p}\left[1 + j\omega C_p R_p\left(1 - \frac{1}{\omega^2 L_p C_p}\right)\right] \\
&= \frac{1}{R_p}\left[1 + j\omega C_p R_p\left(1 - \frac{(\omega_0)^2}{\omega^2}\right)\right] \\
&= \frac{1}{R_p}\left(1 + j\omega C_p R_p \frac{\omega^2 - (\omega_0)^2}{\omega^2}\right) \\
&= \frac{1}{R_p}\left(1 + jC_p R_p \frac{\omega^2 - (\omega_0)^2}{\omega}\right) \\
&= \frac{1}{R_p}\left(1 + j\omega_0^2 C_p R_p \frac{\left(\frac{\omega}{\omega_0}\right)^2 - 1}{\omega}\right) \\
&= \frac{1}{R_p}\left(1 + j\omega_0 C_p R_p \frac{\left(\frac{\omega}{\omega_0}\right)^2 - 1}{\omega/\omega_0}\right)
\end{aligned}$$

The quantity Q_p, the imaginary part of the admittance divided by the real part of the admittance at resonance, is defined in an analogous fashion to the series *R–L–C* circuit definition. Thus:

$$Q_p = \omega_0 C_p/(1/R_p) = \omega_0 C_p R_p \qquad \textbf{(Eq. 9.29)}$$

At resonance the admittance of the capacitor is equal to the admittance of the inductor. Substitution into Equation 9.29 produces another expression for the quality factor of the circuit. Thus:

$$Q_p = R_p/\omega_0 L_p \qquad \textbf{(Eq. 9.30)}$$

Using Equation 9.28 we can eliminate the dependence of Q_p on the resonant frequency and find that we can write Q_p as:

$$Q_p = R_p\sqrt{C/L} \qquad \textbf{(Eq. 9.31)}$$

Notice that the equations for the quality factor for a parallel *R–L–C* circuit are the reciprocals of the equations for the series circuit (see Eqs. 9.8, 9.9, and 9.12).

Using Equation 9.29 we can write the admittance of the parallel circuit as:

$$Y(\omega) = \frac{1}{R}\left(1 + jQ_p\frac{\left(\frac{\omega}{\omega_0}\right)^2 - 1}{\omega/\omega_0}\right) \qquad \textbf{(Eq. 9.32)}$$

Taking the reciprocal of Equation 9.32 produces the impedance $Z(\omega)$.

$$Z(\omega) = \frac{R_p}{1 + jQ_p\dfrac{\left(\frac{\omega}{\omega_0}\right)^2 - 1}{\omega/\omega_0}} \qquad \textbf{(Eq. 9.33)}$$

Converting the expression into polar form produces:

$$= \frac{|Z(\omega)|}{R_p} = \frac{1}{\sqrt{1 + Q_p^2\dfrac{\left(\left(\frac{\omega}{\omega_0}\right)^2 - 1\right)^2}{(\omega/\omega_0)^2}}} \qquad \textbf{(Eq. 9.34)}$$

$$\theta(\omega) = -\arctan\left(Q_p\frac{\dfrac{\omega^2}{(\omega_0)^2} - 1}{\omega/\omega_0}\right) \qquad \textbf{(Eq. 9.35)}$$

If $|Z(\omega)|$ is multiplied by the magnitude of the current source, the result is the output voltage across the parallel circuit for any value of the normalized frequency ω/ω_0. The value of the output at resonance is given as IR_p because the impedance equals R_p. Equation 9.34 then represents the output voltage at any frequency divided by the output voltage at resonance. This is the magnitude of the transfer function of the circuit drawn in Fig. 9.11. Equation 9.35 yields the phase-angle performance.

Comparing Equations 9.16 and 9.17 to Equations 9.33 and 9.34, we see that the form of the equations is identical. This means that a parallel *R–L–C* circuit driven by a current source produces the same kind of output voltage versus frequency curves as the series *R–L–C* circuit driven by a voltage source and shown in Fig. 9.8. Since the two circuits have identical phase-angle behavior, the consequences in regard to the cutoff frequencies are the same. Thus Equations 9.20 through 9.24 are exactly the same provided we replace Q_s in the formulas by the value of Q_p.

The lower cutoff frequency for a parallel circuit is given as:

$$\omega_1 = \omega_0\left(-\frac{1}{2Q_p} + \sqrt{\left(\frac{1}{2Q_p}\right)^2 + 1}\right)\text{rad/s} \qquad \textbf{(Eq. 9.36)}$$

$$f_1 = f_0\left(-\frac{1}{2Q_p} + \sqrt{\left(\frac{1}{2Q_p}\right)^2 + 1}\right)\text{Hz} \qquad \textbf{(Eq. 9.36a)}$$

The upper cutoff frequency for a parallel circuit is given as:

$$\omega_2 = \omega_0\left(-\frac{1}{2Q_p} + \sqrt{\left(\frac{1}{2Q_p}\right)^2 + 1}\right)\text{rad/s} \qquad \textbf{(Eq. 9.37)}$$

$$f_2 = f_0\left(1/2Q_p + \sqrt{\left(\frac{1}{2Q_p}\right)^2 + 1}\right)\text{Hz} \qquad \textbf{(Eq. 9.37a)}$$

The bandwidth of the parallel circuit can be found from the formula:

$$\omega_2 - \omega_1 = \frac{\omega_0}{Q_p} \qquad \textbf{(Eq. 9.38)}$$

from which one more definition for Q_p appears, namely:

$$Q_p = \frac{\omega_0}{\omega_2 - \omega_1} \qquad \textbf{(Eq. 9.39)}$$

This formula, which relates the ratio of resonant frequency to bandwidth, can also be written as:

$$Q_p = \frac{f_0}{f_2 - f_1}\text{ Hz} \qquad \textbf{(Eq. 9.39a)}$$

where frequencies are expressed in hertz instead of radians per second. Equations 9.39 and 9.39a can be used to experimentally measure the actual value of Q_p for a circuit and are therefore very valuable in circuit design. Another interesting relationship between the cutoff frequencies and the resonant frequency ω_0 can be obtained by forming the product of ω_1 and ω_2 as given by Equations 9.20 and 9.21. We discover that for all values of Q_p:

$$\omega_1 \cdot \omega_2 = (\omega_0)^2 \qquad \textbf{(Eq. 9.40)}$$

$$f_1 \cdot f_2 = (f_0)^2 \qquad \textbf{(Eq. 9.40a)}$$

EXAMPLE 9.6

A parallel circuit consists of a current source of 1 mA set to the resonant frequency of the circuit, a resistor of 10 k Ω in parallel with an inductor of 2 mH, and a capacitor of 0.005 μF.

For this circuit find:

(a) The resonant frequency
(b) The value of Q_p
(c) The bandwidth of the circuit
(d) The output voltage
(e) The current in each element

Solution The resonant frequency is found from Equation 9.28 as:

$$\omega_0 = 1/\sqrt{L_pC_p} = 1/\sqrt{2\cdot 10^{-3}\cdot 5\cdot 10^{-9}} = 3.16\cdot 10^5 \text{ rad/s} = 50.3 \text{ kHz} \quad \textbf{(Eq. 9.28)}$$

The value of Q_p can be found from Equation 9.29 as:

$$Q_p = \omega_0 C_p R_p = 3.16\cdot 10^5\cdot 5\cdot 10^{-9}\cdot 10^4 = 15.8$$

The bandwidth can be found from Equation 9.39a as:

$$f_2 - f_1 = \frac{f_0}{Q_p} = \frac{50.3}{15.8} = 3.19 \text{ kHz}$$

The output voltage at resonance is the source current times the resistance because the reactive components of current cancel:

$$V = IR = 1\cdot 10^{-3}\cdot 10^4 = 10 \text{ V}$$

The currents through the capacitor or inductor are given as the output voltage times the admittance.

$$I_C = VY_C = 10\cdot 1.58\cdot 10^{-3} = 15.8 \text{ mA}$$

Note that the current through the capacitor (or inductor) is Q_p times as large as the source current. This is the parallel equivalent of the series voltage boost.

The simple circuit of Fig. 9.11 does not properly represent the resistance that is a part of all inductors. In the case of the series circuit, the resistance of the inductor simply adds to the series resistance already in the circuit without changing the configuration of the circuit. Figure 9.12 shows a circuit that is a close approximation to a real parallel circuit. In this circuit there is a parallel resistance in the circuit labeled R_p as well as a resistance in series with the inductor labeled R_L.

FIGURE 9.12

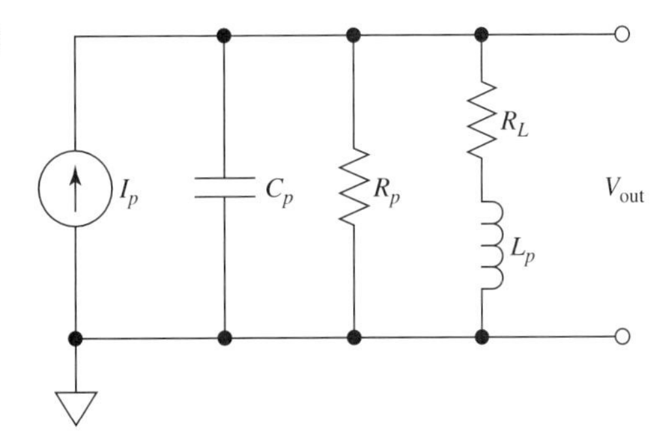

The manufacturer, as one of its properties, usually specifies the value of Q for an inductor. This value of Q is valid over a wide range of frequencies because the actual resistance of the wire is not constant but actually increases with frequency. Thus, rather than specify a single value at a single frequency, the manufacturer provides a graph showing the behavior of the inductor for the frequency range in which it is designed to operate. A typical graph of Q versus frequency is shown in Fig. 9.13.

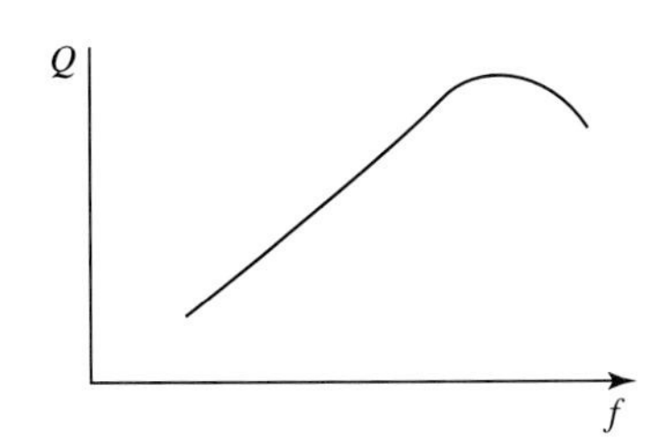

FIGURE 9.13
Q versus frequency

The solution to the circuit of Fig. 9.12 is best carried out by a conversion of the series inductor and resistor into an equivalent parallel combination of elements. We will now perform this operation and show that the series elements can be replaced by the parallel elements. With this approach the solution to the parallel circuit already performed becomes still valid.

An inductor L_s in series with a resistor R_L has an impedance $Z(\omega) = R_L + j\omega L_s$. Its corresponding admittance is given as:

$$Y_1(\omega) = \frac{1}{R_L + j\omega L_s} = \frac{R_L - j\omega L_s}{R_L^2 + (\omega L_s)^2}$$

$$= \frac{R_L}{(R_L)^2 + (\omega L_s)^2} - j\frac{\omega L_s}{(R_L)^2 + (\omega L_s)^2}$$

An inductor L_p in parallel with a resistor R_x has an admittance of

$$Y_2(\omega) = (1/R_x) - j(1/\omega L_p)$$

Equating $Y_1(\omega)$ and $Y_2(\omega)$ yields a relationship between the series elements and the parallel elements. Thus:

$$R_x = \frac{(R_L)^2 + (\omega L_s)^2}{R_L}$$

$$= R_L\left(1 + \frac{(\omega L_s)^2}{(R_L)^2}\right) \qquad \textbf{(Eq. 9.41)}$$

$$R_x = R_L\left[1 + (Q_s)^2\right]$$

and

$$L_p = \frac{(R_L) + (\omega L_s)^2}{\omega^2 L_s}$$

$$L_p = L_s\frac{(1 + Q_s)^2}{Q_s^2} \qquad \textbf{(Eq. 9.42)}$$

Through the use of Equations 9.41 and 9.42 we can convert the circuit of Fig. 9.13 into the form of the circuit shown in Fig. 9.12. For values of Q_s greater than 10, the values of L_p and L_s can be considered to be identical because $(Q_s)^2$ is so much greater than 1. The value of R_x can be found from Equation 9.41. With the substitution of these values into Fig. 9.12 instead of the series combination the new value of parallel resistance becomes the parallel combination of R_x and the original R_p. With these substitutions the circuit reduces to the circuit of Fig. 9.11 and the equations already developed can be used.

EXAMPLE 9.7

A 1 mH coil has a $Q = 50$ at a frequency of 1 MHz. Find:
(a) The resistance of the coil
(b) The parallel combination that can replace the coil and its series resistance

Solution Using the formula for $Q_s = (\omega_0 L)/R$ we find that

$$R = 2\pi \cdot 10^6 \cdot 1 \cdot 10^{-3}/50 = 125.66\ \Omega$$

From Equation 9.41 we find that the parallel resistance is equal to:

$$R_x = R_L\left[1 + (Q_s)^2\right] = 125.66\,(2501) = 314\ \text{k}\Omega$$

From Equation 9.42 we find that the equivalent inductance is:

$$L_p = L_s\left[\frac{1 + (Q_s)^2}{(Q_s)^2}\right] = 10^{-3}\frac{2501}{2500} = 1.0004\ \text{mH}$$

■ SUMMARY

- A circuit consisting of a resistor, an inductor, and a capacitor in series is called a series R–L–C circuit.
- Series R–L–C circuits have a property called resonance. At a frequency called the resonance frequency, which is determined by the value of the inductor and capacitor, the circuit behaves as if the inductor and the capacitor were not in the circuit. The frequency of resonance is given as:

$$f_0 = \frac{1}{2\pi\sqrt{LC}}$$

- The resonance property allows us to select a small band of frequencies from the spectrum of all frequencies that are possible.
- The bandwidth of the selected band of frequencies depends on a property of the circuit denoted by the letter Q and called the quality factor of the R–L–C circuit. The value of Q depends on the values of R, L, and C and can be shown to be equal to

$$Q = \frac{1}{R}\sqrt{L/C}$$

- The larger the value of Q the narrower is the bandwidth of the band of frequencies selected. Q can be expressed as the resonant frequency divided by the bandwidth of the circuit.
- Bandwidth is determined by the frequencies for which the reactive component of the impedance of the circuit is equal to or less than the resistance value of the circuit. The lower frequency value is called f_1 and the upper frequency value is called f_2. The bandwidth is defined as $f_2 - f_1$.
- At resonance the voltage across the resistor equals the applied voltage. The capacitor and the inductor each have a voltage across them that is Q times the applied voltage.
- Parallel R–L–C circuits display resonance phenomena that are similar to series R–L–C circuits. In a parallel R–L–C circuit the admittance is a minimum at the resonant frequency.
- Parallel R–L–C circuits exhibit similar bandwidth properties to series R–L–C circuits.
- For parallel resonance the applied current equals the resistor current. Both the capacitor and the inductor have currents passing through them that are Q times the value of the applied current.

■ PROBLEMS

9.1 Find the resonant frequency in both hertz and radians per second for a series circuit made from the following elements:
(a) $R = 50\ \Omega$, $L = 5$ mH, $C = 0.05\ \mu$F
(b) $R = 10\ \Omega$, $L = 25$ mH, $C = 0.25\ \mu$F
(c) $R = 330\ \Omega$, $L = 300\ \mu$H, $C = 0.001\ \mu$F
(d) $R = 25\ \Omega$, $L = 15\ \mu$H, $C = 300$ pF

9.2 A 50-mH coil is to be used for a series R–L–C circuit that will have a resonant frequency of 1000 Hz. A 5-V source whose frequency is 1000 Hz is attached to the circuit.
(a) What value of capacitor will be needed to produce resonance?
(b) What value of resistance should be placed in series with the source to limit the current to 200 mA?
(c) What voltage will appear across the capacitor at resonance?
(d) What voltage will appear across the inductor at resonance?
(e) How is it possible that the voltage across the capacitor is so large?

9.3 A 0.001 μF capacitor is to be series tuned to a resonant frequency of 20 kHz.
(a) What value of inductor will be needed to tune the circuit?
(b) If the voltage across the capacitor at resonance is 20 times the applied voltage, what value of resistance is in the circuit?
(c) If a 1-V 2-kHz voltage source is applied to the circuit, what will the voltage across the capacitor be?
(d) For the conditions of part (c), what will the voltage across the inductor be?

9.4 A series circuit has the values $R = 20\ \Omega$,

$$L = 500\ \mu\text{H, and } C = 0.047\ \mu\text{F.}$$

(a) What is the value of Q_s for this circuit?
(b) What is the resonant frequency of the circuit in hertz?
(c) What is the bandwidth of the circuit in hertz?
(d) Find the lower and upper cutoff frequencies expressed in hertz for this circuit.

9.5 For a series circuit with an L/C ratio of 400 and a resonant frequency of 2 mHz:
(a) Find the values of L and C.
(b) If the series resistance is to be adjusted so that the Q of the circuit is 20, what is the value of the resistance needed?
(c) Calculate the upper- and lower-frequency cutoff points.

9.6 Repeat problem 9.5 for the case that $L/C = 25$.

9.7 A series resonant circuit has a bandwidth of 100 Hz and a resonant frequency of 2000 Hz.
(a) What is the Q for the circuit?
(b) If the series resistance is 10 Ω, what is the value of the inductor being used?
(c) What is the size of the capacitor?
(d) Calculate the lower and upper cutoff frequencies.

9.8 A series resonant circuit has an upper cutoff frequency of 2000 Hz and a lower cutoff frequency of 1800 Hz.
(a) What is the resonant frequency of the circuit?
(b) What is the Q of the circuit?
(c) If the series resistance is 5 Ω, what is the impedance of the inductor?
(d) Find the values of L and C for this circuit.

9.9 Design a series resonant circuit with an input voltage of 2 V, a bandwidth of 400 Hz, and a maximum current of 250 mA at resonance, with a resonant frequency of 10 kHz.
(a) Find the value of the quality factor.
(b) Find the value of the inductance used.
(c) Find the value of the capacitance used.
(d) Find the upper and lower cutoff frequencies.

9.10 An inductor with a $Q = 50$ and an inductance of 500 μH is to be used in a parallel resonant circuit at 200 kHz.
(a) Find the parallel equivalent inductance and resistance of the inductor.

(b) Find the capacitor required for resonance.
(c) Find the value of the resistance in parallel with the coil so that the Q of the circuit is 10.
(d) Find the bandwidth of the circuit.

9.11 A parallel resonant circuit with a resistor of 10 Ω and a $Q = 10$ uses a 25-μH coil to resonate at 2 MHz.
(a) What is the value of the needed capacitor?
(b) Calculate the upper and lower cutoff frequencies.
(c) Calculate the bandwidth of the circuit.
(d) If a 10 mA current source is attached to the circuit, find the values of output voltage and the currents in the circuit.

9.12 A 2 mH coil with a $Q = 80$ is resonant at 50 kHz and has a bandwidth of 10 kHz.
(a) What value of resistance must be placed in parallel with the coil to create the circuit?
(b) What value of capacitance must be used to resonate the coil?
(c) If instead of a parallel resistance we add resistance in series with the coil, what should its value be to achieve proper operation?

9.13 Design a parallel resonant circuit with an input current of 20 mA, a bandwidth of 400 Hz, and a maximum voltage of 50 V at resonance with a resonant frequency of 10 kHz.
(a) Find the value of the quality factor.
(b) Find the value of the inductance used.
(c) Find the value of the capacitance used.
(d) Find the upper and lower cutoff frequencies.

9.14 For a parallel circuit with an L/C ratio of 400 and a resonant frequency of 2 MHz:
(a) Find the values of L and C.
(b) If the parallel resistance is to be adjusted so that the Q of the circuit is 20, what is the value of the resistance needed?
(c) Calculate the upper- and lower-frequency cutoff points.

CHAPTER **10**

MAGNETIC INDUCTION AND TRANSFORMERS

INTRODUCTION

We have so far been studying the characteristics of the three basic passive circuit elements: the resistor, capacitor, and inductor. In this chapter we introduce an additional circuit element called a transformer.

Many of us may already be familiar with one of the primary applications of a transformer, changing voltage magnitude. The electric power company, which delivers electricity to homes and industrial organizations, ***creates*** power at extremely high ac voltages (in the kilovolt range or more). We, as ***users*** of electric power, require voltages of much lower magnitude. Most homes, in fact, utilize voltages of 120 V or 220 Vrms. Transformers allow us to convert the high ac voltages created by power utilities to the lower voltages we require in our homes.

In addition to changing the magnitude of voltages, two other important applications of the transformer will be explored in this chapter: ***impedance matching*** and ***circuit isolation.***

Inductors used to make transformers are generally referred to as coils or windings because each inductor is made by winding electrically conducting wire around a magnetic core in what resembles a spring, or coil, in shape. In referring to the individual inductors that comprise a transformer, therefore, we will use the terms ***coil*** and ***winding*** interchangeably.

Furthermore, each time a conductor is wound around the magnetic core, a small "circle" of wire, called a ***turn,*** is formed. The total number of turns in each winding is, as we will see later on, an important parameter and is always given by the manufacturer of the transformer.

10.1 REVIEW OF MAGNETIC INDUCTION

Transformers are basically made by arranging inductors in a manner that makes use of a physical law known as magnetic induction. This law, discussed in detail in Appendix A, is actually the basis on which the inductor itself operates.

The Law of Magnetic Induction may be simply stated in two parts as follows:

1. A turn of wire placed in a ***time-varying*** magnetic field, such that the magnetic lines of flux are positioned at right angles to the area of the turn, will cause a voltage to appear at the extreme ends of the turn of wire.

The voltage specified in this law is said to be magnetically induced, which is why it is called the Law of Magnetic Induction.

The voltages developed in each turn of wire in a coil may be added to give the total voltage appearing at the terminals of the winding. This leads to part 2 of our statement of the Law of Magnetic Induction:

2. The voltage induced in a coil is proportional to the rate of change of the magnetic flux and to the total number of turns in the coil.

Note the reference to "rate of change" in part 2. This explains why we represent an inductor as a short circuit in a dc circuit.

In a ***dc circuit,*** the magnitudes of all voltages and currents are constant. This means that the rate of change of any magnetic flux appearing in an inductor as a result of the dc currents is zero. Since the rate of change of magnetic flux is zero in a dc circuit, there can be no voltage induced in a coil, and we therefore represent the coil as a short.

The Law of Magnetic Induction may be mathematically stated as follows:

$$e = N(d\phi/dt) \text{ V} \qquad \textbf{(Eq. 10.1)}$$

where N = total number of series turns in a winding (units: turns)
ϕ = total magnetic flux "cutting" the winding (i.e., at right angles to the winding) (units: webers)
$d\phi/dt$ = differential rate of change of magnetic flux (units: webers/second)
e = voltage developed across the winding (units: volts)

The term $d\phi/dt$ will probably not be familiar to those readers who have not taken calculus. We will not utilize this term in our calculations but will make use of it in the development of certain transformer equations.

10.2 IDEAL TWO-WINDING TRANSFORMER

The basic two-winding transformer consists of two individual windings wound around the same magnetic core. In an ***ideal*** transformer the following conditions are assumed to exist:

1. The wire used to form the winding is an ideal conductor. This means the resistance of the winding is zero.
2. ***All*** the magnetic flux, ϕ, that cuts one winding cuts the other as well. This effectively means that there are no magnetic losses in the ideal transformer.

An ideal two-winding transformer is shown in Fig. 10.1. In any two-winding transformer one winding has a voltage source connected to it and is known as the ***primary*** winding. The other winding has the load connected to it and is called the ***secondary*** winding.

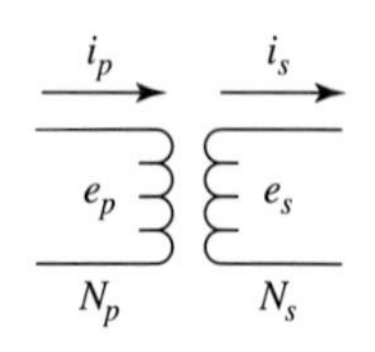

FIGURE 10.1
Ideal two-winding transformer

The parameters used to describe the primary winding have the subscript p and those used to describe the secondary winding have the subscript s. Therefore, in Fig. 10.1:

N_p = number of turns in the primary winding
N_s = number of turns in the secondary winding

e_p = voltage across the primary winding
e_s = voltage across the secondary winding
i_p = current in the primary winding
i_s = current in the secondary winding

From Equation 10.1:

$$e_p = N_p(d\phi_p/dt) \quad \textbf{(Eq. 10.2)}$$

$$e_s = N_s(d\phi_s/dt) \quad \textbf{(Eq. 10.2a)}$$

where ϕ_p = total magnetic flux cutting the *primary* winding
ϕ_s = total magnetic flux cutting the *secondary* winding

From assumption 2 for the ideal transformer:

$$\phi_p = \phi_s \quad \textbf{(Eq. 10.3)}$$

Dividing Equation 10.2 by Equation 10.2a and using Equation 10.3:

$$e_p/e_s = N_p/N_s \quad \textbf{(Eq. 10.4)}$$

Equation 10.4 is the basic transformer equation. It says that the voltages developed across each winding are directly proportional to the number of turns in each winding.

The transformer industry defines a parameter known as the transformation ratio of a transformer by the following equation:

$$a = N_p/N_s = \text{transformation ratio} \quad \textbf{(Eq. 10.5)}$$

We may therefore write Equation 10.4 as Equation 10.6:

$$e_p/e_s = a \quad \textbf{(Eq. 10.6)}$$

Equations 10.4 and 10.6 are ***instantaneous*** equations. If, however, the voltages involved are sinusoidal, then these equations are valid when using the rms values of the sinusoidal voltages. Equation 10.6 may then be written as Equation 10.7:

$$E_p/E_s = a \quad \textbf{(Eq. 10.7)}$$

where E_p and E_s are the rms values of the primary and secondary sinusoidal voltages, respectively.

From our assumption of zero losses in an ideal transformer, the apparent power delivered to the primary winding by the source must be the same as the apparent power delivered to the load by the secondary winding (see Chapter 11). This may be mathematically stated as follows:

$$E_pI_p = E_sI_s \quad \textbf{(Eq. 10.8)}$$

Rearranging terms in Equation 10.8 and using Equation 10.5, we may write:

$$I_p/I_s = N_s/N_p = 1/a \quad \textbf{(Eq. 10.9)}$$

Equation 10.9 indicates that the currents through the transformer windings are inversely proportional to the number of turns in each winding.

10.2.1 The Step-Up and Step-Down Transformers

The first application of transformers we will consider is the one already mentioned in our introduction. This was the changing of a voltage magnitude.

Recall from Equation 10.7 that the voltages across the windings are ***directly proportional*** to the number of turns in each winding. To ***reduce*** a voltage, we must apply it to the primary of a transformer in which the number of primary winding turns is greater than the number of secondary winding turns. Such a transformer is known as a ***step-down*** transformer because it is "stepping down" the voltage magnitude.

For a ***step-down*** transformer, $N_p > N_s$.

To ***increase*** a voltage, we must apply it to the primary of a transformer in which the number of primary turns is less than the number of secondary turns. This type of transformer is known as a step-up transformer.

For a ***step-up*** transformer, $N_p < N_s$.

Equations 10.7 and 10.9 indicate that the voltages and currents in a transformer act just the opposite of each other. ***If the secondary voltage is smaller than the primary voltage then the secondary current will be greater than the primary current and vice versa.*** It is therefore important to remember that the terms ***step-down*** and ***step-up refer to the voltages only*** and that the current will always do the opposite.

The following examples will demonstrate the opposite behavior of transformer voltages and currents.

Note that the letter T will represent ***turns*** in all the following. Furthermore, we are assuming that the polarity of voltages across the primary and secondary windings are known. The determination of polarity will be discussed briefly in a later section.

EXAMPLE 10.1

In the ideal transformer of Fig. 10.2 $N_p = 1000$ T, $N_s = 100$ T, $E_p = 100$ Vrms, and $R_a = 2\ \text{k}\Omega$.

(a) Is this a step-up or a step-down transformer?
(b) Find the secondary rms voltage.
(c) Find the secondary rms current.
(d) Find the primary rms current.

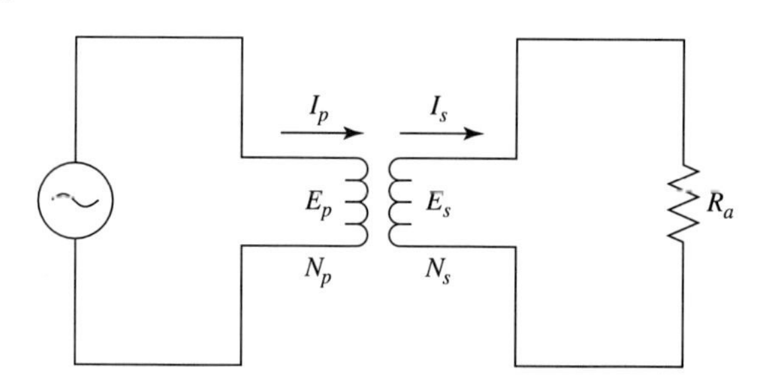

FIGURE 10.2
Circuit using an ideal two-winding transformer

Solution

(a) Since $N_p > N_s$, this must be a step-down transformer.

Answer Step-down transformer.

(b) From Equation 10.5, the transformation ratio is:

$$a = N_p/N_s = 1000/100 = 10$$

Using this value in Equation 10.7 and solving for E_s:

$$E_s = E_p/a = 100/10 = 10$$

Answer $E_s = 10$ Vrms.

(c) From Ohm's Law:

$$I_s = E_s/R_a = 10/2\text{K} = 5\ \text{mA}$$

Answer $I_s = 5$ mA.

(d) Using the value of a in Equation 10.9 and solving for I_p:

$$I_p = I_s/a = 5 \times 10^{-3}/10 = 0.5\ \text{mA}$$

Answer $I_p = 0.5$ mA.

Note how the voltage was ***reduced*** from 100 V to 10 V while the current was ***increased*** from 0.5 mA to 5.0 mA. This was as expected because in part (a) we first determined that this was a ***step-down*** transformer.

In the next example, we will demonstrate the opposite effects of a ***step-up*** transformer.

EXAMPLE 10.2

The parameters of the circuit in Fig. 10.2 have been changed as follows: $N_p = 50$ T, $N_s = 2500$ T, $E_s = 200$ Vrms, and $R_a = 400\ \Omega$.

(a) Is this a step-up or a step-down transformer?
(b) Find the primary rms voltage.
(c) Find the secondary rms current.
(d) Find the primary rms current.
(e) Find the power dissipated in the load resistor, R_a.

Solution

(a) Since $N_p < N_s$, this must be a step-up transformer.

Answer Step-up transformer.

(b) From Equation 10.5:

$$a = N_p/N_s = 50/2500 = 0.02$$

Using this value of a in Equation 10.7 and solving for E_p:

$$E_p = aE_s = 0.02(200) = 4$$

Answer $E_p = 4$ Vrms.

(c) From Ohm's Law:

$$I_s = E_s/R_a = 200/400 = 0.5 \text{ A}$$

Answer $I_s = 0.5$ Arms.

(d) Using the values of a and I_s in Equation 10.9 and solving for I_p:

$$I_p = I_s/a = 0.5/0.02 = 25 \text{ A}$$

Answer $I_p = 25$ Arms.

(e) The power dissipated (see Chapter 11) in the load resistor is given by:

$$P = R_a(I_{srms})^2 = 400(0.5)^2 = 100 \text{ W}$$

Answer $P = 100$ W.

Note how in Example 10.1 the voltage was ***increased*** from 4 V to 200 V while the current was ***decreased*** from 25 A to 0.5 A, as expected with a step-up transformer.

Both examples demonstrate the first application of transformers: changing voltage magnitudes. The next section deals with a second application of transformers: ***isolation.***

10.2.2 The Isolation Transformer

Recall that according to the law of magnetic induction, a ***change*** in magnetic flux is required to induce a voltage across either winding of a transformer. Since a dc current produces a magnetic flux that is constant, no voltage will be induced by this dc magnetic flux.

This property of a transformer allows us to physically separate circuits into two sections. The primary section, or circuit, is connected to the primary winding, and the secondary circuit is connected to the secondary winding. A dc voltage or current in the primary side will have no effect on the secondary circuit and vice versa.

A transformer used primarily to separate two circuits in the manner just described is called an ***isolation transformer.*** Although all circuits separated by a transformer have this property of isolation the term ***isolation transformer*** is generally reserved for one that has a transformation ratio equal to one. This is expressed mathematically in Equation 10.10:

$$a = N_p/N_s = 1 \quad \text{for an isolation transformer} \qquad \textbf{(Eq. 10.10)}$$

Equation 10.10 indicates that $N_p = N_s$ for an isolation transformer. This means that the primary and secondary winding voltages are equal. This fact reflects the previous statement that the primary purpose of an isolation transformer is the separation of two circuits and not the change of voltage magnitude.

EXAMPLE 10.3

The transformer in the circuit of Fig. 10.2 is an isolation transformer, $E_p = 120$ Vrms, and $I_s = 40$ mA.

(a) Find the transformation ratio.
(b) Find the secondary voltage.
(c) Find the load resistance, R_a.
(d) Find the primary current.
(e) Find the power dissipated in the load resistor, R_a.

Solution

(a) From Equation 10.10 the transformation ratio must be unity.

Answer $a = 1$.

(b) Since $a = 1$, the secondary and primary winding voltages must be equal.

Answer $E_s = 120$ Vrms.

(c) From Ohm's Law:

$$R_a = E_s/I_s = 120/40 \times 10^{-3} = 3\ \text{k}\Omega$$

Answer $R_a = 3\ \text{k}\Omega$.

(d) Again, since $a = 1$ the primary and secondary winding currents are equal.

Answer $I_p = I_s = 40$ mA.

(e) The power dissipated in the load resistor is given by:

$$P = R_a(I_{srms})^2 = (3\ \text{K})(40 \times 10^{-3})^2 = 4.8\ \text{W}$$

Answer $P = 4.8$ W.

Note, in the last example, that the load power of 4.8 W must be equal to the total power supplied by the primary, or

$$E_p(I_p) = (120)(40 \times 10^{-3}) = 4.8\ \text{VA}$$

The next section deals with the third primary application of a transformer, ***impedance matching.***

10.2.3 The Impedance Matching Transformer

Before discussing this third application of a transformer we must familiarize ourselves with a transformer property known as ***reflection.*** To do this consider the circuit of Fig. 10.3(a). From Ohm's Law:

$$Z_s = E_s/I_s \qquad \textbf{(Eq. 10.11)}$$

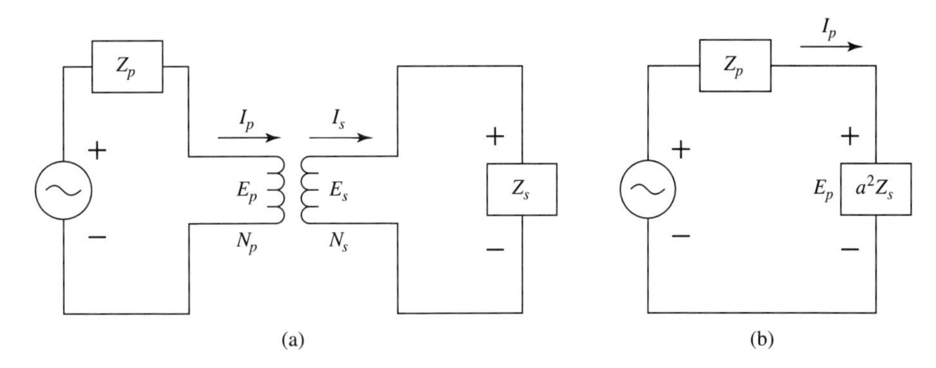

FIGURE 10.3
Circuit to demonstrate reflection

If we let Z_r be the impedance seen at the input terminals to the primary winding, then:

$$Z_r = E_p/I_p \quad \textbf{(Eq. 10.12)}$$

From Equations 10.7 and 10.9 we may solve for E_p and I_p as follows:

$$E_p = aE_s \quad \textbf{(Eq. 10.13)}$$

$$I_p = I_s/a \quad \textbf{(Eq. 10.13a)}$$

Substituting Equations 10.13 and 10.13a into Equation 10.12:

$$Z_r = aE_s/(I_s/a) = a^2E_s/I_s \quad \textbf{(Eq. 10.14)}$$

Substituting Equation 10.11 into Equation 10.14:

$$Z_r = a^2Z_s \quad \textbf{(Eq. 10.15)}$$

Equation 10.15 shows that the secondary load impedance is ***reflected*** into the primary circuit by multiplying it by a^2. This equation is interpreted as follows:

> Any impedance in the secondary circuit is reflected into the primary circuit by a factor of a^2.

Similarly, impedances in the secondary may be reflected as follows:

$$Z_r = Z_p/a^2 \quad \textbf{(Eq. 10.16)}$$

Equation 10.16 is interpreted as follows:

> Any impedance in the primary circuit is reflected into the secondary circuit by a factor of $1/a^2$.

These properties allow us to look at the entire transformer circuit from the point of view of either the primary circuit or the secondary circuit. Using Equation 10.15, the entire circuit of Fig. 10.3(a) may be redrawn as shown in Fig. 10.3(b).

The reflected impedance property allows us to use the transformer to match a load impedance to a source impedance. From the Maximum Power Transfer Theorem (see Chapter 11) this condition would result in maximum power being transferred to the load.

EXAMPLE 10.4

Consider the circuit of Fig. 10.4(a). In this circuit: $R_L = 50\ \Omega$, $V_i = 40$ Vrms, and $R_o = 800\ \Omega$.

(a) Find the power dissipated in the load resistance, R_L.
(b) Find the power drawn by the source resistance, R_o.

(c) Modify the circuit of Fig. 10.4(a) using an ideal two-winding transformer to maximize the power drawn by the 50-Ω load resistance.
(d) Calculate the new power delivered to the load resistance.
(e) Calculate the new power dissipated in the source resistance.

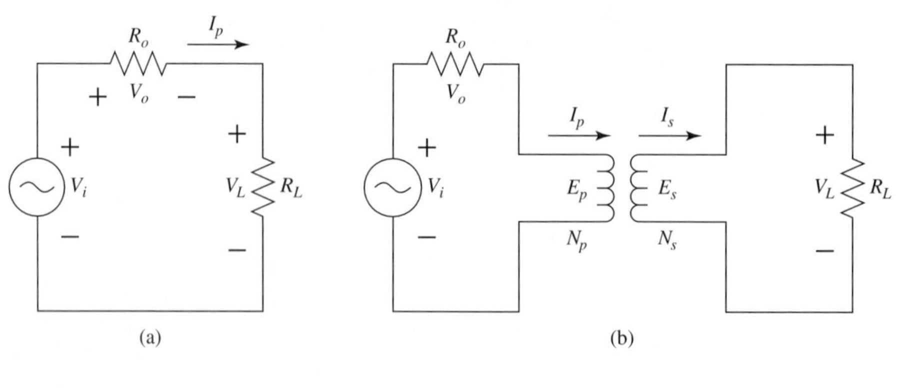

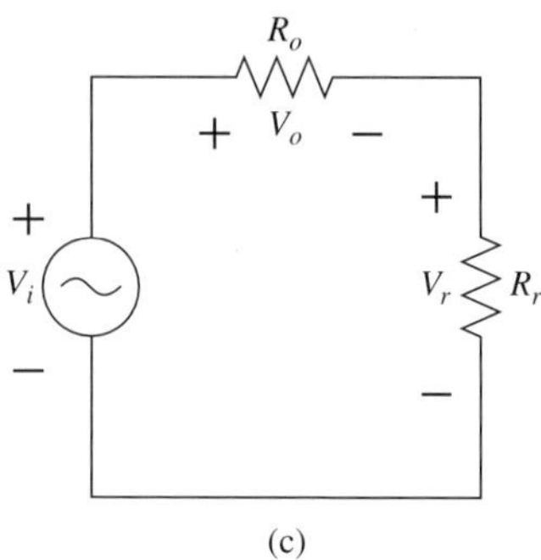

FIGURE 10.4
Use of reflection in impedance matching

Solution

(a) By voltage division, the voltage across the load resistance is:

$$V_L = R_L V_i/(R_o + R_L) = 50(40)/850 = 2.35 \text{ Vrms}$$
$$P_L = (V_L)^2/R_L = (2.35)^2/50 = 0.111 \text{ W}$$

Answer $P_L = 0.111$ W.

(b) The voltage across the source resistance is:

$$V_o = V_i - V_L = 40 - 2.35 = 37.65 \text{ Vrms}$$
$$P_o = (V_o)^2/R_o = (37.65)^2/800 = 1.77 \text{ W}$$

Answer $P_o = 1.77$ W.

(c) Note that the power dissipated in the ***source*** resistance is approximately ***16 times*** the power drawn by the ***load*** resistance. To correct this situation, we will use the impedance matching properties of the transformer to make the reflected load impedance "seen" by the source equal to the source resistance. From the Maximum Power Transfer Theorem, this will maximize the power drawn by the 50-Ω load resistance.

The new circuit is drawn in Fig. 10.4(b). In this circuit, we want the reflected load impedance to equal the source resistance. Accordingly:

$$R_r = a^2 R_L = R_o \quad \text{or} \quad a^2(50) = 800 \quad \text{or} \quad a^2 = 16$$

Answer The circuit of Fig. 10.4(b) with $a = 4$.

(d) The voltage across the reflected load resistance is:

$$V_r = R_r(V_i)/(R_r + R_o) = 800(40)/1600 = 20 \text{ Vrms}$$

The power drawn by the reflected load resistance is:

$$P_r = (V_r)^2/R_r = 20^2/800 = 0.5 \text{ W}$$

Since the ideal transformer does not detract from the power into the primary winding, this is also the power drawn by the actual 50-Ω load resistance.

Answer $P_L = 0.5$ W.

(e) Since the reflected load resistance and the source resistance are now equal, they both draw the same power.

Answer $P_o = 0.5$ W.

The last example demonstrated how the impedance matching property of the transformer could be used to take advantage of the maximum power transfer theorem and maximize the power delivered by a given source to a fixed load. Note that the load power in the last example was increased by nearly 500 percent simply by modifying the circuit to include the two-winding transformer.

10.3 MULTIWINDING TRANSFORMERS

All our transformer discussions thus far have been limited to the ***ideal two-winding*** transformer. Many circuits require the use of transformers that include ***more than one*** secondary winding. These are generally referred to as ***multiwinding*** transformers.

The analysis of multiwinding transformers is facilitated by the fact that each secondary winding may be treated independently. That is, the ratio equations developed for the ideal two-winding transformer may be applied to each secondary in relation to the primary winding without considering the presence of the other secondary windings. The ***total primary winding voltage,*** however, must include the effects of all secondary winding.

EXAMPLE 10.5

The circuit of Fig. 10.5 includes an ideal transformer with three secondary windings. Subscript 1 refers to the primary winding and subscripts 2, 3, and 4 refer to the three secondary windings in the following parameters: $N_1 = 2800$ T, $N_2 = 100$ T, $N_3 = 200$ T, $N_4 = 400$ T, $E_1 = 2000$ Vrms, $R_2 = 1$ kΩ, $R_3 = 2$ kΩ, and $R_4 = 4$ kΩ.

(a) Find E_2, the voltage across secondary winding 2.
(b) Find E_3, the voltage across secondary winding 3.
(c) Find E_4, the voltage across secondary winding 4.
(d) Find I_2, the current in secondary winding 2.
(e) Find I_3, the current in secondary winding 3.
(f) Find I_4, the current in secondary winding 4.
(g) Find I_1, the current in the primary winding.

Solution In each of the following cases, transformation ratio Equations 10.7 and 10.9 are employed with:

a_{12} = transformation ratio of primary to secondary winding 2
a_{13} = transformation ratio of primary to secondary winding 3
a_{14} = transformation ratio of primary to secondary winding 4

$$a_{12} = N_1/N_2 = 2800/100 = 28$$
$$a_{13} = N_1/N_3 = 2800/200 = 14$$
$$a_{14} = N_1/N_4 = 2800/400 = 7$$

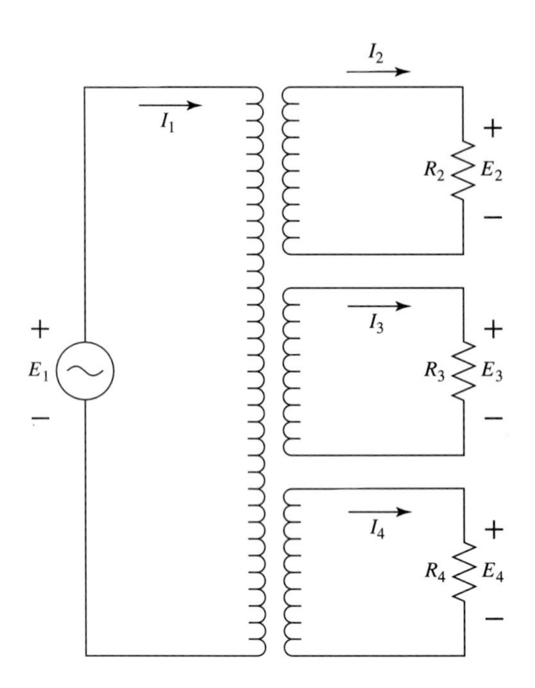

FIGURE 10.5
Multiwinding transformer circuit

(a) $E_2 = E_1/a_{12} = 2000/28 = 71.4$ V

Answer $E_2 = 71.4$ Vrms.

(b) $E_3 = E_1/a_{13} = 2000/14 = 142.8$ V

Answer $E_3 = 142.8$ Vrms.

(c) $E_4 = E_1/a_{14} = 2000/7 = 285.7$ V

Solution 285.7 Vrms.

(d) $I_2 = E_2/R_2 = 71.4/1\text{K} = 71.4$ mA

Answer $I_2 = 71.4$ mA.

(e) $I_3 = E_3/R_3 = 142.8/2\text{K} = 71.4$ mA

Answer $I_3 = 71.4$ mA.

(f) $I_4 = E_4/R_4 = 285.7/4\text{K} = 71.4$ mA

Answer $I_4 = 71.4$ mA.

(g) Since all three secondary loads are resistive, all the primary currents resulting from the three secondary windings are in phase and can be added algebraically.

$$I_1 = I_{12} + I_{13} + I_{14}$$

where I_{12}, I_{13}, and I_{14} are the primary currents due to secondary windings 2, 3 and 4, respectively.

$$I_{12} = I_2/a_{12} = 71.4 \times 10^{-3}/28 = 2.55 \text{ mA}$$
$$I_{13} = I_3/a_{13} = 71.4 \times 10^{-3}/14 = 5.1 \text{ mA}$$
$$I_{14} = I_4/a_{14} = 71.4 \times 10^{-3}/7 = 10.2 \text{ mA}$$
$$I_1 = (2.55 \times 10^{-3}) + (5.1 \times 10^{-3}) + (10.2 \times 10^{-3}) = 17.85 \text{ mA}$$

Answer $I_1 = 17.85$ mA.

10.4 NONIDEAL TRANSFORMERS

Practical, or nonideal, transformers fall primarily into two general categories, which describe the type of material used as the transformer core. The core is the material around which the primary and secondary coils are wound.

One category is known as the ***iron core transformer.*** In this category the coils are wound around a magnetic core made of some compound of iron. The iron core serves to ensure that virtually all the magnetic flux linking one winding also links the other winding.

The other category is known as an ***air core transformer.*** In this category, no magnetic core is used. Instead, the coils must be wound so that they all physically enclose the same magnetic flux.

Iron core transformers are shown with two vertical lines between the windings to represent the magnetic core. Air core transformers have no vertical lines between the windings in order to signify the absence of a physical core (see Fig. 10.6). The symbol we have been using in this chapter is that of an air core transformer. In both cases, however, the transformer characteristics differ from those of an ideal transformer by only a few percent except in some special cases.

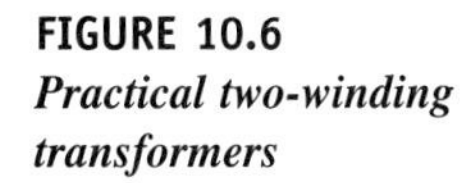

FIGURE 10.6
Practical two-winding transformers

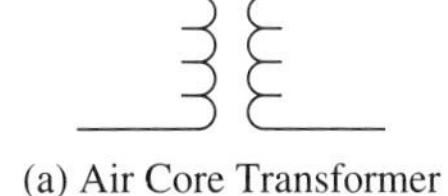

(a) Air Core Transformer

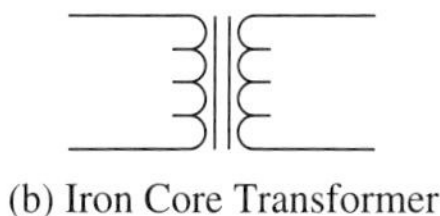

(b) Iron Core Transformer

10.4.1 Iron Core Transformers

Recall that the two primary assumptions made in analyzing an ideal transformer are the absence of any winding resistance, which would represent an electrical power loss, and the total cutting of all magnetic flux by all windings, the lack of which would represent a magnetic loss. In the iron core transformer, we may include the effects of these two losses by introducing a resistance and reactance into the primary and secondary windings as shown in Fig. 10.7(a):

where R_p = dc resistance of primary winding
X_p = leakage reactance of primary winding
R_s = dc resistance of secondary winding
X_s = leakage reactance of secondary winding
R_L = load impedance

Note that additional losses, such as hysteresis and eddy current losses, need only be taken into account in special cases.

For convenience, the circuit shown in Fig. 10.7(a) is often simplified by reflecting all impedances to either the primary or secondary circuit. Using Equation 10.15, the circuit of Fig. 10.7(a) may be redrawn as shown in Fig. 10.7(b). Alternatively, using equation 10.16, the circuit of Fig. 10.7(a) may be redrawn as in Fig. 10.7(c).

In Fig. 10.7(b):

$$R_{e_p} = R_p + a^2R_s \qquad \textbf{(Eq. 10.17)}$$

$$X_{e_p} = X_p + a^2X_s \qquad \textbf{(Eq. 10.18)}$$

In Fig. 10.7 (c):

$$R_{e_s} = R_s + R_p/a^2 \qquad \textbf{(Eq. 10.19)}$$

$$X_{e_s} = X_s + X_p/a^2 \qquad \textbf{(Eq. 10.20)}$$

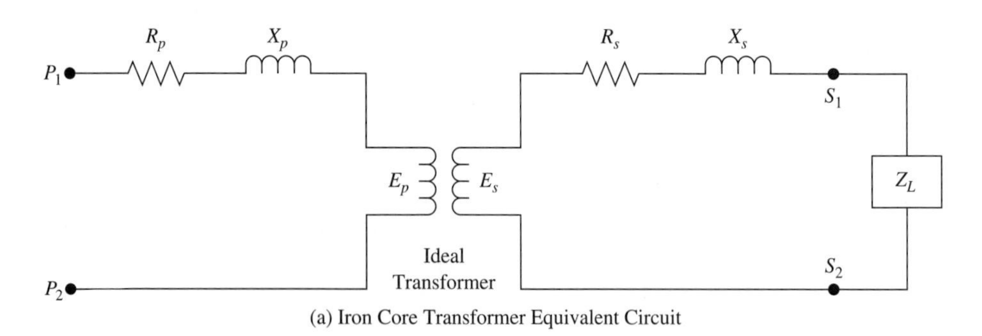

(a) Iron Core Transformer Equivalent Circuit

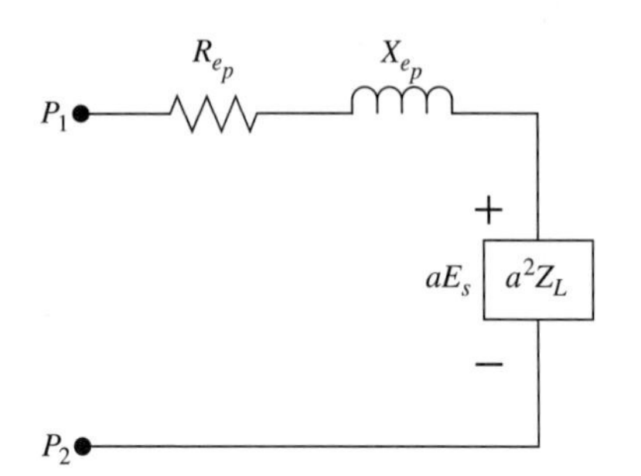

(b) Equivalent Circuit Reflected to the Primary

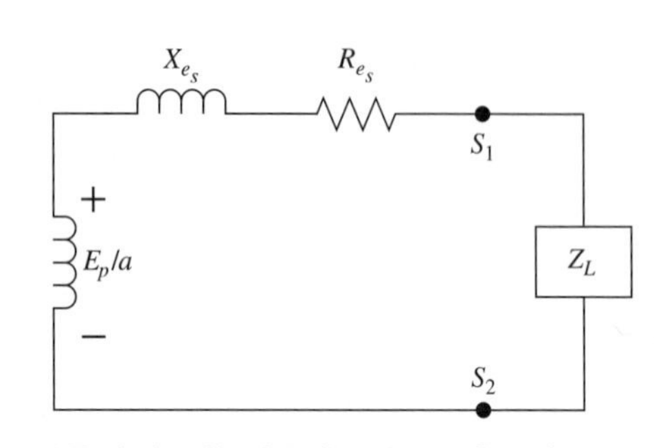

(c) Equivalent Circuit Reflected to the Secondary

P_1P_2 = Primary winding input terminals
S_1S_2 = Secondary winding input terminals

FIGURE 10.7
Total transformer equivalent circuits

where R_{e_p} = total transformer equivalent resistance reflected to the primary
X_{e_p} = total transformer equivalent reactance reflected to the primary
R_{e_s} = total transformer equivalent resistance reflected to the secondary
X_{e_s} = total transformer equivalent reactance reflected to the secondary

EXAMPLE 10.6

The iron core transformer of Fig. 10.8(a) has the following parameters: V_i = 120 Vrms, $R_p = 4\ \Omega$, $X_p = 4\ \Omega$, $R_s = 2\ \Omega$, $X_s = 3\ \Omega$, $R_L = 20\ \Omega$, $a = 4$.

(a) Draw the two-winding equivalent circuit of the transformer.
(b) Draw the total transformer equivalent circuit reflected to the primary.
(c) Draw the total transformer equivalent circuit reflected to the secondary.
(d) Find the rms primary current.
(e) Find the rms secondary current.
(f) Find the rms load voltage.

Solution

(a) The two-winding iron core transformer equivalent circuit was drawn in Fig. 10.7(a). Using this circuit, we may draw the two-winding transformer equivalent circuit for our example as shown in Fig. 10.8(b).

Answer Fig. 10.8(b).

(b) The total transformer equivalent circuit reflected to the primary may be drawn using the circuit of Fig. 10.7(b). This is shown in Fig. 10.8(c). In this circuit:

$$R_{e_p} = R_p + a^2R_s = 4 + 4^2(2) = 36\ \Omega$$
$$X_{e_p} = X_p + a^2X_s = 4 + 4^2(3) = 52\ \Omega$$
$$a^2R_L = 4^2(20) = 320\ \Omega$$

Answer Fig. 10.8(c).

(a) Iron Core Transformer Circuit

(b) Two-Winding Transformer Equivalent Circuit

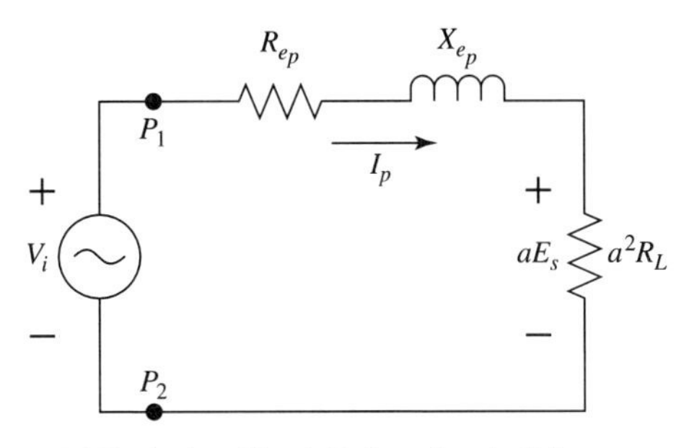

(c) Equivalent Circuit Reflected to the Primary

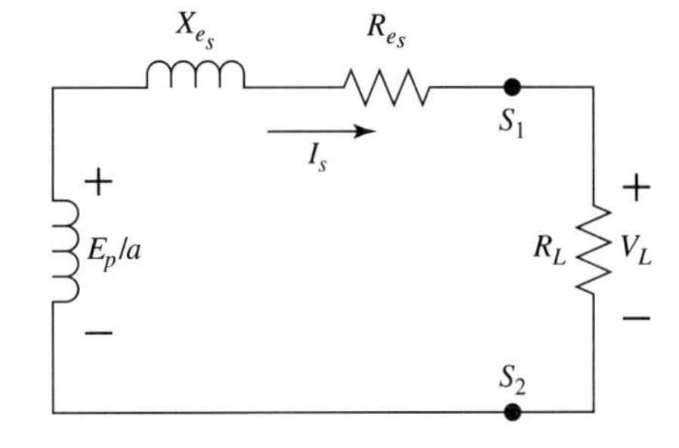

(d) Equivalent Circuit Reflected to the Secondary

FIGURE 10.8
Transformer circuits for example 10.6

(c) The total transformer equivalent circuit reflected to the secondary may be drawn using the circuit of Fig. 10.7(c). This is shown in Fig. 10.8(d). In this circuit:

$$R_{e_s} = R_s + R_p/a^2 = 2 + 4/4^2 = 2.25\ \Omega$$
$$X_{e_s} = X_s + X_p/a^2 = 3 + 4/4^2 = 3.25\ \Omega$$

Answer Fig. 10.8(d).

(d) The rms primary current may be found using the total transformer equivalent circuit reflected to the primary drawn in Fig. 10.8(c). In this circuit, the total input impedance seen by the voltage source is found by adding all the series impedances shown.

$$Z_{\text{in}} = R_{e_p} + jX_{e_p} + a^2RL = 36 + j52 + 320 = 356 + j52\ \Omega$$

From Ohm's Law the rms primary current is:

$$I_p = V_i/|Z_{\text{in}}| = 120/360 = 0.333\ \text{A}$$

Answer $I_p = 0.333$ Arms.

(e) The rms secondary current may be found by using the two-winding equivalent circuit of the transformer in Fig. 10.8(b).

$$I_p/I_s = 1/a \quad \text{or} \quad I_s = aI_p = 4(0.333) = 1.33\ \text{A}$$

Answer $I_s = 1.33$ Arms.

(f) The rms load voltage may be found from Ohm's Law:

$$V_L = R_L(I_s) = 20(1.33) = 26.6\ \text{V}$$

Answer $V_L = 26.6$ Vrms.

It is instructive to compare the load voltage in this example with the value of load voltage we would have obtained if we had assumed the iron core transformer was ideal. In this case, V_L would be equal to E_s as follows:

$$V_L = V_i/a = 120/4 = 30\ \text{V}$$

The 3.2-V difference between the two values of V_L represents about a 12 percent error. This error is higher than what would typically result from the assumption of an ideal transformer.

10.4.2 Air Core Transformers

As previously mentioned, the air core transformer uses no magnetic core. The coils are physically placed so that any flux present in one winding will also be present in the other winding. The equivalent circuit of the air core transformer is similar to that of the iron core transformer.

10.5 DETERMINING PARAMETERS OF IRON CORE TRANSFORMERS

It is important, particularly in high-power circuits, to determine the elements that make up the equivalent circuit of the iron core transformer. The total transformer equivalent resistance represents an electrical power loss while the total transformer equivalent reactance represents a magnetic loss.

Transformers are generally given ***rated values.*** These are values of primary and secondary voltages, volt amperes, and frequency that define the operating point that results in ***maximum efficiency*** of the transformer.

The efficiency of any device is generally defined by the following equation:

$$\eta = (P_o/P_i) \times 100\% \qquad \textbf{(Eq. 10.21)}$$

where P_o = power output from the device
P_i = power input to the device
η = efficiency in percent

The difference between the power output and power input will, of course, be the power losses. In the case of the iron core transformer, these are the electrical and magnetic losses previously mentioned.

Once the total transformer equivalent resistance and reactance are determined, the power losses can be calculated at any point. In this way, the values of voltage and volt amperes that result in maximum efficiency can be determined.

Two tests are performed on transformers to measure their equivalent resistances and reactances. These tests are known as the ***short circuit test*** and the ***open circuit test.***

In both tests, power is measured by a device called a ***wattmeter.*** This device basically measures current and voltage and uses a multiplier circuit that takes power factor into account in order to determine the power being dissipated in the circuit. (See Chapter 11.)

10.5.1 Short Circuit Test

In the short circuit test, use is made of the fact that electrical power dissipation in a transformer will depend on the current drawn. It is desirable, therefore, to perform the short circuit test at ***rated current.***

It is necessary to short one winding while running this test, however. This is to ensure that ***only*** the transformer resistance is included in the circuit. A voltage source is applied to the other winding. A wattmeter is included in the circuit to measure the power being dissipated.

Generally, this test is run with the ***primary*** winding shorted and a voltage source applied to the ***secondary*** winding (see Fig. 10.9a). This is done to enable us to obtain rated current with a substantially reduced voltage.

If P_{sc} is the power measured by the wattmeter during the short circuit test, then the total transformer equivalent resistance is given by:

$$R_{e_s} = P_{sc}/I_{sr}^2 \qquad \textbf{(Eq. 10.22)}$$

where R_{e_s} = total transformer equivalent resistance reflected to the secondary
I_{sr} = rated secondary current
P_{sc} = power dissipated in the transformer at *rated current*

The wattmeter measures power but cannot differentiate between the ***types*** of power it measures. Since the total power loss in a transformer is primarily ***electrical*** and ***magnetic,*** we must somehow eliminate the magnetic losses while running the short circuit test to ensure that the wattmeter is measuring ***only*** electrical losses. Otherwise, Equation 10.22 would be invalid.

This is accomplished by running the test at substantially reduced voltage by shorting the primary as previously described. Since the magnetic losses are proportional to ***voltage,*** this makes the magnetic losses negligible compared to the electrical losses and Equation 10.22 is valid.

Once the electrical power dissipated at rated current is known, the electrical power dissipated at any other secondary current may be calculated by:

$$P_e = (I_s/I_{s_r})^2 P_{sc_r} \qquad \textbf{(Eq. 10.23)}$$

where P_e = electrical power dissipated at a secondary current of I_s.

From Fig. 10.9(a), the total impedance seen by the voltage source connected to the secondary winding is:

$$Z_{e_s} = E_s/I_{s_r} \qquad \textbf{(Eq. 10.24)}$$

From the total transformer equivalent circuit reflected to the secondary:

$$Zes^2 = (R_{e_s})^2 + (X_{e_s})^2 \qquad \textbf{(Eq. 10.25)}$$

Since R_{e_s} and Z have already been determined (by equations 10.22 and 10.24), X_{e_s} can be determined by solving equation 10.25:

$$X_{e_s} = \sqrt{Zes^2 - (R_{e_s})^2} \qquad \textbf{(Eq. 10.25a)}$$

10.5.2 Open Circuit Test

This test is run to determine the ***magnetic*** losses in the transformer. It must be run at rated voltage but ***no current*** to ensure that all the power measured by the wattmeter is magnetic in nature and not electrical. To accomplish this, the primary winding is energized with a voltage source at rated voltage while the secondary winding is opened (see Fig. 10.9b).

A wattmeter measures the total magnetic power being lost in the transformer. If E_{pr} is the rated primary voltage at which the magnetic losses, P_{oc}, were measured then the magnetic power losses at any other voltage may be determined by:

$$P_m = (E_p/E_{pr})P_{oc} \qquad \textbf{(Eq. 10.26)}$$

where P_m = magnetic power loss at primary voltage E_p
P_{oc} = magnetic power loss at rated primary voltage E_{pr}

EXAMPLE 10.7

An iron core transformer is rated as follows:

Primary voltage = 1200 V

Secondary voltage = 120 V

Volt amperes = 12 kVA

Load power factor = 0.95

The transformer is subjected to the short circuit and open circuit tests using the circuits of Fig. 10.9 with the following results:

Short Circuit Test Results	*Open Circuit Test Results*
Power dissipation = 600 W	Power dissipation = 300 W
V_i = 10 Vrms	V_i = 1200 Vrms

(a) Find R_{e_s}, the total transformer equivalent resistance reflected to the secondary.
(b) Find X_{e_s}, the total transformer equivalent reactance reflected to the secondary.
(c) Find the transformer efficiency at rated conditions.
(d) Find P_e, the electrical power dissipated in the transformer if it is used at a secondary current of 50 Arms.
(e) Find P_m, the magnetic power lost at a primary input voltage of 900 Vrms.
(f) If the transformer is supplying a load of 3 kW under the conditions stated in parts (d) and (e), calculate the new efficiency of the transformer.

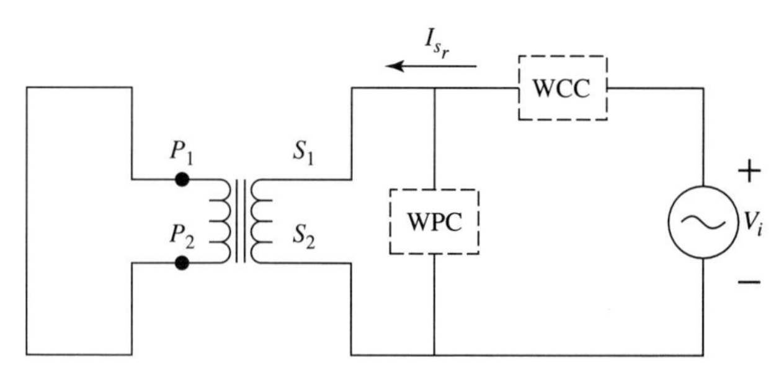

(a) Short Circuit Test

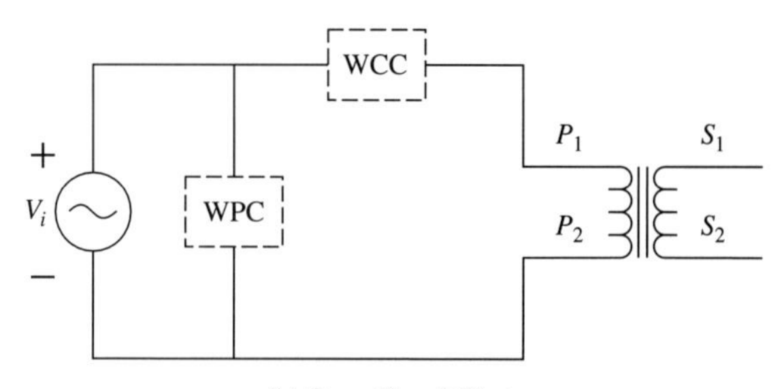

(b) Open Circuit Test

WPC = Wattmeter potential coil, used to measure circuit voltage
WCC = Wattmeter current coil, used to measure circuit voltage

FIGURE 10.9
Determining transformer equivalent circuit parameters

Solution

(a) Since the short circuit test is run at rated current, we must first ***calculate*** the rated secondary current. This is done using the rated apparent power given.

$$I_{s_r} = \text{VA}/\text{V}_s = 12\text{K}/120 = 100 \text{ A}$$

Using Equation 11.22:

$$R_{e_s} = P_{sc}/(I_{s_r})^2 = 600/(100)^2 = 0.06\ \Omega$$

Answer $R_{e_s} = 0.06\ \Omega$.

(b) The total transformer equivalent impedance reflected to the secondary may now be found using Equation 10.24.

$$Z_{e_s} = E_s/I_{s_r} = 10/100 = 0.1\ \Omega$$

Using Equation 10.25a:

$$X_{e_s} = (0.1^2 - 0.06^2)^{1/2} = 0.08\ \Omega$$

Answer $X_{e_s} = 0.08\ \Omega$.

(c) When a transformer is assigned a rated apparent power, this number always represents ***output.*** At rated conditions the output, or load, power will be:

$$P_o = S(\text{fp}) = 12\text{K}(0.95) = 11.4\ \text{kW}$$

The total losses at rated conditions will be the sum of the electrical power loss, as measured by the short circuit test, and the magnetic power loss as measured by the open circuit test.

$$P_{\text{loss}} = P_{sc} + P_{oc} = 600 + 300 = 900\ \text{W}$$

The total input power requirement is the sum of the output power plus the power losses.

$$P_{\text{in}} = P_o + P_{\text{loss}} = 11{,}400 + 900 = 12{,}300\ \text{W}$$

From the definition of efficiency:

$$\eta = 100(P_o/P_{\text{in}}) = 100(11.4\text{K}/12.3\text{K}) = 92.683\%$$

Answer $\eta = 92.68\%$.

(d) Since the transformer is now being run at a current that is different from the rated current, we can use Equation 10.23 to calculate the new electrical power loss.

$$P_e = (I_s/I_{s_r})^2(P_{sc}) = (50/100)^2(600) = 150$$

Answer $P_e = 150$ W.

(e) Since the transformer is now being run at a voltage that is different from the rated voltage, we can use Equation 10.26 to calculate the new magnetic power loss.

$$P_m = (E_p/E_{p_r})P_{oc}) = (900/1200)(300) = 225\ \text{W}$$

Answer $P_m = 225$ W.

(f) The input power under these conditions must be:

$$P_{\text{in}} = P_o + P_{\text{loss}} = 3000 + 150 + 225 = 3375\ \text{W}$$
$$\eta = 100(3000/3375) = 88.888\%$$

Answer $\eta = 88.89\%$.

Note that, as expected, running the transformer at a load that is different from the rated load resulted in a lower transformer efficiency.

10.6 TRANSFORMER WINDING POLARITY

It is impossible to ascertain the voltage polarity of a secondary winding by inspection. To enable users of transformers to determine polarity, a dot convention is utilized.

Transformer manufacturers place a dot at one of the terminals of both windings. The dot convention stipulates as follows:

If a current ***enters*** a ***dotted*** terminal in the ***primary*** winding, a ***positive*** polarity will be induced at the ***dotted*** terminal of the ***secondary*** winding.

Similarly: If a current enters the ***undotted*** terminal in the primary winding, a positive polarity will be induced in the ***undotted*** terminal of the secondary winding.

The above dot convention may be applied equally for current in the secondary winding inducing a voltage in the primary winding.

■ SUMMARY

- Transformers are devices that consist basically of inductors, known as windings or coils, all cutting the same magnetic flux.
- Voltages are induced in the transformer windings using the principle of magnetic induction.
- A ***turn*** of wire is the name given to each individual circle of wire that is wound around a core to form the transformer windings.
- In an ideal two-winding transformer there is no winding resistance and no winding leakage reactance. This means that there are no electrical or magnetic power losses.
- The transformer winding that is connected to a voltage source is called the ***primary*** winding. The winding that is connected to a load is called the ***secondary*** winding.
- The voltages across the windings are directly proportional to the number of turns in each winding. The currents in the windings are ***inversely*** proportional to the number of turns in each winding.
- The transformation ratio, a, is defined as the ratio of the number of turns in the primary winding to the number of turns in the secondary winding.
- The three primary applications of the transformer are:

  Voltage magnitude changing (step up and step down)

  Circuit isolation

  Impedance matching
- Iron core transformer equivalent circuits are useful in analyzing transformer performance.
- Equivalent circuit parameters are measured using two tests:

  The ***short circuit*** test to measure R_{e_s}, X_{e_s} and the electrical loss at rated conditions.

  The ***open circuit*** test to measure the magnetic loss at rated conditions.
- The short circuit and open circuit test results enable us to predict the electrical and magnetic transformer losses at any load conditions.
- Transformer ratings represent the operating conditions that will result in maximum transformer efficiency.

■ PROBLEMS

10.1 Name and describe the three primary applications of a transformer.

10.2 What are the two main assumptions about an ***ideal*** transformer?

10.3 What is the difference between a step-up and a step-down transformer?

10.4 The primary winding of an ideal two-winding transformer is connected to a 360-Vrms voltage source. If the secondary voltage is measured to be 12 Vrms:
(a) Is this a step-up or a step-down transformer?
(b) What is the transformation ratio of this transformer?

10.5 The secondary winding of an ideal two-winding transformer is connected to a 40-Ω resistor. The voltage across this resistor is measured at 50 Vrms. The primary winding has 1000 T, and the transformation ratio is 8.
(a) Is this a step-up or a step-down transformer?
(b) How many turns are there in the secondary winding?
(c) What is the voltage across the primary winding?
(d) What is the secondary rms current?
(e) What is the primary rms current?

10.6 The secondary winding of an ideal two-winding transformer is connected to a 100-Ω resistor. The current through this resistor is 1 Arms. The primary current is 0.05 Arms.
(a) Find a = transformation ratio.
(b) Is this a step-up or step-down transformer?
(c) Find the secondary winding rms voltage.
(d) Find the primary winding rms voltage.

10.7 In the multiwinding ideal transformer of Fig. P10.7, $N_1 = 1600$ T, $N_2 = 160$ T, $N_3 = 400$ T, $N_4 = 200$ T, $R_2 = 80\ \Omega$, $R_3 = 160\ \Omega$, $R_4 = 100\ \Omega$, and $V_i = 2400$ Vrms.
(a) Find the rms voltage V_2.
(b) Find the rms voltage V_3.
(c) Find the rms voltage V_4.
(d) Find the current I_2.

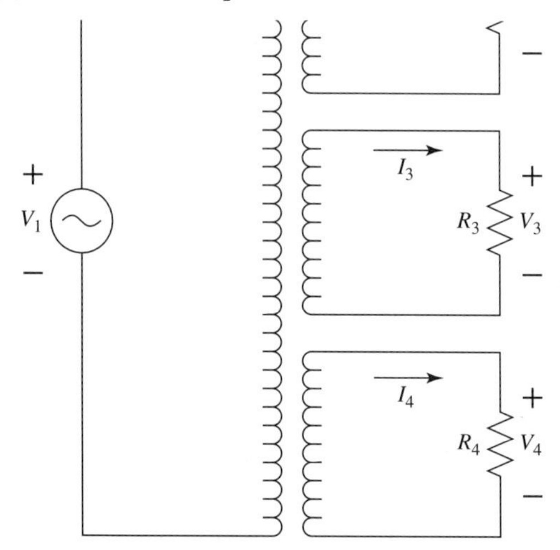

FIGURE P10.7
Multiwinding transformer circuit for problem 10.7

(e) Find the current I_3.
(f) Find the current I_4.
(g) Find the current I_1.
(h) Find the impedance seen by the primary winding.

10.8 Repeat (a) through (h) of problem 10.7 if the following changes are made to the circuit of Fig. P10.7:

R_2 is replaced by an 80-Ω inductive reactance.

R_4 is replaced by a 100-Ω inductive reactance.

10.9 What should the transformation ratio of a two-winding transformer be to match a 20-Ω resistive load to a 1800-Ω source resistance?

10.10 An iron core transformer circuit is shown in Fig. P10.10. In this circuit $R_p = 1\ \Omega$, $X_p = 2\ \Omega$, $R_s = 1.5\ \Omega$, $X_s = 0.5\ \Omega$, $a = 4$, $Z_o = 36\underline{/45^\circ}\ \Omega$, and $V_i = 100$ Vrms.
(a) Draw the equivalent circuit reflected to the primary.
(b) Draw the equivalent circuit reflected to the secondary.

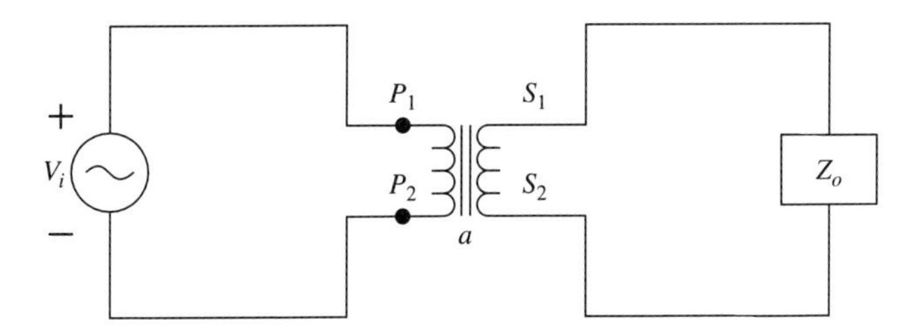

FIGURE P10.10
Transformer circuit for problem 10.10

10.11 For the circuit of problem 10.10:
(a) Find the rms primary current.
(b) Find the rms secondary current.
(c) Find the rms load voltage.

10.12 What is meant by a transformer's rating?

10.13 A two-winding iron core transformer is rated as follows:

Primary voltage = 2400 V

Secondary voltage = 240 V

Volt amperes = 24 KVA

Load power factor = 0.96

The transformer is subjected to the short circuit and open circuit tests using the circuit of Fig. P10.13 with the following results:

Short Circuit Test Results

Power dissipation = 800 W
V_i = 20 Vrms

Open Circuit Test Results

Power dissipation = 400 W
V_i = 2400 Vrms

(a) Find R_{es}.
(b) Find X_{es}.
(c) Find η if the transformer is run at rated conditions.
(d) Find the electrical power dissipated in the transformer if it is used at a secondary current of 100 Arms.
(e) Find the magnetic power lost at a primary input voltage of 1800 Vrms.
(f) If the transformer is supplying a load of 6 kW under the conditions stated in parts (d) and (e), calculate the new efficiency of the transformer and compare it to the value found in part (c). Comment on the results.

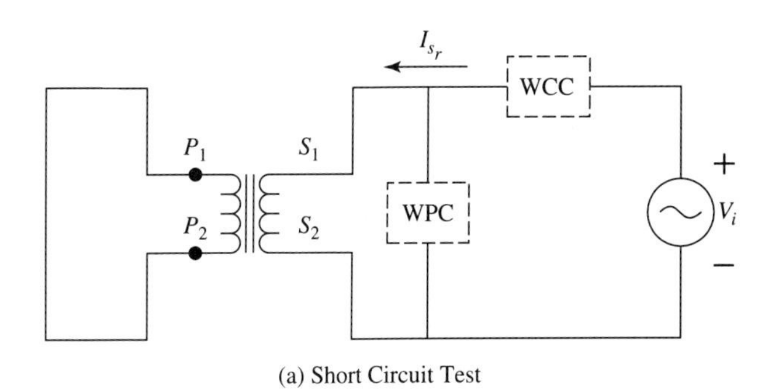

(a) Short Circuit Test

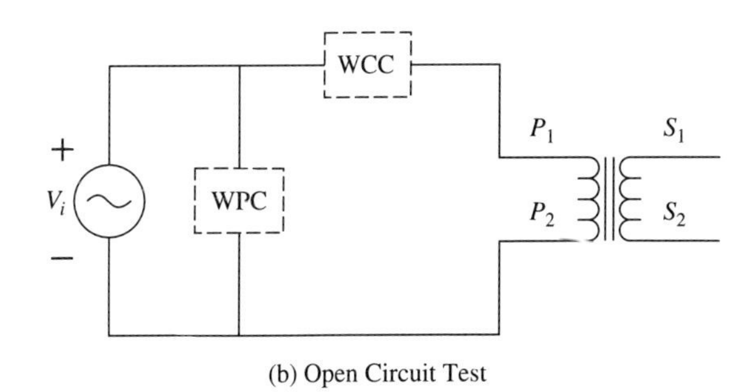

(b) Open Circuit Test

FIGURE P10.13
Circuits for problem 10.13

CHAPTER 11

POWER AND ENERGY

INTRODUCTION

The term ***power*** is important in all disciplines: electrical, mechanical, chemical, and so on. It is defined as the rate at which energy is either ***produced*** by a device or ***used*** by a device.

In electrical circuits, therefore, power is defined as the rate at which energy is ***delivered*** by a power source, or circuit, or the rate at which energy is ***used*** by a circuit. Since virtually all power generated is sinusoidal in nature, in this chapter we will investigate the power characteristics of circuits energized by sinusoidal sources.

11.1 INSTANTANEOUS POWER

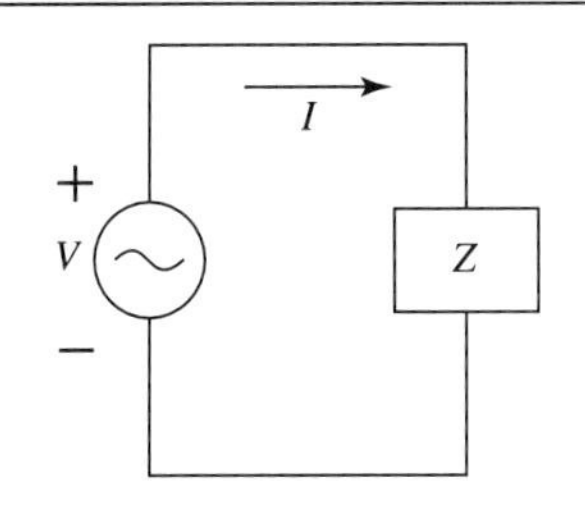

FIGURE 11.1

Consider the simple circuit of Fig. 11.1. A sinusoidal voltage source is delivering power to a circuit of impedance Z. Using the definition of instantaneous power, p_i, may be shown to be given by the following equation:

$$p_i = vi \quad \textbf{(Eq. 11.1)}$$

where v = instantaneous voltage across impedance Z
i = instantaneous current through impedance Z
p_i = instantaneous power delivered by the voltage source to the impedance

If the unit of v is volts and the unit of i is amperes, then the unit of p_i is watts (W).

Assume that the voltage source is given by the equation:

$$v = V_p \sin \omega t$$

and the impedance is given by the polar expression:

$$Z = Z\angle\theta$$

Based on the foregoing and Ohm's Law, the current in the circuit will be given by:

$$i = I_p \sin(\omega t - \theta)$$

Using these expressions for voltage and current in Equation 11.1:

$$p_i = (V_p \sin \omega t)[I_p \sin(\omega t - \theta)] = V_p I_p (\sin \omega t) \sin(\omega t - \theta) \quad \textbf{(11.2)}$$

This last equation for instantaneous power may be utilized to determine the average power delivered by a source to a circuit. In addition, it will be used to analyze the behavior of passive circuit elements when subjected to the voltage source as shown in Fig. 11.1.

11.2 POWER IN PASSIVE CIRCUIT ELEMENTS

The average power delivered by a source to a load as shown in Fig. 11.1 may be found by integrating Equation 11.2 over a period and dividing by the period. Since it is assumed that most students using this text do not have a background in calculus, we will approach the problem of finding average power a little differently. In the process of determining average power we will introduce two new concepts: ***reactive power*** and ***apparent power.***

11.2.1 Power in the Resistor

Consider the situation when the impedance of Fig. 11.1 is a pure resistor. This is shown in Fig. 11.2(a). Since we know that θ is 0 for a resistor, Equation 11.2 becomes

$$p_i = p_r - V_p I_p \sin^2 \omega t \quad \textbf{(11.3)}$$

Equation 11.3 is plotted in Fig. 11.2(b) in ladder form under the waveforms for voltage and current. From these waveforms we may note that the value of p_r is ***always positive.*** This could also be ascertained from Equation 11.3 directly because any term that is squared yields a positive result. This, in turn, means that the integration process, which basically gives us the area under the p_i curve, always gives us a positive value. The physical interpretation of this is the following rule:

All power delivered by a source to a resistance is dissipated by the resistance.

In other words, a resistance utilizes all the power delivered to it. This may seem obvious, but its importance will be seen when we discuss the behavior of reactive elements, the inductor and capacitor, under the same circuit conditions.

The positive average value obtained under the integration process is given the name ***real*** or ***average power.*** We will use uppercase P to denote the real or average power.

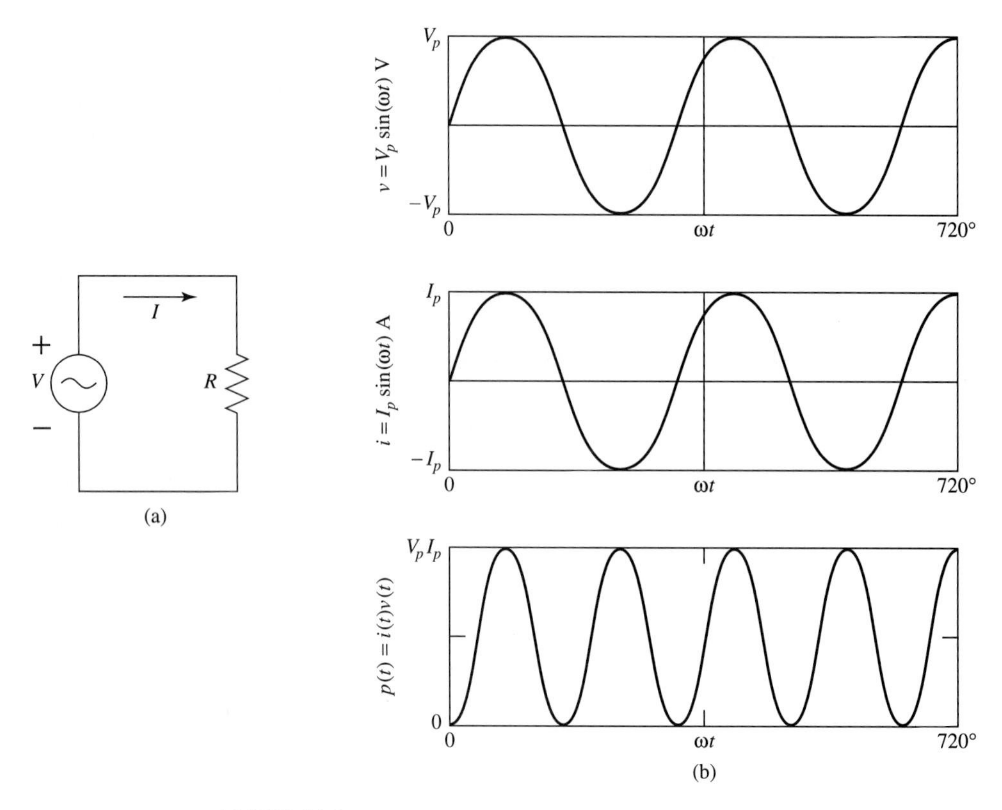

FIGURE 11.2
Voltage, current, and power waveforms for a resistor

When the integration process is carried out, the following equation is obtained for P:

$$P_R = V_{\text{rms}} I_{\text{rms}} \qquad \textbf{(Eq. 11.4)}$$

where V_{rms} = rms voltage across the resistor (V)
I_{rms} = rms value of current through the resistor (A)
P_R = average power (or real power) dissipated in the resistor (W)

Several alternate forms of Equation 11.4 may be developed by considering Ohm's Law. From Ohm's Law:

$$V_{\text{rms}} = RI_{\text{rms}}$$

Substituting this expression for V_{rms} in Equation 11.4, we get:

$$P_R = R(I_{\text{rms}})^2 \qquad \textbf{(Eq. 11.5)}$$

If Ohm's Law is solved for I_{rms} as follows:

$$I_{\text{rms}} = V_{\text{rms}}/R$$

and this result for I_{rms} is substituted into Equation 11.4, we get:

$$P_R = (V_{\text{rms}})^2/R \qquad \textbf{(Eq. 11.6)}$$

Equations 11.4, 11.5, and 11.6 are equivalent expressions for the power dissipated in a resistor.

11.2.2 Power in the Inductor

Consider the situation when the impedance of Fig. 11.1 is a pure inductor. This is shown in Fig. 11.3(a). For an inductor, the current lags the voltage by 90°. The waveforms for current and voltage are shown in ladder form in Fig. 11.3(b). The waveform for instantaneous power developed in the inductor is also included in Fig. 11.3(b) in ladder form with the voltage and current waveforms.

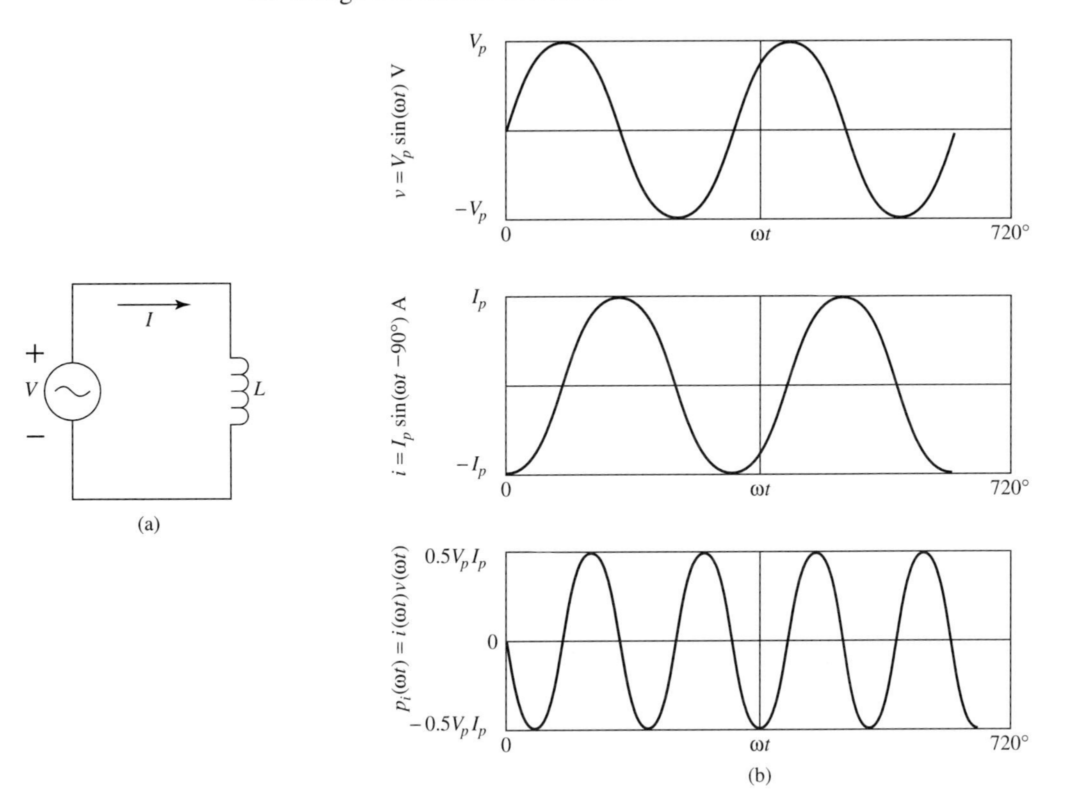

FIGURE 11.3
Voltage, current, and power waveforms for an inductor

Integrating the power equation for the inductor to find its average value includes finding the area under the instantaneous power curve of Fig. 11.3(b). Examining this waveform we note the following: During any one cycle, the magnitude of all the ***positive*** area under the instantaneous power curve is equal to the magnitude of all the ***negative*** area under the instantaneous power curve. The total average power is therefore ***zero.*** This is interpreted as follows:

> None of the power delivered to an inductor is ***used*** or ***dissipated.*** The power delivered to an inductor during the positive portion of a cycle is stored in the magnetic field of the inductor. All this stored power is ***returned*** to the source of the power during the negative portion of a cycle.

Since an inductor does not dissipate power it is said to have no average, or real, power. Mathematically this may be written as:

$$P_L \equiv 0 \qquad \textbf{(Eq. 11.5)}$$

To account for the power delivered to an inductor we introduce the concept of ***reactive power.*** We consider that the power delivered to an inductor during one portion of the cycle (and returned to the power source during the rest of the cycle) is reactive power, denoted by the letter Q_L. To assure that it is not considered as real power, we assign a different unit to reactive power.

The expression for reactive power associated with an inductor is given by Equation 11.6:

$$Q_L = V_{\text{rms}} I_{\text{rms}} \quad \textbf{(Eq. 11.6)}$$

where V_{rms} = rms voltage across the inductor in units of volts
I_{rms} = rms current through the inductor in units of amps
Q_L = reactive power associated with the inductor in units of VAR (reactive volt-amperes)

Alternate forms of Equation 11.6 may be developed by considering Ohm's Law for the inductor. Recall that:

$$V_{\text{rms}} = X_L I_{\text{rms}} \quad \text{or} \quad I_{\text{rms}} = V_{\text{rms}}/X_L$$

Substituting each of these expressions in turn into Equation 11.6 we get the following two alternate forms of Equation 11.6:

$$Q_L = (I_{\text{rms}})^2 X_L \quad \textbf{(Eq. 11.7)}$$

$$Q_L = (V_{\text{rms}})^2 / X_L \quad \textbf{(Eq. 11.8)}$$

EXAMPLE 11.1

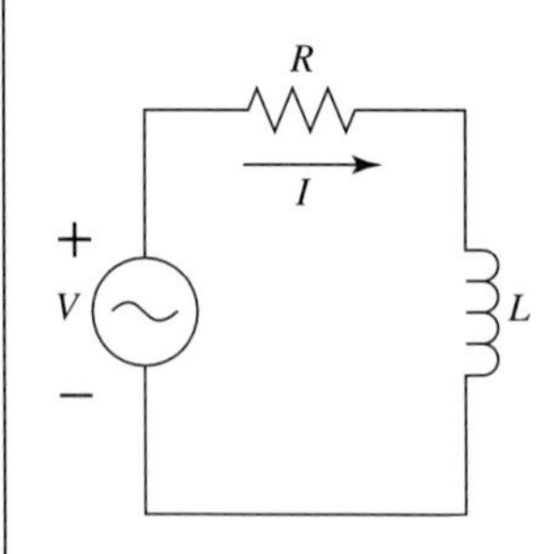

FIGURE 11.4

The voltage source in Fig. 11.4 is given by:

$$v = 12\sin(2000t)\text{ V} \qquad R = 8\ \Omega \qquad L = 6\text{ mH}$$

Find:
(a) P_R, the average power dissipated by the resistor
(b) Q_L, the reactive power associated with the inductor

Solution From the given voltage source:

$$V_{\text{rms}} = 0.707(12) = 8.484\text{ V}$$

$$Z_s = R + j\omega L = 8 + j(2000)(6 \times 10^{-3}) = 8 + j6 = 10\angle 36.9^\circ\ \Omega$$

Using Ohm's Law:

$$I_{\text{rms}} = V_{\text{rms}}/|Z_s| = 8.484/10 = 0.8484 = 848.4\text{ mA}$$

Using this value of I_{rms} in Equations 11.5 and 11.7:

$$P_R = R(I_{\text{rms}})^2 = 8(848 \times 10^{-3})^2 = 5.75\text{ W}$$

$$Q_L = (I_{\text{rms}})^2 X_L = (848 \times 10^{-3})^2(6) = 4.314\text{ VAR}$$

Answer $P_R = 5.75$ W, $Q_L = 4.314$ VAR.

11.2.3 Power in the Capacitor

If the impedance of Fig. 11.1 is a pure capacitor, we have the circuit shown in Fig. 11.5(a). For a capacitor, the current leads the voltage by 90°. The waveforms for current and voltage are shown in ladder form in Fig. 11.5(b). The waveform for instantaneous power developed by the capacitor is also included in Fig. 11.5(b) in ladder form with the voltage and current.

As mentioned previously, integrating the instantaneous power equation to find the average power involves determining the area under the instantaneous power curve. Examining the power waveform, we note, as in the case of the inductor, that the magnitude of all the ***positive*** area under the power curve is equal to the magnitude of all the

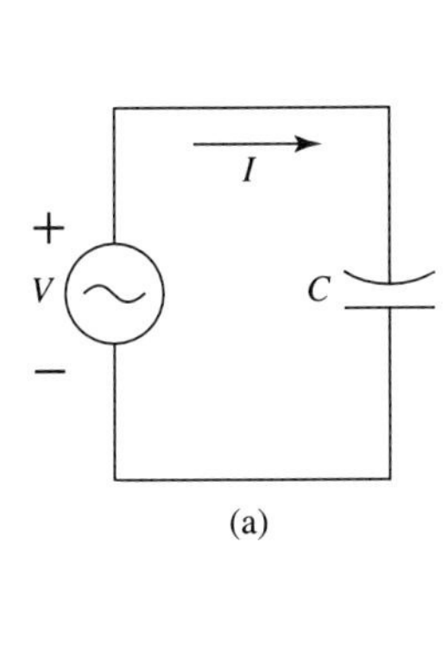

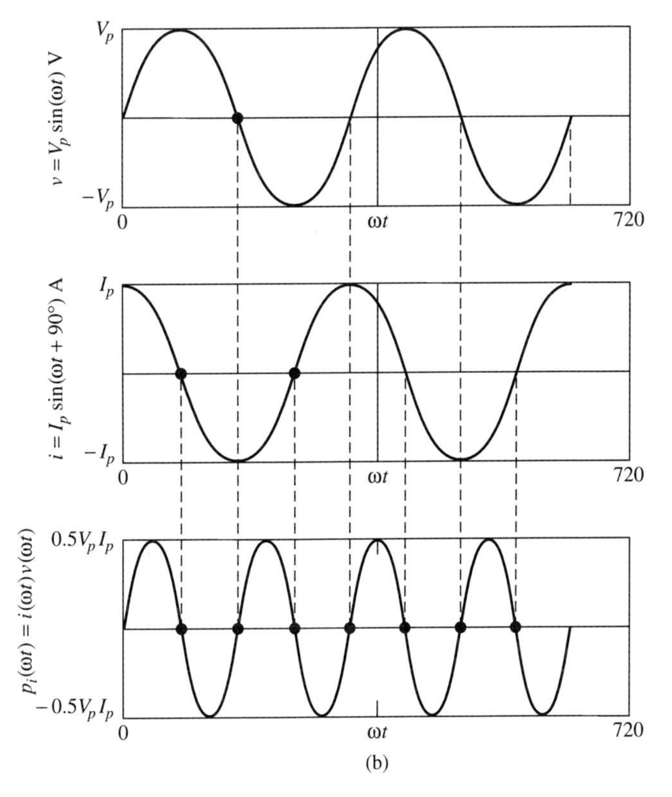

FIGURE 11.5
Voltage, current, and power waveforms for a capacitor

negative area under the power curve. The total area, or the average power, is once again equal to zero. This is interpreted in the same way as for the inductor:

> None of the power delivered to a capacitor is ***used*** or ***dissipated.*** The power delivered to a capacitor during the positive portion of a cycle is stored in the electric field of the capacitor. All this stored power is ***returned*** to the source of the power during the negative portion of a cycle.

Since a capacitor does not dissipate power, it, like the inductor, is said to have no average, or real power. Mathematically:

$$P_c \equiv 0 \qquad \textbf{(Eq. 11.9)}$$

To account for the power delivered to a capacitor we again use the concept of ***reactive power.*** We consider that the power delivered to a capacitor during one portion of a cycle (and returned to the power source during the rest of the cycle) is reactive power, denoted by the letter Q_C. We again assign a different unit to capacitive reactive power and its equation is:

$$Q_C = V_{\text{rms}} I_{\text{rms}} \qquad \textbf{(Eq. 11.10)}$$

where V_{rms} = rms voltage across the capacitor in volts
I_{rms} = rms current through the capacitor in amps
Q_C = reactive power associated with the capacitor in units of VARS (reactive volt-amperes)

Alternate forms of Equation 11.10 using Ohm's Law as in the case of the inductor:

$$Q_C = (I_{\text{rms}})^2 X_C \qquad \textbf{(Eq. 11.11)}$$

$$Q_C = (V_{\text{rms}})^2 / X_C \qquad \textbf{(Eq. 11.12)}$$

11.3 POWER FACTOR, APPARENT POWER, AND THE POWER TRIANGLE

We have seen that the resistor, inductor, and capacitor behave differently when connected to a source of power. These differences lead to the concept of reactive power and average, or real, power. We further note that, when the reactive power of the capacitor is compared to that of the inductor, they appear to behave in the same way. There is a subtle difference, however, as may be seen by the fact that reactive power in an inductor results from multiplying a voltage by a current that ***lags*** that voltage by 90° whereas reactive power in a capacitor results from multiplying a voltage by a current that ***leads*** that voltage by 90°. To account for these differences and ease the mathematical process of power computation, we will introduce two new concepts: the ***power factor*** and the ***power triangle.*** This will lead to the concept of ***apparent power.***

11.3.1 Power Factor

The concept of power factor may be introduced by recalling that when the voltage and current phasors associated with an impedance are plotted in the complex plane, the angle between these phasors is the impedance angle, θ_Z. The value of this angle was:

For a resistor: $\theta_R = 0°$

For an inductor: $\theta_L = 90°$

For a capacitor: $\theta_C = -90°$

We define power factor as follows:

Power factor is the cosine of the impedance angle, θ_Z.

This definition alone is not sufficient to distinguish between Q_L and Q_C because $\cos(90°) = \cos(-90°)$. We therefore add the suffix ***leading*** to $\cos\theta_C$ to represent the fact that the capacitive current leads the capacitive voltage and ***lagging*** to the $\cos\theta_L$ to represent that the inductive current lags the inductive voltage. Accordingly, denoting power factor by fp:

For a resistor: $\text{fp} = \cos(0°) = 1$

For an inductor: $\text{fp} = \cos(90°) = 0$ lagging (0 lag)

For a capacitor: $\text{fp} = \cos(-90°) = 0$ leading (0 lead)

The different power types may now be represented by complex numbers. Real power is represented by a complex number with a 0° phase angle, inductive reactive power by a complex number with a 90° phase angle, and capacitive power by a complex number with a −90° phase angle. The phase angle, when used in the context of power, will be referred to as a "power angle" θ_p. We should remain aware, however, that they are equal.

$$\theta_p = \theta_Z \qquad \textbf{(Eq. 11.13)}$$

The power associated with the three passive elements will now be represented as follows:

$$P_R = P\angle 0°\ \text{W} \qquad Q_L = Q_L\angle 90°\ \text{VAR} \qquad Q_C = Q_C\angle -90°\ \text{VAR} \qquad \textbf{(Eq. 11.14)}$$

Since real power is always associated with a resistor, we will drop the subscript R when writing P and understand we are referring to real power dissipated by a resistor. We will continue, of course, to use the subscripts C and L when referring to reactive power.

11.3.2 Apparent Power and the Power Triangle

In circuits that include more than a single element type, we will encounter both real and reactive power. This was demonstrated in Example 11.1, where a series R–L circuit (Fig. 11.4) resulted in both real and reactive powers. When these powers are plotted on the complex plane, we get the result shown in Fig. 11.6(a). Note how, as expected, the power and reactive power are perpendicular to one another.

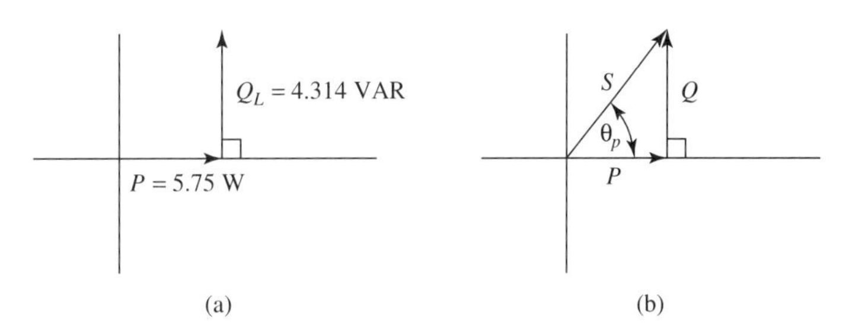

FIGURE 11.6

If the resultant of these two complex numbers is drawn a right triangle results as shown in Fig. 11.6(b). This right triangle is called the ***power triangle.*** The resultant is called ***apparent power*** and represents the sum of the real and reactive powers. Apparent power is represented by the complex number S and is given, mathematically as follows:

$$S = I^*V = S\angle\theta_P \qquad \textbf{(Eq. 11.15)}$$

When I^* (the conjugate of I) is given in amperes and V is given in volts, then the unit for S is the volt-ampere (VA). The power triangle is interpreted as follows:

> The power source delivers an apparent power, S, to the circuit. This apparent power is partially dissipated in the circuit resistance as real or average power, and the remainder is stored in the circuit reactive elements as reactive power.

The ***power angle*** of a circuit is the angle between the real power and the apparent power. This is shown in Fig. 11.6(b) as θ_p. The power factor of the circuit would be the cosine of this power angle.

EXAMPLE 11.2

For the circuit of Fig. 11.4, calculate the following:

(a) The circuit power factor, θ_p

(b) Apparent power of the circuit, S

Solution

(a) The power triangle for this circuit was drawn in Fig. 11.6(b). From this triangle we may find the power angle, θ_p, as follows:

$$\theta_p = \tan^{-1}(Q_L/P) = \tan^{-1}(4.314/5.75) = 36.9°$$
$$\text{fp} = \cos\theta_p = \cos(36.9°) = 0.8$$

Note that the power angle, θ_p, is the same as the angle between the source voltage and circuit current, θ_Z, as expected. Just as the current in an inductive circuit is said to lag the voltage, the power in an inductive circuit is said to lag the apparent power. We must therefore attach the label ***lagging*** to the power factor to indicate an inductive circuit.

Answer fp = 0.8 lag.

(b) The circuit apparent power may be found from the power triangle in any of the following ways:

$$S = Q_L/\sin\theta_p = 4.314/\sin(36.9°) = 7.19 \text{ VA}$$
$$S = P/\cos\theta_P = 5.75/\cos(36.9°) = 7.19 \text{ VA}$$
$$S = \sqrt{P^2 + (Q_L)^2} = \sqrt{4.314^2 + 5.75^2} = 7.19 \text{ VA}$$

Answer $S = 7.19$ VA.

EXAMPLE 11.3

In the R–C circuit of Fig. 11.7, $R = 4\text{ k}\Omega$, $C = 0.4\ \mu\text{F}$, and $v = 360\sin(417t)$ V. Find:
(a) P = power dissipated in the circuit
(b) Q_C = reactive power associated with the circuit
(c) θ_p = circuit power angle
(d) fp = circuit power factor
(e) S = apparent power

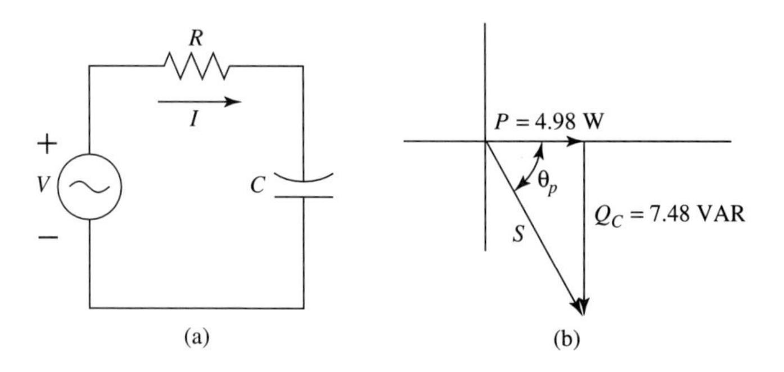

FIGURE 11.7

Solution
(a) $Z_C = X_C\angle{-90°} = (1/\omega C)\angle{-90°} = [1/(417 \times 0.4 \times 10^{-6}]\angle{-90°}$
$= 6\text{K}\angle{-90°}\ \Omega$

$$Z_s = R + Z_C = 4\text{K} + (6\text{K}\angle{-90°}) = 7.21\text{K}\angle{-56.3°}\ \Omega$$
$$V_{\text{rms}} = 0.707(360) = 254.5 \text{ V}$$
$$I_{\text{rms}} = V_{\text{rms}}/|Z_s| = 254.5/7.21\text{K} = 35.3 \text{ mA}$$
$$P = R(I_{\text{rms}})^2 = 4 \times 10^3(35.3 \times 10^{-3})^2 = 4.98 \text{ W}$$

Answer $P = 4.98$ W.
(b) $Q_C = (I_{\text{rms}})^2 X_C = (35.3 \times 10^{-3})^2(6 \times 10^3) = 7.48$ VAR

Answer $Q = Q_C = -7.48$ VAR.
(c) The power triangle is drawn in Fig. 11.7(b). From this triangle:

$$\theta_p = \tan^{-1}(-7.48/4.98) = -56.3°$$

Note that the circuit power angle is exactly the same as the impedance angle, θ_Z, as expected.

Answer $\theta_p = -56.3°$.
(d) $\cos(-56.3°) = 0.55$, fp $= \cos(-56.3°)$ lead

Answer fp = 0.55 lead.
(e) $S = |Q_C/\sin(-56.3°)| = 7.48/\sin(-56.3°) = 9.0$ VA

Answer $S = 9.0$ VA.

Note that the apparent power found was the magnitude only, not the complex number. When the apparent power is requested, it will always be assumed to be the magnitude only unless the complex apparent power is specifically requested.

When a circuit consists of both inductors and capacitors, the circuit will have a leading or lagging power factor depending on which reactive power has the greater value.

EXAMPLE 11.4

In the circuit of Fig. 11.8(a), $R = 500\ \Omega$, $L = 100$ mH, $C = 0.2\ \mu$F, and $v = 4\sin(10{,}000t)$ V.

Find:

(a) Power dissipated in the circuit
(b) Total circuit reactive power
(c) Circuit power factor
(d) Circuit apparent power

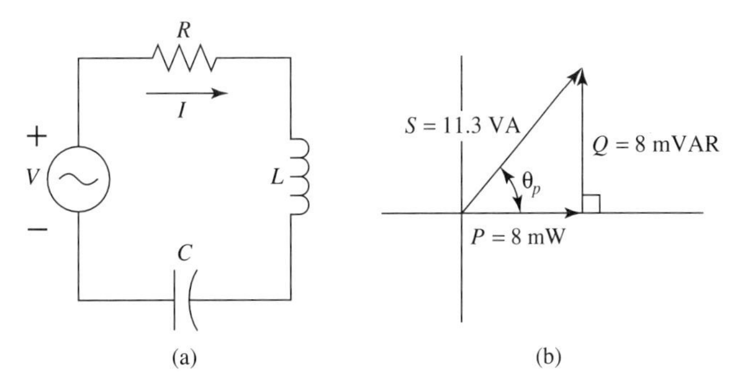

FIGURE 11.8

Solution

(a) $Z_S = R + j(X_l - X_C)$

$X_L = \omega L = (10{,}000)(0.1) = 1000\ \Omega \qquad X_C = 1/\omega C$
$= 1/(10{,}000 \times 0.2 \times 10^{-6}) = 500\ \Omega$

$Z_s = 500 + j(1000 - 500) = 500 + j500 = 707\angle 45^\circ\ \Omega$

$V_{\text{rms}} = 0.707(4) = 2.8\ V \qquad I_{\text{rms}} = V_{\text{rms}}/Z_s = 2.8/707 = 4.0$ mA

$P = R(I_{\text{rms}})^2 = 500(4 \times 10^{-3})^2 = 8 \times 10^{-3}$ W

Answer $P = 8$ mW.

(b) $Q = Q_L - Q_C$, $Q_L = (I_{\text{rms}})X_L = (4 \times 10^{-3})^2(1000) = 16 \times 10^{-3}$ VAR

$Q_C = (I_{\text{rms}})^2 X_C = (4 \times 10^{-3})^2(500) = 8 \times 10^{-3}$ VAR

$Q = 16 \times 10^{-3} - 8 \times 10^{-3} = 8 \times 10^{-3}$ VAR

Answer $Q = 8$ mVAR.

(c) The power triangle for this circuit is drawn in Fig. 11.8(b). From this triangle:

$$\theta_p = \tan^{-1}(Q/P) = \tan^{-1}(500/500) = 45^\circ$$
$$\cos\theta_p = \cos 45^\circ = 0.707 \qquad \text{fp} = \cos\theta_p \text{ lag}$$

Answer fp = 0.707 lag.

(d) $S = \sqrt{P^2 + Q^2} = \sqrt{(8 \times 10^{-3})^2 + (8 \times 10^{-3})^2} = 11.3 \times 10^{-3}$

Answer $S = 11.3$m VA.

In this example, the circuit power factor was lagging because the inductive reactive power was greater than the capacitive reactive power. If the reverse had been true, then the circuit power factor would have been leading.

11.3.3 Circuit Power Calculations

In any given circuit there may be many resistors, inductors, and capacitors. Fortunately, the calculation of total circuit power, reactive power, and apparent power is simplified by the fact that they are all represented by complex numbers. Accordingly, the following rules apply to circuit power calculations:

> The total power dissipated in a circuit is the sum of the individual powers dissipated by individual circuit impedances.

This may be mathematically expressed as:

$$P_T = \Sigma P_n \qquad \textbf{(Eq. 11.16)}$$

where P_n represents the power dissipated in impedance Z_n.

> The total reactive power associated with a circuit is the sum of the individual reactive powers associated with the individual impedances.

Mathematically:

$$Q_T = \Sigma Q_n \qquad \textbf{(Eq. 11.17)}$$

where Q_n represents the reactive power associated with impedance Z_n.

No such simple summation may be performed to determine the total circuit apparent power. This is obvious from the fact that the apparent power is the polar resultant of the real and reactive powers, and complex number magnitudes cannot be added simply. Therefore, finding the total circuit apparent power is found only after the total real and reactive powers have been found.

$$S_T = \sqrt{(P_T)^2 + (Q_T)^2} = P_T/\cos \theta_p = Q_T/\sin \theta_p \qquad \textbf{(Eq. 11.18)}$$

Note that these calculations are further simplified by the fact that they are the same regardless of whether the circuit under consideration is series, parallel, series–parallel or otherwise connected.

EXAMPLE 11.5

For the circuit of Fig. 11.9(a) find the following:

(a) The power dissipated in each individual impedance (load)
(b) The reactive power associated with each individual load
(c) The power factor associated with each individual load
(d) The apparent power associated with each individual load
(e) The total circuit power
(f) The total circuit reactive power
(g) The total circuit power factor
(h) The total circuit apparent power

Solution

$$V_{\text{rms}} = 0.707 \times 75 = 53 \text{ V}$$

If we let Z represent the total circuit impedance:

$$Z = Z_1 + Z_2 \| Z_3 = 4\text{K}\angle 30^\circ + \frac{(4\text{K}\angle -45^\circ)(6\text{K}\angle 60^\circ)}{(4\text{K}\angle -45^\circ) + (6\text{K}\angle 60^\circ)}$$

$$Z = 7.4\text{K}\angle 11.9^\circ\ \Omega$$

$$I_{1,\text{rms}} = V_{\text{rms}}/Z = 53/7.4\text{K} = 7.16 \text{ mA}$$

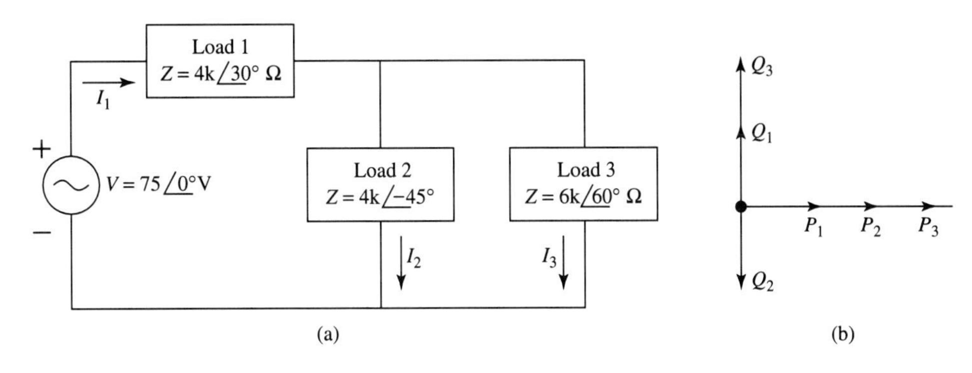

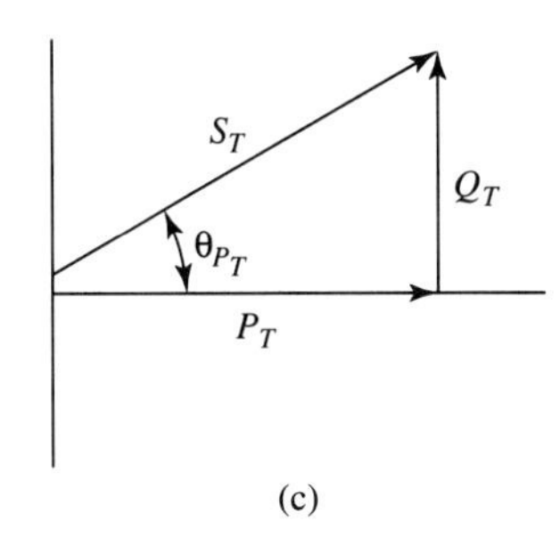

FIGURE 11.9

Using current division:

$$I_{2,\text{rms}} = |Z_3 I_{1,\text{rms}}/(Z_2 + Z_3)| = \left|\frac{(6\text{K}\angle 60°)(7.16 \times 10^{-3})}{(4\text{K}\angle -45°) + (6\text{K}\angle 60°)}\right| = 6.83 \text{ mA}$$

$$\begin{aligned} I_{3,\text{rms}} &= |Z_2 I_{1,\text{rms}}/(Z_2 + Z_3)| \\ &= |(4\text{K}\angle -45°)(7.16 \times 10^{-3})/(4\text{K}\angle -45° + 6\text{K}\angle 60°)| = 4.55 \text{ mA} \end{aligned}$$

(a) $P_1 = R_1(I_{1,\text{rms}})^2 = \text{Re}(Z_1)(I_{1,\text{rms}})^2 = 4\text{K}\cos(30°)(7.16 \times 10^{-3})^2$
$= 177.6 \times 10^{-3}$ W

$P_2 = R_2(I_{2,\text{rms}})^2 = \text{Re}(Z_2)(I_{2,\text{rms}})^2 = 4\text{K}\cos(-45°)(6.83 \times 10^{-3})^2$
$= 131.9 \times 10^{-3}$ W

$P_3 = R_3(I_{3,\text{rms}})^2 = \text{Re}(Z_3)(I_{3,\text{rms}})^2 = 6\text{K}\cos(60°)(4.55 \times 10^{-3})^2$
$= 62.1 \times 10^{-3}$ W

Answer $P_1 = 177.6$ mW, $P_2 = 131.9$ mW, $P_3 = 62.1$ mW.

(b) $Q_1 = (I_{1,\text{rms}})^2 X_1 = (I_{1,\text{rms}})^2 \text{Im}(Z_1) = (7.16 \times 10^{-3})^2 4\text{K}\sin(30°)$
$= 102.5 \times 10^{-3}$ VAR

$Q_2 = (I_{2,\text{rms}})^2 X_2 = (I_{2,\text{rms}})^2 \text{Im}(Z_2) = (6.83 \times 10^{-3})^2 4\text{K}\sin(-45°)$
$= -131.9 \times 10^{-3}$ VAR

$Q_3 = (I_{3,\text{rms}})^2 X_3 = (I_{3,\text{rms}})^2 \text{Im}(Z_3) = (4.55 \times 10^{-3})^2 6\text{K}\sin(60°)$
$= 107.6 \times 10^{-3}$ VAR

Answer $Q_1 = 102.5$ mVAR, $Q_2 = -131.9$ mVAR, $Q_3 = 107.6$ mVAR.

(c) $\text{fp}_1 = \cos\theta_{p_1} = \cos(30°) = 0.866$ lag, since θ_{p_1} is positive

$\text{fp}_2 = \cos\theta_{p_2} = \cos(-45°) = 0.707$ lead, since θ_{p_2} is negative

$\text{fp}_3 = \cos\theta_{p_3} = \cos(60°) = 0.50$ lag, since θ_{p_3} is positive

Answer $\text{fp}_1 = 0.866$ lag, $\text{fp}_2 = 0.707$ lead, $\text{fp}_3 = 0.50$ lag.

(d) $S_1 = \sqrt{(P_1)^2 + (Q_1)^2} = \sqrt{(177.6 \times 10^{-3})^2 + (102.5 \times 10^{-3})^2} = 205.1$ mVA

$S_2 = \sqrt{(P_2)^2 + (Q_2)^2} = \sqrt{(131.9 \times 10^{-3})^2 + (131.9 \times 10^{-3})^2}$
$= 186.5$ mVA

$S_3 = \sqrt{(P_3)^2 + (Q_3)^2} = \sqrt{(62.1 \times 10^{-3})^2 + (107.6 \times 10^{-3})^2} = 124.2$ mVA

Answer $S_1 = 204.1$ mVA, $S_2 = 186.5$ mVA, $S_3 = 124.2$ mVA.
(e) Using equation 11.16:

$$P_T = P_1 + P_2 + P_3 = 177.6 \text{ mW} + 131.9 \text{ mW} + 62.1 \text{ mW} = 371.6 \text{ mW}$$

Answer $P_T = 371.6$ mW.
(f) Using Equation 11.17:

$$Q_T = Q_1 + Q_2 + Q_3 = 102.5 \text{ mVAR} - 131.9 \text{ mVAR} + 107.6 \text{ mVAR}$$
$$= 78.2 \text{ mVAR}$$

Answer $Q_T = 78.2$ mVAR.
(g) $\theta_{PT} = \tan^{-1}(Q_T/P_T) = \tan^{-1}(78.2 \times 10^{-3}/371.6 \times 10^{-3}) = 11.9°$

$f_{PT} = \cos(\theta_{P_T}) = \cos(11.9°) = 0.979$ lag since θ_{P_T} is positive

Answer $f_{PT} = 0.979$ lag.
(h) Using Equation 11.18:

$$S_T = \sqrt{(P_T)^2 + (Q_T)^2} = \sqrt{(371.6 \times 10^{-3})^2 + (78.2 \times 10^{-3})^2} = 379.7 \text{ mVA}$$

Answer $S_T = 379.7$ mVA.

The power triangle for this circuit is drawn in Fig. 11.9(c). Note that, as expected, the circuit has a positive power angle (11.9°).

The previous example involved power calculations of individual circuit components. The basic technique is the same even if the individual impedances are not known, because Equations 11.16 through 11.18 and their associated rules depend only on knowing the individual power components, not the impedances. This is demonstrated in Example 11.6, which involves knowledge strictly of the individual power components.

EXAMPLE 11.6

For the circuit of Fig. 11.10 find:
(a) Total circuit power
(b) Total circuit reactive power
(c) Total circuit power factor
(d) Total circuit apparent power
(e) Source current I

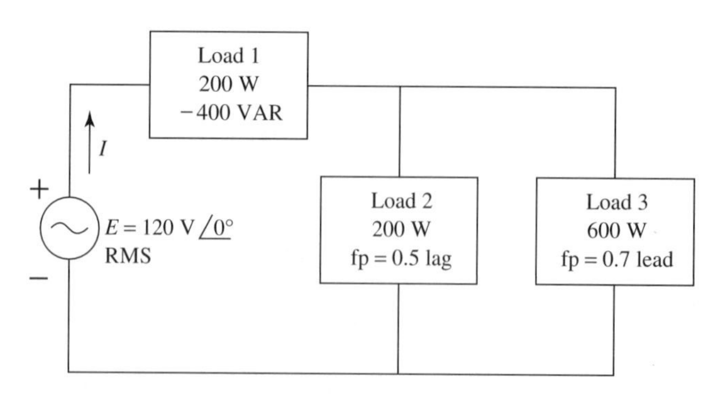

FIGURE 11.10

Solution
(a) $P = P_1 + P_2 + P_3 = 200 + 200 + 600 = 800$

Answer $P = 800$ W.

(b) $Q = Q_1 + Q_2 + Q_3$

$\mathbf{Q_1 = -400\ VAR}$

$\theta_2 = \cos^{-1}(0.5) = 60°$ $\quad Q_2 = P_2 \tan\theta_2 = 200\tan(60°)$

$\mathbf{Q_2 = 346\ VAR}$ (positive, since the fp is lagging)

$\theta_3 = \cos^{-1}(0.7) = -45.57°$ $\quad Q_3 = P_3 \tan\theta_3 = 600\tan(-45.57°)$

$\mathbf{Q_3 = -612\ VAR}$ (negative, since the fp is leading)

$Q = -400 + 346 - 612 = -666$

Answer $Q = -666$ VAR

(c) $\text{fp} = \cos\theta$

$\theta = \tan^{-1}(Q/P) = \tan^{-1}(-666/800) = -39.78°$ $f_p = \cos(-38.78°) = 0.780$

Answer fp = 0.780 leading (since θ is negative)

(d) $S = \sqrt{P^2 + Q^2} = \sqrt{800^2 + 666^2} = 1040.94$ VA

Answer $S = 1040.94$ VA.

(e) $S = EI$ or $I = S/E = 1040.94/120 = 8.675$ A

Answer $I = 8.675$ A

Note that in the preceding example the power parameters could be found even though the individual load components were not specifically known. We do know that loads 1 and 3 are capacitive because they have negative reactive powers and leading power factors. Load 2 is inductive because it has a lagging power factor.

11.4 MAXIMUM POWER TRANSFER THEOREM

A load may receive power from a single voltage or current source with a single impedance or from a circuit containing several sources and impedances. In any case, it is often useful to know what conditions will result in the load receiving the maximum power from the source (or source circuit). This may be determined from the Maximum Power Transfer Theorem. This theorem may be stated as follows:

MAXIMUM POWER TRANSFER THEOREM

The condition under which a load receives maximum power from a source is that the load impedance is equal to the conjugate of the Thevenin impedance. The value of this maximum power is the Thevenin voltage squared divided by the Thevenin resistance.

$$Z_L = (Z_{\text{Th}})^* \quad \text{for maximum power transfer} \qquad \textbf{(Eq. 11.19)}$$

$$P_m = (E_{\text{Th}})^2/4R_{\text{Th}} = \text{value of maximum power} \qquad \textbf{(Eq. 11.20)}$$

where E_{Th} = rms value of the Thevenin voltage

EXAMPLE 11.7

(a) Find the value of the load impedance, Z_L, that will result in maximum power being transferred from the voltage source and its series impedance in Fig. 11.11.

(b) Find the value of the maximum power being transferred.

Solution

(a) This circuit is already in the form of Thevenin's equivalent circuit. From Equation 11.19: $Z_L = (Z_S)^* = 480\underline{/-36.9°}$

Answer $Z_L = 480\underline{/-36.9°}\ \Omega$.

FIGURE 11.11

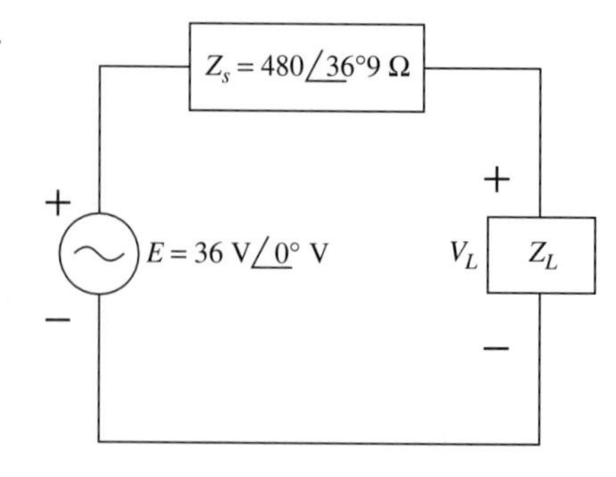

(b) From Equation 11.20:

$$P_m = E_{Th}^2/4\,R_{Th}$$
$$Z_L = (383.85 - j288.20)\ \Omega$$
$$R_{Th} = 383.85\ \Omega$$
$$E_{Th} = 0.707(36) = 25.45\ V_{rms}$$
$$P_m = (25.45)^2/4(383.85) = 0.422\ \text{W}$$

Answer $P_m = 422$ mW.

EXAMPLE 11.8

In the circuit of Fig. 11.12(a), Z_L is the load receiving power from the circuit to the left of terminals a and b. Find:

(a) The value of Z_L that will result in its receiving ***maximum*** power from the circuit to the left of terminals a and b.

(b) The value of the maximum power transferred.

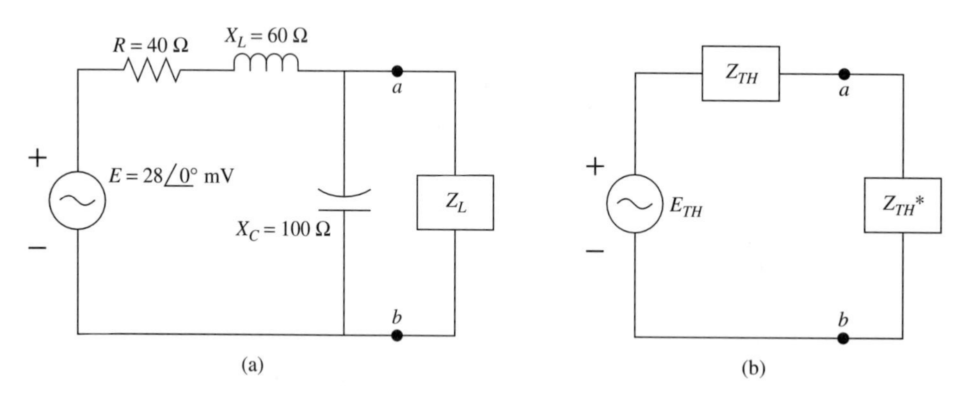

FIGURE 11.12

Solution Before we can find the values requested in parts (a) and (b), we must find Thevenin's equivalent circuit with respect to terminals a and b. The circuit of Fig. 11.12(a) is redrawn in Fig. 11.12(b) with Thevenin's equivalent circuit replacing the original circuit to the left of terminals a and b. In Fig. 11.12(b):

$$Z_{Th} = \frac{(100\angle -90°)(40 + j60)}{40 + j60 - j100} = 127.46\angle 11.3°$$

$$Z_L = (Z_{Th})^*$$

Answer $Z_L = 127.46\angle -11.3^\circ\ \Omega$.

(b) $Z_{Th} = 127.46\angle 11.3^\circ = 125.0 + j25.0$

$R_{Th} = 125\ \Omega$

$$E_{Th} = \frac{(100\angle -90^\circ)(28 \times 10^{-3}\angle 0^\circ)}{40 + j60 - j100}$$

$$= \frac{(100\angle -90^\circ)(28 \times 10^{-3}\angle 0^\circ)}{(40 - j40)} = 49.50\angle -45^\circ \text{ mV}$$

$$E_{Th,rms} = 0.707(49.50) \times 10^{-3} = 35.0 \times 10^{-3} = 35.0 \text{ mVrms}$$

From Equation 11.20:

$$P_m = (35.0 \times 10^{-3})^2/4(125) = 2.45\ \mu\text{W}$$

Answer $P_m = 2.45\ \mu$W.

In working with dc circuits, the Maximum Power Transfer Theorem is modified in two ways:

1. Replace the word ***impedance*** with ***resistance.***
2. Replace the term ***rms value*** (of Thevenin's voltage) with ***Thevenin's voltage.***

The following example demonstrates use of the Maximum Power Transfer Theorem with a dc circuit.

EXAMPLE 11.9

In the dc circuit of Fig. 11.13(a) find:

(a) The value of load resistance that will result in maximum power transfer from the circuit to the left of terminals *a* and *b*

(b) The value of the maximum power transferred

Solution The circuit of Fig. 11.13(a) is redrawn in Fig. 11.13(b) using Thevenin's equivalent circuit of the portion to the left of terminals *a* and *b*. In this circuit:

$$R_{Th} = R_1 \| R_2 = (1\text{K})(3\text{K})/(4\text{K}) = 750\ \Omega$$
$$E_{Th} = R_2(E)/(R_1 + R_2) = (3\text{K})(80)/4\text{K} = 60 \text{ V}$$

Solution

(a) From Equation 11.19:

Answer $R_L = 750\ \Omega$.

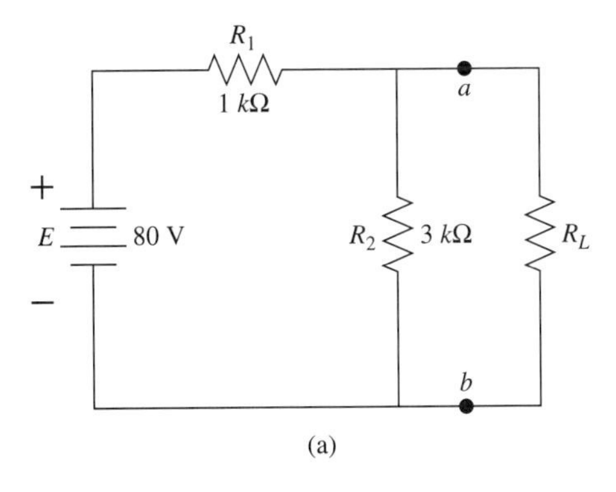

(a)

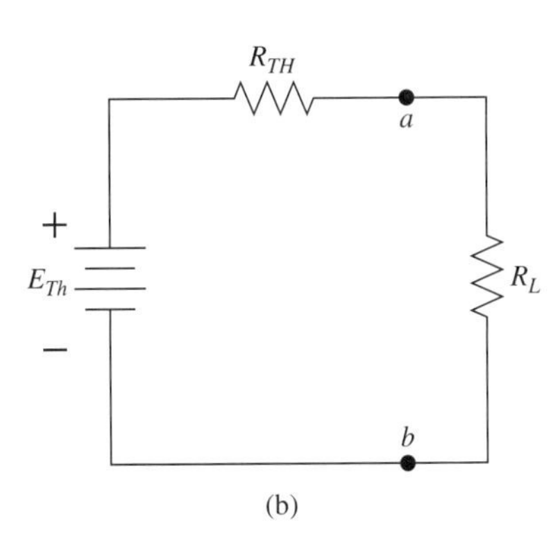

(b)

FIGURE 11.13

(b) Using Equation 11.20:

$$P_m = (60)^2/(4 \times 750) = 1.2 \text{ W}$$

Answer $P_m = 1.2$ W.

The Maximum Power Transfer Theorem may also be stated in terms of Norton's equivalent circuit as follows:

> The condition under which a load receives maximum power from a source is that the load impedance is equal to the conjugate of the Norton impedance. The value of this maximum power is the Norton resistance multiplied by the Norton current squared divided by 4.

$$Z_L = (Z_N)^* \quad \text{for maximum power transfer} \qquad \textbf{(Eq. 11.21)}$$

$$P_m = R_N(I_N)^2/4 = \text{value of maximum power} \qquad \textbf{(Eq. 11.22)}$$

where I_N = rms value of Norton current

EXAMPLE 11.10

In Fig. 11.14(a):

(a) Find the value of load impedance required for maximum power transfer from the source circuit to the left of terminals 1 and 2.

(b) Using Norton's equivalent circuit, find the value of the maximum power transferred to the load.

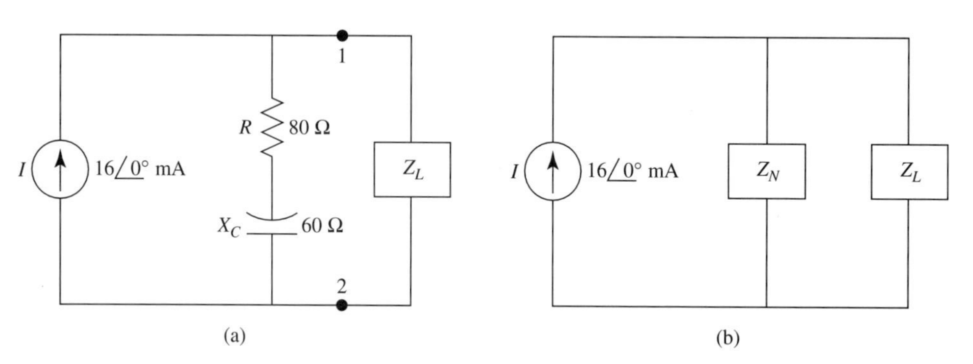

FIGURE 11.14

Solution

(a) The circuit of Fig. 11.14(a) is already in the form of a Norton circuit. This is emphasized by drawing this circuit in the form shown in Fig. 11.14(b). In this circuit:

$$Z_N = 80 - j60 = 100\underline{/-36.9^\circ}$$

Therefore, from Equation 11.21:

Answer $Z_L = 100\underline{/36.9}\ \Omega$.

(b) In this circuit, $I_N = I = 16\underline{/0^\circ}$ mA. $I_{N,\text{rms}} = 0.707(16) = 11.31$ mA

Using Equation 11.22:

$$P_m = (80)(11.31 \times 10^{-3})^2/4 = 2.56 \times 10^{-3} \text{ W}$$

Answer $P_m = 2.56$ mW.

In some cases, either the real or imaginary part of the load impedance cannot be adjusted for maximum power transfer. This may be due to any of a number of reasons such as conflicts with other requirements or simply inability to vary the impedance. In such a case, as a rule of thumb, adjust the load impedance as closely as possible to the ideal value. This will yield the maximum power that can be transferred under the given conditions.

11.5 POWER FACTOR CORRECTION

Recall from discussions in the previous sections that the only passive component that dissipates (uses) real, or average, power is the resistor. Both the inductor and capacitor store energy during one portion of a periodic power cycle and return that same energy to the source during the remainder of the cycle. This is very important in power circuits, especially when large amounts of power are involved.

Companies that use power supplied by power utilities pay only for the ***real*** power that they use. The power utility, however, must not only supply the power required by a company but also the power that is lost in the transmission lines due to i^2R losses.

Power lost in the transmission lines is directly proportional to the square of the current drawn by a company. This current, however, can be reduced without sacrificing any real power required by the company. Reducing the current will obviously result in a reduction in the i^2R losses that the utility must supply over and above the actual power required by the company. Power utilities require the company to employ a technique known as power factor correction in order to reduce the current flowing through the transmission line. The power factor of any circuit can be altered without changing the real power utilized by the circuit.

Recall that the inductive reactive power of a circuit, Q_L, has a 90° power angle while the capacitive reactive power of a circuit has a −90° power angle. The total power angle of a circuit is determined by summing the real and reactive powers and determining the apparent power as described in the previous sections. The resulting power factor of the circuit is the cosine of the total power angle.

The foregoing indicates that the total power factor of a circuit may be modified by adding either inductive or capacitive reactive power without changing the real power drawn by the circuit. If the reactive element used to change the circuit power factor is placed in parallel with the circuit, the voltage across the circuit remains the same but the total current drawn is changed. The following examples demonstrate this procedure.

EXAMPLE 11.11

In the circuit of Fig. 11.15(a):

$E = 400$ Vrms is the voltage supplied by a power company

$Z = 2\ \Omega\angle 60°$ is the total circuit impedance of a customer

(a) Calculate the power factor of the circuit shown in Fig. 11.15(a).
(b) Draw the power triangle for this circuit.
(c) Calculate the line current, I, drawn by this circuit.
(d) Modify the circuit in Fig. 11.15(a) to yield a power factor of unity by placing the appropriate element in parallel with the circuit.
(e) Calculate the line current drawn by the modified circuit and draw the new power triangle.
(f) Assuming the total transmission line resistance of the wiring connecting the load to the voltage source is 0.5 Ω, calculate the difference in power dissipated in the transmission line between the original circuit and the modified circuit.

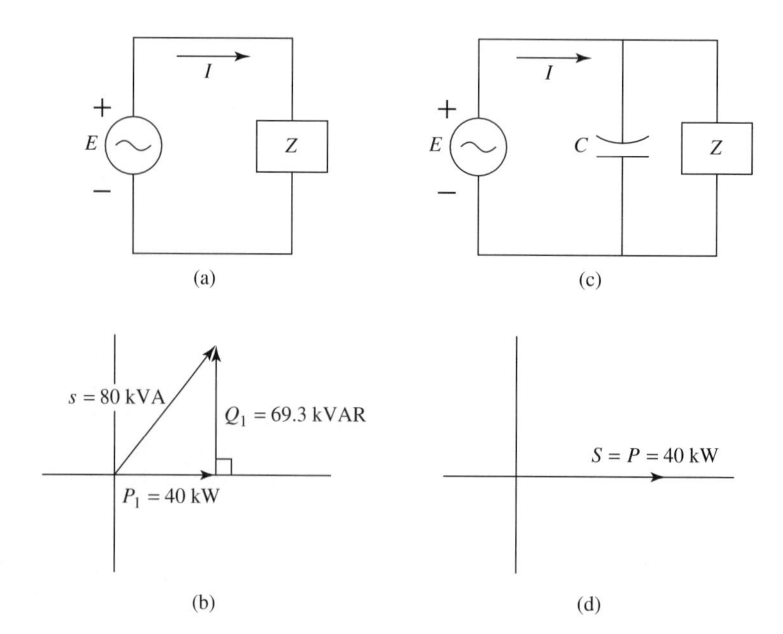

FIGURE 11.15

Solution We will use condition 1 to represent the conditions before modifying the circuit and condition 2 to represent the conditions after modifying the circuit.
(a) Recall that the power angle of a circuit is the same as its phase angle. Therefore:

$$\text{fp}_1 = \cos\theta_1 = \cos(60°) = 0.5$$

Answer $\text{fp}_1 = 0.5$ lag.
(b) From Equation 11.15: $S = IE$
From Ohm's Law: $I_1 = E/Z = 400/2 = 200$ Arms

$$S_1 = 200(400) = 8 \times 10^4 = 80 \text{ kVA}$$

Answer The power triangle is drawn in Fig. 11.15(b). In this figure:

$$P_1 = 80\text{K}\cos(60°) = 80\text{K}(0.5) = 40 \text{ kW}$$
$$Q_1 = 80\text{K}\sin(60) = 80\text{K}(0.866) = 69.3 \text{ kVAR}$$

Note that P_1 and Q_1 are the real and reactive powers associated with this circuit.
(c) The line current was calculated in part (b).

Answer $I_1 = 200$ A.
(d) Consider the power triangle shown in Fig. 11.15(b). To attain a unity power factor we must make the power angle zero because cos (0°) is 1. To make the power angle zero, we must add negative reactive power to balance the positive reactive power that exists.

Since negative reactive power is associated with a capacitor, we must place a capacitor in parallel with our circuit to ***add*** negative reactive power $-Q_C$, where:

$$Q_C = Q_1 = 69.3 \text{ kVAR} \quad \text{to the circuit}$$

The reactance of this capacitor may be found from Equation 11.12:

$$X_C = E^2/Q_C = (400)^2/69.3\text{ K} = 2.3\ \Omega$$

Answer The modified circuit is drawn in Fig. 11.15(c).
(e) The power triangle of the modified circuit is drawn in Fig. 11.15(d). In this triangle:

$$S_2 = P = 40 \text{ kVA} \qquad I_2 = S/E = 40\text{K}/400 = 100 \text{ A}$$

Answer $I_2 = 100$ A.

(f) The power lost in the transmission lines ***before*** modifying the circuit is:

$$P_{L_1} = (I_1)^2(0.5) = (200)^2(0.5) = 20 \text{ kW}$$

The power lost in the transmission lines ***after*** modifying the circuit is:

$$P_{L_2} = (I_2)^2(0.5) = (100)^2(0.5) = 5 \text{ kW}$$
$$\Delta P_L = P_{L_1} - P_{L_2} = 20 \text{ kW} - 5 \text{ kW} = 15 \text{ kW}$$

Note that the line losses were reduced from 20 kW to 5 kW, a 75% reduction, without a loss in the amount of power used by the company.

Answer $\Delta P_L = 15$ kW.

In Example 11.11 the capacitive reactance required to increase the power factor to unity was calculated. The value of the capacitance was not calculated because the frequency was not specified.

EXAMPLE 11.12

Calculate the value of the capacitance required to achieve unity power factor in Example 11.11 if the frequency of the power is 60 Hz.

Solution From the equation for capacitive reactance:

$$C = 1/(2\pi f X_C) = 1/(2\pi \times 60 \times 2.3) = 1.15 \times 10^{-3} \text{ F}$$

Answer $C = 1.15$ mF.

In practice, it is almost impossible to attain a unity power factor. This would require that the element added in parallel with the original circuit introduce absolutely no resistive component. Since all physical devices (see Chapter 13) contain some resistance, the actual power factor achieved cannot be 1.

Power factors very close to 1, however, are very possible. Power factors of 0.98 and better are not uncommon, and the resulting reduction in line losses is usually significant.

11.6 POWER IN DC CIRCUITS

Recall that in dc circuits inductors are effectively shorts and capacitors are effectively opens. Power calculations in dc circuits therefore involve only real power dissipated in the resistors. Power dissipated in a resistance R with a current I flowing through it may be shown to be given by the following equation:

$$P_{\text{dc}} = I^2R \qquad \textbf{(Eq. 11.21)}$$

Since, from Ohm's Law, the voltage across a resistor is IR, then Equation 11.21 may be written as:

$$P_{\text{dc}} = V^2/R \qquad \textbf{(Eq. 11.22)}$$

where V is the voltage across the resistor carrying a current I.

EXAMPLE 11.13

Calculate the total power dissipated in the circuit of Fig. 11.16(a). In this figure, $R_A = 4\ \text{k}\Omega$, $R_B = 6\ \text{k}\Omega$, $R_C = 12\ \text{k}\Omega$, $L = 12$ mH, and $C = 4\ \mu$F.

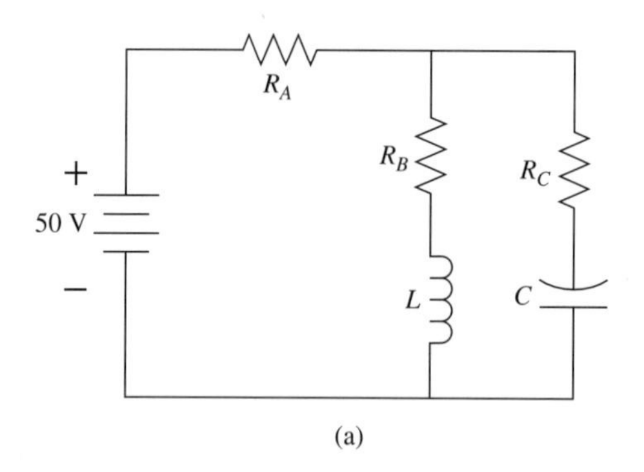

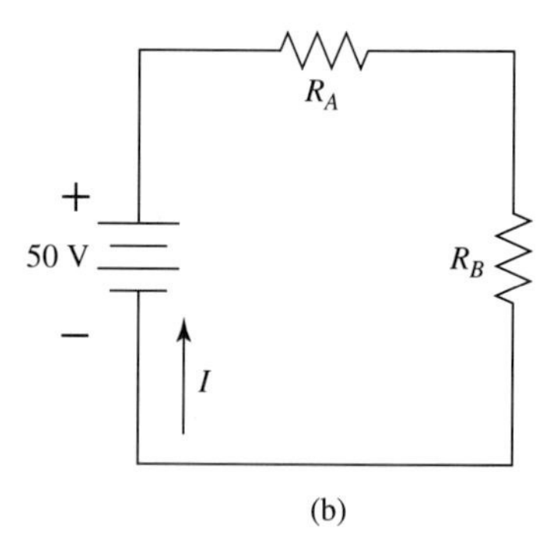

FIGURE 11.16

Solution Using the fact that capacitors are opens and inductors are shorts in dc circuits, the circuit of Fig. 11.16(a) may be redrawn as shown in Fig. 11.16(b). In this figure, note that we have a series dc circuit. Therefore:

$$I = E/(R_A + R_B) = 50/(10\text{K}) = 5\ \text{mA}$$

Using Equation 11.21:

$$P_A = I^2(R_A + R_B) = (5 \times 10^{-3})^2(10\text{K}) = 2.5 \times 10^{-1}\ \text{W}$$

Answer $P = 250$ mW.

■ SUMMARY

- Power is defined as the rate at which energy is utilized.
- Power in ac circuits must be separated into real, reactive, and apparent power.
- Real power (P) is dissipated in a resistor.
- Reactive power in an inductor (Q_L) is considered positive.
- Reactive power in a capacitor (Q_C) is considered negative.
- The power angle is equal to the phase angle of an impedance.
- The power factor of a circuit is defined as the cosine of the power angle.
- The power factor of an inductor and of a capacitor is zero.
- The power factor of a capacitive circuit is labeled ***leading.***
- The power factor of an inductive circuit is labeled ***lagging.***
- The total real power dissipated in a circuit is the sum of all individual powers dissipated by all elements in a circuit ***regardless*** of how these elements are connected.
- The total reactive power associated with a circuit is the sum of all individual reactive powers associated with each reactive element, regardless of how they are connected.
- A power triangle is the diagram of the real and reactive powers of a circuit treated as complex numbers. Real power is always drawn along the positive real axis, inductive reactive power along the positive imaginary axis, and capacitive reactive power along the negative imaginary axis. The complex sum of all the real and reactive powers is called the apparent power.
- Only real power is associated with a dc circuit.
- Maximum power is drawn from a source circuit when the load impedance is equal to the conjugate of the source circuit Thevenin impedance.
- The value of the maximum power drawn is the Thevenin voltage squared divided by four times the Thevenin resistance.

■ PROBLEMS

11.1 Describe the difference between real and reactive powers.

11.2 What is apparent power?

11.3 Which passive element(s) utilize real power and which ones utilize reactive power?

11.4 What is meant by the ***power factor*** of a circuit?

11.5 A series R–L circuit is shown in Fig. P11.5. In this figure, $R = 12$ k, $L = 40$ mH, and $v = 24\sin(400t)$ V.
(a) Find the rms value of current in the circuit.
(b) Find the power dissipated in the circuit.
(c) Find the reactive power associated with the circuit.
(d) Find the apparent power of the circuit.
(e) Find the circuit power factor.

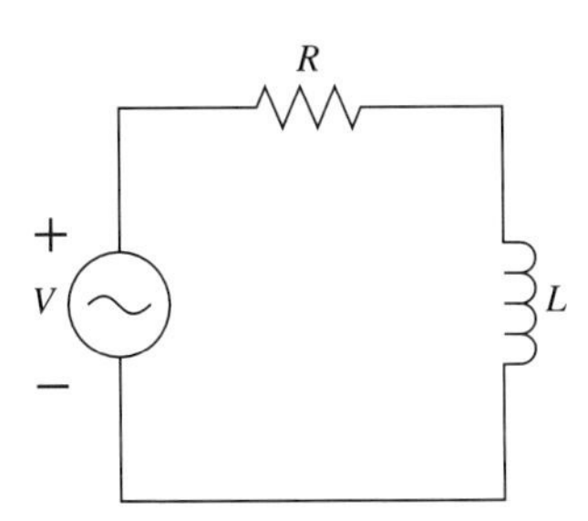

FIGURE P11.5

11.6 A series R–L–C circuit is shown in Fig. P11.6. Given $R = 360\ \Omega$, $X_C = 600\ \Omega$, and $X_L = 240\ \Omega$:

(a) Find the real circuit power.
(b) Find the reactive circuit power.
(c) Find the apparent circuit power.
(d) Find the circuit power factor.
(e) Is this circuit "capacitive" or "inductive" in nature? Why?

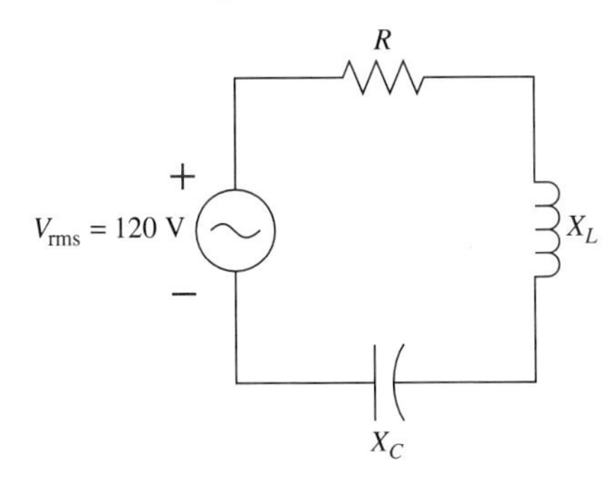

FIGURE P11.6

11.7 **(a)** Find the value of load impedance, Z_L, in Fig. P11.7 that will result in drawing maximum power from the source circuit between terminals a and b.
(b) Find the value of the maximum power transferred in part (a).

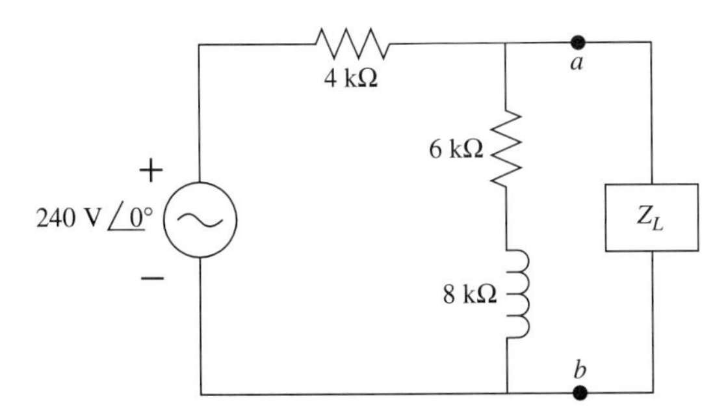

FIGURE P11.7

11.8 It is known that a load impedance $Z_L = 24 - j36\ \Omega$ results in maximum power transfer from a source circuit. It is also known that the value of the maximum power transferred is 1200 W.

(a) What is the Thevenin impedance of the source circuit?
(b) What is the value of the rms Thevenin voltage?

11.9 **(a)** What value of load resistance, R_L, will result in maximum power transfer from the source circuit of Fig. P11.9?
(b) What is the value of the maximum power transferred under the conditions of part (a)?

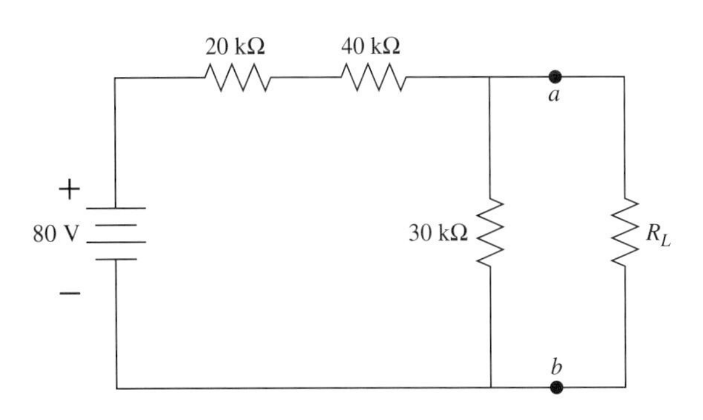

FIGURE P11.9

11.10 The dc power supply in Fig. P11.10 is $E = 40$ V. The components are $R_a = 100\ \Omega$, $R_b = 140\ \Omega$, $R_c = 180\ \Omega$, $C_a = 40\ \mu$F, $L_b = 30$ mH and $L_c = 60$ mH.

(a) Find the value of load resistance that must be placed between terminals a and b to draw maximum power from the source circuit.
(b) Find the value of the maximum power drawn from the source circuit.

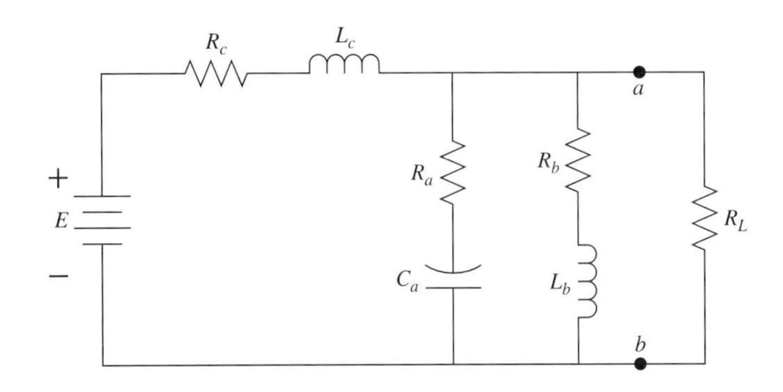

FIGURE P11.10

11.11 **(a)** What is meant by ***power factor correction***?
(b) Why is power factor correction important?

11.12 For the circuit of Fig. P11.12:

(a) Find the total circuit average power.
(b) Find the total circuit reactive power.
(c) Find the reactive load impedance that must be placed in parallel with the circuit of Fig. P11.12 to make the total circuit power factor unity.

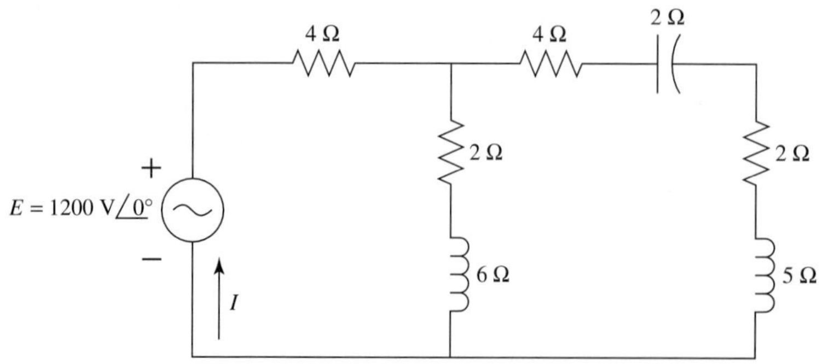

4 Ω
4 Ω
2 Ω
2 Ω
2 Ω
6 Ω
5 Ω
+
E = 1200 V/0°
−
I

FIGURE P11.12

11.13 In the circuit of Fig. P11.12:

(a) Find the rms current delivered to the circuit by the power source ***before*** power factor correction.

(b) Find the rms current delivered to the circuit by the power source ***after*** power factor correction.

(c) Assuming that the total line resistance carrying the circuit current is 0.8 Ω, find the power dissipated in the line resistance ***before*** and ***after*** power factor correction.

(d) Compare the line loss in the circuit before and after power factor correction by finding the difference in power between the two conditions.

11.14 If the frequency of the power delivered to the circuit of Fig. P11.12 is 12 kHz, find the value of the capacitance required for power factor correction to unity.

11.15 **(a)** Find the total circuit power factor for the circuit in Fig. P11.15.

(b) Find the value of capacitance that must be placed in parallel with the circuit to adjust the total circuit power factor to unity.

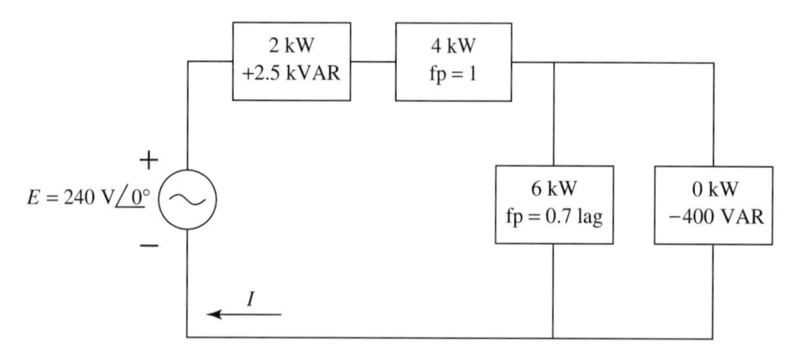

FIGURE P11.15

11.16 For the circuit of Fig. P11.15 find the difference in circuit currents between the two conditions (before and after power factor correction).

11.17 The complex power diagram for a circuit is shown in Fig. P11.17.

(a) Find the circuit power factor.

(b) Find the reactive power that must be added to the circuit to adjust the circuit power factor to unity.

(c) If the voltage across the circuit is 120 Vrms and the frequency is 60 Hz, find the value of capacitance that will result in the desired unity power factor.

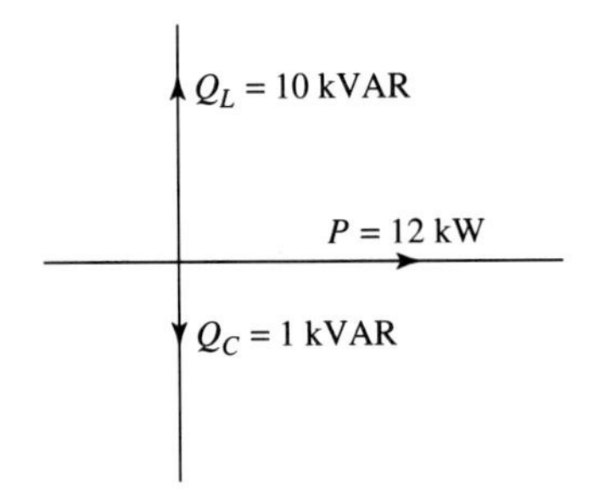

FIGURE P11.17

11.18 **(a)** Draw the complex power diagram for a circuit with the following specifications: $P = 38$ kW, $Q_L = 30$ kVAR, $Q_C = -80$ kVAR.

(b) Find the circuit power angle.

(c) Find the circuit power factor.

(d) Find the reactive power required to make this circuit power factor unity.

CHAPTER 12

TRANSIENT ANALYSIS OF CIRCUITS

INTRODUCTION

In earlier chapters of this text we have analyzed circuits whose steady state solutions do not change with time. For the case of dc circuits we considered that all capacitors were open and all inductors were shorted. For ac circuits, where the voltage itself varies with time, we solved circuit problems by introducing the concept of phasors, thereby suppressing the time variation of the voltages. We used phasor notation and the concept of impedance to analyze the sinusoidal behavior of the circuit. This enabled us to calculate the peak values of voltages and currents. It is true that the instantaneous value of the voltages in the circuit varies with time, but the calculated peak values of the voltages in the circuit are time independent.

The solution to any circuit analysis problem describes how the circuit voltages and currents go from their initial values to their steady state final values. When a change is made in a circuit, such as a switch closing or opening, the voltages and currents change from their initial values (before the switch is operated) to their final values at a speed and in a manner determined by the circuit. This first part of a circuit analysis solution is called the transient part of the solution. It represents how the circuit switches from its initial state to its final state. The second part of the complete solution deals with voltage values that exist after the transient mode has vanished. This part of the solution is called the steady state condition or mode of operation of the circuit. ***In the steady state condition the peak voltage and current amplitudes and phases are time independent even though the voltage itself may vary instantaneously in the circuit.***

Analysis of circuits during the time that they change from their initial voltage and current state to their final voltage and current state is called ***transient circuit analysis.*** The techniques needed to solve problems of this type differ from the techniques of previous chapters but still rely very heavily on algebraic calculations. The tools required for transient circuit analysis will be developed in this chapter.

12.1 OHM'S LAW REVISITED

The voltage versus current relationship that exists for resistors is called Ohm's Law and is stated as:

$$V = IR \qquad \textbf{(Eq. 12.1)}$$

Note that this equation is independent of the properties of the waveform that describes the voltage source or the resultant current flow in the circuit. In fact, for any purely resistive network, since the resistance is a constant and independent of frequency, the voltage waveform and the current waveform must be identical in shape and differ only in magnitude. No matter what the voltage waveform looks like, the current waveform is identical in shape for all instants of time. Thus if the voltage waveform is a triangular wave, then the current is likewise a triangular shape of exactly the same appearance and shape but of different magnitude. For a purely resistive network there is no transient solution. Voltages and currents change instantly in response to changes in the input conditions. The total solution for a resistive network is the steady state solution.

12.1.1 Ohm's Law for Sinusoidal Sources in the Steady State Mode

In previous chapters we discussed the analysis of ac circuits in the steady state mode. It was assumed that the circuit had been energized for a long time and that transient conditions no longer existed. Capacitors and inductors in an ac circuit under these conditions have a voltage versus current characteristic that is a function of the frequency of the source.

The relationship between the voltage across the capacitor and the current through it is given by the formula:

$$V_C = [1/(j\omega C)]I_C \qquad \textbf{(Eq. 12.2)}$$

In this formula, V_C is the voltage across the capacitor, I_C is the current through the capacitor, and the factor $1/(j\omega C)$ is the impedance Z_C of the capacitor. This identification allows us to write the voltage vs. current relationship as:

$$V_C = I_C Z_C \qquad \textbf{(Eq. 12.2a)}$$

$$Z_C = 1/j\omega C \qquad \textbf{(Eq. 12.2b)}$$

In Equation 12.2b the factor $j = \sqrt{-1}$, $\omega = 2\pi f$ (called the radian frequency), and C is the value of the capacitance in farads. Capacitors in an ac circuit can thus be analyzed with a simple voltage versus current relationship.

The relationship between the voltage across the inductor and the current through it is given by:

$$V_L = (j\omega L)I_L \qquad \textbf{(Eq. 12.3)}$$

As we did earlier for the capacitor, we can identify the impedance of the inductor to be $Z_L = j\omega L$, where $j = \sqrt{-1}$, $\omega = 2\pi f$ (called the radian frequency), and L is the inductance expressed in henries. With this identification Equation 12.3 can be rewritten as:

$$V_L = I_L Z_L \qquad \textbf{(Eq. 12.3a)}$$

With the impedance for a capacitor and an inductor defined as they are here, the voltage versus current relationships for resistors, capacitors, and inductors take similar forms in the steady state mode of analysis. These relationships, which apply ***only for the case of sinusoidal waveforms*** allow us to write a general "***impedance***" form of Ohm's Law. It must be strongly stressed that the concept of "impedance" applies only to the case where the voltage and/or current sources in the circuit are sinusoidal in nature (i.e., sine or cosine waveforms) and are all of the exact same frequency. If they are not of the same frequency, then we can use superposition to find a solution that satisfies the circuit values and conditions for each unique frequency separately.

The generalization of Ohm's Law for the ac case as shown in Equation 12.4 becomes:

$$V = IZ \qquad \textbf{(Eq. 12.4)}$$

where Z is the total impedance of the circuit and both V and I are phasor voltages and currents of the same frequency. This approach works because the equations that describe the behavior of capacitors and inductors in the presence of single-frequency sinusoidal voltage and/or current waveforms in the steady state mode of operation can be represented by complex algebraic expressions.

12.1.2 Physical Laws for Capacitors and Inductors

The actual equations that describe the instantaneous behavior of capacitors and inductors are differential and integral equations. To solve these equations requires the use of techniques of differential and integral calculus. If the capacitance and inductance values are constant and independent of frequency, the complex analysis process can be simplified. These simplified equations can be solved with an algebraic approach. The results of the algebraic analysis can then be used to express all solutions in their time-dependent form.

The relationship between the instantaneous voltage across a capacitor and the current through it can be stated as:

$$i(t) = C\frac{dv(t)}{dt} \qquad \textbf{(Eq. 12.5)}$$

or

$$v(t) = \frac{1}{C}\int i(t)\,dt + v(0^+) \qquad \textbf{(Eq. 12.6)}$$

where I is the current through the capacitor, C is the capacitance of the capacitor expressed in farads and $V(t)$ is the voltage across the capacitor. These equations are valid independent of the shape of the voltage waveform. Equation 12.6 reveals that the voltage across a capacitor as a function of time is equal to the integral of the current through it plus the initial value of the voltage at time $t = 0^{+}$. These equations require the use of techniques of calculus for their solution.

The relationship that describes the behavior of an inductor may be written as:

$$v(t) = L\frac{di(t)}{dt} \qquad \textbf{(Eq. 12.7)}$$

or

$$I(t) = (1/L)\int V(t)\,dt + I(0^+) \qquad \textbf{(Eq. 12.8)}$$

where $V(t)$ is the voltage across the inductor, L is the inductance of the inductor, and $I(t)$ is the current through the inductor. No matter what the form of the current waveform is, the equation for the inductor circuit is valid. As was the case for the equations for the capacitor, the equations for the inductor require differential and integral calculus for their solution.

With the use of Laplace transforms, a technique that we will discuss in this chapter, Equations 12.5 through 12.8 will be converted into algebraic equations. These equations will then be solved with algebraic techniques rather than calculus techniques.

12.1.3 Kirchhoff's Laws Revisited

Equations 12.1 and 12.5 through 12.8 express the electrical properties of the basic passive circuit elements that are used in linear circuit analysis. These elements along with voltage and current sources are the building blocks of all passive electrical circuits.

Analysis of circuits relies on two fundamental theorems due to Kirchhoff, namely, Kirchhoff's Voltage Law (KVL) and Kirchhoff's Current Law (KCL). These two laws express the physicality of all electric circuits.

Kirchhoff's Voltage Law states that the sum of the voltages around the circuit equals zero for any and all instants of time. We can implement this theorem by writing the sum of the voltages around the circuit. We use the equations for each element to relate the voltage across the element to the current that flows through the circuit. The equations thus generated produce mesh current equations, not unlike those that are used for the steady state analysis done previously. While the equations written for transient analysis look different, they have the same form as the mesh current equations previously studied.

Kirchhoff's Current Law states that the sum of currents entering a node of a circuit must equal zero. This law is related to conservation of charge and is true for every instant of time, as is KVL. We can use the equations for the elements to express the current in terms of the voltages around the circuits in the same way that steady state node voltage analysis was carried out earlier. As was the case for KVL, the form of the equations does not look exactly like the form developed for steady state analysis, but the equations have the same form as the node current equations developed for steady state analysis.

The equations developed from Equations 12.5 through 12.8 with the use of Kirchhoff's laws are called integrodifferential equations. Somehow these complex equations must be converted into a simpler set of equations that can be manipulated with normal algebraic techniques. To perform this conversion we must convert equations that are time dependent into equations that are frequency dependent. This conversion will transform the equations into algebraic equations that are easier to handle. The method employed uses something called a Laplace transform. This transform replaces complex derivatives and integrals with simple algebraic expressions.

12.1.4 Transient Properties of Electrical Circuits

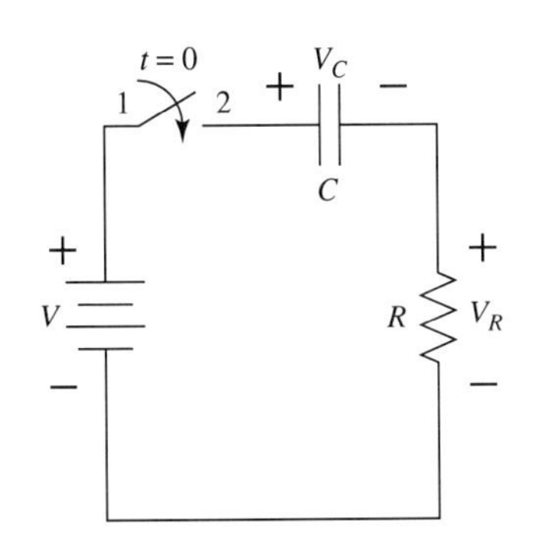

FIGURE 12.1

Before we discuss the use of Laplace transforms, we should discuss the type of response that circuits exhibit. Consider the case of the circuit shown in Figure 12.1. It consists of a battery in series with a capacitor and a resistor. Let us suppose that at time $t = 0$ we close the switch and ask what happens in the circuit. We will assume that the capacitor is initially uncharged and that there is therefore no voltage across it. Equation 12.9 relates the charge on a capacitor to the voltage across it. Thus:

$$Q = C\,V \qquad \textbf{(Eq. 12.9)}$$

where Q is the net charge in coulombs, C is the value of the capacitor expressed in farads, and V is the voltage across the capacitor expressed in volts. After the switch is closed, the application of Kirchhoff's Voltage Law reveals:

$$V = V_C + V_R$$

Initially, the value of V_C, the voltage across the capacitor, is zero because there is no initial charge on the capacitor and the charge takes time to build up as the current flows through the circuit. This means that all of the battery voltage is across the resistor. Using Ohm's Law we find that the current flowing through the resistor is given by:

$$I = V_R/R = V/R$$

For a given resistance value, the current is therefore determined by only the applied voltage and the resistance. If the resistance is increased, the current in the circuit decreases. Current is charge in motion, and as the current flows in the circuit it piles up charge on the capacitor. This in turn causes a voltage to appear across the capacitor as given by Equation 12.9. The bigger the capacitor value is, the smaller the voltage across it will be for a given amount of charge. Voltage across the capacitor is therefore related to the capacitance value and the resistance value. Increasing either one of them decreases the voltage across the capacitor for a given amount of time.

If there is a voltage drop across the capacitor, then the voltage across the resistor must be smaller, because the total voltage applied to the circuit is constant. This means that as the voltage develops across the capacitor, the current through the circuit becomes smaller because the voltage across the resistor is less and the current is related to the voltage by Ohm's Law. If we wait long enough, the voltage across the capacitor approaches the applied voltage V and the current through the resistor is zero. Thus the steady state condition of the circuit is described as $V_C = V$, $V_R = 0$, and $I = 0$.

How long the circuit takes to get to its steady state values is clearly a function of R and C. The larger either of them is made, the longer it takes for the circuit to get to the steady state. When the voltage across the resistor is equal to the voltage across the capacitor, we may write:

$$V = IR = Q/C$$

from which it follows that:

$$IRC = Q$$

But current is charge per unit time, so for this equation to make sense it follows that RC must have units of time. Notice that RC controls the time it takes for the current I to deliver a given charge to the capacitor. From this it follows that the voltage across the capacitor is some function of time and the value of RC. For this reason ***the value of RC is called the time constant of the circuit.***

It will be proven later in this chapter that the value of the voltage across the resistor as a function of time is given by the expression:

$$V_R = Ve^{-t/RC} \qquad \textbf{(Eq. 12.10)}$$

Where e is the base of the natural logarithm system (discussed in Chapter 3), a number that is equal to approximately 2.71828. Equation 12.10 is called an exponential function because the variable t is the exponent of the base e.

With the aid of Ohm's Law and Kirchhoff's laws we can solve for the expressions relating the current and the capacitor voltage to time as:

$$I = \frac{V}{R}e^{-t/RC} \qquad \textbf{(Eq. 12.11)}$$

$$V_C = V(1 - e^{-t/RC}) \qquad \textbf{(Eq. 12.12)}$$

t/RC	$e^{-t/RC}$
0	1
1	0.368
2	0.135
3	0.049
4	0.018
5	0.007

When we examine the behavior of the expression $e^{-t/RC}$ as a function of time, we find that at $t = 0$ the exponential function $e^{-t/RC}$ equals 1. As time increases the value of $e^{-t/RC}$ approaches 0. How fast the function approaches zero is related to the value of the time constant RC. The larger the time constant is, the slower the exponential function approaches zero. Figure 12.2 shows a plot of $e^{-t/RC}$ versus t/RC.

Instead of considering time expressed in terms of seconds, it is more useful to think of time in terms of the number of time constants of the circuit. We see from the preceding table and Fig. 12.2 that after three time constants the value of the exponential term is only 4.9% of its original value. After five time constants it is only 0.7% of the original value. At this point the transient term can be said to have vanished.

If we use the table of values for the time constant on the equations of Figure 12.1, we find that for $t = 0$ the initial value of the resistor voltage is V. The initial value of the current is given by Ohm's Law as V/R. The initial value of the voltage across the capacitor is zero. After five time constants, the voltage across the capacitor is V and the voltage across the resistor is zero. This demonstrates that the equations produce the right values for large time and for time equal to zero.

It is in the nature of electrical circuits that the initial values of the voltages and currents in the circuit always change exponentially. We can expect that no matter what the circuit, its transient will die out in an exponential manner. That is, for all of the simple circuits that we analyze there will always be a factor of $e^{-t/\tau}$ in the solution, where ***τ is the characteristic time constant*** of the circuit. The expression for τ is dependent on the

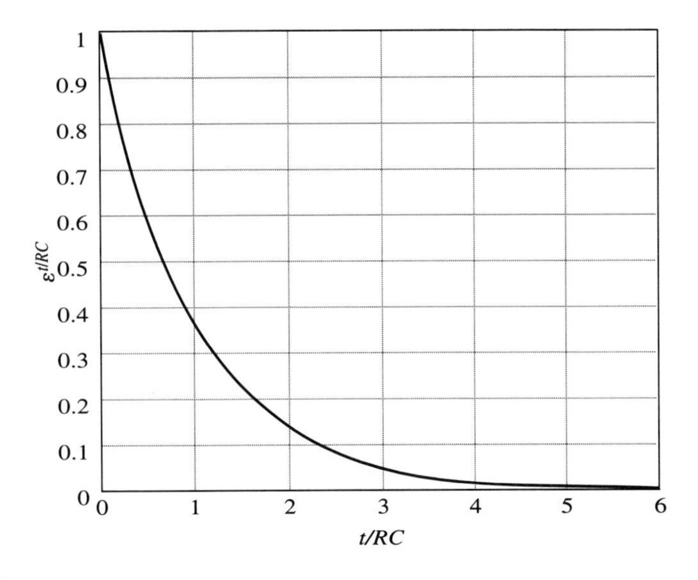

FIGURE 12.2

circuit and may not always be of the form *RC*. Nevertheless, its dimensions are always expressed in units of time.

EXAMPLE 12.1

In Figure 12.1, if $V = 25$ V, $R = 3.3$ kΩ, and $C = 10\ \mu$F:

(a) Calculate the time constant of the circuit.

(b) What is the voltage across the resistor at $t = 20$ ms?

The time constant of the circuit is given as *RC*, which equals:

$$RC = 3.3 \times 10^3 \times 10 \times 10^{-6} = 33 \text{ ms}$$

The voltage across the resistor is given by Equation 12.10 as:

$$V_R = Ve^{-t/RC}$$

from which

$$V_R = 25e^{-t/0.033}$$

At $t = 20$ ms the voltage across the resistor equals:

$$V_R = 25e^{-0.02/0.033} = 25(0.545) = 13.64 \text{ V}$$

To reach equilibrium will take five time constants or 165 ms. At that time the voltage across the resistor will be approximately zero (actually 0.175 V).

12.2 LAPLACE TRANSFORMS: AN INTRODUCTION

Laplace transforms are a complex mathematical technique that is used to convert integral and differential equations into algebraic equations. While the mathematical proof of their properties is beyond the level of this text, we can use Laplace transforms by resorting to tables that list conversions between the original equations and their algebraic equivalents. After solving the algebraic equations, we can convert them, through use of the same tables, back into real-time functions that express the way that the circuit values change with time. With this approach we can use the tool without understanding its details of construction.

The first step in the use of Laplace transforms is to understand that the solutions obtained are only good for time greater than $t = 0$. In our equations we will include the voltages and currents that flow in the circuit at time $t = 0$. These voltages and currents are called the ***initial conditions*** of the circuit. We shall represent these initial conditions as voltage or current sources added in series or parallel with circuit elements that have initial conditions associated with them. Thus a capacitor may have a voltage across it due to charge on the capacitor at time $t = 0$. Similarly, an inductor may have an initial current flowing through it at time $t = 0$. All of these voltage sources representing these initial conditions are called initial condition sources or initial condition generators. The terms are used interchangeably in the literature.

We begin by defining some terms in the Laplace domain. In the real world the independent variable that expresses all voltage variations is time. We use the letter t to denote time and define t only for values greater than or equal to zero. We write expressions that can be used to find the voltage at any point of a circuit in terms of time. If we know the time we can evaluate the expression and then get the voltage at that instant of time. The corresponding variable in the Laplace world is called s and is thought of as a complex frequency variable. That is, it has real and imaginary parts. We solve for functions of s in the equations we generate, and these expressions can in turn be converted back into expressions in time.

If $f(t)$ represents a time-varying signal, then its corresponding function in the Laplace domain is called $\boldsymbol{F}(s)$. We shall use boldface italicized capital letters here to represent functions of the variable s to distinguish them from the time-variable signals, which we will represent with lower case roman letters. Thus the Laplace transform of $\mathrm{v}(t)$ is equal to $\boldsymbol{V}(s)$. Equation 12.13 expresses the relationship between the time domain derivative of a function $\mathrm{f}(t)$ and its equivalent in the algebraic Laplace domain $\boldsymbol{F}(s)$:

$$\mathcal{L}d\mathrm{f}(t)/dt = s\boldsymbol{F}(s) - f(0^+) \qquad \textbf{(Eq. 12.13)}$$

This equation says that the Laplace transform of the time derivative of a function of time that we label $\mathrm{f}(t)$ can be replaced by the expression $s\boldsymbol{F}(s) - \mathrm{f}(0^+)$, where $\boldsymbol{F}(s)$ is the Laplace transform of the function of time $\mathrm{f}(t)$, and $\mathrm{f}(0^+)$ represents the value of the time function $\mathrm{f}(t)$ at time $t = 0^+$. The notation 0^+ is used to indicate that the time is the instant after the switch is closed.

An integral in the time domain, $\mathrm{g}(t) = \int \mathrm{f}(t)\,dt$, converts to an algebraic expression in the Laplace domain as

$$\mathcal{L}\mathrm{g}(t) = \mathcal{L}\int \mathrm{f}(t)\,dt = (1/s)[\boldsymbol{F}(s) + \mathrm{g}(0^+)] \qquad \textbf{(Eq. 12.14)}$$

This equation expresses the relationship between the integral of a function $\mathrm{f}(t)$ and its corresponding Laplace transform.

12.2.1 Representing Passive Elements in the Laplace Domain

Through the use of Equations 12.13 and 12.14, we can convert the capacitance and inductance defining equations given earlier as Equations 12.5 through 12.8 as follows:

If $\quad i(t) = Cdv(t)/\,dt$

then $\quad \boldsymbol{I}(s) = sC\,v(s) - C\,\mathrm{v}(0^+) \qquad$ **(Eq. 12.15)**

and if $\quad v(t) = (1/C)\int i(t)\,dt + \mathrm{v}(0^+)$

then $\quad \boldsymbol{V}(s) = (1/sC)[\boldsymbol{I}(s) + C\mathrm{v}(0^+)] \qquad$ **(Eq. 12.16)**

Similarly if $\quad v(t) = L\dfrac{di(t)}{dt}$

then $\quad \boldsymbol{V}(s) = sL\,\boldsymbol{I}(s) - L\mathrm{i}(0^+) \qquad$ **(Eq. 12.17)** and if $\quad i(t) = \dfrac{1}{L}\displaystyle\int v(t)\,dt - i(0^+)$

then $\quad \boldsymbol{I}(s) = (1/sL)[\boldsymbol{V}(s) + L\,\mathrm{i}(0^+)$ **(Eq. 12.18)**

where the boldface capital letters **I** and **V** are functions of time and the italicized boldface capitals are the corresponding functions of complex frequency s. Note that Equation 12.16 can be derived from Equation 12.15 by algebraic manipulation. Similarly, Equation 12.18 can be obtained from Equation 12.17 by algebraic manipulation. The Laplace transform thus converts derivatives and integrals into algebraic functions of complex frequency s.

Consider Equation 12.15:

$\boldsymbol{I}(s) = sC\,\boldsymbol{V}(s) - C\mathrm{v}(0^+)$ **(Eq. 12.15)**

The equation states that the current in the circuit is equal to the current through the capacitor minus a current equal to $C\mathbf{V}(0^+)$. This current represents the effect of the existence of charge on the capacitor at time $t = 0^+$ just after the switch in the circuit is closed. From Equation 12.15 we see that it is dimensionally a current because the left side of the equation is expressed as a current. ***Symbolically the current*** $C\mathrm{v}(0^+)$ ***is represented as a current source and because it represents the effect of the initial charge on the capacitor, it is called an initial condition current generator.*** The complete equivalent circuit for this equation is shown in Fig. 12.3.

FIGURE 12.3

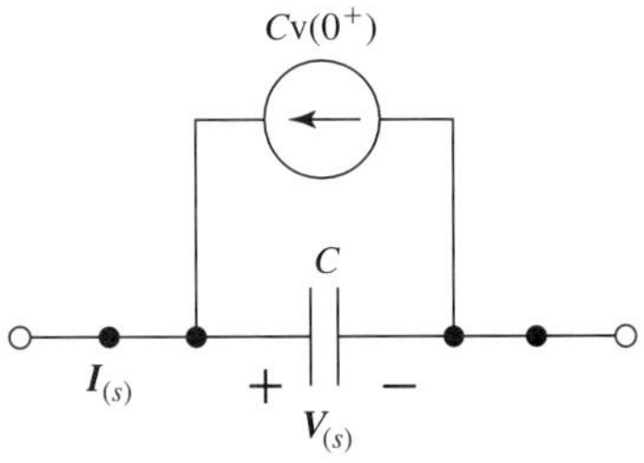

$C\mathrm{v}(0^+)$

C

$\boldsymbol{I}_{(s)}$ + − $\boldsymbol{V}_{(s)}$

A current source in parallel with an element can be converted to a voltage source in series with the element by multiplying the value of the current source by the impedance of the element. For the Laplace transform the operational impedance of the capacitor is given as $1/sC$. The value of the voltage source thus becomes $(1/sC)\,[C\mathbf{V}(0^+)]$, which equals $\mathbf{V}(0^+)/s$ and is in series with the capacitor. This conversion is shown in Fig. 12.4.

FIGURE 12.4

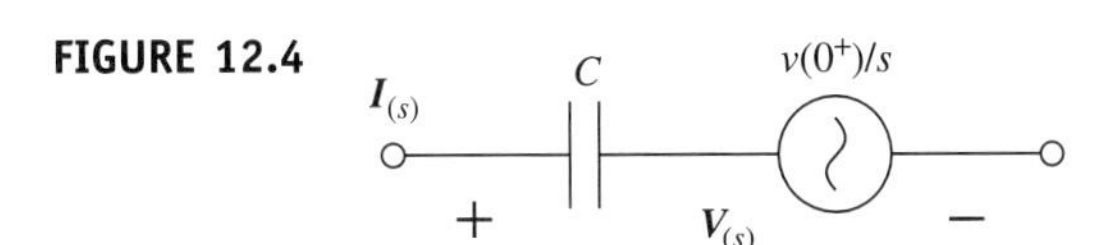

We can write the corresponding equation for this configuration as:

$\boldsymbol{V}(s) = (1/sC)\,\boldsymbol{I}(s) + \mathrm{v}(0^+)/s$

When we compare this equation generated from the equivalent circuit to Equation 12.16, we see that they are identical. With the aid of the two representations for a capacitor that has an initial charge on it (series with voltage source and parallel with current source), any capacitor circuit can be studied using either mesh current analysis or node voltage analysis.

Next let us consider Equation 12.17:

$\boldsymbol{V}(s) = sL\,\boldsymbol{I}(s) - L\,i(0^+)$ **(Eq. 12.17)**

This equation says that the voltage across an inductor is equal to the voltage drop across the inductor minus a voltage source represented in the s domain as equal to $L\,i(0^+)$. This voltage source is due to the initial current flowing in the inductor at time $t = 0^+$. Once again, it is an example of a source that is due to the initial condition of the current through the inductor. It is therefore an example of an ***initial condition voltage generator.*** The series circuit representing this equation is shown in Fig. 12.5.

FIGURE 12.5

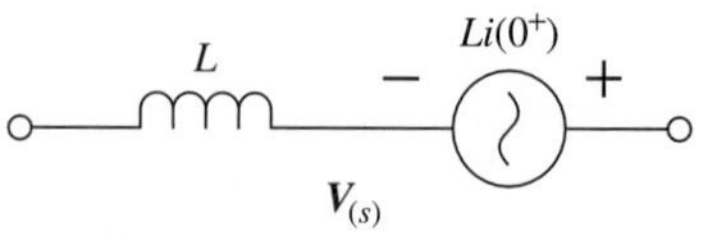

If we start with Equation 12.18 rather than 12.17, we can draw a parallel inductor and current source representation. Thus:

$$\boldsymbol{I}(s) = (1/sL)[\boldsymbol{V}(s) + L\,i(0^+)] \qquad \textbf{(Eq. 12.18)}$$

can be represented by an inductor in parallel with a current source equal to $\mathbf{I}(0^+)/s$. The circuit is shown in Fig. 12.6 below.

FIGURE 12.6

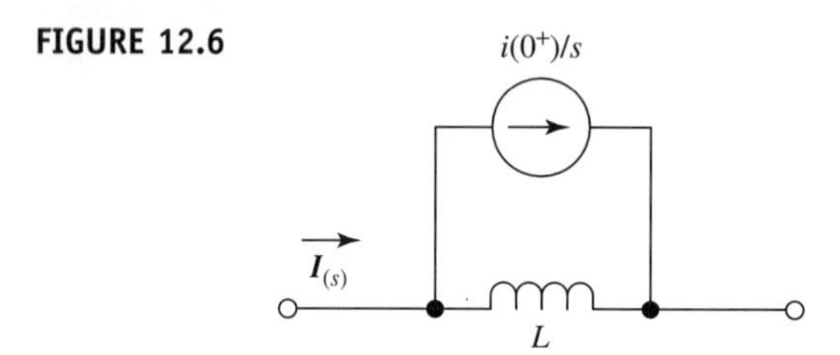

We have shown that capacitors and inductors can be represented by a series circuit representation with the element and an initial condition voltage source in series. A parallel representation with a combination of an initial condition current source and the element can also be used. The transition between voltage and current source is performed in the same way as was done for the steady state conversions covered earlier. We further note that if the values of the voltage or current sources are zero, then Equations 12.15 through 12.18 represent the same equations developed for the steady state solution provided we replace s by $j\omega$. The Laplace transform is somehow an extension of the steady state solution method with the addition of sources for the initial conditions of capacitors and inductors.

TABLE 12.1
Laplace transform representation of passive circuit elements

$f(t)$ for $t > 0$	$F(s)$	Equation Number
$v(t) = i(t)R$	$\boldsymbol{V}(s) = \boldsymbol{I}(s)R$	**Eq. 12.1**
$i_C(t) = C\dfrac{dv_C(t)}{dt}$	$\boldsymbol{I}_C(s) = sC\,\boldsymbol{V}_C(s) - C\boldsymbol{V}_C(0^+)$	**Eq. 12.15**
$v_C(t) = 1/C\int i_C(t)\,dt$	$\boldsymbol{V}_C(s) = \dfrac{\mathbf{I}_C(s)}{sC} + \dfrac{v_C(0^+)}{s}$	**Eq. 12.16**
$v_L(t) = \dfrac{L\,di_L}{dt}$	$\boldsymbol{V}_L(s) = sL\,\boldsymbol{I}_L(s) - Li_L(0^+)$	**Eq. 12.17**
$\mathbf{I}_L(t) = 1/L\int \mathbf{V}_L(t)\,dt$	$\boldsymbol{I}_L(s) = \dfrac{\boldsymbol{V}_L(s)}{sL} + \dfrac{i_L(0^+)}{s}$	**Eq. 12.18**

Table 12.1 shows a summary of the equations for passive elements expressed in both time and Laplace equivalents. With this table plus Kirchhoff's laws, any circuit containing passive elements can be converted into a set of equations that can be solved using algebraic techniques.

EXAMPLE 12.2

For the circuit shown at the below, find an expression for the total current passing through the parallel combination of elements.

Solution

$$\mathrm{i}_T = \mathrm{i}_R + \mathrm{i}_C + \mathrm{i}_L$$

From Table 12.1 we find $\mathscr{L}\,\mathrm{i}_T = \boldsymbol{I}_T(s)$:

$$\boldsymbol{I}_T(s) = \frac{\boldsymbol{V}(\mathrm{s})}{R} + \mathrm{s}C\boldsymbol{V}(\mathrm{s}) - C\,\mathrm{v}(0^+) + \frac{\boldsymbol{V}(\mathrm{s})}{\mathrm{s}L} + \frac{\mathrm{i}_L(0^+)}{\mathrm{s}}$$

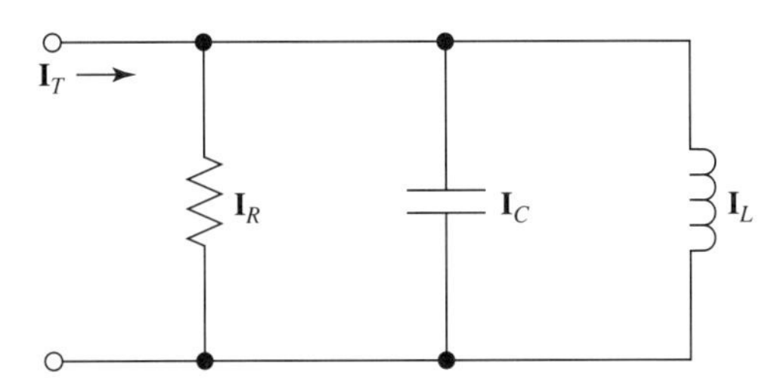

12.2.2 Time Dependent Sources in the Laplace Domain

We have previously employed dc as well as ac sinusoidal sources in the analysis of the steady state response of circuits. Our analysis was limited to just those two voltage waveforms. With the use of Laplace transforms in transient analysis we can express any realizable voltage and current waveforms as functions of s. Through the use of Laplace transform integral, any waveform that can be written as a function of time can also be expressed as a function of the Laplace variable s. It is beyond the scope of this text to calculate functions of s for a given function of time t. It is also unnecessary because many tables exist that show the relationship between the time domain function and the corresponding Laplace domain expression. We shall tabulate several useful functions of time and their corresponding expressions in s. This approach makes it possible to solve transient analysis problems without the use of high-powered mathematics.

Table 12.2 lists some of the more useful waveforms and their equivalences in the Laplace domain. These waveforms, which were generated with Laplace transform integrals, can be used for voltage sources and current sources. We need not be able to derive the expressions from scratch in order to be able to use them. Once we have solved the Laplace domain equations for the circuit, this table will be used to convert the solution back into functions of time. This table therefore represents one key element in the solution of transient problems in circuit analysis.

The combination of Tables 12.1 and 12.2 allows us to convert circuits into the equations needed to solve for the transient condition. We now have equations linking the voltage and current through any circuit element as well as Laplace functions for the time-variable source waveforms. Therefore, we may apply Kirchhoff's laws to develop a set of equations that can be solved for the current and voltage responses of the circuit for all time. We will solve this set of equations using the determinant techniques previously demonstrated in the solution of steady state solutions. The resultant solution is an expression for each voltage and current in the circuit in terms of R, L, C, their initial conditions, and the applied voltage and current sources. ***The solution is always expressible as a ratio of two polynomials in powers of s.*** This property derives from the linear nature of the equations and the determinant methodology. Corresponding to these polynomials in powers of s are expressions that are functions of time. These are the solutions to the circuit problem. The transient solution is obtained by looking up the inverse transform in Table 12.2.

TABLE 12.2
Laplace transform representation of sources

$f(t)$ for $t > 0$	$F(s)$	Equation Number
$a_1 f(t)$	$a_1 F(s)$	**Eq. 12.19**
$u(t) = 1$	$\dfrac{1}{s}$	**Eq. 12.20**
$\sin \omega t$	$\dfrac{\omega}{s^2 + \omega^2}$	**Eq. 12.21**
$\cos \omega t$	$\dfrac{s}{s^2 + \omega^2}$	**Eq. 12.22**
$e^{-\alpha t}$	$\dfrac{1}{s + \alpha}$	**Eq. 12.23**
$e^{-\alpha t} f(t)$	$F(s + \alpha)$	**Eq. 12.24**
$e^{-\alpha t} \sin \omega t$	$\dfrac{\omega}{(s + \alpha)^2 + \omega^2}$	**Eq. 12.25**
$e^{-\alpha t} \cos \omega t$	$\dfrac{s + \alpha}{(s + \alpha)^2 + \omega^2}$	**Eq. 12.26**
t	$\dfrac{1}{s^2}$	**Eq. 12.27**
$tf(t)$	$F(s^2)$	**Eq. 12.28**
$\dfrac{t^n}{n!}$ (n a positive integer)	$\dfrac{1}{s^{n+1}}$	**Eq. 12.29**

12.3 PROPERTIES OF THE SOLUTION EQUATIONS

An nth order polynomial in s can be written as:

$$s^n + A_{n-1}s^{n-1} + A_{n-2}s^{n-2} + \cdots + A_0 = 0 \qquad \textbf{(Eq. 12.30)}$$

By definition, a root of Equation 12.30 is a value of s that satisfies the equation. That is, if a value of s is substituted into the equation and makes the equation equal to zero, then that value of s is called a root of the polynomial. It can be shown that the number of roots of an equation is equal to the highest exponent of the equation. Thus an equation whose highest power is n, where n is any positive integer, has n roots. Applying this concept to Equation 12.30, we can rewrite it as:

$$(s - s_1)(s - s_2)\cdots(s - s_n) = 0 \qquad \textbf{(Eq. 12.30a)}$$

where the roots are denoted by the values of s for which $s = s_i$. Each of these roots represents a value for s that satisfies the original Equation 12.30. If we were to multiply out the terms of Equation 12.30a, we would find that we have written Equation 12.30.

Since the solution to any circuit analysis problem is a ratio of polynomials, it follows that both the denominator and the numerator may be written in the same way as Equation 12.30a. Factors that appear in both numerator and denominator will cancel. The roots of the numerator equation are called ***zeros,*** because a root of the numerator polynomial will make the numerator and therefore the entire ratio equal to zero. The roots of the denominator polynomial are called ***poles,*** because a root of the denominator polynomial will make the value of the ratio equal to infinity. It is the poles of the denominator polynomial that actually determine the response of the circuit in the time domain.

For the circuits that we analyze, the ***denominator polynomial is always of higher order than the numerator polynomial.*** The actual difference in order is dependent on the circuit and the Laplace transform of the waveform that is applied to the circuit. A sample solution might look like:

$$\mathbf{V}(s) = \frac{(s - z_1)(s - z_2)}{(s - p_1)(s - p_2)\cdots(s - p_n)} \qquad \textbf{(Eq. 12.31)}$$

where the zeros of the numerator are written as z_i and the poles of the denominator are designated as p_i.

The denominator of Equation 12.31 is written as a product of terms of the form $(s - p_i)$. In arithmetic we learned that when we had many fractions to add we could produce a common denominator by multiplying the denominator of each of the fractions together to form one common denominator. We performed the same operations in algebra when we had algebraic fractions to add. The denominator of Equation 12.31 will be treated as if it is one common denominator that is the result of adding many algebraic fractions together. This combination process can be reversed and the original fractions can be determined. The process of reversal is called ***partial fraction expansion.*** This technique will be explained and demonstrated next.

12.4 PARTIAL FRACTION EXPANSION

Let us assume that Equation 12.31 results from an algebraic combination of the following fractions:

$$\mathbf{V}(s) = \frac{c_1}{(s - p_1)} + \frac{c_2}{(s - p_2)} + \cdots + \frac{c_n}{(s - p_n)} \qquad \textbf{(Eq. 12.32)}$$

Clearly each of the fractions must have nonzero values for all of the constants (c) because the common denominator contains a factor from all of the simple fractional terms listed. Equation 12.32 is called a ***partial fraction expansion*** of Equation 12.31. Values of c_1, c_2, etc. exist that will make Equation 12.32 equal to Equation 12.31. The coefficients for the partial fraction expansion of Equation 12.31 can be found by setting Equation 12.31 equal to Equation 12.32. Doing this we find:

$$\frac{(s - z_1)(s - z_2)}{(s - p_1)(s - p_2)\cdots(s - p_n)} = \frac{c_1}{(s - p_1)} + \frac{c_2}{(s - p_2)} + \cdots + \frac{c_n}{(s - p_n)}$$

Multiplying both sides of the resultant equality by $(s - p_1)$ we find:

$$\frac{(s - z_1)(s - z_2)}{(s - p_2)\cdots(s - p_n)} = c_1 + \frac{c_2(s - p_1)}{(s - p_2)} + \cdots + \frac{c_n(s - p_1)}{(s - p_n)}$$

Setting $s = p_1$ yields a value for c_1. Thus:

$$\frac{(p_1 - z_1)(p_1 - z_2)}{(p_1 - p_2)\cdots(p_1 - p_n)} = c_1$$

In similar fashion we can determine values for each of the unknown coefficients c.

Another way to solve for the values that make this equation an identity is to multiply both sides of the equation by the denominator of the left side of the equation. This produces the same polynomial on both sides of the equation. On the left side each of the powers of s has a coefficient that is a number. On the right side the things multiplying the powers of s are expressions involving the unknown coefficients. If we equate powers of s on both sides of the equal sign, we can solve the equations we get for the values of the unknowns. We will now demonstrate each of the described methods.

EXAMPLE 12.3

Find the partial fraction expansion for:

$$\mathbf{V}(s) = \frac{(s+1)(s+3)}{(s+2)(s+4)(s+6)}$$

Solution Setting up the identity we find:

$$\frac{(s+1)(s+3)}{(s+2)(s+4)(s+6)} = \frac{c_1}{(s+2)} + \frac{c_2}{(s+4)} + \frac{c_3}{(s+6)}$$

If we multiply both sides of the equation by the factor $(s + 2)$ we find that:

$$\frac{(s+1)(s+3)}{(s+4)(s+6)} = c_1 + \frac{c_2(s+2)}{(s+4)} + \frac{c_3(s+2)}{(s+6)}$$

Setting $s = -2$ we find that $c_1 = -1/8$. Notice that both c_2 and c_3 are multiplied by the factor $(s + 2)$ and are therefore multiplied by 0 when $s = -2$. If we multiply the original equation by $(s + 4)$ and set $s = -4$, we find that $c_2 = -3/4$. Finally, multiplying the original equation by $(s + 6)$, we find that $c_3 = 15/8$. The identity can therefore be written as:

$$\frac{(s+1)(s+3)}{(s+2)(s+4)(s+6)} = \frac{-1/8}{(s+2)} + \frac{-3/4}{(s+4)} + \frac{15/8}{(s+6)}$$

Starting with the identity again we multiply both sides of the identity by the denominator $(s + 2)(s + 4)(s + 6)$. This produces the following identity:

$$(s+1)(s+3) = c_1(s+4)(s+6) + c_2(s+2)(s+6) + c_3(s+2)(s+4)$$

Simplifying by multiplying through and grouping like terms we find:

$$s^2 + 4s + 3 = (c_1 + c_2 + c_3)s^2 + (10c_1 + 8c_2 + 6c_3)s + (24c_1 + 12c_2 + 8c_3)$$

From which we find by equating powers of s:

$$(c_1 + c_2 + c_3) = 1 \qquad (10c_1 + 8c_2 + 6c_3) = 4 \quad \text{and} \quad (24c_1 + 12c_2 + 8c_3) = 3$$

The solution of these three equations yields the same answers as those provided by the first technique demonstrated.

On occasion the polynomial ratio that represents the solution may contain two or more poles of the same value. This is called a ***higher-order pole.*** Under these conditions, the partial fraction expansion must contain all possible powers of that pole. Thus if the pole is of second order, both first- and second-order poles must be represented in the partial fraction expansion. This factor changes the way in which the constants must be evaluated. One method of evaluation is demonstrated in Example 12.3, where the factor $(s + 4)^2$ appears as a pole of the solution.

EXAMPLE 12.4

Find the partial fraction expansion for:

$$\mathbf{V}(s) = \frac{(s+1)(s+3)}{(s+2)(s+4)^2}$$

Solution Since $(s + 4)$ appears as a pole of order 2, both linear and squared factors of $(s + 4)$ must appear as poles of the partial fraction expansion. We write this as:

$$\mathbf{V}(s) = \frac{c_1}{(s+2)} + \frac{c_2}{(s+4)} + \frac{c_3}{(s+4)^2}$$

Thus:

$$\frac{(s+1)(s+3)}{(s+2)(s+4)^2} = \frac{c_1}{(s+2)} + \frac{c_2}{(s+4)} + \frac{c_3}{(s+4)^2}$$

Multiplying both sides by $(s+2)$ produces:

$$\frac{(s+1)(s+3)}{(s+4)^2} = c_1 + \frac{c_2(s+2)}{(s+4)} + \frac{c_3(s+2)}{(s+4)^2}$$

Setting $s = -2$ yields $c_1 = -1/4$.

Multiplying both sides of the identity by $(s+4)^2$ produces:

$$\frac{(s+1)(s+3)}{(s+2)} = \frac{c_1(s+4)^2}{(s+2)} + c_2(s+4) + c_3$$

setting $s = -4$ produces $c_3 = -3/2$.

If we now multiply both sides of the identity by the denominator and equate powers of s:

$$s^2 + 4s + 3 = (c_1 + c_2)s^2 + (8c_1 + 6c_2 + c_3)s + (16c_1 + 8c_2 + 2c_3)$$

we find that $(c_1 + c_2) = 1$, $(8c_1 + 6c_2 + c_3) = 4$, and $(16c_1 + 8c_2 + 2c_3) = 3$. Now, since we have found values for c_1 and c_3 we can use any of the three coefficients of the powers of s to find c_2. Using the expression for the coefficient of s^2, $(c_1 + c_2) = 1$, and substituting for c_1 we find that $c_2 = 5/4$. Therefore:

$$\frac{(s+1)(s+3)}{(s+2)(s+4)^2} = \frac{-1/4}{(s+2)} + \frac{5/4}{(s+4)} - \frac{3/2}{(s+4)^2}$$

We can carry out the partial fraction expansion by use of the methods demonstrated in Examples 12.3 and 12.4 for any combination of poles. Sometimes the pole has a complex value or a purely imaginary value. We must realize that if a complex pole exists then its complex conjugate also exists because the coefficients of s are all real. We shall now demonstrate how to handle complex poles. Consider the case of Example 12.5.

EXAMPLE 12.5

Find the partial fraction expansion for:

$$\mathbf{V}(s) = \frac{(s+1)(s+3)}{(s+2)(s^2+4)}$$

We notice that there are poles at $s = j2$ and at $s = -j2$. We can write the partial expansion as:

$$\mathbf{V}(s) = \frac{c_1}{(s+2)} + \frac{c_2 s + c_3}{(s^2+4)}$$

$$\frac{(s+1)(s+3)}{(s+2)(s^2+4)} = \frac{c_1}{(s+2)} + \frac{c_2 s + c_3}{(s^2+4)}$$

multiplying by $(s+2)$ on both sides and setting $s = -2$ we find $c_1 = -1/8$. Multiplying by (s^2+4) on both sides and setting $s^2 = -4$ and $s = j2$ we find:

$$c_2 = 9/8 \quad \text{and} \quad c_3 = 7/4$$

Alternatively we can multiply the denominator of the left side and equate powers of s, which yields:

$$s^2 + 4s + 3 = (c_1 + c_2)s^2 + (2c_2 + c_3)s + (4c_1 + 2c_3)$$

Solving as before for the values of c_i we find $c_1 = -1/8$, $c_2 = 9/8$, and $c_3 = 7/4$. So:

$$\mathbf{V}(s) = \frac{-1/8}{(s+2)} + \frac{(9/8)s + 7/4}{(s^2+4)}$$

Thus far we have learned how to write equations for circuit analysis that are valid for all values of time greater than 0. Through the use of the Laplace transform we have converted these values into a set of algebraic equations with real coefficients. These equations with real coefficients are linear equations. A set of linear equations may be solved using determinants through the use of Kramer's rules. We have taken the solutions thus obtained and used partial fraction expansions to reduce the equation to a sum of simple fractions. The only remaining task is the conversion of these simple fractions to their corresponding functions of time. This will require the use of Table 12.2.

12.5 LAPLACE TRANSFORMS: FINDING THE CIRCUIT'S BEHAVIOR IN TIME

It is assumed that all voltage and/or current sources are introduced into the circuit at time $t = 0$. Before time $t = 0$ the value of all voltage and current sources is unknown. As we examine the equations contained in Table 12.1 we see that the first five equations represent the behavior of the circuit elements R, C, and L as functions of time and as functions of the Laplace transform variable s. When solving circuit problems we shall always use the Laplace representation, keeping in mind that the time behavior of the circuit is given by the inverse transform.

The sixth equation in the table reveals that multiplication by a scalar quantity in the time domain is the same operation in the frequency domain. This means that, when we take the transform of a time domain function in the table, the constant multiplying the function also multiplies its frequency domain by the same factor.

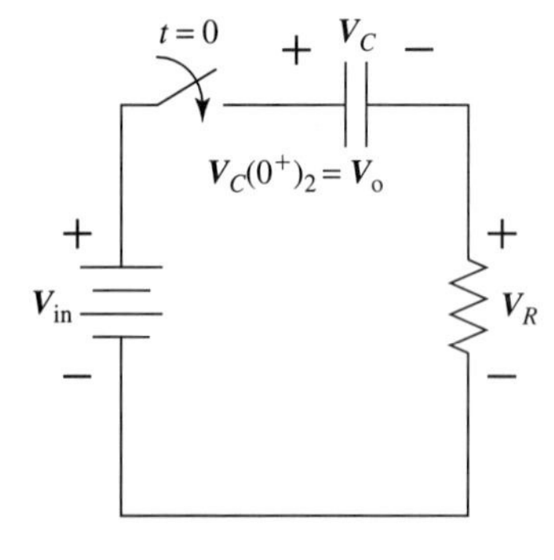

FIGURE 12.7

Equations 12.20 through 12.29 represent transforms of particular waveforms. Each of these waveforms can have any value before $t = 0$. That is, the expressions are only valid for time greater than zero. The first of these equations is called the unit step equation. It is equal to 1 for all time greater than zero and equal to zero for all time less than zero. This function enables us to represent a voltage source as turning on at time $t = 0$ and equal to zero before that time. Thus in Fig. 12.7 the input voltage to the circuit is given as V_{in}/s. This follows from application of Equation 12.19 to the unit step equation (12.20).

12.5.1 The Series *R–C* Circuit

Consider the circuit shown in Fig. 12.7. It consists of a battery, a switch, and a series combination of a resistor and a capacitor. We will solve the circuit for the circuit voltages and currents as functions of time. At time $t = 0$, the switch is closed and the capacitor has an initial voltage across it of V_0 volts. From Kirchhoff's Voltage Law the sum of the voltages around the circuit must equal to the applied voltage. Thus:

$$v_{in}(t) = v_C(t) + v_R(t) \qquad \textbf{(Eq. 12.33)}$$

From Table 12.1, Equation 12.16 we find the Laplace transform of $v_C(t)$ to be:

$$V_C(s) = \frac{I_C(s)}{sC} + \frac{v_C(0^+)}{s}$$

which becomes:

$$V_C(s) = \frac{I_C(s)}{sC} + \frac{\mathbf{V}_0}{s}$$

From Equation 12.1 we find $\boldsymbol{V}_R = \boldsymbol{I}(s)\,R$. Finally, the applied voltage V_{in} is represented as $\mathbf{V}_{\text{in}}/s$ because of the switch closing at $t = 0$. Substituting these values into Kirchhoff's equation, we find:

$$\frac{\mathbf{V}_{\text{in}}}{s} = \boldsymbol{V}_C(s) + \boldsymbol{V}_R(s) = \frac{\boldsymbol{I}_C(s)}{sC} + \frac{\mathbf{V}_0}{s} + \boldsymbol{I}(s)\,R \qquad \textbf{(Eq. 12.34)}$$

Since this is a series circuit $\boldsymbol{I}_C(s) = \boldsymbol{I}(s)$. Solving the algebraic equation for $\boldsymbol{I}(s)$ we find:

$$\boldsymbol{I}(s) = \frac{\mathbf{V}_{\text{in}} - \mathbf{V}_0}{s\left(R + \dfrac{1}{sC}\right)} = \frac{(\mathbf{V}_{\text{in}} - \mathbf{V}_0)/R}{s + (1/RC)} \qquad \textbf{(Eq. 12.35)}$$

This solution is of the form $a_1/(s + \alpha)$ where a_1 is a constant equal to $(\mathbf{V}_{\text{in}} - \mathbf{V}_0)/R$ and $\alpha = 1/RC$. Using Equation 12.23 in Table 12.2 we can write the current in the time domain as:

$$\mathbf{i}(t) = \frac{(\mathbf{V}_{\text{in}} - \mathbf{V}_0)}{R}\,e^{-t/RC} \quad \text{for} \quad t > 0. \qquad \textbf{(Eq. 12.36)}$$

The voltage across the resistor is given by Ohm's Law as $V_R = IR$, from which:

$$\mathrm{v}_R(t) = (\mathbf{V}_{\text{in}} - \mathbf{V}_0)e^{-t/RC} \quad \text{for} \quad t > 0. \qquad \textbf{(Eq. 12.37)}$$

Finally, the voltage across the capacitor can be found from:

$$\boldsymbol{V}_C(s) = \frac{\boldsymbol{I}_C(s)}{sC} + \frac{\mathbf{V}_0}{s} \qquad \textbf{(Eq. 12.16)}$$

By substitution of $\boldsymbol{I}_C(s)$ into the equation:

$$\boldsymbol{V}_C(s) = \frac{(\mathbf{V}_{\text{in}} - \mathbf{V}_0)/R}{sC[s + (1/RC)]} + \frac{\mathbf{V}_0}{s}$$

$$\frac{(\mathbf{V}_{\text{in}} - \mathbf{V}_0)/R}{sC[s + (1/RC)]} = \frac{(\mathbf{V}_{\text{in}} - \mathbf{V}_0)/RC}{s[s + (1/RC)]} = \frac{a_1}{s} + \frac{a_2}{s + (1/RC)}$$

Evaluating the constants we find that $a_1 = \mathbf{V}_{\text{in}} - \mathbf{V}_0$ and $a_2 = \mathbf{V}_0 - \mathbf{V}_{\text{in}}$. Therefore:

$$\boldsymbol{V}_C(s) = \frac{\mathbf{V}_{\text{in}} - \mathbf{V}_0}{s} + \frac{\mathbf{V}_0 - \mathbf{V}_{\text{in}}}{s + (1/RC)} + \frac{\mathbf{V}_0}{s} = \frac{\mathbf{V}_{\text{in}}}{s} + \frac{\mathbf{V}_0 - \mathbf{V}_{\text{in}}}{s + (1/RC)}$$

from which

$$\mathbf{V}_C(t) = \mathbf{V}_{\text{in}} + (\mathbf{V}_0 - \mathbf{V}_{\text{in}})e^{-t/RC} \quad \text{for} \quad t > 0. \qquad \textbf{(Eq. 12.38)}$$

This problem was considered in Section 12.1.4, where the solution was stated without proof. In Section 12.1.4 we discussed the role of the time constant when the capacitor had no initial voltage across it. If we set $\mathbf{V}_0$ equal to zero in our current solution, the answer given here matches the solution previously quoted there. Note the appearance of the time constant RC in the expressions for the transient part of the solution.

12.5.2 The Parallel *R–C* Circuit

Figure 12.8 shows a circuit that has a current source in parallel with a resistor and a capacitor. At time $t = 0$ the switch is closed. The capacitor has an initial voltage across it equal to $\mathbf{V}_0$ volts. We will now solve this circuit for its response as a function of time.

Application of Kirchhoff's Current Law to the circuit results in the equation:

$$I_{\text{in}} = I_R + I_C$$

FIGURE 12.8

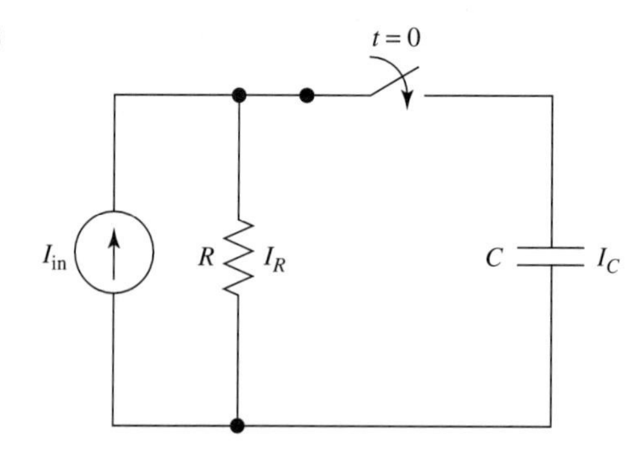

Using the equations of Table 12.1 we convert this to the algebraic equation:

$$\frac{\mathbf{I}_{\text{in}}}{s} = \frac{V(s)}{R} + sC\,V(s) - C\,V_0$$

Solving for $V(s)$ we find:

$$V(s) = \frac{(I_{\text{in}}/s) + C\mathbf{V}_0}{\frac{1}{R} + sC} = \frac{\mathrm{I}_{\text{in}}(1/C)}{s(s + 1/RC)} + \frac{\mathbf{V}_0}{s + \frac{1}{RC}}$$

Using partial fraction expansion we find:

$$V(s) = \frac{\mathbf{I}_{\text{in}}R}{s} - \frac{\mathbf{I}_{\text{in}}R}{s + (1/RC)} + \frac{\mathbf{V}_0}{s + (1/RC)}$$

Table 12.1 provides the inverse as:

$$\mathrm{v}(t) = I_{\text{in}}R + (\mathbf{V}_0 - \mathrm{I}_{\text{in}}R)e^{-t/RC} \quad \text{for} \quad t > 0$$

Note that for t approaching infinity $\mathbf{V}(t) = I_{\text{in}}R$, which is the value we would obtain from a steady state calculation.

12.5.3 A Series *R–L* Circuit Solution

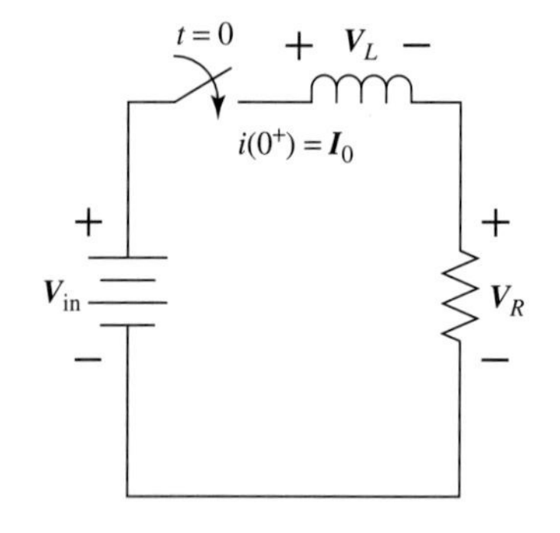

FIGURE 12.9

We have shown solutions for a capacitor and resistor in series and in parallel with the initial voltage on the capacitor equal to V_0. In the case of an inductor, the current through the inductor can not change instantly. Therefore the solution for an R–L circuit will contain a term due to the initial value of the current through the inductor. The solution for the transient behavior follows the same form as the solution for the R–C circuit. Figure 12.9 shows a series R–L circuit. We shall now solve this circuit when there is an initial current through the inductor of I_0 amperes. We start with Kirchhoff's Voltage Law, which states that the sum of the voltages is equal to the input voltage. Thus:

$$\mathbf{V}_{\text{in}}(t) = \mathbf{V}_R(t) + \mathbf{V}_L(t)$$

Using Table 12.1 as a starting point and converting everything to the Laplace domain we obtain:

$$V_{\text{in}}(s) = V_R(s) + V_L(s)$$

$$\frac{\mathrm{V}_{\text{in}}}{s} = I(s)R + sLI_L(s) - Li_L(0)$$

But since it is a series circuit $I(s) = I_L(s)$, from which it follows that:

$$\frac{V_{\text{in}}}{s} = I(s)R + sLI(s) - Li(0^+)$$

$$\boldsymbol{I}(s) = \frac{\mathbf{V}_{in}}{s(sL + R)} + \frac{L\mathrm{I}_0}{(sL + r)}$$

$$= \frac{\mathbf{V}_{in}/L}{s[s + (R/L)]} + \frac{\mathrm{I}_0}{s + (R/L)}$$

Partial fraction expansion applied to this produces:

$$\boldsymbol{I}(s) = \frac{\mathrm{V}_{in}/R}{s} - \frac{\mathrm{V}_{in}/R}{s - (R/L)} + \frac{\mathrm{I}_0}{s + (R/L)}$$

From which using Table 12.1 we find:

$$\mathbf{I}(t) = \mathrm{V}_{in}/R + [\mathrm{I}_0 - (\mathrm{V}_{in}/R)]e^{-Rt/L} \quad \text{for} \quad t > 0$$

We can find $\mathbf{V}_R(t)$ by Ohm's Law as:

$$\mathbf{v}_R(t) = \mathbf{V}_{in} + (\mathbf{I}_0 R - \mathbf{V}_{in})e^{-Rt/L} \quad \text{for} \quad t > 0$$
$$\mathbf{v}_L(t) = \mathbf{V}_{in} - \mathbf{v}_R(t) = (\mathbf{V}_{in} - \mathbf{I}_0 R)e^{-Rt/L} \quad \text{for} \quad t > 0$$

Notice that for this circuit the time constant is equal to L/R.

12.5.4 A Parallel *R–L* Circuit Solution

For the circuit shown in Figure 12.10 we can write Kirchhoff's Current Law as:

$$I_{in} = I_R + I_L$$

FIGURE 12.10

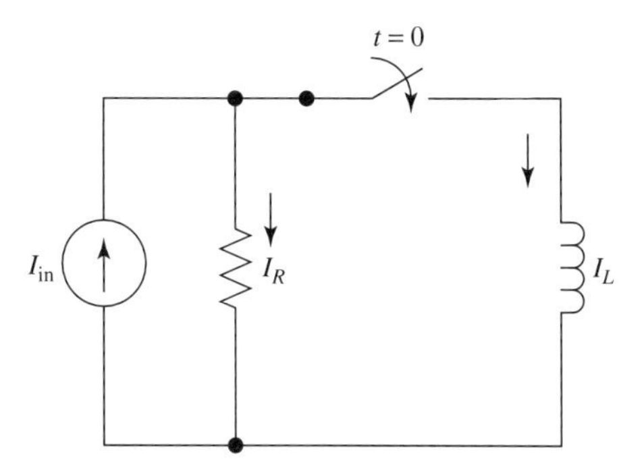

From Table 12.1 this becomes:

$$\boldsymbol{I}_{in}(s) = \boldsymbol{I}_R(s) + \boldsymbol{I}_L(s) = \frac{\boldsymbol{V}}{R} + \frac{\boldsymbol{V}}{sL} + \frac{\mathbf{I}(0^+)}{s}$$

Where $\mathbf{i}(0^+) = \mathbf{I}_0$, which is the initial current through the inductor at time $t = 0$. Substituting $\mathbf{I}_{in}/s$ for $\boldsymbol{I}_{in}(s)$ we obtain:

$$\frac{\mathbf{I}_{in} - \mathbf{I}_0}{s} = \boldsymbol{V}\left(\frac{1}{R} + \frac{1}{sL}\right)$$

from which we obtain an expression for $\boldsymbol{V}(s)$ as:

$$\boldsymbol{V}(s) = \frac{\mathbf{I}_{in} - \mathbf{I}_0}{s}\frac{sLR}{(sL + R)} = \frac{(\mathbf{I}_{in} - \mathbf{I}_0)(LR)}{sL + R} = \frac{(\mathbf{I}_{in} - \mathbf{I}_0)R}{s + (R/L)}$$

Using Table 12.2 to convert this back to the time domain we find:

$$\mathbf{V}(t) = (\mathbf{I}_{in} - \mathbf{I}_0)Re^{-t/(L/R)}$$

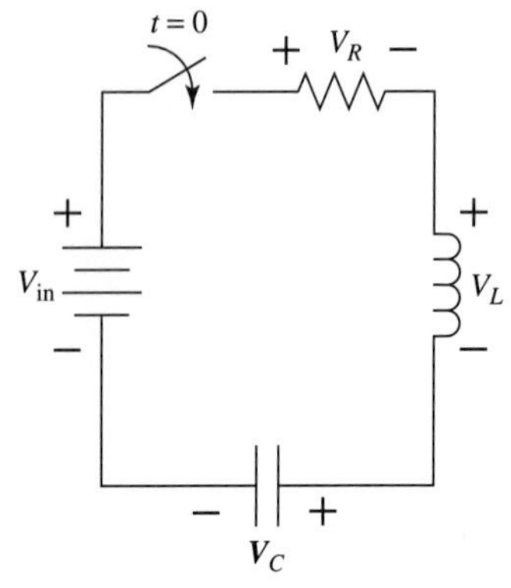

12.5.5 Solutions for a Series R–L–C Circuit

FIGURE 12.11

We shall now demonstrate the solutions that may arise when we consider a circuit that contains resistance, inductance, and capacitance at the same time. It will be demonstrated that the form of the solution will depend on the resistance in the circuit. As the value of the resistance changes the form of the solution will change.

For the circuit shown in Fig. 12.11 we can write Kirchhoff's Voltage Law as:

$$V_{in} = V_R + V_L + V_C$$

If the initial current in the circuit is given as I_0 and the voltage across the capacitor at time $t = 0^+$ is given as V_0, we may write the Laplace transform of the circuit equation as:

$$V_{in}(s) = I(s)R + sLI(s) - L\mathbf{i}(0^+) + \frac{I(s)}{sC} + \frac{\mathbf{v}(0^+)}{s}$$

If we substitute in the initial values the equation becomes:

$$V_{in}(s) = I(s)R + sLI(s) - L\mathbf{I}_0 + \frac{I(s)}{sC} + \frac{\mathbf{V}_0}{s}$$

But $V_{in}(s) = \frac{\mathbf{V}_{in}}{s}$ from which the equation becomes:

$$\frac{\mathbf{V}_{in}}{s} = I(s)R + sLI(s) - L\mathbf{I}_0 + \frac{I(s)}{sC} + \frac{\mathbf{V}_0}{s}$$

Rearranging terms we find:

$$\frac{\mathbf{V}_{in}}{s} + L\mathbf{I}_0 - \frac{V_0}{s} = I(s)\left(R + sL + \frac{1}{sC}\right)$$

From which we can find:

$$I(s) = \frac{\mathbf{V}_{in} + sL\,\mathbf{I}_0 - \mathbf{V}_0}{s\left(R - sL + \frac{1}{sC}\right)}$$

If we multiply through by s/L top and bottom we find:

$$I(s) = \frac{(\mathbf{V}_{in} + sL\,\mathbf{I}_0 - \mathbf{V}_0)}{\left(\frac{sR}{L} + s^2 + \frac{1}{LC}\right)}(1/L)$$

Rewriting the expression we find:

$$I(s) = \frac{(\mathbf{V}_{in} - \mathbf{V}_0)(1/L) + s\,\mathbf{I}_0}{\left(s^2 + \frac{sR}{L} + \frac{1}{LC}\right)}$$

Examination of the denominator of the expression for $I(s)$ reveals that this is a second order polynomial in s. In the event that $R = 0$ the solution for $\mathbf{I}(t)$ can be found to be a combination of sine and cosine terms because the roots are purely imaginary. The form of $I(s)$ matches Equations 12.21 and 12.22 in Table 12.2. If we define $1/LC = (\omega_0)^2$, then the solution of the equation in the time domain can be written as:

$$\mathbf{I}(t) = \frac{(\mathbf{V}_{in} - \mathbf{V}_0)}{\omega_0 L}\sin \omega_0 t + \mathbf{I}_0 \cos \omega_0 t$$

For the condition where R is not equal to zero we must evaluate the roots of the denominator. Consider the equation:

$$s^2 + (R/L)s + (1/LC) = 0$$

The roots of this equation are given as:

$$S_{1.2} = -(R/2L) \pm \sqrt{(R/2L)^2 - (1/LC)}$$

Case 1: Complex Conjugate Roots

In the event that $R/2L < 1/LC$ the roots of the equation are complex conjugates. Under these conditions:

$$\begin{aligned} s^2 + (R/L)s + (1/LC) &= s^2 + (R/L)s + (R/2L)^2 - (R/2L)^2 + (1/LC) \\ &= [s + (R/2L)]^2 + (1/LC) - (R/2L)^2 \end{aligned}$$

If we define ω^2 as $(1/LC) - (R/2L)^2$ then the denominator can be written as:

$$[s + (R/2L)]^2 + \omega^2$$

Rewriting $\boldsymbol{I}(s)$ with the new denominator we find:

$$\boldsymbol{I}(s) = \frac{(\mathbf{V}_{\text{in}} - \mathbf{V}_0)(1/L) + s\,\mathbf{I}_0}{[s + (R/2L)]^2 + \omega^2}$$

Which can be rewritten as:

$$\boldsymbol{I}(s) = \frac{(\mathbf{V}_{\text{in}} - \mathbf{V}_0)(1/L) + (\mathbf{I}_0R/2L) - \mathbf{I}_0R/2L + s\,\mathbf{I}_0}{[s + (R/2L)]^2 + \omega^2}$$

which becomes:

$$\boldsymbol{I}(s) = \frac{(\mathbf{V}_{\text{in}} - \mathbf{V}_0)(1/L) - (\mathbf{I}_0R/2L)}{[s + (R/2L)]^2 + \omega^2} + \frac{(\mathbf{I}_0R/2L) + s\,\mathbf{I}_0}{(s + R/2L)^2 + \omega^2}$$

From Table 12.2 we find the time domain equation to be:

$$\mathbf{I}(t) = \left(\frac{\mathbf{V}_{\text{in}} - \mathbf{V}_0}{L} - \frac{\mathbf{I}_0R}{2L}\right)e^{-tR/2L}\sin \omega t + \mathbf{I}_0e^{-tR/2L}\cos \omega t$$

Case 2: Two Equal Roots

In the event that $R/2L = 1/LC$ the roots of the equation are equal. Under these conditions:

$$s^2 + (R/L)s + (1/LC) = s^2 + (R/L)s + (R/2L)^2 = [s + (R/2L)]^2$$

From which:

$$\boldsymbol{I}(s) = \frac{(\mathbf{V}_{\text{in}} - \mathbf{V}_0)(1/L) + s\,\mathbf{I}_0}{(s + R/2L)^2} = \frac{(\mathbf{V}_{\text{in}} - \mathbf{V}_0)(1/L) - (\mathbf{I}_0R/2L)}{[s + (R/2L)]^2} + \frac{(\mathbf{I}_0R/2L) + s\,\mathbf{I}_0}{[s + (R/2L)]^2}$$

Comparing with Table 12.2 we find the proper expression to be:

$$i(t) = \mathbf{I}_0e^{-Rt/2L} = \left(\frac{(\mathbf{V}_{\text{in}} - \mathbf{V}_0)}{L} - \frac{\mathbf{I}_0R}{2L}\right)te^{-Rt/2L}$$

Case 3: Two Unequal Real Roots

In the event that $R/2L > 1/LC$ the roots of the equation are both negative, are real, and are unequal. Under these conditions the roots of this equation are given as:

$$S_{1.2} = -(R/2L) \pm \sqrt{(R/2L)^2 - (1/LC)}$$

Under this condition the expression for $\boldsymbol{I}(s)$ can be written as:

$$\boldsymbol{I}(s) = \frac{(\mathbf{V}_{\text{in}} - \mathbf{V}_0)(1/L) + s\,\mathbf{I}_0}{(s + s_1)(s + s_2)}$$

With the aid of partial fraction expansions we can find:

$$I(s) = \frac{(\mathbf{V}_{\text{in}} - \mathbf{V}_0)(1/L) - s_2\,\mathbf{I}_0}{(s + s_1)} + \frac{(\mathbf{V}_{\text{in}} - \mathbf{V}_0)(1/L) - s_1\,\mathbf{I}_0}{(s + s_2)}$$

With the aid of Table 12.2 we find the expression for $i(t)$. Thus:

$$i(t) = [(\mathbf{V}_{\text{in}} - \mathbf{V}_0)(1/L) - s_2\,\mathbf{I}_0]e^{-s1t} + [(\mathbf{V}_{\text{in}} - \mathbf{V}_0)(1/L) - s_1\,\mathbf{I}_0]e^{-s_2t}$$

12.6 OTHER SOLUTIONS

Although we have considered only dc sources in our discussion of various types of circuits, it should be pointed out that our solutions for the dc case may be repeated for any input voltage as long as the Laplace transform for the voltage is known. Substitution of any waveform will yield the proper results for that waveform even if the waveform is not one of those specified in our limited table. Tables of Laplace transforms can be found in most engineering libraries.

■ SUMMARY

- Transient analysis is the study of how circuit response varies from the time the switch is closed on a circuit until the circuit values of voltage and current no longer change.
- The total solution for a circuit consists of the transient solution and the steady state solution.
- The actual laws controlling how inductors and capacitors behave are differential equations. In the case of sinusoidal voltages the relationship between the current through and the voltage across an element can be shown to be algebraic for steady state but more complex for the transient condition.
- The time that a transient takes until it disappears is related to the values of the elements that make up the circuit. Circuits respond exponentially to changes in operating condition. These exponentials are controlled by circuit properties called time constants.
- The time constant of a simple R–C circuit is given as the product of R in ohms and C in farads. This product has units of seconds. The corresponding expression for a simple R–L circuit is L/R, where L is in henries and R is in ohms.
- After five time constants the size of the transient is less than 1% of its original value.
- Laplace transforms can be used to convert a differential equation into an algebraic one.
- The algebraic equations generated by Laplace transforms are solved using tables of inverse transforms.

■ PROBLEMS

12.1 Convert the following sources in the time domain to the Laplace domain.

(a) $v(t) = 10 + \sin(20t)$
(b) $i(t) = 20e^{-5t} + 10e^{-10t}$
(c) $v(t) = 3t\cos(200t) + 2t^2$
(d) $i(t) = 15\cos(30t + 1.4)$

12.2 Convert the following solutions in the Laplace domain into solutions in the time domain.

(a) $\dfrac{s + 6}{(s + 4)s}$ **(b)** $\dfrac{s + 3}{(s + 4)(s + 2)s}$

(c) $\dfrac{s + 7}{s^2 + 2s + 10}$ **(d)** $\dfrac{s + 10}{(s + 3)^2}$

12.3 For the circuit shown below:

(a) Find the time domain equation for the circuit.
(b) Find the Laplace equation for the circuit.
(c) Find the current in the circuit as a function of time.
(d) Sketch the voltage across the capacitor as a function of time.

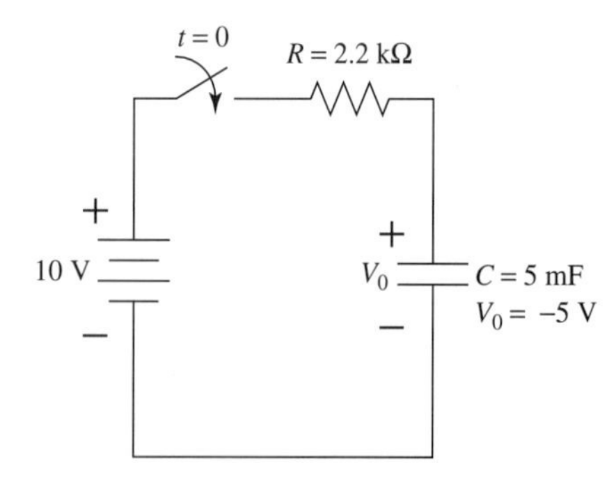

FIGURE P12.3

12.4 Repeat problem 12.3 for $R = 100$ kΩ.

12.5 For the circuit of problem 12.3 find the voltage across the resistor as a function of time. At what time is the voltage across the resistor equal to 3 V?

12.6 For the circuit of problem 12.3 the voltage source is replaced by a source $v_{\text{in}}(t) = 10\sin 100t$. Find:

(a) The voltage across the resistor as a function of time

(b) The voltage across the capacitor as a function of time
(c) The time that it takes the circuit to reach steady state

12.7 Repeat problem 12.3 for $v(t) = 10e^{-2t}$.

12.8 For the circuit shown in Fig. P12.8 the initial voltage on the 100 μF capacitor is 50 V. At time $t = 0$ the switch is closed.
(a) Find the voltage $v_R(t)$ across the resistor as a function of time.
(b) Find the voltage $v_e(t)$ across the 10-μF capacitor as a function of time.
(c) What is the steady state voltage across each element?

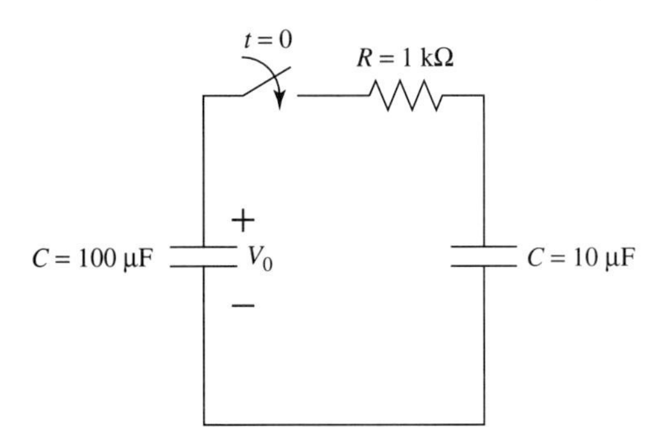

FIGURE P12.8

12.9 Repeat problem 12.8 if the value of R is changed to 10 kΩ.

12.10 At time $t = 0$ the switch is closed. The initial voltage on the capacitor is 50 V.
(a) For $R = 10\ \Omega$, find the current in the circuit as a function of time.
(b) What value of R will produce two equal roots in the denominator polynomial?
(c) Find the current in the circuit for this value of R.
(d) Find the voltage across the capacitor as a function of time for part (c).
(e) Sketch the waveform for the voltage across the resistor and the inductor.

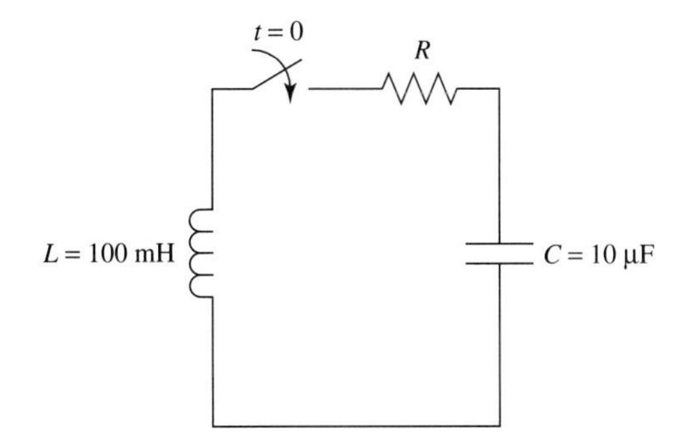

FIGURE P12.10

12.11 A 20-V battery is connected to a series circuit consisting of a 2.2-kΩ resistor and a 5-mH inductor. If the initial current through the inductor is 5mA, find:
(a) The time constant of the circuit
(b) The voltage across the inductor as a function of time
(c) The time for which the voltage across the inductor is 7 volts

12.12 For the circuit of problem 12.11, find the voltage across the resistor as a function of time. At what time is the voltage across the resistor equal to 15 V?

12.13 For the circuit of problem 12.11, the voltage source is replaced by a source = 10 sin 100t. Find:
(a) The voltage across the resistor as a function of time
(b) The voltage across the inductor as a function of time
(c) The time that it takes the circuit to reach steady state

12.14 For the circuit of problem 12.11 the voltage source is replaced by a source $v(t) = 20e^{-10t}$. Find:
(a) The voltage across the resistor as a function of time
(b) The voltage across the inductor as a function of time
(c) The time that it takes the circuit to reach steady state

CHAPTER **13**

COMPONENT PROPERTIES

INTRODUCTION

In this chapter we will consider the differences between real and idealized components. Through a detailed understanding of the origins of physical properties you will develop an awareness of the difference between actual components and the idealized components used in solving circuit equations. This will enable you to understand why results of calculations often differ from measurements in circuits when real components are used to build a previously calculated circuit.

13.1 LINEAR AND PASSIVE ELEMENTS

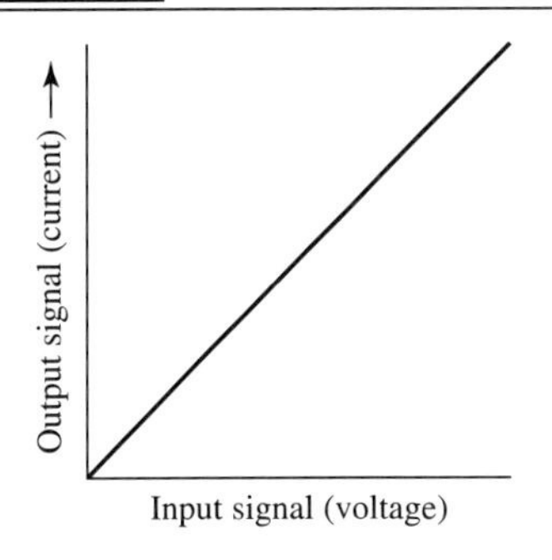

FIGURE 13.1
Linear element characteristic

The resistors, capacitors, and inductors that are discussed in this chapter are all considered to be linear elements whose properties are constant and do not vary with time. By definition, a ***linear element*** is one whose input versus output characteristic is a straight line. If we consider the voltage across an element to be its input, then the current through the element would be its output. Conversely, if we consider the current through the element to be its input then the voltage across the element is its output. Figure 13.1 exhibits a typical input/output characteristic plot for a linear element.

For a linear element we consider that the resistance, capacitance, and inductance values associated with the element are constant and are independent of the magnitude and polarity of the applied voltage. We can formally define a linear element as follows:

(If $v_1(t)$ across the element produces $i_1(t)$ through the element and $v_2(t)$ across the element produces $i_2(t)$ through the element then the element is linear if $(v_1(t) + v_2(t))$ produces $(i_1(t) + i_2(t))$ through the element.)

A ***passive element*** is an element that does not need a source of power to establish its electrical properties. The resistors, capacitors, and inductors considered in this chapter are all passive devices. With these restrictions (linear and passive), the concept of linear superposition is applicable to any circuit. That is, the solutions generated by the application of linear superposition to such circuits are valid.

13.2 PROPERTIES OF MATTER: TEMPERATURE

All matter is formed as the result of interactions between basic chemicals that scientists call elements. These elements combine to form substances we call compounds. We experience these compounds and elements through our physical senses—touch, taste, sight, smell, and sound. Matter is therefore a physical quantity that occupies a finite volume. This concept is discussed further in Chapter 1.

All matter possesses a property called temperature. Temperature is our way of describing the amount of heat energy contained in a piece of matter. When a rock, for example, has a lot of heat energy in it, we say that the rock feels hot. If we somehow remove some of the heat energy from the rock, then the rock becomes colder or less hot. Heat energy, as a concept, is based on the observation that the chemical elements contained in a piece of matter are all in a state of constant motion. That is, the atoms that make up the physical matter vibrate and move in a random and constantly changing way. The hotter the matter is, the greater the size of the atomic vibrations and motion.

13.2.1 The Celsius Temperature Scale

Temperature is a direct measure of the heat energy content of matter. The more heat energy a substance contains, the higher its temperature is said to be. Temperature is measured in units called degrees. Several different scales are used for measurement of temperature. The most common scale is the Celsius scale, which is established by the freezing point and boiling point of water. The freezing point is set equal to zero degrees (0°) Celsius, and the boiling point of water is set equal to 100 degrees (100°) Celsius. Thus 100 Celsius degrees span the temperature difference between freezing water and boiling water.

The Celsius scale can be indefinitely extended above 100 degrees for things that are hotter than boiling water. Substances that contain more heat energy than boiling water have a temperature greater than 100 degrees Celsius. Substances that have a lower temperature than the freezing point of water have a negative Celsius temperature. Dry ice, for example, has a temperature of −40 degrees Celsius (−40°C). It is 40 degrees colder than the freezing point of water (0°C).

13.2.2 Thermal Equilibrium

It is important to understand that temperature is an average value rather than a value applied to each of the atoms in the body being measured. Some atoms vibrate with more energy and some vibrate with less. The atoms continually collide and interchange kinetic energy with each other, one getting "hotter" and the other getting "colder." When a body with colder atoms comes into contact with a hotter body, the atoms from the hotter body transfer some of their excess kinetic energy to the atoms of the colder body. This causes the colder body to warm and the hotter body to cool. This process continues until the two substances are at the same temperature. When the two temperatures are the same, the substances are said to be in thermal equilibrium. Each body transfers energy to the other body, but the net flow of energy between the bodies is zero.

13.2.3 Temperature Measurements—Absolute Zero

The heat content of substances is measured with a device called a thermometer. Many different types of thermometers exist to cover the wide range of temperatures that exist in the universe. These thermometers depend on the physical properties of the thermometer as well as the heat content of the body being measured. The physical changes of the thermometer allow us to measure the temperature of unknown substances through the use of known substances and their properties.

When we consider the operation of removing heat energy from any substance, we realize that all substances have a finite amount of heat energy. As we remove heat energy, the temperature of the substance decreases. When we have removed all of the heat energy that can be removed we reach a temperature called ***absolute zero.*** This temperature is approximately equal to $-273.16°C$ (273.16 degrees below zero). At this temperature no more heat energy can be removed from any substance. In the real world we cannot reach absolute zero, although we can get to within a small fraction of a degree of absolute zero.

Scientists use a temperature scale related to the Celsius scale. In this scale there is a lowest temperature point: absolute zero. In this new scale ***all matter exists at temperatures that are greater than zero.*** This new temperature scale is called the ***Kelvin*** scale and is derived from the Celsius scale by adding 273.16 degrees to every Celsius temperature reading.

$$\text{Kelvin temperature} = \text{Celsius temperature} + 273.16 \quad \textbf{(Eq. 13.1)}$$

13.3 PROPERTIES OF MATTER: CONDUCTIVITY

Certain chemical elements have the property that some of their electrons are not tightly bound to their individual atoms. These loosely held electrons are called ***free electrons*** because at normal temperatures they are free to drift at random throughout the material. Their interaction with both free and bound electrons as well as positively charged atoms causes the overall charge on any finite region of matter to average out to zero. That is, there is as much positive charge in any region as there is negative charge, so no net charge exists in any region except momentarily. These free electrons drift at random from atom to atom as they absorb and release kinetic energy (heat) in their travels.

A material that contains many free electrons is called an ***electrical conductor.*** For example, copper and silver both contain many free electrons and are therefore good electrical conductors. When a battery or other voltage source is applied across a conductor, it produces an electromotive force (an electron moving force) across the conductor in response to the voltage difference applied across the conductor. Note that only voltage differences can be produced because all voltage sources have two terminals—one positive and one negative. We read the voltage difference between these two points, not the voltage at any one point by itself. All voltmeters have two leads used to measure voltage.

The free electrons in the conductor move in the direction of the more positive voltage terminal of the source. The effect of the applied voltage difference is added to the random motion that the electrons already possess and the applied voltage accelerates (or speeds up) the electrons, which thereby move in the direction of the positive voltage terminal. The accelerated electrons strike other electrons and atoms, and lose energy to them. This process continues as the electron moves through the conductor. Energy, gained from the voltage source, is transferred into the atoms that make up the conductor. Thus the power supplied to the electrons by the voltage source ultimately goes into giving the atoms more kinetic energy. This increases the temperature of the conductor because the energy absorbed from the accelerated electrons passing through the material is transferred to the atoms, thereby heating them.

It is important to realize and understand that the voltage difference applied across the conductor affects every free electron within the conductor. Thus a material that has many free electrons will have a large flow of electrons in response to the applied voltage while a material with few free electrons will have a small flow of electrons. Consider a conductor of cross-sectional area A whose electrons move under the influence of a voltage source in a direction perpendicular to the cross-sectional area, (see Fig. 13.2). The electrons flowing through the cross-sectional area produce an ***electrical current.***

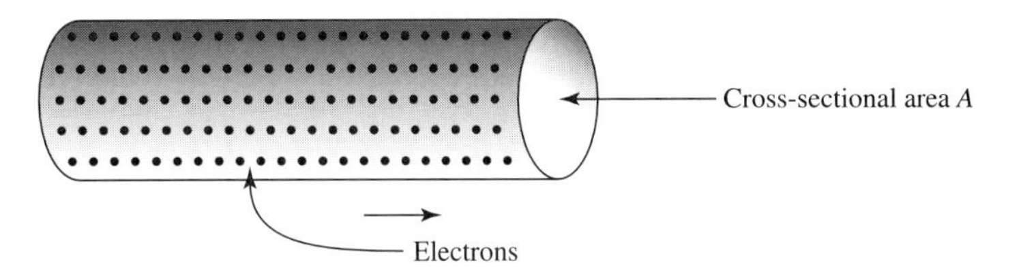

FIGURE 13.2
Electron flow in a conductor

The electrical current is defined as the total charge passing through the cross-sectional area per second. Since the charge on each electron is identical, the greater the number of electrons passing through the area per second, the larger is the current. If there are n free electrons per cubic meter each with a charge e and a velocity v meters per second, the current flowing through the area A is:

$$I = nevA \qquad \textbf{(Eq. 13.2)}$$

where A is the cross-sectional area of the conductor.

TABLE 13.1

Metal	Electron Concentration (electrons/m^3)
Lithium	4.60×10^{28}
Sodium	2.50×10^{28}
Potassium	1.34×10^{28}
Rubidium	1.08×10^{28}
Cesium	8.60×10^{27}
Copper	8.50×10^{27}
Silver	5.76×10^{27}
Gold	5.90×10^{27}

Table 13.1 lists the electron concentration n for several metals. Using this value in Equation 13.2 we can calculate the speed of electrons in a wire as a function of the current passing through it.

EXAMPLE 13.1

A copper wire with a diameter of 1 mm is carrying a current of 1 A. What is the velocity of the electrons passing through the wire?

Solution From Equation 13.2:

$$I = nevA$$
$$1\text{ A} = 8.5 \times 10^{27} \times 1.602 \times 10^{-19} \times v \times \pi \times (5 \times 10^{-4})^2$$
$$v = 9.35 \times 10^{-4}\text{ m/s}$$

As the calculation shows, electrons in a metal move very slowly because of the great number of electrons in a metal.

The velocity of the moving electrons is controlled in part by the obstacles in the path of the electrons as they travel. More obstacles means lower average velocity and therefore lower current. As we increase the voltage across a given conductor, we increase the speed of the electrons traveling through it and therefore the current flowing. Every substance presents different conditions for the free electrons, and ***therefore the net velocity of the electrons is directly linked to the properties of the material through which they flow.***

13.3.1 Conductivity

Consider a conductor with a constant cross-sectional area A and a given length L. Such a conductor is called an electrical wire or simply a wire. When we apply a voltage across the ends of the wire, a current flows. The electrons in the wire move from the negative terminal of the voltage source to the positive terminal of the voltage source. Electrons enter the wire at the negative terminal to replace the electrons that have moved out of the wire into the positive terminal of the voltage source. Since the cross-sectional area of the wire is constant, the number of electrons moving at any point is also constant, and therefore ***the current in the wire is the same at every point.***

For a given applied voltage, a longer length of a wire will have less voltage available per unit length than a short wire. The smaller the voltage per unit length is, the slower the speed of the flowing electrons because the voltage accelerating them over that distance is smaller. The larger the voltage per unit length is, the greater the acceleration and therefore the greater the speed of the electrons and thus the current flow. For a given wire geometry, the number of free electrons that can move is proportional to the cross-sectional area of the wire. The greater the number of electrons, the greater is the current that can flow. We thus see that the current that flows is proportional to the area of the wire and inversely proportional to the length of the wire. Finally, the actual current that flows is a function of the number of electrons available to flow and their velocity. We call this property of the wire its ***conductivity*** and say that a greater conductivity means a greater current for a given applied voltage.

The actual current that flows in a wire is directly proportional to the conductivity, the area of the wire, and the voltage applied across the wire. The current is inversely proportional to the length of the wire. We summarize this relationship by the equation:

$$\text{Current} = \frac{(\text{Conductivity})(\text{Cross-sectional area of the wire})(\text{Voltage})}{\text{Length of the wire}} \qquad \textbf{(Eq. 13.3)}$$

Using the letter I to represent the current and the letter V to represent the voltage we can write:

$$I = \frac{(\text{Conductivity})(\text{Area})}{\text{Length}} V \qquad \textbf{(Eq. 13.3a)}$$

The expression multiplying the voltage is called the ***conductance*** of the wire. We see that it is a function of the physical dimensions of the wire as well as its conductivity. Representing conductivity by the Greek letter sigma (σ), cross-sectional area by the letter A, and length by the letter L, we may write:

$$\text{Conductance} = G = \sigma A/L$$

This results in the formula:

$$I = GV \quad \textbf{(Eq. 13.4)}$$

13.3.2 Resistance

By definition, the reciprocal of conductance G is called ***resistance.*** Resistance is designated by the letter R and measured in units called ***ohms.*** Thus:

$$R = 1/G \quad \textbf{(Eq. 13.5)}$$

Substituting resistance for conductance into Equation 13.4 produces

$$I = V/R \quad \textbf{(Eq. 13.4a)}$$

We can rewrite this equation as:

$$V = IR \quad \textbf{(Eq. 13.4b)}$$

which we recognize as ***Ohm's Law.*** If we use the letter rho (ρ) to represent the reciprocal of the conductivity sigma (σ) we may write an expression for the resistance of a wire as:

$$R = \rho L/A \quad \textbf{(Eq. 13.6)}$$

where rho is the ***resistivity*** ($1/\sigma$) of the material, L is the length of the conductor, and A is its cross-sectional area.

Resistivity is therefore a reflection of the fact that as the electrons move through any material they interact with atoms and other electrons in their path. The electrons are alternately sped up by the voltage applied to the conductor and slowed down by collisions with other electrons and atoms. When electrons finally arrive at the other end of the conductor from which they started, they have used up the energy that the voltage has provided to them. This energy that was lost in collisions with the material was converted into greater atomic vibration. The greater vibration is recognized as an increase in the temperature of the material.

13.3.3 Resistance and Wire Size

The wires that interconnect electrical circuits are usually thought of as having zero resistance. This approximation is true if the resistance in the circuit is large in comparison to the resistance in the connecting wires. We can check the accuracy of this approximation by examination of Table 13.2, which gives the resistance of a wire as a function of the size of the wire. This table, which lists American Wire Gauge (AWG) sizes and dimensions versus the resistance per unit length, is constructed from Equation 13.6, which gives the resistance of a wire in terms of its physical dimensions. The table is based on the resistivity of copper, which is given as:

$$\rho = 1.7241 \times 10^{-6}\ \Omega\ \text{cm at } 20^\circ\text{C}$$

TABLE 13.2

AWG B & S Gauge	Diam- eter in Mile	Cross Section circular mile	square inches	Ohms per 1000 ft at 20°C (68°F)
0000	460.0	211,600	0.1662	0.04901
000	409.6	167,800	0.1318	0.06180
00	364.8	133,100	0.1045	0.07793
0	324.9	105,500	0.08289	0.09827
1	289.3	83,690	0.06573	0.1239
2	257.6	66,370	0.05213	0.1563
3	229.4	52,640	0.04134	0.1970
4	204.3	41,740	0.03278	0.2485
5	181.9	33,100	0.02600	0.3133
6	162.0	26,250	0.02062	0.3951
7	144.3	20,820	0.01635	0.4982
8	128.5	16,510	0.01297	0.6282
9	114.4	13,090	0.01028	0.7921
10	101.9	10,380	0.008155	0.9989
11	90.74	8,234	0.006467	1.260
12	80.81	6,530	0.005129	1.588
13	71.96	5,178	0.004067	2.003
14	64.08	4,107	0.003225	2.525
15	57.07	3,257	0.002558	3.184
16	50.82	2,583	0.002028	4.016
17	45.26	2,048	0.001609	5.064
18	40.30	1,624	0.001276	6.385
19	35.89	1,288	0.001012	8.051
20	31.96	1,022	0.0008023	10.15
21	28.46	810.1	0.0006363	12.80
22	25.35	642.4	0.0005046	16.14
23	22.57	509.5	0.0004002	20.36
24	20.10	404.0	0.0003173	25.67
25	17.90	320.4	0.0002517	32.37
26	15.94	254.1	0.0001996	40.81
27	14.20	201.5	0.0001583	51.47
28	12.64	159.8	0.0001255	64.90
29	11.26	126.7	0.00009953	81.83
30	10.03	100.5	0.00007894	103.2
31	8.928	79.70	0.00006260	130.1
32	7.950	63.21	0.00004964	164.1
33	7.080	50.13	0.00003937	206.9
34	6.305	39.75	0.00003122	260.9
35	5.615	31.52	0.00002476	329.0
36	5.000	25.00	0.00001964	414.8
37	4.453	19.83	0.00001557	523.1
38	3.965	15.72	0.00001235	659.6
39	3.531	12.47	0.000009793	831.8
40	3.145	9.888	0.000007766	1049.0

13.3.4 Resistance and Power

The total energy absorbed by a conductor is equal to the energy absorbed per electron times the number of electrons that make the trip. If N coulombs of charge are moved through a conductor by a voltage of V volts, then the number of joules of energy used to perform this transfer is given by:

$$\text{Energy used} = N \cdot V \text{ joules} \qquad \textbf{(Eq. 13.7)}$$

Power is defined as the ***number of joules per second*** that a conductor absorbs. The ***units*** used are used expressed as ***watts.*** Thus ***one watt is defined as one joule of energy used in one second.*** If the total energy is used to transport N coulombs of charge in T seconds, then the power associated with this charge transfer is given as:

$$\text{Power} = \frac{N \cdot V}{T} \text{ watts} \qquad \textbf{(Eq. 13.8)}$$

But N/T is equal to the current that flows in the conductor (I). From this we find that the power dissipated by a conductor with a voltage drop V across it is given as:

$$P(\text{power}) = I \cdot V \text{ watts} \qquad \textbf{(Eq. 13.9)}$$

Equation 13.9 represents the power dissipated by any conductor that has current passing through it and a voltage drop across it. Since every conductor has some resistance, it follows that every electrical circuit has some unintended added resistance as well as some unintended power loss. Combining Ohm's Law (Eq. 13.4b) with the power loss equation allows us to write the power loss in terms of the current through the circuit and the resistance as:

$$\text{Power} = I^2 R \qquad \textbf{(Eq. 13.10)}$$

Alternatively, we may also write the power loss as:

$$\text{Power} = V^2/R \qquad \textbf{(Eq. 13.11)}$$

13.3.5 Resistance and Temperature

In a conductor the atoms making up the conductor pick up energy as the electrons moving through the material interact with them and transfer kinetic energy to the atoms in the form of heat. This interaction results in the material becoming hotter, which in turn affects the properties of the interaction between the electrons and the atoms. As a general rule, the hotter the atoms are, the greater the interference with the flow of electrons and therefore the higher the resistance of the conductor appears. Since the dimensions of the conductor do not change appreciably with temperature, it follows that the resistivity must be responsible for the change in resistance with temperature. As temperature increases, the resistivity increases, which in turn increases the resistance of the conductor.

The resistivity versus temperature behavior curve for a conductor is shown in Fig. 13.3. Note that this curve is not quite a straight line. For every temperature there is a unique value of resistivity. Since the resistivity is a function of temperature we can express the resistivity as a function of temperature as:

$$\rho(T) = \rho_0[1 + \alpha(T - T_0) + \beta(T - T_0)^2 + \cdots] \qquad \textbf{(Eq. 13.12)}$$

where ρ_0 is the resistivity at $T = T_0$ and α, β, and so on are constants experimentally determined from the shape of the actual curve. These constants allow us to express the resistivity versus temperature curve analytically and to solve problems related to a change of temperature in a circuit. The values of α and β are small compared to 1, and α is usually much larger than β. For small temperature changes the resistivity curve is almost a

FIGURE 13.3
Resistivity versus temperature curve

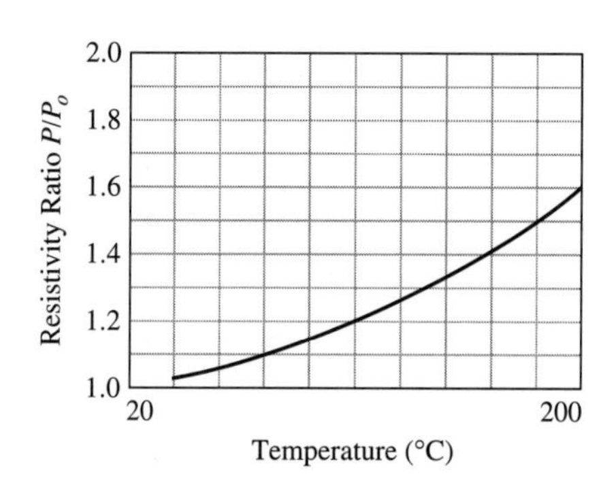

straight line and we can assume that β and all higher-order constants are zero. Under this condition Equation 13.12 reduces to:

$$\rho(T) = \rho_0[1 + \alpha(T - T_0)] \qquad \textbf{(Eq. 13.13)}$$

The quantity α is called the temperature coefficient of resistivity. Table 13.3 lists the resistivity and the temperature coefficient of resistivity for several conductors.

TABLE 13.3

Material	Resistivity (μ Ω cm)	Temperature Coefficient (Ω per Ω per °C)
Aluminum	2.828	0.0039
Constantin	49.05	0.0002
Copper	1.724	0.00393
Gold	3.14	0.0034
Iron	9.65	0.0052–0.0062
Manganin	44.8	0.00002
Nichrome	112	0.00017
Silver	1.638	0.0038

EXAMPLE 13.2

Find the resistance of a copper wire that is 1000 m long and has a diameter of 1 mm. If the temperature goes to 50°C, what is the resistance of the wire?

Solution From Equation 13.6 we find $R = \rho L/A$, from which, using Table 13.2, we find:

$$R = \frac{1.724 \times 10^{-6}\ \Omega\ \text{cm} \times 1000\ \text{m}}{\pi(0.5 \times 10^{-1})^2\ \text{cm}^2}$$
$$= \frac{1.724 \times 10^{-6} \times 10^5}{\pi(0.25 \times 10^{-2})}$$
$$= 21.95\ \Omega$$

A temperature change from 20°C to 50°C represents a rise of 30°C and produces a resistance change given by Equation 13.13

$$\rho(T) = \rho_0[1 + \alpha(T - T_0)]$$
$$\rho(50°) = 1.724 \times 10^{-6}[1 + 0.00393(50 - 20)]$$
$$= 1.724 \times 10^{-6} \times 1.1179$$
$$= 1.927 \times 10^{-6}\ \Omega\ \text{cm}$$

Using this value for resistivity we find from Equation 13.6 that:

$$R(50) = 24.54\ \Omega$$

13.4 RESISTORS

Resistance is an unavoidable consequence of current flowing through real conductors. It is also a property of an electrical circuit that, when used properly, can be exploited to our advantage. It can be used to turn a current flow into a voltage drop, which can be used to generate heat and light (to name just some of its useful purposes).

Many electrical circuits make use of resistance for proper circuit operation. When resistance is required in a circuit, elements called resistors are used. Resistors are made with only a few specified standard values. The standard resistances range in value from a fraction of an ohm to many millions of ohms. The resistance needed for a particular purpose is calculated with the aid of Ohm's Law. Once the required value is determined, a resistor with a standard value close to the calculated value is used. This approach allows resistors to be off-the-shelf items rather than special values used for each new circuit. Most circuits do not require that the resistance be exactly as calculated but can operate properly over a range of values. Thus standard resistor values make economic sense. Figure 13.4 shows a picture of several different types of resistors.

FIGURE 13.4
Several types of resistors

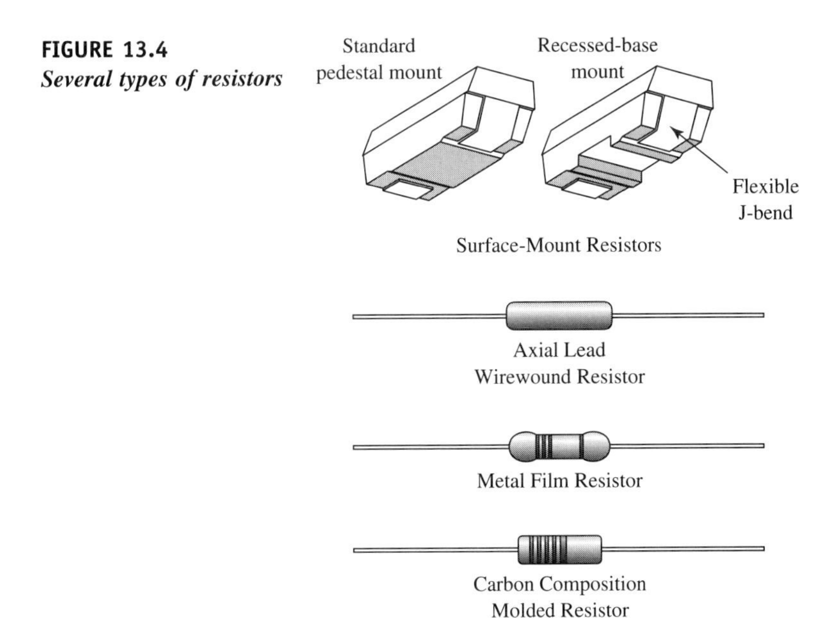

A common type of resistor used in electrical circuits is called a carbon composition resistor. Carbon is an element that conducts electricity readily. When ground into a powder it can be mixed with nonconducting powders (insulators) to create materials with higher resistivity than plain carbon. This mixture may be compressed into a solid and used to form resistors that vary over an extremely wide range of values. By varying the percentage of each component in the mixture resistors of several ohms to several megohms (millions of ohms) can be routinely fabricated. Figure 13.5 shows a cutaway view of a carbon composition resistor. The mixed powder is compressed into a hollow phenolic tube, and metal leads are attached to each end of the tube to form the resistor.

FIGURE 13.5
Cutaway of a carbon composition resistor

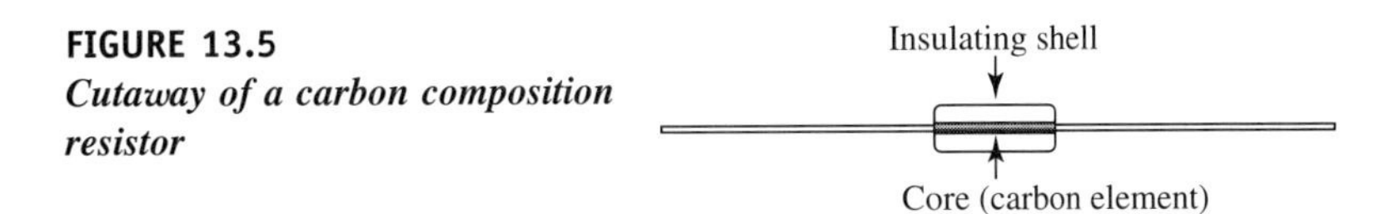

Carbon composition resistors have the advantage of being small in size and very inexpensive to manufacture. Their physical size, which has been standardized, is based solely on the power dissipation (wattage) rating of the resistor and is independent of the ohmic value of the resistance. They are commonly available in $^1/_8$-W, $^1/_4$-W, $^1/_2$-W, 1-W, and 2-W sizes. Billions of these carbon composition resistors are manufactured each year for use in electrical circuits. Figure 13.6 shows the relative size of the different wattage carbon composition resistors. Resistors that dissipate more energy (higher wattage) are correspondingly larger than those that dissipate less.

Wattage	Ohms	Body Dimensions (in./mm)		Lead diameter (in./mm)
		Length	Diameter	
0.125	2.7–22M	0.160/4.1	0.066/1.7	0.018/0.46
0.250	2.7–22M	0.265/7.7	0.098/2.5	0.027/0.69
0.500	1.0–22M	0.406/10.3	0.148/3.8	0.035/0.89
1.00	1.0–22M	0.593/15.1	0.233/5.9	0.043/1.09
2.00	1.0–22M	0.719/18.3	0.320/8.1	0.048/1.22

FIGURE 13.6
Size of resistors

The wattage rating of resistors is based upon their ability to safely dissipate the heat that is generated within as the current passes through them. For the case of carbon composition resistors, the wattage rating is based on a maximum ambient temperature of 70°C. Temperature ratings for resistors are determined so that the resistor will not be damaged, provided the resistor is used at or below the maximum operating temperature. If a resistor dissipates more power than its rating, the resistor will overheat and may even be permanently damaged.

EXAMPLE 13.3

A 470-Ω, $^1/_4$-W resistor is connected across a power supply. What is the maximum voltage that can be placed across the resistor before its power dissipation is exceeded?

Solution For any resistor $P = V^2/R$, from which we find that:

$$V^2 = P \cdot R$$

For the 470-Ω, $^1/_4$-W resistor we find:

$$V^2 = (^1/_4)(470) = 117.5$$
$$V = 10.84 \text{ V}$$

Carbon composition resistors are marked using an industry-standard color marking code. This code allows for the rapid identification of a resistor's value. Figure 13.7 and Table 13.4 shows the scheme used for color-coding resistors. There are usually four bands of color painted on the body of the resistor. The color nearest the edge of the resistor is the first band and indicates the first significant figure of the resistance value in ohms. The second band shows the second significant figure of the resistance value in ohms. The third band is a decimal multiplier that multiplies the value of the first two bands by some power of ten, thereby establishing the value of the resistance in ohms to two significant figures. The fourth band, if it exists, reveals the tolerance of the resistor. That is, it expresses how close to its indicated value the resistor must be. This marking code, which was established by RETMA (Radio-Electronics-Television Manufacturers Association), is used worldwide to identify resistor values as well as the value of many other electronic devices.

FIGURE 13.7
Carbon composition molded resistor

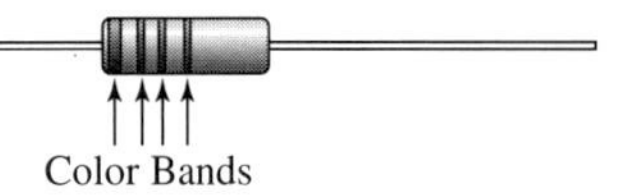

TABLE 13.4

Color	Significant Figures	Decimal Multiplier	Tolerance in Percent
Black	0	1	
Brown	1	10	
Red	2	100	
Orange	3	1,000	
Yellow	4	10,000	
Green	5	100,000	
Blue	6	1,000,000	
Violet	7	10,000,000	
Gray	8		
White	9		
Gold		0.1	±5
Silver		0.01	±10
No Band			±20

13.4.1 Standard Resistance Values

Along with the resistor color-coding scheme, RETMA established a set of standard resistance values that are used worldwide. The standard value chart is shown in Table 13.5. The values of the standards were established so that the number of values needed at every tolerance level was minimized. For the 20% tolerance range only 7 values per decade are used. At the 10% tolerance there are 13 values per decade, while at the 5% level 25 values are used. This is because larger tolerances require fewer resistors to represent the decade span of values.

Only resistors whose value is some integer power of 10 times those listed in the table are standard values of resistance. Any value needed for a circuit is selected from those in the table by using the closest RETMA value listed.

TABLE 13.5
RETMA significant figure values

20% Tolerance	10% Tolerance	5% Tolerance	20% Tolerance	10% Tolerance	5% Tolerance
10	10	10	47	47	47
		11			51
	12	12		56	56
		13			62
15	15	15	68	68	68
		16			75
	18	18		82	82
		20			91
22	22	22	100	100	100
		24			
	27	27			
		30			
33	33	33			
		36			
	39	39			
		43			

EXAMPLE 13.4

What are the color bands for a 7500-Ω, 5% tolerance resistor?

Solution From Table 13.4 we find:

Violet represents the number 7.
Green represents the number 5.
Two zeros are represented by the color red.
A 5% tolerance is represented by a gold band.
The resistor should be marked violet, green, red, gold.

13.4.2 Other Types of Resistors

Carbon film resistors are the modern version of carbon composition resistors and have superior properties. Instead of using a mixture of carbon and some insulating material, these resistors are fabricated by depositing carbon vapor on a ceramic form until the desired resistance value is produced. In this process a more stable resistance element is obtained with an accuracy that is easier to control. It is so easy to manufacture these resistors with a 5% tolerance that lower-tolerance units are not even made. Figure 13.8 shows the characteristic shape of these resistors.

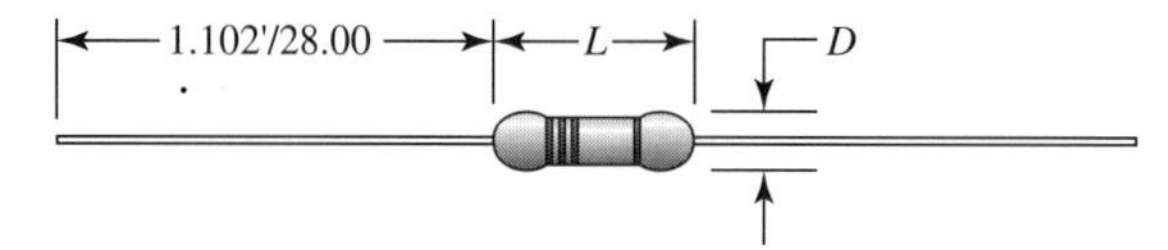

Wattage	Ohms	Dimensions (in./mm) Max. Length	Max. Diameter	Lead guage	Lead diameter (in./mm)
0.125	1.0–1M	0.138/3.5	0.073/1.85	24	.0201 / 0.51
0.250	1.0–10M	0.268/6.8	0.099/2.5	22	.0254 / 0.65
0.500	1.0–10M	0.355/9.0	0.118/3.0	22	.0254 / 0.65
1.00	1.0–10M	0.473/12.0	0.197/5.0	20	.032 / 0.81
2.00	1.0–10M	0.630/16.0	0.217/5.5	20	.032 / 0.81

FIGURE 13.8
Carbon film resistors, 5% tolerance

Modern circuits call for resistors that are extremely stable and extremely accurate. For resistors of this type metal vapor is used instead of carbon vapor, and the resistor is carefully trimmed to its final value with a laser etcher. Resistors of this type can be produced with accuracies of 0.01% or better. Such resistors are not marked with the color bands used for carbon film resistors but rather have their values printed on the body of the resistor.

When a circuit needs a resistor to dissipate more than 2 W of energy, wirewound resistance elements are used. These elements consist of high-resistivity metals wound on an insulating core. These so-called wirewound resistors can be designed to dissipate many watts of power. The power rating needed and the resistance value required determine their size and configuration. Their maximum resistance is usually limited to several thousand ohms.

13.5 MAGNETIC FIELDS AND INDUCTANCE

Consider a wire in which no current is flowing. When we attach a voltage source to the wire, the electrons start to move in the direction of the positive voltage source terminal under the influence of the applied voltage. As the electrons move, their motion produces a current. It takes a finite time for the current in the wire to reach its final value. This final value cannot occur instantly because the electrons must accelerate from zero velocity to their final velocity under the influence of the voltage applied, which comes from the voltage source. This energy transfer takes a finite amount of time because the acceleration is finite. As the electrons start to move the current increases from zero to some final value and a magnetic field appears around the wire. The magnetic field changes from zero to some final value and is determined by the magnitude of the current flowing in the circuit. The strength of the magnetic field is linearly dependent on the current. As the current in the circuit increases and decreases, the magnetic field that surrounds the wire likewise increases and decreases. Moving electrons represent energy, and this energy is stored in the magnetic field. When the current stops, the magnetic field vanishes.

It is important to understand that there is no difference in the properties of a magnetic field produced by a wire carrying a dc current and those of a magnetic field surrounding a permanent magnet. These fields are identical in all of their properties, and no test can differentiate between them. A magnet reacts to another magnet by either repelling it or attracting it. A fixed wire near a suspended permanent magnet can therefore be used to move the permanent magnet by passing a current through the fixed wire. This means that the magnetic field of the wire can do work. Since the magnetic field around the wire can do work, this is further evidence that the field must contain energy of some sort. This energy enables the wire to do work.

Just as a current in a wire produces a magnetic field, so a permanent magnet moving near a wire produces a voltage. The faster the motion of the magnet with respect to the wire, the larger the voltage generated across the wire will be. If the wire is connected to a circuit, a current will flow in the circuit in response to the voltage produced. Electric motors and generators are based on this interaction between a magnetic field and a conductor carrying current.

A wire carrying an electric current produces a magnetic field around the wire. If the current level decreases, the magnetic field around the wire will likewise decrease. This variation in the magnetic field produces a voltage across the wire. The polarity of the voltage produced is in such a direction as to attempt to maintain the current in the circuit. The faster the current is changing, the larger the voltage produced to counteract that change in current. We can express this fact by saying that the voltage generated is proportional to the change of current with time. Mathematically, this can be written as:

$$V = L\, dI/dt \quad \textbf{(Eq. 13.14)}$$

The proportionality constant in this equation, symbolized by the letter L, is called inductance and is measured in units called henries. For a given change in current over a given amount of time, the voltage generated is expressed by Equation 13.14. This equation is a linear differential equation that relates the change in current through the wire to the voltage created across the wire.

Suppose that the wire is formed into a series of loops as shown in Fig. 13.9. Such a configuration is called a coil of wire. The magnetic field caused by the current through a loop of wire interacts with the adjacent loops of wire, but if the magnetic field of a loop interacts with its neighbor then the neighbor's magnetic field interacts with the first loop. The effect of this interaction is to increase the value of the proportionality constant L. Thus the inductance of a coil of wire increases as the number of turns on the coil increases. Inductance plays a vital role in the operation and design of electrical circuits. By adjusting the number of turns on the coil, the inductance of the coil can be set to any value desired.

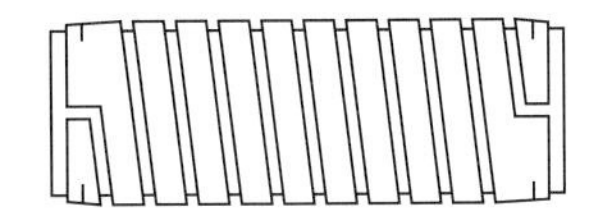

FIGURE 13.9
Inductor (coil of wire)

If a coil of wire carrying a current is placed around a piece of iron, an electromagnet is created. The electrons in the iron line up in such a way as to increase the size of the magnetic field generated within the coil. The increased magnetic field contains more energy than was present in the generated magnetic field when the iron was not inside the coil. This greater field causes the value of the inductance L to be larger in value. The iron somehow acts as a magnifier of the magnetic energy and is said to be a magnetic material. Other elements have properties similar to the iron, but no other single element has the

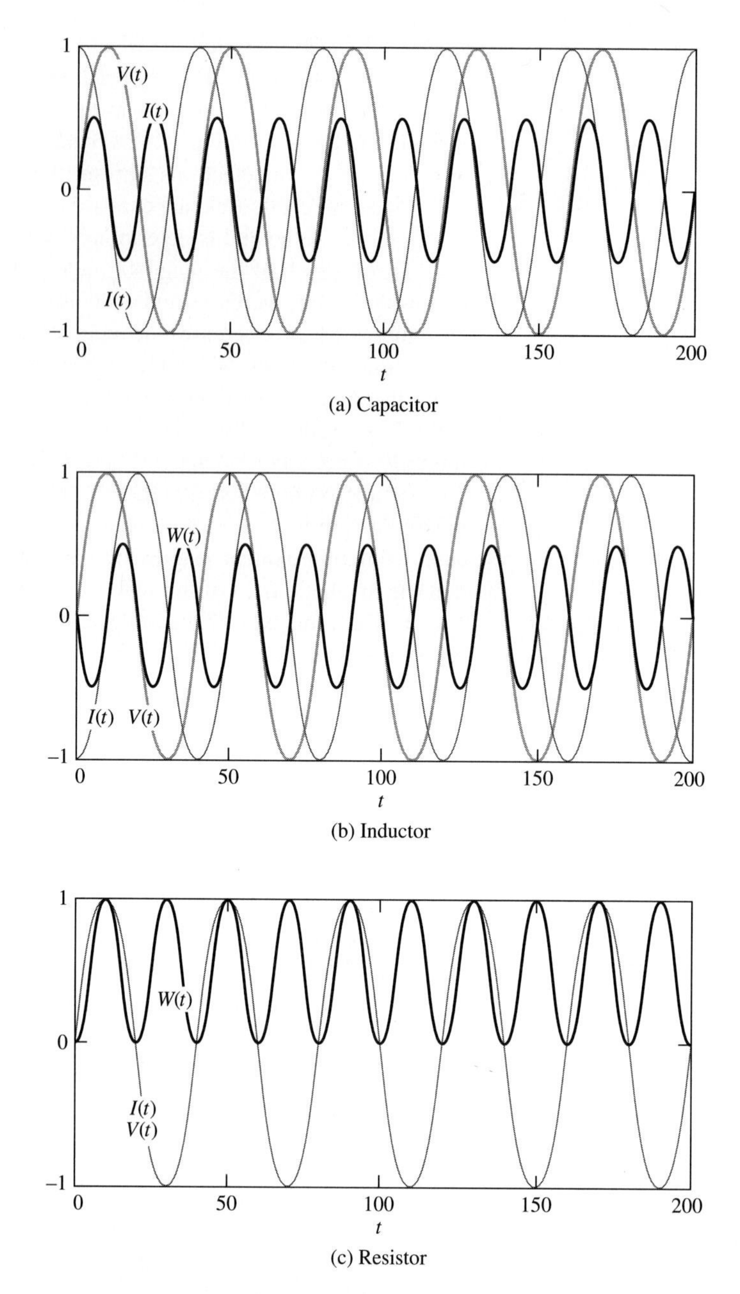

FIGURE 13.10
Curves of voltage, current, power versus time

same large effect on the inductance of the coil. In addition to iron, certain compounds have a large effect on the value of the inductance. Each of these compounds has a slightly different set of properties. Through the use of these properties and the original coil of wire, permanent magnets can be produced. These are magnets that retain their magnetism after the current is removed from the coil surrounding them.

If a coil is connected to a sinusoidal voltage source, then as the current in the coil varies the magnetic field also varies. The sinusoidal voltage waveform applied to the coil gives rise to a sinusoidal current. In fact, for the sinusoidal waveform, Equation 13.14 reduces to:

$$V = j\omega LI \qquad \textbf{(Eq. 13.15)}$$

which is a phasor equation that expresses the relationship between the voltage applied and the current that results. This property is discussed more fully in Chapter 12.

As the current is increasing, the magnetic field is absorbing energy from the voltage source. When the current starts to decrease, the magnetic field also decreases. The energy stored in the magnetic field cannot simply disappear. The energy returns to the voltage source that provided it in the first place. If we look at a graph of the voltage applied to the coil versus time and the current through the coil versus time, we observe that the product of voltage and current is sometimes a positive quantity and sometimes a negative quantity. When the product is positive, the voltage source is supplying power to build the magnetic field. When the product is negative, the magnetic field is returning the power taken from the voltage source. In the absence of resistance, these powers must be equal to each other. We thus arrive at a conclusion that inductors sometimes absorb power from a source and sometimes return power to the source. The net power absorbed in any cycle must be zero because the power absorbed is completely returned to the source. Figure 13.10(b) shows a plot of voltage current and instantaneous power versus time for a sinusoidal waveform.

Like resistance, inductance is the consequence of the behavior of moving electrons. All electrical circuits contain some inductance because of the creation of a magnetic field and its properties. Unlike resistance, inductance does not result in power loss in a circuit. When inductance is used in an ac circuit, the higher the frequency the greater the voltage across the inductance element. Thus, as the frequency increases the ratio of the voltage across the inductance to the current through the inductance element increases. Like resistors, inductance elements are packaged into standard sizes, called inductors. Inductors respond differently as the frequency of operation changes. This change in properties with frequency makes the inductor a valuable circuit element in circuits where the frequency varies.

An inductor is made from wire, so it always has some resistance associated with the wire that composes it. This resistance affects the performance of the inductor, making it an element that does not quite return as much power to the source as it absorbs in establishing its magnetic field. The frequency of operation along with the resistance of the inductor determines the circuit performance of an inductor.

13.6 CAPACITANCE AND CAPACITORS

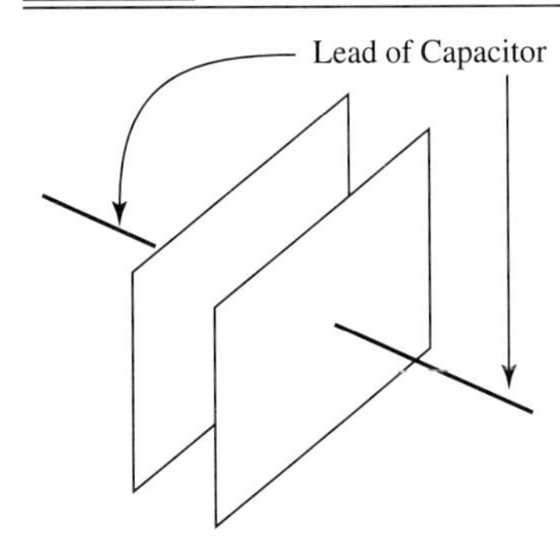

FIGURE 13.11
Two metal plates spaced a distance d apart form a parallel plate capacitor

Consider two metal plates that are arranged to be parallel to each other and spaced a distance *d* apart (see Fig. 13.11). If we attach one plate to the positive terminal of a voltage source and the other plate to the negative voltage source, then a voltage will exist between the two plates. Clearly no current can flow between the plates because there is no conducting connection between the plates. Since the space between the plates is not a conductor, it must be an insulator. An insulator is something that does not conduct electricity. Air is an insulator. Even a vacuum is an insulator. The insulator between the two plates of a capacitor is called a ***dielectric.***

When the plates were first connected to the voltage source, both plates were electrically neutral and there was no voltage difference between them. To establish a voltage difference between the plates, electrons from the plate connected to the positive terminal of the source must flow to the plate attached to the negative terminal of the source. The first plate becomes positively charged, and the second becomes negatively charged. Since no current flows when the voltage across the plates equals the applied voltage, the source that produced this condition can be removed. The electrons stay on the negatively charged plate, and the positively charged plate has a deficiency of electrons equal to the charge on the negatively charged plate.

The charge imbalance between the plates produces the voltage difference between the two plates and remains even when the source is removed. If the two plates are connected together through a resistor, the excess electrons flow through the resistor and the voltage between the two plates goes to zero. Thus the two plates store energy between them that can be used to produce current flow in a resistor. Energy originally placed on the plates by the voltage source is turned into heat as electrons move through the resistor. An electrical component that stores charge in a set of plates is called a ***capacitor,*** and the property the component possesses is called capacitance.

The electrical size of a capacitor is proportional to the area of the plates used to construct it. The larger the area of each plate, the greater is the capacitance value. If we move the plates closer together, the charge that can be stored in the capacitor is found to increase. Thus, a capacitor with plates closer together has a larger capacitance. Finally, the size of the capacitance is a function of the dielectric (insulating material) between the two plates. If a vacuum exists between the two plates, the capacitance can be expressed as:

$$C = \varepsilon_0 A/d \qquad \textbf{(Eq. 13.16)}$$

where $\varepsilon_0 = 8.85 \times 10^{-12}$ F/m, A is the area of a plate expressed in square meters, and d is the distance between plates expressed in meters. For capacitors with a dielectric other than vacuum, the value of the capacitance is larger. We can express the capacitance of these capacitors as:

$$C = \varepsilon_r \varepsilon_0 A/d \qquad \textbf{(Eq. 13.17)}$$

where ε_r is called the relative dielectric constant and is the ratio of the capacitance with the dielectric to the capacitance with a vacuum dielectric. It expresses the factor by which the capacitance is increased. Table 13.6 lists some dielectric constants for materials usually used to construct capacitors.

TABLE 13.6

Dielectric Material	Dielectric Constant
Air	1.00054
Mylar	3.1
Paper	3.7
Paraffin	2.0–2.5
Polystyrene	2.5
Teflon	2.1
Vacuum	1.000
Water	80

EXAMPLE 13.5

A capacitor is formed from two metal plates that each have an area of 1 m^2. The distance between the plates is 1 mm. What is the capacitance of the capacitor if the dielectric is vacuum?

Solution From Equation 13.17 we find:

$$C = \varepsilon_r \varepsilon_0 A/d = 1 \times 8.85 \times 10^{-12} \times (1/10^{-3}) = 8850 \text{ pF}$$

For a given capacitor, the larger the voltage applied to the plates, the larger is the amount of charge stored. The relationship between the charge and the voltage is a linear one. We may write that:

$$Q = CV \qquad \textbf{(Eq. 13.18)}$$

In this equation the charge is measured in ***coulombs*** and the voltage is measured in ***volts.*** The proportionality constant C is measured in farads, which is a unit of capacitance.

As the voltage across the capacitor increases, electrons move between the positive plate and the negative plate, building charge on the negative plate. The flowing charge is a current that is storing energy in the capacitor. ***A decreasing voltage across the capacitor also causes a redistribution of electrons, but in this case the current flow is opposite in direction and represents energy being transferred back to the voltage source.*** Like the inductor, energy is not lost when a capacitor is charged but is returned when the capacitor is discharged. Both capacitors and inductors are thus energy storage devices, unlike a resistor, which is an energy-absorbing device. See Fig. 13–10(a).

We see from this description that the current flow into and out of a capacitor is related to the changing of the voltage across the capacitor. If the voltage changes with time, the current exists; if the voltage does not change with time, the current is zero. We can write an equation that reflects this fact as:

$$I = C\,dV/dt \qquad \textbf{(Eq. 13.19)}$$

This equation is related to Equation 13.16, which expresses the relationship between charge and voltage by theorems of calculus. As was the case for the inductor, when a sinusoidal voltage source is attached to a capacitor, the resultant current is also a sinusoidal waveform. Equation 13.17 then becomes a phasor equation:

$$I = j\omega C\,V \qquad \textbf{(Eq. 13.20)}$$

If we fill the space between the two plates of the capacitor with an insulator, we find that the amount of charge that exists on each plate increases. This can be explained if we assume that some kind of energy is stored between the plates of the capacitor. When an insulator is inserted between the plates, it stores energy within its atomic structure in addition to the energy already stored. This allows more energy to be stored between the plates of the capacitor. The insulator placed between the plates of a capacitor is called a dielectric. From Equation 13.16 we see that the dielectric effect is to increase the charge for a given voltage or to effectively increase the value of the capacitor by some amount. The factor by which the capacitor is increased is called the dielectric constant. This constant is 1 for a vacuum and larger than 1 for any insulator.

The properties of insulators used in the construction of capacitors vary over a wide range. Some, such as glass, are such good insulators that a charge placed on the capacitor will last for many days with no loss of charge. Other insulators have some large but finite resistance associated with them that causes the charge to be dissipated rather rapidly when the voltage source is removed. Each of these insulators has a use. For example, if we want large capacitance in a small package, we may be willing to sacrifice leakage for capacitance. An electrolytic capacitor is a capacitor with an insulator that has a dielectric constant that is extremely high. In this case we sacrifice leakage for capacitance. Mylar has a dielectric constant that is very stable over a wide temperature range. Thus capacitors made with mylar are temperature insensitive.

An understanding of the properties of various types of capacitors can be obtained by a study of their properties as listed in data sheets available from the manufacturers. As in the case of inductors, capacitors come in many shapes and sizes. Like resistors, capacitors have standard values that follow Table 13.4.

13.7 REAL COMPONENTS

All real components are a combination of resistance, inductance, and capacitance. An equivalent circuit that shows all of their properties, not just the dominant element, best represents the components. How closely an element matches an ideal element depends on the frequency range over which the element is used. Through use of more exact equivalent circuits, taking all properties of a component into consideration, a better match is found between theory and actual measurement.

SUMMARY

- A linear element is one whose input–output characteristic is a straight line independent of the magnitude of the signal applied.
- All material properties are temperature sensitive. Electrical properties can change drastically with temperature.
- The Celsius scale is used to measure temperature in the International System of Units. It has a lower limit of −273.16 degrees. This temperature is called absolute zero. No temperature can be colder than this.
- The Kelvin scale is derived from the Celsius scale by adding 273.16 degrees to every Celsius temperature. This means that all Kelvin temperatures are positive values.
- Resistance is an electrical property of matter. When current flows through a resistance, heat is developed and the material increases in temperature.
- Resistances are manufactured in standard values and measured in units called ohms. The heat that they generate is measured in units called watts. Resistors are rated by the maximum amount of wattage they can safely dissipate.
- Inductors are electrical devices that store energy in a magnetic field. Unlike resistors, inductors do not convert current flow to heat. When the current is decreased, the energy stored is returned to the circuit. Inductors are measured in units called henries.
- Capacitors are electrical devices that store charge in the form of electrons. Like the inductor they do not convert current flow to heat but rather return the stored charge to the circuit when the circuit is deenergized. Capacitors are measured in units called farads.

PROBLEMS

13.1 A copper wire with a diameter of 0.010 in. melts when a current of 10.2 A flows through it.

(a) What is the velocity of the electrons when the wire is just about to melt?

(b) If the resistance of the wire is 100 Ω, how long is the wire?

(c) How many watts per meter does this current represent?

13.2 What is the velocity of a silver wire with a diameter of 0.01 cm that carries 2 A?

13.3 What is the diameter of a silver wire that has a resistance of 0.1 Ω/m?

13.4 What is the resistance of a copper wire with a diameter of 0.1 cm that circles the earth at the equator? Assume that the distance around the earth is 40,000 km.

13.5 Repeat problem 13.3 for a nichrome wire.

13.6 What is the diameter of a nichrome wire that has a resistance of 100 Ω/m?

13.7 The nichrome wire of problem 13.6 is heated to 200°C from 20°C. By what factor does the resistance increase due to the temperature change?

13.8 A constantin wire changes its resistance by 1% from its value at 20°C. What is the temperature change that caused this change?

13.9 What is the power dissipation per meter of the wire of problem 13.2?

13.10 The following resistors are identified by their color bands. What resistance does each of them represent?

(a) Red, violet, yellow, silver

(b) Blue, gray, green, gold

(c) Yellow, violet, orange, silver

(d) Brown, black, black

13.11 A capacitor is made up of plates that are 0.25 mm apart. What cross-sectional area is needed to form a capacitor of 500 pF?

13.12 A mylar capacitor has a capacitance of 0.01 μF. If the area of each plate is 0.1 m^2, what is the spacing between the plates?

CHAPTER 14

INSTRUMENTATION AND LAB SIMULATION

INTRODUCTION

After a circuit is built it is necessary to measure the circuit to determine that it works as expected. To accomplish this task we use electrical test equipment that can measure the various properties of the circuit. Only through measurement can we be sure that the circuit was built properly and that all of the parts of the circuit work together to meet the specifications they were designed to meet. This chapter deals with the proper operation of test equipment in the analysis of circuits and the use of software to simulate the use of test equipment.

14.1 TYPES OF MEASUREMENT

In normal operation all electrical circuits develop voltages across elements and currents through elements. These voltages and currents are what cause the circuit to behave as it does. When a circuit is not performing as expected, it is because somewhere in the circuit the voltages and currents are wrong. This may be due to a defective part or to incorrect wiring. The purpose of making measurements on a circuit is to ensure that it is operating as expected. If we record the results of measurements when the circuit is performing properly, we can use the recorded data to analyze the circuit if it ever fails to operate properly. Thus we can repair the circuit if it ever malfunctions. This is the theory behind all circuit repair, generally known as "troubleshooting" or "debugging."

14.2 DC CURRENT MEASUREMENTS

When we wish to measure the dc current through a circuit we use a device that measures current, called an ammeter. Current is expressed in units of amperes, hence the name ammeter. For most measurements the current being measured is a small fraction of an ampere. If the current being measured is in the range of thousandths of an ampere (0.001 A), the device that we use is called a milliammeter; if the current being measured is in millionths of an ampere (0.000001 A) the device is called a microammeter. ***Microammeters, milliammeters, and ammeters are all the same type of device. They all measure current in amperes.*** The only difference between them is in their sensitivity to the current passing through them.

The simplest form of an ammeter consists of a coil of wire mounted in a magnetic field. Figure 14.1 shows the construction details of such a meter. Current for the moving coil passes through two identical control springs that are mounted on the coil's rotation. The control springs are arranged so that they balance the force created when the magnetic field of the coil interacts with the fixed magnetic field. Thus, greater currents produce greater forces that are balanced by the greater torque produced by the springs. A pointer attached to the coil moves in response to the current passing through the coil. A scale is placed behind the pointer, and the larger the current through the coil, the larger the deflection of the pointer and the reading on the scale. Depending on the meter's sensitivity, the scale is calibrated directly in units of amperes, milliamperes, or microamperes.

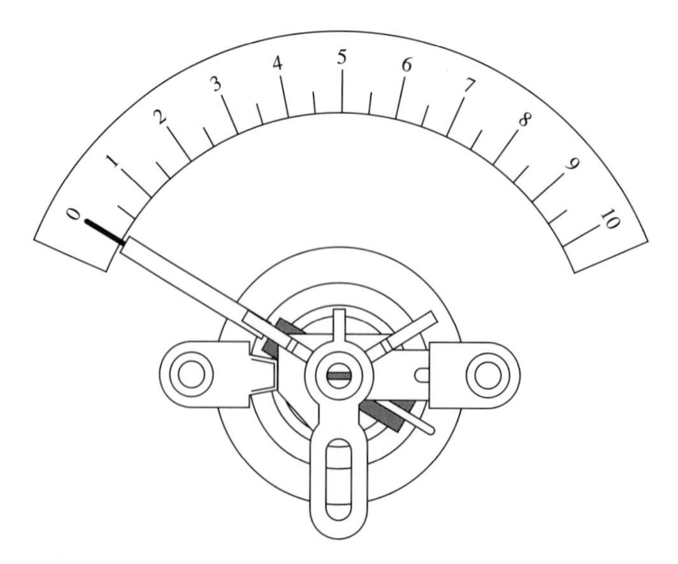

FIGURE 14.1
Ammeter with Pointer and Scale

The coil in an ammeter is made from ordinary copper wire; therefore, it has resistance. The size of the resistance is related to the sensitivity of the ammeter. Greater sensitivity of the meter calls for more turns of wire on the coil, which in turn means more resistance in the coil. Since the ammeter has resistance, we must be sure that when we place it in a circuit its resistance is small compared to the resistance already in the circuit. If the resistance of the ammeter is too large, the current in a circuit will be affected by the ammeter coil resistance. The percentage error in the measured current is related to the ratio of the ammeter resistance divided by the circuit resistance. For good measurements the ammeter resistance should be less than 1% of the circuit resistance. For low-current measurements the ammeter resistance is a very important factor.

The current in a circuit can vary over several orders of magnitude. It is therefore important that current-measuring equipment have ranges of measurement so that the meter deflection will be adequate and not too large for the current being measured. Current ranges are created by using a sensitive meter in parallel with a fixed resistor. Figure 14.2 shows a schematic diagram of a two-range ammeter. With the switch open, the meter reads the current passing through it. However, when the switch is closed only a portion of the current passes through the meter. The balance of the current passes through the resistor labeled R_1. If the resistance of R_1 is chosen equal to 1/9 the microammeter resistance, it follows that most of the current flows through it and the current passing through the ammeter is 1/10 the total current. R_1 therefore causes the meter to be able to respond to 10 times its normal current, and the reading on the meter is simply multiplied by 10 to find the total current in the ammeter circuit. With application of the proper scale factor we can read 10 times the original current. Thus the ammeter indicated has two different ranges.

FIGURE 14.2
Two-range ammeter

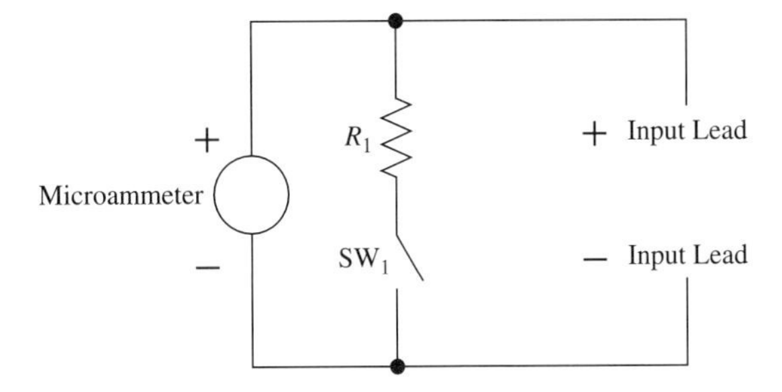

EXAMPLE 14.1

For the ammeter circuit of Fig. 14.2, the microammeter has a full-scale deflection of 50 μA and a resistance of 1500 Ω. What should the value of R_1 be if the full-scale reading of the meter is to be 1 mA when the switch is closed?

Solution If the full-scale deflection is to be 1 mA = 1000 μA, then the current through R_1 should be 1000 μA − 50 μA = 950 μA. At full-scale deflection the voltage across the meter is equal to 50 μA times 1500 Ω = 0.075 V. From Ohm's Law we find

$$R_1 = 0.075/0.000950 = 78.95\ \Omega \text{ (approximately)}$$

Ammeters are polarity-sensitive devices. As such their inputs are marked positive and negative (or common). When placed in series with an element whose current you are trying to measure, the positive lead of the ammeter always goes closer to the positive terminal of the power supply. Current passing through the ammeter must pass from the positive terminal of the meter, through the meter, and then out the negative terminal. Failure to hook the ammeter up with the proper polarity will cause the pointer to try to move in the downscale direction, and the meter will not properly indicate the current through the circuit.

The ammeter that has just been described is called an analog measuring device because the amount of current is indicated by a moving pointer on a scale and the angle the pointer moves through can have any value, not just discrete values. Many inexpensive ammeters use this type of meter because it is inexpensive. Modern ammeters produce a reading that is digital rather than analog. One big advantage of a digital ammeter is that it indicates the current with a negative sign when the meter is accidentally hooked up backward in a circuit. The negative sign merely indicates that the current is flowing in the opposite direction to that assumed. One must be very careful in putting an ammeter into a circuit that it is hooked up in the right way. ***In order to read the current through an element in a circuit it is important to place the ammeter in series with the element whose current is desired. Remember:*** An ammeter ***always*** goes ***in series*** with the element whose current is being measured.

14.3 DC VOLTAGE MEASUREMENTS

The dc microammeter just described can be used in series with a resistor to make voltage measurements. When placed across an unknown voltage, the microammeter measures the current through the combined resistance of its series resistor and the microammeter resistance. Since we know the current and the total resistance, we know the value of the unknown voltage that the meter is connected to by multiplying the current reading of the meter times the total combined resistance of the meter and its series resistor. With full-scale current passing through the microammeter, the voltage across the series combination is equal to the product of the full-scale current reading of the meter and the total resistance of the combination. This reading is called the ***full-scale reading*** of the voltmeter. Through the use of the series resistor the same meter can be used to read both current and voltage. Figure 14.3 shows the basic schematic diagram of a two-range voltmeter based on a microammeter and resistor series combination.

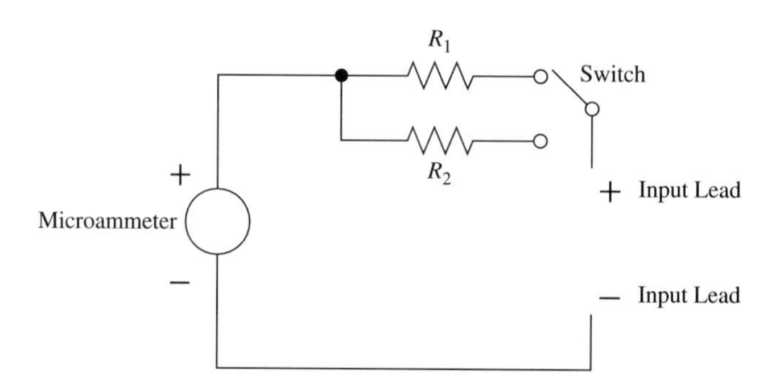

FIGURE 14.3
Two-range voltmeter

Figure 14.3 shows the voltmeter circuit. With the switch in the position to connect the positive input lead to the series combination of R_1 and the microammeter, the voltage needed to produce a full-scale meter reading is given by:

$$V_{\text{full scale}} = I_{\text{full scale}}(R_m + R_1) \quad \textbf{(Eq. 14.1)}$$

where $I_{\text{full scale}}$ is the full-scale current rating of the meter and R_m is the internal resistance of the microammeter.

If the switch is placed in the other position, the R_2 is in series with the meter and the full-scale reading is given by:

$$V_{\text{full scale}} = I_{\text{full scale}}(R_m + R_2) \quad \textbf{(Eq. 14.2)}$$

Through proper choice of R_1 and R_2, two independent voltage measurement scales can be created.

To read the voltage across any element in a circuit, the voltmeter is placed in parallel with the element. The microammeter can be calibrated in terms of the voltage across its terminals by using the equation:

$$V_{\text{indicated}} = I_{\text{indicated}}(R_m + R) \quad \textbf{(Eq. 14.3)}$$

where I is the indicated current reading of the microammeter and R is the value of the series resistor for the circuit. (For the circuit of Fig. 14.3 that is R_1 or R_2, depending on the switch setting.) ***Notice that the operation of the voltmeter circuit requires that some current formerly flowing through the circuit under test is diverted to flow through the meter.*** If the resistance of the voltmeter circuit is too low compared to the resistance of the circuit elements that it is connected across, then the reading of the voltmeter will be in error. ***For proper operation of the voltmeter circuit, the voltmeter resistance should be at least 100 times as large as the value of the resistance of the elements that it is in parallel with.***

Typical values for the resistance of a voltmeter circuit depend on the voltage range that the meter is set on. They usually are in the range of 20,000 Ω per volt of full-scale reading. Thus a meter with a full-scale reading of 10 V would have a resistance of 200,000 Ω. Notice that as the full-scale voltage increases, the resistance of the voltmeter increases.

EXAMPLE 14.2

A two-range (0–2 V and 0–10 V) voltmeter is to be built using the circuit of Fig. 14.3. If the microammeter has a full-scale deflection of 30 μA and a resistance of 2000 Ω, what should be the value of the two resistors needed for the two ranges?

Solution For the 2-V range: If 2 V is to allow a current of 30 μA to flow in the circuit, then the resistance of the voltmeter should be:

$$R_T = R_m + R_1 = 2\ \text{V}/(30\ \mu\text{A}) = 66{,}666\ \Omega$$

$$R_m = 2000\ \Omega \quad \text{therefore} \quad R_1 = 64{,}666\ \Omega$$

For the 10-V range the current is once again 30 μA and therefore:

$$R_T = R_m + R_2 = 10/(30\ \mu\text{A}) = 333{,}333\ \Omega$$

from which

$$R_2 = 331{,}333\ \Omega$$

Modern digital microammeters require much smaller measurement currents than analog-type microammeters. As a consequence, the digital style of voltmeter uses a high-input-resistance microammeter and a voltage divider to make measurements. A typical circuit diagram is shown in Fig. 14.4. This diagram is a simplified version of the real circuit. It helps us understand the operation of the modern digital voltmeter.

With the switch as shown in Fig. 14.4, R_1, R_2, and R_3 form a voltage divider. The current into the box labeled High-Input-Resistance ammeter is a small fraction of the input current. Since the resistance of the divider is much lower than the microammeter resistance, the microammeter does not affect the division ratio of the divider. Through proper choice of resistors, multiscale voltmeters can be constructed. The voltage at the input is equal to the reading on the meter multiplied by the reciprocal of the division ratio. If the division ratios are set to powers of 10, the reading on the digital meter can be adjusted by lighting the proper decimal point on the display. Because the voltage divider is made up of a set of resistors, the input resistance of the meter is constant and independent of

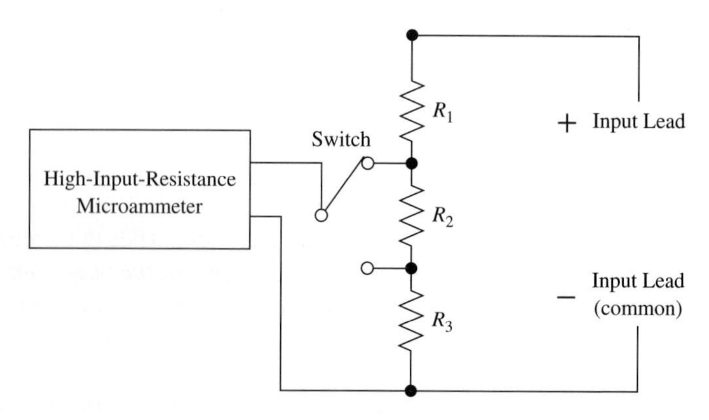

FIGURE 14.4
Modern digital voltmeter circuit

the scale being used for reading. Modern digital voltmeters have an input resistance of 10 MΩ independent of the scale being used. This high value is usually enough to prevent interaction with circuits under test.

EXAMPLE 14.3

Figure 14.4 shows a two-range digital voltmeter circuit. If the input-resistance of the voltmeter is 10 MΩ and its microammeter range is 0–200 mV, find R_1, R_2, and R_3 such that the input resistance of the meter is 10 MΩ and the scales are 0–10 V and 0–100 V.

Solution For the 10-V range, the division ratio must be chosen such that when there is 10 V on the input of the divider, the output voltage will be 200 mV. Thus $(R_2 + R_3)/(R_1 + R_2 + R_3)$ must equal 1/50. But since $R_T = R_1 + R_2 + R_3$ must equal 10 MΩ, it follows that $R_2 + R_3 = 200{,}000\ \Omega$ and $R_1 = 9.8\ \text{M}\Omega$.

For the 100-V range, R_3 must be 1/500 times 10 MΩ, or $R_3 = 20{,}000$ ohms. Thus $R_2 = 180{,}000\ \Omega$.

14.4 AC MEASUREMENTS: VOLTMETERS AND AMMETERS

Instruments used for ac measurements operate by adding a conversion circuit to the basic dc instrument. The simplest conversion circuit is a rectifier circuit, which converts the ac measurement into a related dc measurement. More expensive instruments use a circuit that generates a dc signal proportional to the true rms value of the ac input. Most ac meters assume that the input waveform is a sinusoid and the meters are arranged to display the rms value of the ac sinusoid. ***If the ac signal being measured is not actually a sinusoidal wave, then the readings indicated for the ac signal may be completely valueless. All ac voltmeters and ammeters are calibrated in rms of an equivalent sinusoid rather than peak value of the ac waveform.***

The problem with ac measurements is that the reading is very sensitive to the shape of the waveform. In the case of nonsinusoidal voltages the readings have little or no meaning. Even the use of a more expensive meter with a true rms converter provides little real information about the waveform and its properties. Multimeters that read both ac and dc values are useful as a tool if the waveform is sinusoidal, but they provide poor accuracy compared to other methods that can be used to measure ac properties of circuits.

14.5 RESISTANCE MEASUREMENT CONCEPTS

All linear circuits possess values of resistance, inductance, and capacitance, which control their operation. These circuit values are independent of the voltages placed on the circuit and are a consequence of the way in which the circuit elements making up the circuit are assembled as well as their values. When testing a circuit, the first measurements made should be the values of resistance measured at each node with respect to a reference node (generally chosen as circuit ground). By comparing the measured and expected values, many variations in the properties of the circuit will be detected. Wiring errors and component value errors will result in readings that do not match the expected readings. Only after the readings match the expected readings should a circuit be energized. In this way damage to the circuit may be avoided and many simple errors may be detected. Manufacturers of electrical equipment generally include resistance values for each node of the circuit so that problems in defective equipment may often be detected by measuring the circuit resistance properties. In this way the circuit can be tested without power being applied.

14.5.1 Measuring Resistance: Ohmmeters

Devices used to measure resistance are called ***ohmmeters.*** Since the value of measured resistance may vary over many orders of magnitude, ohmmeters generally are constructed with several ranges so that accurate readings may be obtained. The letter used to identify resistance in a circuit is a capital R. Thus, on the dial controlling the range of the ohmmeter, the letter R identifies a resistance range. Very often we find markings such as $R \times 1$, or $R \times 10$ on the range switch. These markings mean that ***the value of a measured resistance is determined from the meter reading by multiplying the meter reading by the factor following the letter R.*** For the examples noted here the indicated value of resistance is to be multiplied by 1 or 10 depending on the scale that the ohmmeter is set to. It is very important to remember to check the range of the ohmmeter when reading resistance so that the right multiplication factor is used in the measurement. Some modern test equipment uses the symbol Ω (the Greek letter ***omega***) as the symbol to mark the resistance scale rather than the letter R, since Ω is the international symbol used to symbolize ohms. In this case the range switch might be marked $\Omega \times 1$ or $\Omega \times 10$.

14.5.2 Analog Ohmmeters

A simple ohmmeter circuit consists of a dc voltage source V in series with a limiting resistor (R_2) and a shunted microammeter. The circuit for the ohmmeter is shown in Fig. 14.5. Shunt resistor R_1 is adjusted so that the microammeter has full-scale deflection when the leads of the ohmmeter are shorted together. This corresponds to a 0-Ω resistor connected between the test leads of the ohmmeter. When the leads are open, the meter reads zero current. Zero current corresponds to an infinite resistance between the open leads. The ohmmeter scale therefore indicates zero at the full-scale reading of the meter and infinity at the zero-current point of the meter. Values of resistance between zero and infinity are marked on the scale of the ohmmeter. Figure 14.6 shows a sample scale for the ohmmeter. Resistor R_1 is called the ***zero adjust control*** of the ohmmeter and not only adjusts for full-scale deflection but can be used to adjust for variations in battery voltage as the battery ages.

In operation the ohmmeter of Fig. 14.5 has its leads connected across an unknown resistance. Voltage source V causes a current to flow through the unknown resistance. A resistance value is read directly from the ohmmeter scale. This reading must be multiplied by the resistance factor indicated on the meter range switch to obtain the correct resistance value. It is important to realize that the circuit being measured must not be energized by any voltage source except that of the ohmmeter. ***If the circuit whose resistance is being measured is energized, the ohmmeter readings obtained are meaningless. In addition the ohmmeter may be damaged by the circuit it is being used to test.***

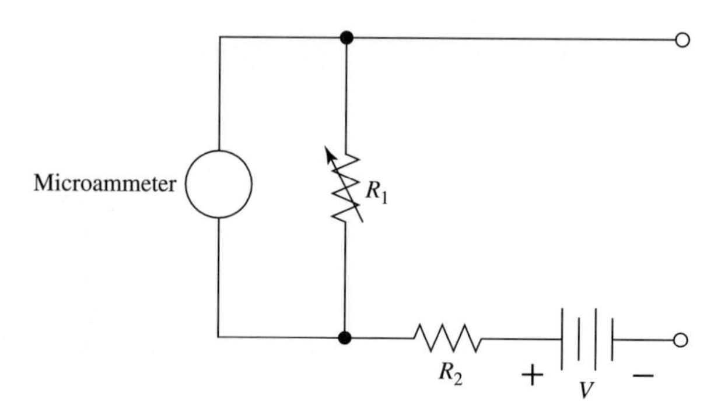

FIGURE 14.5
Simple ohmmeter

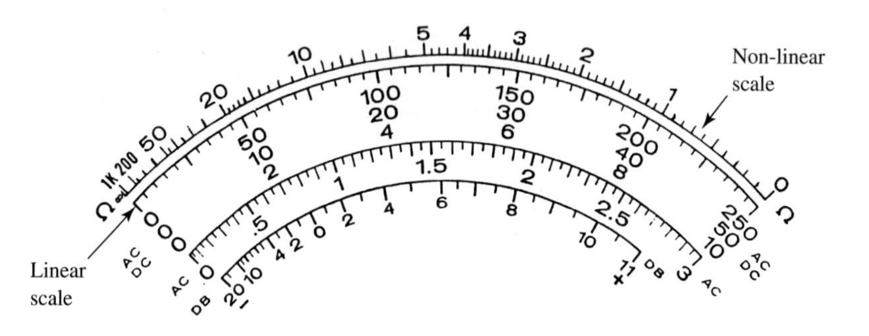

FIGURE 14.6
Ohmmeter scale picture

FIGURE 14.7

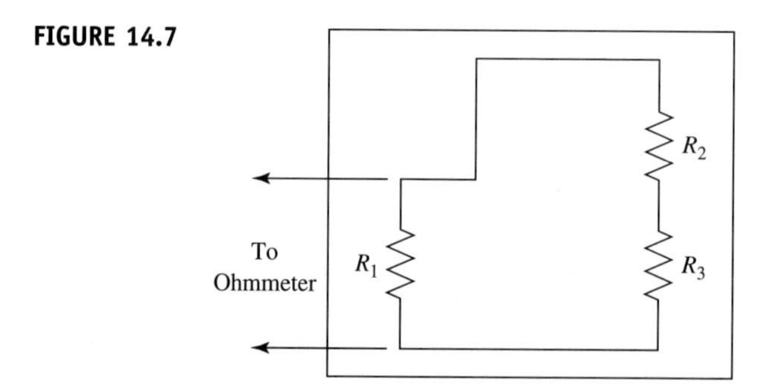

When the ohmmeter is used to measure the resistance in a circuit it measures all of the resistors connected between its terminals. Figure 14.7 shows a circuit containing three resistors R_1, R_2, and R_3. With the ohmmeter leads placed across R_1, the ohmmeter actually reads the resistance of R_1 in parallel with the series combination of R_2 and R_3. We must be careful in making resistance measurements so that readings are meaningful.

14.5.3 Linear Ohmmeter

The analog ohmmeter suffers from a combination of poor sensitivity and poor resolution. In an effort to improve the accuracy of ohmmeters, modern digital voltmeters take advantage of the high input impedance of the voltmeter module. Figure 14.8 shows a sim-

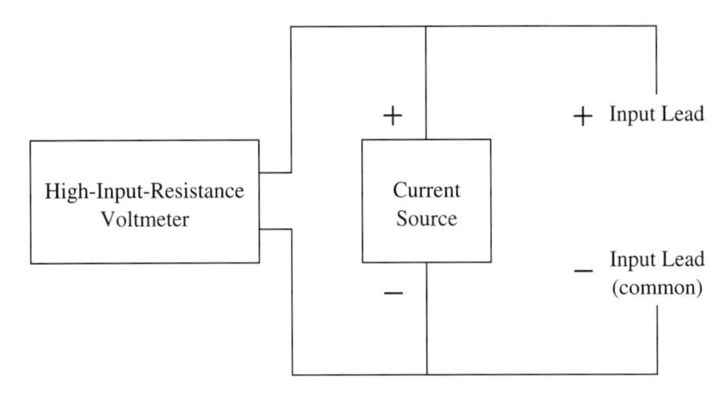

FIGURE 14.8
Linear ohmmeter

plified schematic of a linear ohmmeter. The schematic shows the voltmeter module of Figure 14.4 in parallel with a current source and the test leads.

When a resistor is placed across the test leads, the voltage developed across the resistor is equal to the product of the current from the current source and the resistance of the resistor. The voltmeter in parallel with the resistor reads the voltage across the resistor. If the current source is 1 mA and the resistance is 200 Ω, then the voltmeter reads 200 mV. This reading, which is actually a voltage reading, can be directly interpreted as 200 Ω. Through a series of current sources that differ by a factor of 10 from one another, the 200-mV reading can represent 2000 Ω, 20,000 Ω, and so on. One must merely arrange for the proper decimal point to be displayed along with the unit ohms, kilohms, and so on. Through the use of constant current sources, resistance is directly measured in terms of a voltage. When the leads of the ohmmeter scale are left open, the meter reads overrange because the voltage of the current source exceeds the full-scale reading of the voltmeter. This approach to measuring resistance results in an ohmmeter that can display small differences in resistance much more accurately than the analog type of voltmeter. In addition the voltage across the device being measured is always less than 200 mV.

14.6 MEASUREMENT CONCEPTS

All measurement instruments have at least two leads that connect to the circuit under test. These leads enable the instrument to measure the value of the circuit elements placed between the two leads. These leads are the eyes of the measurement equipment. Without them the instrument cannot make any measurements. For instruments used to measure dc properties the two leads are labeled positive (+) and negative (−). This convention is used to indicate the polarity of the dc voltage that will appear across the instrument in normal use. ***If the polarity of the voltage that appears across the instrument is opposite to the marked polarity, then the instrument may indicate a negative value for the reading or may fail to operate properly at all.*** The negative reading, if it actually appears, indicates that the assumption made for the measured signal polarity is incorrect. Reversal of the leads will result in a correct reading in those instruments. Unlike digital meters, analog meters do not operate when the leads are reversed because the meter is trying to move in the downscale direction and is stopped by a limit pin built into the meter.

Portable test instruments can be used to measure voltages across any two nodes in a circuit. Similarly, portable ammeters can measure current through any part of a circuit. By definition a portable piece of test equipment contains its own power source and is not connected to the ac power line. That is, it either needs no power source (for example, an analog voltmeter or ammeter derives its power from the circuit under test) or derives its power from its own internal power supply (usually batteries). Such meters are very useful because they can be used anywhere without the need for an ac power line.

14.7 VIEWING SIGNALS THAT VARY WITH TIME: THE OSCILLOSCOPE

The voltmeters and ammeters we have discussed are very useful to characterize waveforms whose properties do not vary rapidly with time. DC voltages and currents are time invariant. If we were to display these voltages and currents versus time, the display would be a straight line parallel to the x axis. A display of this type is shown in Fig. 14.9. The display does not provide any more information than the meters previously discussed—a dc voltage is completely specified by its amplitude and polarity.

If we display a waveform that changes with time, we can gain some information in addition to the information obtained from the voltmeters and ammeters. Figure 14.10 shows a voltage waveform displayed versus time. Notice that the waveform is not sinusoidal. This waveshape would not be visible on any of the instruments we have previously discussed, but the display reveals the nature of the waveform. The visual display makes it possible to measure the frequency of the waveform as a result of the time base representation along the x axis.

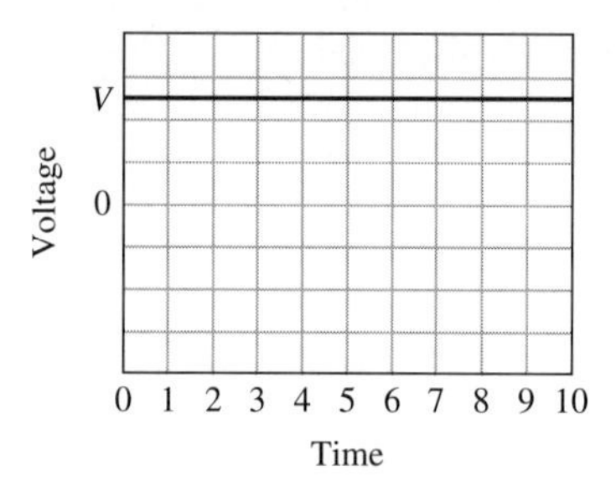

FIGURE 14.9

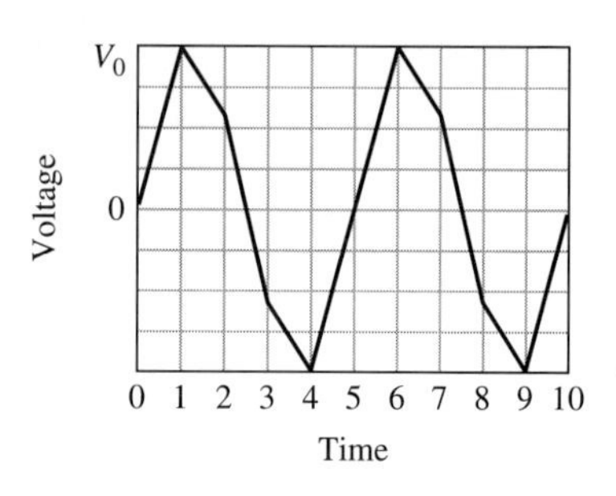

FIGURE 14.10

If we display two different waveforms on the same screen at the same time, we can measure the relationships between them. When the signals are an input and an output waveform, we can observe any difference between the two waves caused by the circuit that produces them. This information adds greatly to our knowledge of the operation of the circuit.

Often the waveforms being observed are really complex in structure. The instrument used to display waveforms is called an ***oscilloscope.*** Basically, an oscilloscope displays waveforms versus time on a display screen. It can be used to see even the smallest differences between closely related waveforms. In its normal mode of operation an oscilloscope displays time as the horizontal axis and the input signal voltage amplitude as a vertical axis. With this visual display we can see how the waveform changes with time. To understand how the oscilloscope produces its display we will consider the operation of each axis separately.

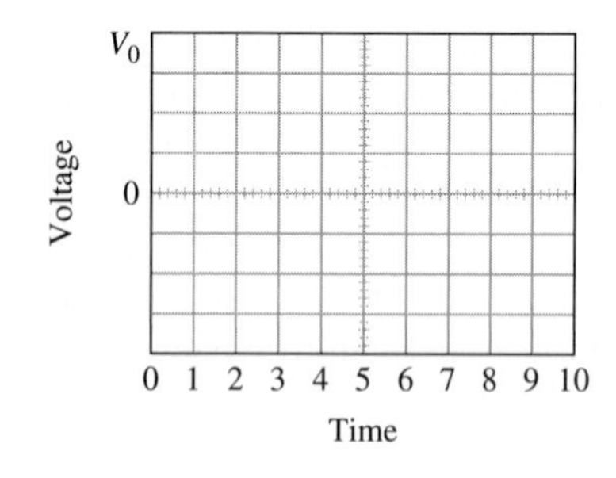

FIGURE 14.11

Figure 14.11 shows the display face of an oscilloscope. It has a display grid that is divided into centimeter boxes in both the horizontal and the vertical directions. Notice that the boxes are further divided along the x and y axes into five equal subdivisions. Thus each subdivision represents 1/5 division or 0.2 divisions. These subdivisions allow the user to read the x axis to 1/10 division, which results in an x axis (or time measurement) accuracy of approximately 1% of the full-scale measurement. Correspondingly, the y axis (or signal voltage) can be measured to 1.25% accuracy. The x and y axes intersect at the middle of the display (commonly called a display screen or screen). The display grid is built into the display device, which is usually a cathode ray tube (or CRT). The display grid can be illuminated for aid in reading results from the display grid.

The display screen is directly behind the display grid and is an integral part of the cathode ray tube. This tube produces a narrow beam of electrons that strike the inner front surface of the tube, which is coated with phosphor to form the display screen. Phosphor is a material that produces light when it is struck by electrons. The duration of this light when the electron beam is removed is dependent on the type of phosphor used. Some phosphors remain visible for a long time after the electron beam is gone; others remain for a very short time. Phosphors come in many different colors. An ordinary television set, for example, uses a cathode ray tube with three different colors of phosphors (red, green, and blue) to produce the picture that we watch on the screen of the cathode ray tube, which makes up the picture tube of our television set.

The CRT used for the oscilloscope contains two sets of metallic plates used to deflect the electron beam as it travels past them. These plates are arranged at right angles to each other. One set of plates can move the beam of electrons from left to right (horizontally), and the other set of plates can move the beam of electrons up and down (vertically). Any voltage placed across either set of plates deflects the electron beam toward the plate with the more positive voltage on it. The beam may be simultaneously moved up and down as well as right to left by voltages applied to each set of plates.

14.7.1 The *x* Axis (Time Base)

The oscilloscope contains a circuit that generates a triangular waveform as shown in Fig. 14.12. The waveform, called a sawtooth waveform, varies linearly with time. The rising portion of the sawtooth waveform is called the ***sweep time.*** After it reaches its maximum positive value, the waveform starts falling to its lowest value. This portion of the waveform is called the ***retrace time.*** This voltage waveform, when applied to the horizontal deflecting plates of the cathode ray tube, causes the electron beam to move or "sweep" across the display screen at a uniform rate. During the rising portion of the sawtooth the beam sweeps from left to right, and during the falling portion of the sawtooth the beam sweeps from right to left. The voltage level of the sawtooth applied to the horizontal deflection plates is set so that the resultant beam deflects the 10-cm length of the screen and covers the entire *x* axis region.

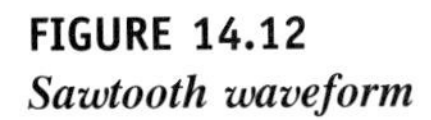

FIGURE 14.12
Sawtooth waveform

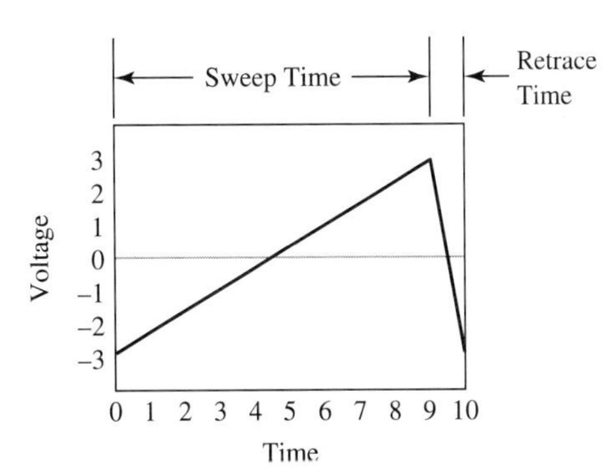

The oscilloscope contains a ***blanking circuit*** to eliminate or blank the electron beam during the time that the electron beam moves from right to left. This blanking circuit eliminates the beam during the right to left sweep, making the reverse trace invisible. Only the left to right sweep portion of the sawtooth produces a visible line on the face of the cathode ray tube. This sweep line moves across the screen from left to right at a speed determined by the sawtooth generator setting. The beam may move very slowly to trace out low-frequency waveforms or rapidly to trace out high-frequency waveforms.

14.7.2 The Triggering Signal: Synchronization

Consider a waveform that has a period of 0.001 s (1 ms). To properly display this signal, the beam must move across the screen in 1 ms. This is too fast for the eye to detect. To make the waveform visible we make use of its periodicity. The waveforms that we study repeat exactly every T seconds. We call the value of T the period of the waveform. Since the waveform is exactly the same from cycle to cycle, we can pick a point on the waveform and start the sweep at that point every time the waveform repeats. Under this condition, all of the waveform displays on the oscilloscope screen will be identical. Our eyes will not detect individual sweeps but rather the average effect of many identical sweeps. As it is in the movies, if the retrace rate is fast enough, the pattern viewed will not flicker from trace to trace. To create this uniform repetition of the input signal, a circuit is needed that examines the input signals and starts the sweep every time the waveform crosses the selected point on the waveform. The circuit that performs this action is called a ***triggering circuit.*** This circuit examines the voltage level and the slope of the input signal, and when the voltage and slope match the preset values the circuit generates a trigger signal that starts the sweep of the beam across the screen. The process of adjusting the sweep starting point is called ***synchronization,*** and the operator must adjust the synchronization so that the oscilloscope pattern displayed does not change, but rather stays fixed in its appearance.

14.7.3 Setting the Input Controls: The Oscilloscope in Operation

There are many different types of oscilloscopes, each one customized for the frequency of operation and types of signals to be observed. What follows here is a brief discussion of the controls on a typical oscilloscope. Refer to the operating manual of any instrument that you use for greater detail about the operation of an individual oscilloscope.

Oscilloscopes generally have two identical signal input channels labeled Channel 1 and Channel 2. These two channels are identical in properties. In operation the oscilloscope can be configured to display one channel at a time or both channels at the same time. This choice is controlled by a switch on the front panel of the scope marked Operating Mode. Among the modes available are:

Channel 1 only

Channel 2 only

Alternate Mode, which switches channels between sweeps. In this mode first one channel is swept completely and then the other is swept completely. The scope continuously alternates between the two channels.

Chopped Mode, which switches rapidly between the two channels and displays the results simultaneously

Add Mode, in which the two signals are added or subtracted from one another to produce the sum or difference of the two input signals

Each input channel has a range switch, which controls the gain of the input channel. Since the full-scale vertical deflection of the CRT is fixed, an amplifier is needed so that any size signal can be displayed at full-scale deflection on the screen. For example, if the deflection sensitivity of the CRT was 10 V/cm, an amplifier with a gain of 10 would be needed for a 1-V signal to appear as a 1-cm deflection. The range switch on each channel is fixed in 12 steps from 1 mV/cm to 5 V/cm in a 1–2–5 sequence. In addition, a vernier control allows any setting per division to be obtained in addition to the preset values. In actual operation it is important to remember that the vernier must be turned off to use any of the preset range switch settings.

In addition to the range switch, each channel has a switch that adjusts the frequency response of the oscilloscope to the input signal. This switch is labeled DC, AC, and Off. In the off position the input signal is removed from the channel and the channel input is grounded. This position is used to set the x axis location of the channel and to adjust the zero voltage level. The DC position of the switch causes the amplifier to respond to both dc and ac signals in displaying the waveform. Finally, the AC position allows just the ac voltage to be displayed on the screen. In the ac position only voltage changes affect the display. This allows small ac variations on large dc signals to be observed.

The highest voltage scale on an oscilloscope is 5 V/cm. The upper range of the oscilloscope can be increased through the use of a calibrated probe. These probes connect to the input of the scope and increase the range of the scope to 50 V/cm. The probes also increase the input impedance of the scope from the standard 1 MΩ to a 10-MΩ input impedance. Auxiliary probes are also available to measure current as well as increase the input range to 500 V/cm.

Scopes are generally single-ended input devices. That is, the input signal is developed between the channel input and ground. This means that the two input channels share a common ground. ***This is important because it means that the ground terminals on the two input channels must be connected to the same point. Failure to understand this ground connection may result in erroneous readings. In addition, damage to the circuit under test or to the scope or both may result from grounding errors.***

14.7.4 Setting the Time Base Controls

The speed with which the electron beam moves across the display screen is set in seconds per division, and like the gain controls, the settings are grouped in factors of 1, 2, and 5. The scope can have values of speed per division that range from as slow as 5 s per division to as fast as 100 ns per division or faster. ***Like the gain-setting controls, the trace speed can be adjusted with the use of a potentiometer to be anything within the limits of the machine that the operator desires. Also, as with the gain-setting controls, one must be sure that the potentiometer is in the calibrated position before any readings of time are taken from the oscilloscope screen.***

In addition to the normal sweep mode, scopes have a switch that will expand the sweep by a fixed factor of 10. This switch is usually placed on the horizontal position control. With this control, the effective sweep width is expanded by a factor of 10 so that small parts of a complex waveform can be examined more closely for their fine structure.

Once the sweep speed and input channel selectors are set, the triggering signal must be adjusted so that the pattern observed is stationary and does not vary from cycle to cycle. For most signals there is a preset level close to the zero voltage level of the signal that can be used to trigger properly. It only requires a setting of the desired input signal slope to finish the adjustment of the trigger signal circuitry. All scopes come with many modes of trigger operation. One can either experiment to find the right trigger setting and mode or refer to the Oscilloscope Operating Manual for the properties of the trigger signal modes for the given instrument.

The operation of test equipment to observe the properties of a circuit is straightforward if the operating concepts of the test equipment are known. To fully understand the measurement process takes a great amount of hands-on practice. There is no substitute for this kind of practice. The details discussed here are just a small amount of the information available on test equipment. Students interested in learning more about test equipment should read all of the operating manuals that come with the instruments. This will strengthen their understanding of the measurement process.

14.8 CIRCUIT ANALYSIS SOFTWARE

Many good commercial software packages currently on the market allow us to utilize the extensive capabilities of the PC to analyze electric circuits for us. One such software package is Electronic Workbench (EWB).

EWB is particularly suited for not only analyzing electric circuits but doing so in a manner that simulates the procedure actually used in laboratories. This is because the circuits drawn in the EWB environment include laboratory instruments in their setup. This gives us the ability to see how these instruments must be connected in our circuit in order to provide us with the desired results.

Accordingly, the rest of this chapter will be devoted to utilizing EWB in the analysis of specific circuits. In doing so, we will be utilizing the proper procedure for connecting the instruments discussed in the first part of this chapter into a circuit.

14.8.1 Analysis Using EWB Version 4.1

The EWB version used in this text is version 4.1. Currently, EWB version 5.0 is being made available to the public. The difference lies primarily in the GUI (Graphical User Interface) used and analysis capabilities. The circuits and instrumentation are virtually the same. The descriptions that follow pertain directly to EWB version 4.1. Each one may be extrapolated to version 5.0 with little modification.

14.8.2 EWB Version 4.1 Techniques

To demonstrate the use of EWB version 4.1, we will utilize some of the circuits analyzed in previous chapters using classical circuit analysis techniques. Combined with the rules for connecting instruments into a circuit described in previous sections of this chapter, the student will obtain a good background into the use of circuit simulation as a valuable tool for analysis.

The first step in using EWB is to activate the software by clicking on the EWB icon in Windows 3.1 or on the EWB program listing in Windows 95. This gives us an open window in which we can create the circuit we wish to analyze.

Circuits are created by obtaining the components and instruments we need from "parts bins," so named to reflect the actual terminology used in laboratory storage areas. Each parts bin is activated by clicking on its associated button in the parts bin toolbar. This causes all the parts contained in that bin to be displayed vertically on the left side of the window.

A part is obtained by clicking on its button, holding down the mouse button, and dragging it into the area where we will create the circuit. Sources and instruments are similarly obtained from their respective parts bins.

Some instruments are not contained in any parts bin, typically because they are larger and are kept in inventory in much fewer numbers than parts. The oscilloscope and function generator are two such instruments. Note that the buttons representing these instruments are located on the top of the window. Only one of each of these instrument types is available for a given circuit. This is in contrast with ammeters and voltmeters, which may be used in any numbers.

Once the components, sources, and instruments are placed in the window, they must be connected with "wires." Wires are drawn by pointing the mouse to the terminal of one component, where a dot representing a connector appears when the mouse button is clicked and held down, and dragging the mouse to the terminal to which the wire is to be terminated. When the dot representing the connector of the second terminal appears, the mouse button is released and the wire remains. Each dot, or connector, can support four wires. If additional wires are needed, a connector may be obtained from the parts bin and inserted into the circuit just as an additional part.

A given instrument may be enlarged by double clicking on it. This enables us to get a better view of a waveform or a reading. To move a component to a different part of the circuit, double click on the component, hold down the mouse button, and drag the component to its new location. To modify the label or value of a component, double click on the component. This causes a dialog box to appear. Follow the directions in the dialog box to obtain a new value. Let us now investigate some specific circuits.

14.8.2.1 A DC Series Circuit

Consider the dc series circuit of Fig. 5.13(a), which is reproduced here as Fig. 14.13(a). The current, I, and the voltage across each resistor, V_1, V_2, and V_3, were found in Chapter 5 by using classical circuit analysis techniques. To verify the solutions found in Chapter

5 we could set up this circuit in the laboratory and check these values using an ammeter and voltmeters. Instead we will save time and money by using EWB to check these values for us.

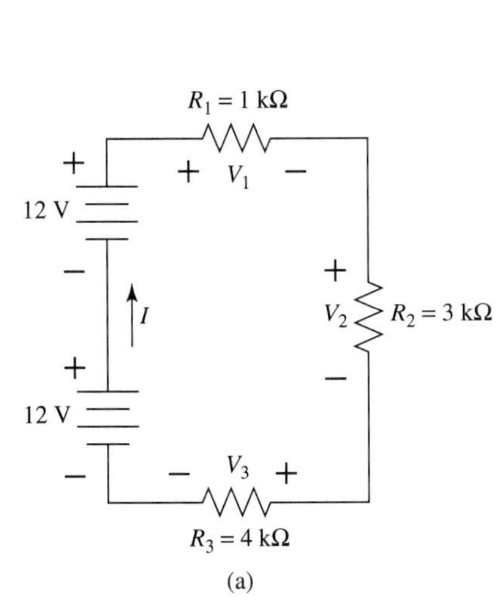

FIGURE 14.13A
(a) Circuit of Fig. 5.13

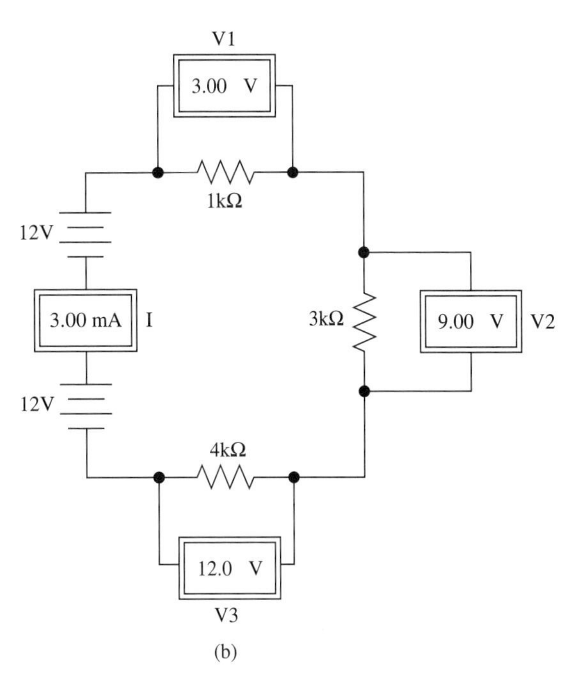

FIGURE 14.13B
(b) EWB solution of circuit in part (a)

Note that the ammeter is connected in ***series*** with the circuit so that the circuit current passes through it. The voltmeters, however, are each connected in ***parallel*** with the voltage to be measured. Recall that the dc ammeter and voltmeter are polarity sensitive. This means that the voltmeter must be connected so that its positive terminal is connected to the positive terminal of the voltage being measured and its negative terminal is connected to the negative terminal of the voltage being measured. The ammeter must be connected so that the current flowing enters the ammeter at its positive terminal and leaves the ammeter at its negative terminal.

If the ammeter, or any voltmeter, were connected with the terminals reversed, the resulting value would be indicated with a negative sign. This would be accurately indicating that the voltmeter was measuring the voltage of the lower-polarity terminal (negative) with respect to the higher-polarity (positive) terminal.

Read the voltages and current from the voltmeters and ammeter connected into the circuit. Compare these values with those obtained in Example 5.10. Note that these values are the same.

14.8.2.2 An AC Series Circuit

Now let us analyze the circuit of Fig. 5.14(a), which is redrawn in Fig. 14.14(a). The sinusoidal current in this circuit and the voltage across each element were found by classical circuit analysis. We will now find these same parameters using EWB.

The EWB circuit, including all the instrumentation, is shown in Fig. 14.14(b). Note that in this circuit we are using an ac ammeter (multimeter) to read current and an oscilloscope to read the voltage across the capacitor.

To compare these results with those obtained in Example 5.11 note the following:

1. The multimeter reads 869 mA. As in the case of actual multimeters, this reading is the *rms* value of the current. To convert this to ***peak*** value we must multiply by the square root of 2 giving us 0.869 × 2 or 1.23 A. Note that this is the same as the peak value obtained in part (a) of Example 5.11.

2. The peak value of the capacitor voltage can be read directly off the oscilloscope waveform. At 5 V/div, the peak value is found by multiplying 5 by 2.45 div, or 12.25 V, which again compares favorably with the results of Example 5.11.
3. The frequency of the capacitor waveform may be determined by reading the time scale as 5 ms/div and multiplying this by the number of divisions in one period. This is simpler to do if we zoom the oscilloscope waveform as shown in Fig. 14.14(c).

In this figure, the markers T_1 and T_2 give us the difference between two consecutive peaks as 12.56 ms. The period is therefore 12.56 ms and the frequency is 1/(12.56 ms), or 79.6 Hz.

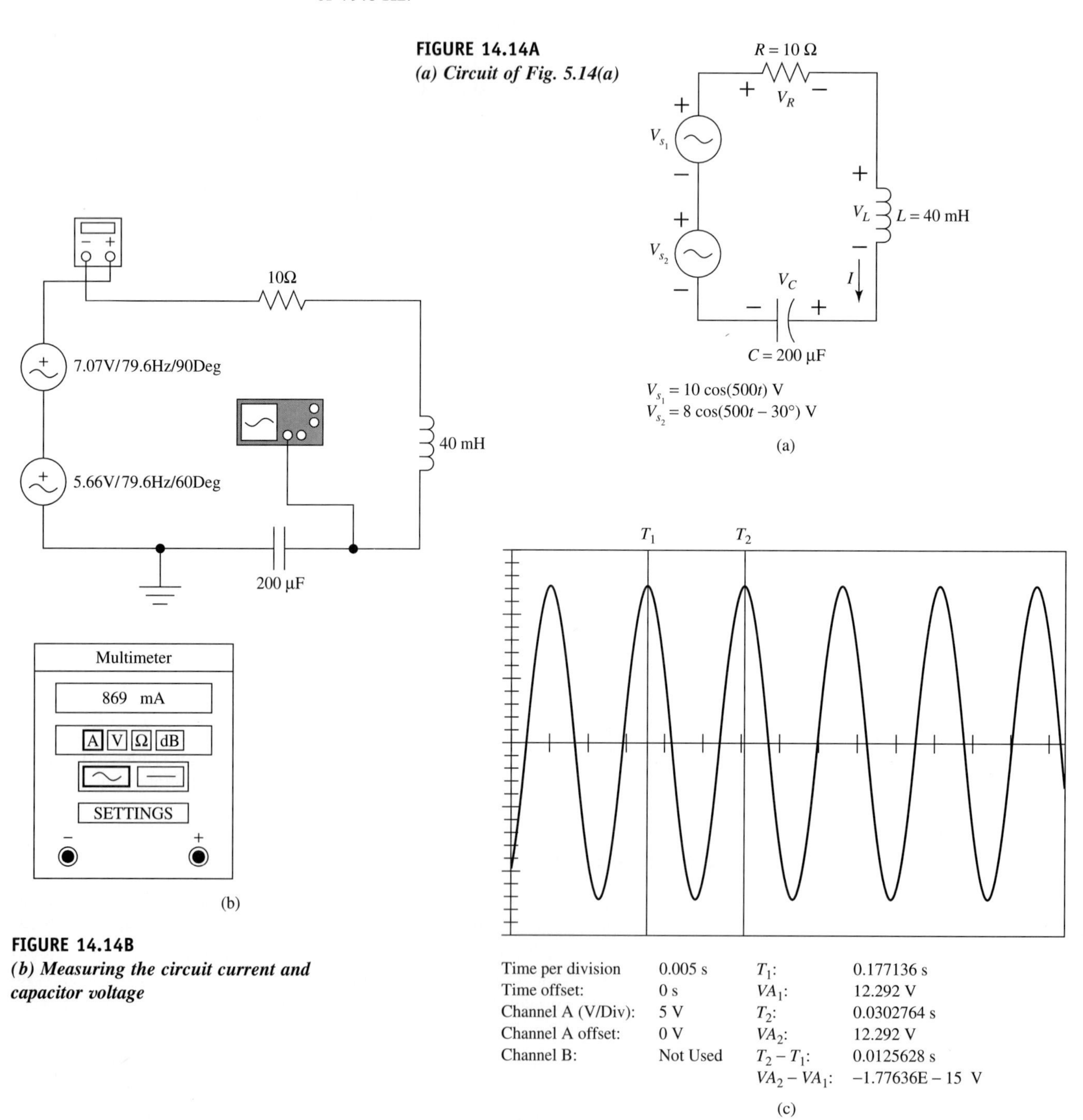

FIGURE 14.14A
(a) Circuit of Fig. 5.14(a)

FIGURE 14.14B
(b) Measuring the circuit current and capacitor voltage

FIGURE 14.14C
(c) Expanding the oscilloscope waveform—voltage across capacitor

14.8.2.3 A Series–Parallel AC Circuit

The series–parallel circuit of Fig. 6.7(a) is reproduced here as Fig. 14.15(a). In Example 6.7, the voltage v_{cd} was calculated. We will now "measure" this voltage using EWB. In the circuit of Fig. 14.15(a), $v = 36 \sin(10{,}000t)$ V, $R_1 = 4$ kΩ, $R_2 = 6$ kΩ, $R_3 = 10$ kΩ, $R_4 = 12$ kΩ, $R_5 = 2$ kΩ, $L_1 = 200$ mH, $L_3 = 1$ H, $C_2 = 0.02$ μF, $C_4 = 0.01$ μF, and $C_5 = 0.04$ μF. The EWB representation of this circuit is shown in Fig. 14.15(b).

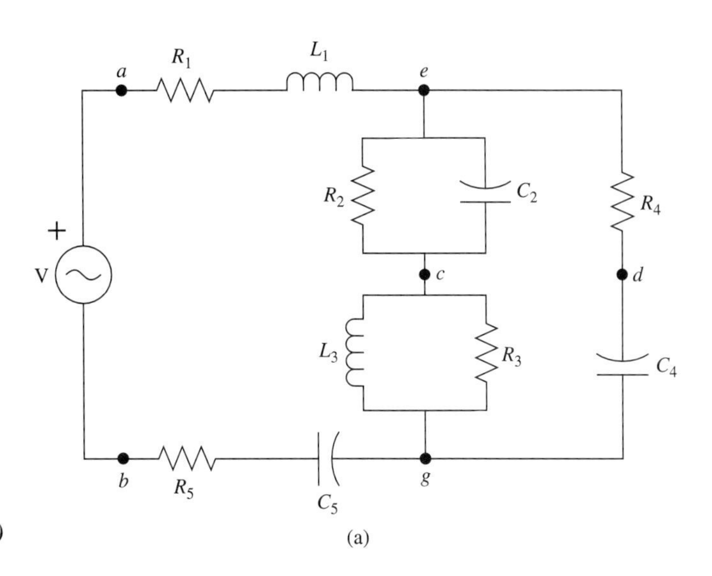

FIGURE 14.15A
(a) Circuit of Fig. 6.7(a)

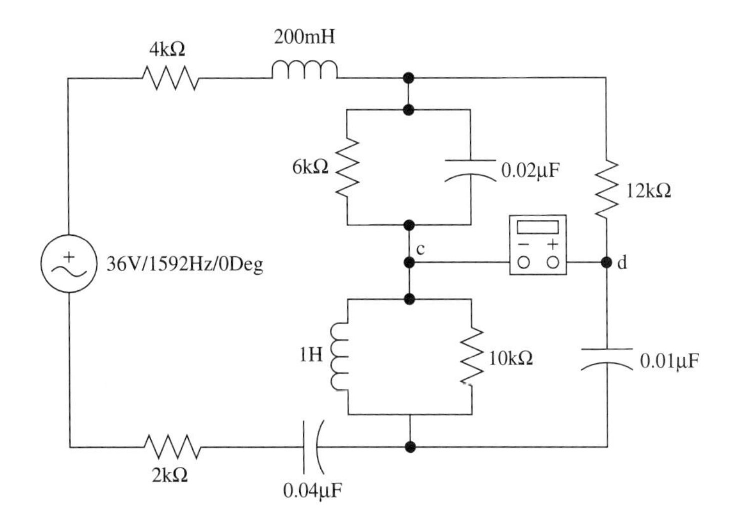

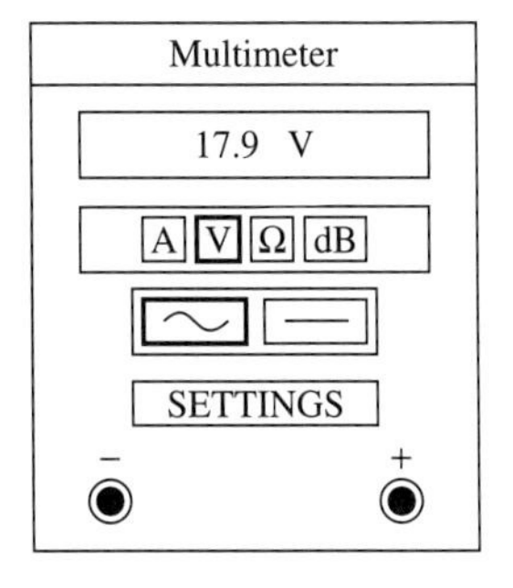

FIGURE 14.15B
(b) Measuring the voltage from c to d in part (a)

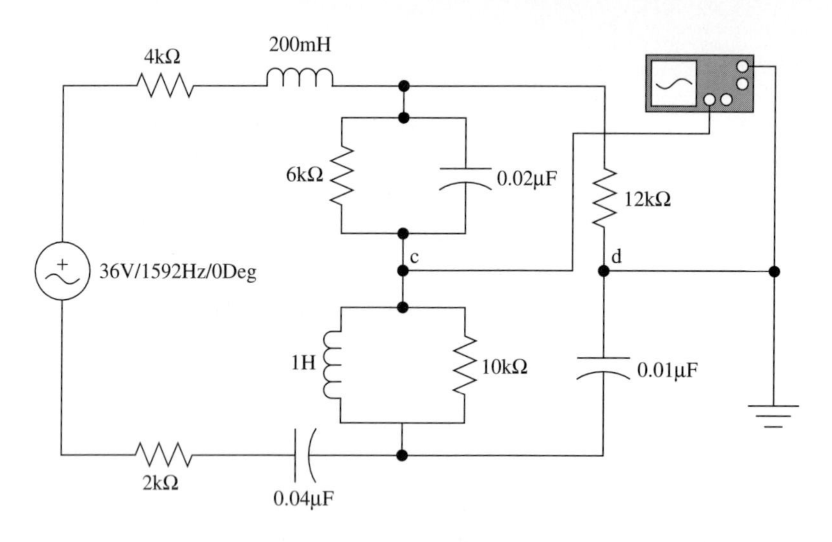

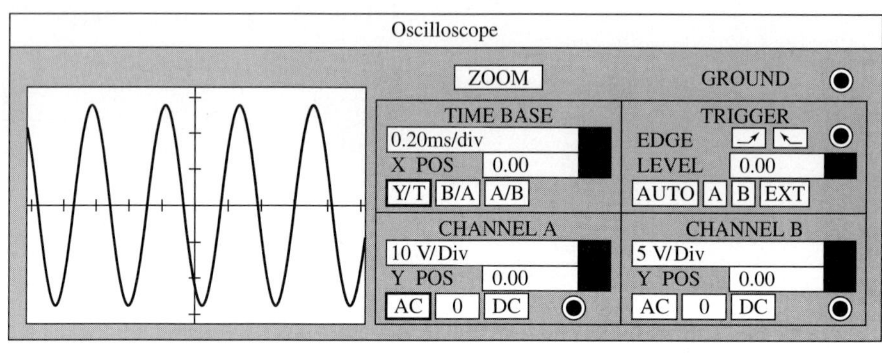

(c)

FIGURE 14.15C
(c) Measuring the voltage from c to d in part (a) using an oscilloscope

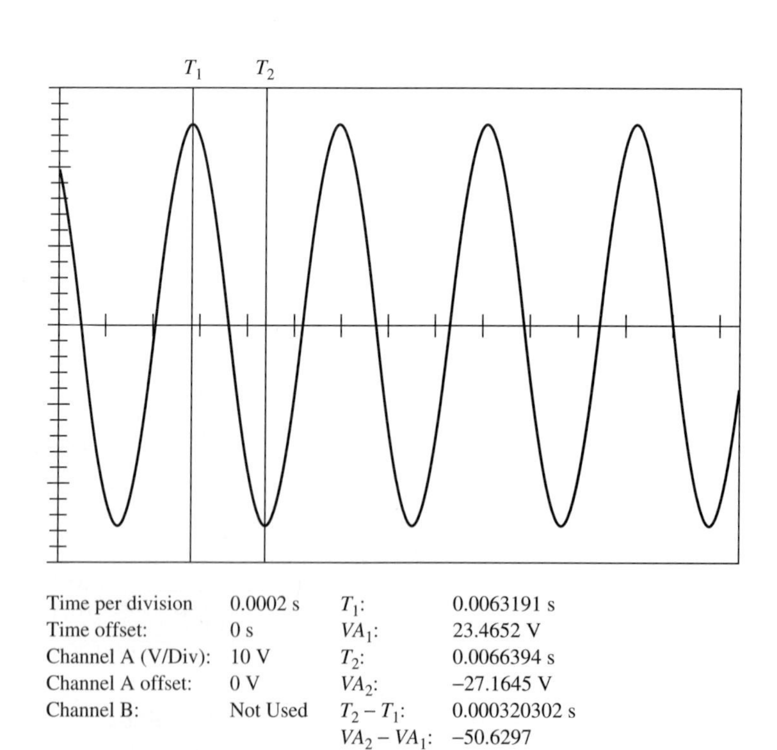

(d)

FIGURE 14.15D
(d) Oscilloscope response

A multimeter is used in this circuit to measure the desired voltage. Note that the multimeter reads 17.9 V. Since the multimeter reads rms values, we must convert this to peak value by multiplying by the square root of 2. This yields a peak value of 25.3 V. This agrees with the value found in Example 6.7.

To view the voltage v_{cd} on an oscilloscope, we can use the circuit of Fig. 14.15(c). The "zoomed" response is shown in Fig. 14.15(d). Note that the peak to peak value can be read from the zoomed response by using the two markers, 1 and 2. The peak to peak value is shown as $V_{A_1} - V_{A_2} = 50.6$ V. Dividing by two yields the same peak voltage of 25.3 V.

14.8.2.4 A Superposition Problem

The circuit of Fig. 7.9(a) was solved using the superposition theorem in Example 7.13. The circuit is reproduced in Fig. 14.16(a). In this figure, we will find the current I_x using EWB.

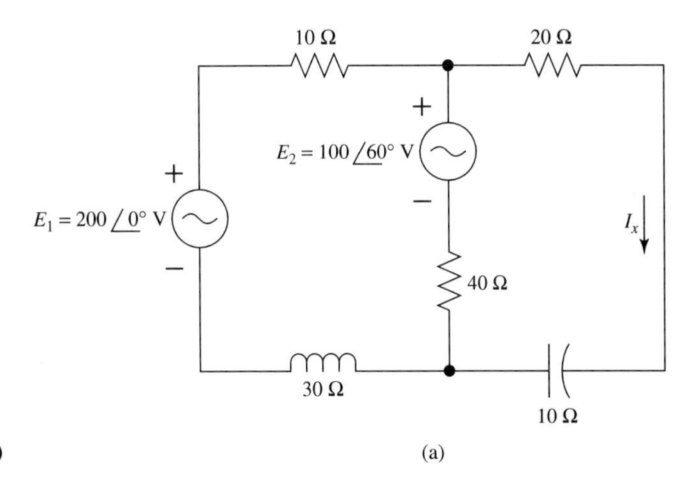

FIGURE 14.16A
(a) Circuit of Fig. 7.9(a)

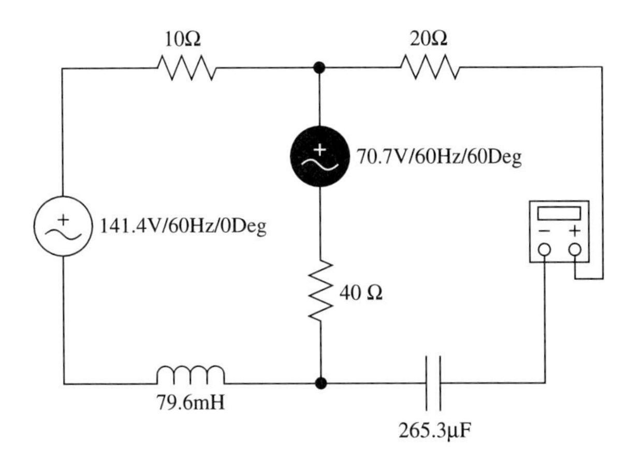

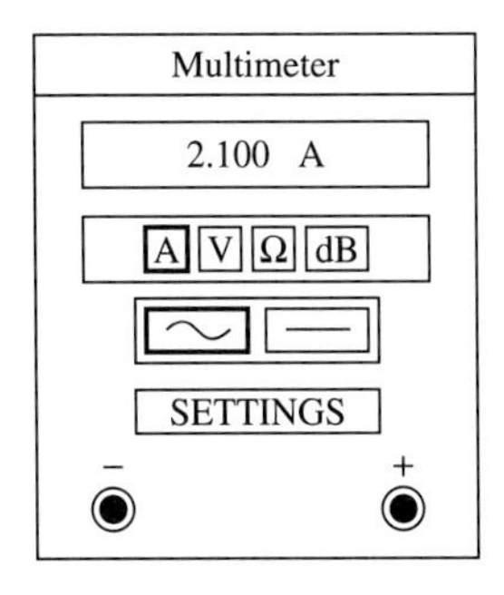

FIGURE 14.16B
(b) Using EWB to find the current I_x in part (a)

The EWB circuit is shown in Fig. 14.6(b). A multimeter is used to "measure" the current I_x. The multimeter reads 2.1 A. In comparing this result with that obtained in Example 7.13 note the following:

1. The peak value of $I_x = 2.94$ A obtained in Example 7.13 can be found from the 2.1 Arms reading of the EWB multimeter by multiplying 2.1 by the square root of 2.
2. The values of the voltages shown in the EWB circuit are rms values. The 141.4 Vrms value of one voltage source corresponds to the 200-V peak value shown in Fig. 14.16(a). The 70.7-Vrms value of the other voltage source corresponds to the 100-V peak value shown in Fig. 14.16(a).
3. The values of the capacitor and inductor shown in the EWB circuit were calculated to yield the reactances of 10 Ω and 30 Ω shown in Fig. 14.16(a) at the source frequency of 60 Hz. Since the frequency of the voltage sources used in EWB must be specified, the value of any reactive element must also be specified.

14.8.3 Other EWB Capabilities

EWB can be used for analyzing many circuits that are beyond the scope of this book: semiconductor circuits, integrated circuits, and digital circuits, among others. Further, curves of amplitude versus frequency, known sometimes as Bode Plots, can be obtained using EWB.

■ SUMMARY

- Measurement equipment is used to confirm that circuits are operating in accordance with their designed properties.
- To measure a current flowing in an electrical circuit we use a device called an ammeter. The device responds to the current flowing through it. An ammeter that measures thousandths of an ampere is called a milliammeter. A device that measures millionths of an ampere is called a microammeter.
- The earliest type of ammeter measured current by using the magnetic interaction between a coil of wire and a permanent magnet. This meter used a calibrated scale to indicate the amount of current flowing through the meter.
- The ammeter can be used to measure dc voltage in a circuit by letting a current flow through a known resistor value. The current that flows is related to the size of the voltage by Ohm's Law for the resistance element.
- Modern electronic measuring equipment uses electronic circuitry to perform the same measurements more rapidly and more accurately. These devices are called digital voltmeters and ammeters.
- To measure ac we first convert the ac waveform to a dc value and then measure the DC value.
- Resistance properties of a circuit are measured using a device called an ohmmeter. This device, which is self-powered, is employed when the circuit is de-energized.
- Ohmmeters operate by putting a known voltage across a circuit and seeing how much current flows in response. The smaller the current is, the larger the resistance. The scale of the ammeter is calibrated with the aid of Ohm's Law. Each value of current corresponds to a unique value of resistance.
- Modern ohmmeters place a known current through a resistance and measure the voltage developed across the circuit. The voltage is directly related to the value of resistance in the circuit.
- Time-varying signals are measured with the aid of a device called an oscilloscope. This device uses a display panel to show how the signals in a circuit vary with time.
- In addition to the available test equipment for measurement, there is software that can simulate the circuit operation. One such piece of software is called Electronic Workbench (EWB). This software makes it possible to test a circuit without ever actually building it first. This speeds up the design and analysis process.

APPENDIX **A**

THE LAW OF MAGNETIC INDUCTION

The law of magnetic induction may be stated as follows:

> An emf (electromotive force) is induced in a winding, or coil, of N turns whenever the magnetic flux lines linking the winding in a direction that is perpendicular to the area of the winding varies with time.

If e is the electromotive force in volts, N is the number of turns in the winding, ϕ is the magnetic flux perpendicular to the winding area in Wb/s and t is time in seconds then:

$$e = -Nd\phi/dt \quad \textbf{(1)}$$

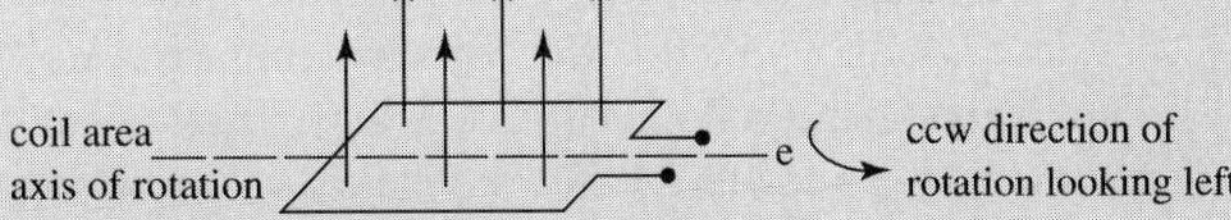

FIGURE A.1
Illustration of the law of magnetic induction.

Equation 1 is known as Faraday's law of induction. Fig. A.1 shows the spatial relationship between ϕ, the area of the coil and direction of rotation that yields an electromotive force e across the two terminals shown.

The significance of this law is that the induced voltage, e, is a function of the ***rate of change*** of the magnetic flux that "cuts" the coil. It does not matter whether the flux is changing because:

(a) The coil is physically moving in a stationary magnetic field or
(b) The coil is stationary but the magnetic flux is changing with time.

The induced emf appears whenever there is ***relative motion*** between the magnetic field and the coil regardless of whether the relative motion is produced as a result of (a) or (b) above.

Condition (a)

Recall that a magnetic field is established whenever an electric current flows. If the current is constant, as in the case of dc, then the magnetic flux lines will be constant and unchanging. The only way an emf can be induced in a coil by this magnetic field is if the coil is moving physically with respect to the field.

This is the situation that exists in a rotating electrical generator. The magnetic field is created by a dc current in a stationary winding located on the *stator,* or the stationary housing of the generator. This dc winding is generally called the "field" winding since its function is to create the magnetic field. The magnetic lines of flux are therefore stationary.

Another winding is physically located on the rotating portion of the generator, or *rotor.* This winding is known as the "armature winding" since the rotating portion of the generator on which it is located is called the "armature." The armature and field windings are designed so that the magnetic lines of flux have a significant component pointing at the optimum angle to the armature winding throughout the 360° of rotation of the armature. This relative motion between the magnetic field lines and armature winding result in an emf being induced in the armature winding according to equation 1.

The coil shown in Fig. A.1 would be the armature winding. The flux shown in Fig. A.1 would be that established by the dc current in the field winding.

Note the minus sign in equation 1. This negative sign is interpreted as meaning that the polarity of the induced emf must be such that it creates a current which will tend to *reduce* the magnetic field that created the emf.

Condition (b)

In Chapter 10 we discussed an application of the law of magnetic induction involving relative motion produced by method (b), the transformer. In this application the coils (primary and secondary windings) are stationary while the flux is changing.

In Chapter 10, we saw how a sinusoidal voltage imposed on one winding, the primary, resulted in a sinusoidal induced voltage in the other winding, the secondary. The flux shown in Fig. A.1 is basically that produced by the primary winding current.

The current in the primary winding is changing in a sinusoidal manner and therefore results in a magnetic flux which is also changing in a sinusoidal manner. The two windings experience the same total magnetic flux since they are both wound around the same core.

This sinusoidal flux in turn produces the sinusoidal voltage in the stationary secondary winding which is the coil shown in Fig. A.1.

The polarity of an emf induced by Faraday's law can be determined by a rule known as the right hand rule. It is impossible, however, to predict the polarity of induced voltage simply by inspection of a transformer. Industry therefore uses something called a "dot convention" to specify the polarity of induced voltage.

A dot is placed on the schematic of each winding. A current flowing into a *dotted* terminal in one winding induces a positive polarity at the *dotted* terminal of the other winding. Similarly, a current flowing into an *undotted* terminal in one winding induces a positive polarity at the ***undotted*** terminal of the other winding.

The phenomenon of induced emf can be studied in greater detail by studying any of the dozens of excellent Physics books available which include electromagnetic fields.

APPENDIX B

NSWERS TO SELECTED ODD-NUMBERED PROBLEMS

CHAPTER 1

1. **(a)** 3.58×10^3
(b) 4.56×10^{-5}
(c) 4.57×10^8
(d) 9.37×10^{-1}
(e) 3.99×10^{12}
(f) 1.36×10^2

3. **(a)** 34.8 kg
(b) 78.6×10^9 μg
(c) 1.47×10^4 mm
(d) 8.99×10^{-4} km

5. **(a)** 4.68×10^{20} protons
(b) 4.68×10^{20} electrons
(c) 6.02×10^{23} protons
(d) Neutrons are electrically neutral; they can't be used for charge.

7. **(a)** 50.5 Ω
(b) 9.5 Ω
(c) 14.2 Ω
(d) 32.5 Ω

9. **(a)** 5.495×10^{30} electrons
(b) 8.8×10^{11} C
(c) 279 years

11. **(a)** 18.75 g
(b) 2.06×10^{28} electrons

13. 20.8 km/s

CHAPTER 2

3. **(a)** 6400 mA
(b) 60°
(c) $\pi/3$ or 1.047 rad
(d) 4000 rad/s
(e) 636.62 Hz
(f) 4525.48 mA

5. **(a)** 360°
(b) 2π rad

7. Through source from − to +.

9. Determined by circuit.

15. **(a)** 12 μs
(b) 83.33 kHz
(c) 100 mA
(d) 0
(e) 100 mA

17. **(a)** i leads v by 65°
(b) v leads i by 30°
(c) v_1 leads v_2 by 10°
(d) i_2 leads i_1 by 10°

■ CHAPTER 3

1. **(a)** 120°
(b) −180°
(c) 126°
(d) 264°
(e) −90°

3. **(a)** 62.8 cm
(b) 37.7 cm
(c) 31.4 cm
(d) 23.6 m
(e) 0.872 cm

5. **(a)** 2005°
(b) 15.8°
(c) 275°
(d) 30°
(e) 12°

7.

	S	*C*	*T*
(a)	0.643	0.766	0.833
(b)	−0.94	0.342	−2.75
(c)	0.891	0.454	1.96
(d)	−0.717	0.697	−1.03
(e)	0.806	−0.591	−1.36

9. **(a)** $3.83 + j3.21$
(b) $-3.5 + j6.06$
(c) $-4 + j0$
(d) $4.35 + j11.2$
(e) $3.96 + j4.51$

11.

	AB	*A/B*
(a)	$12 + j$	$0.072 + j0.588$
(b)	$-70 + j14$	$0.143 + j0.714$
(c)	$-30 - j54$	$0.750 - j0.417$
(d)	$24 + j82$	$0.986 + j0.630$
(e)	$-144 + j12$	$0.466 + j0.414$

13.

	A	*B*
(a)	$5.83\angle 59°$	$9.85\angle -24°$
(b)	$7.21\angle 124°$	$9.90\angle 45°$
(c)	$7.28\angle 74.1°$	$8.49\angle -45°$
(d)	$10.0\angle 53.1°$	$8.54\angle 20.6°$
(e)	$9.49\angle -71.6°$	$15.2\angle -113°$

15.

	AB	*A/B*
(a)	$85\angle 0.7$ rad	$0.294\angle -0.24$ rad
(b)	$156\angle 0.517$ rad	$0.270\angle 1.6$ rad
(c)	$5.66\angle -1.68$ rad	$3.35\angle 1.4$ rad
(d)	$115\angle 0.7$ rad	$6.35\angle -1.27$ rad
(e)	$22.9\angle 0.5$ rad	$0.510\angle 0.183$ rad

17.

	ω (rad/s)	Period (s)
(a)	1260	5.00×10^{-3}
(b)	7540	8.33×10^{-4}
(c)	13,800	4.55×10^{-4}
(d)	1.26×10^6	5.00×10^{-6}
(e)	126	5.00×10^{-2}

■ CHAPTER 4

1. **(a)** 200 mV $\angle 0°$
(b) 23.5 A $\angle 30°$
(c) 15 V $\angle -60°$
(d) 2 μA $\angle 135°$
(e) 4 μV $\angle 15°$
(f) 34 mV $\angle -120°$
(g) 66 μA $\angle -120°$

3. **(a)** $8.75 \sin(1000t + 101.5°)$ V
(b) $91.65 \sin(377t - 40.9°)$ A
(c) $12.39 \sin(200t + 73.8°)$ mV
(d) $50 \sin(400t + 53.1°)$ V
(e) $4.47 \sin(1000t - 26.6°)$ μA

5. **(a)** Resistor
(b) 1 A $\angle 30°$
(d) $100 \sin 1000t$ V, $1 \sin 1000t$ A

7. **(a)** Capacitor
(b) 0.5 μA $\angle 135°$
(d) $20 \sin(2\pi \times 100t + 45°)$ mV, $0.5 \sin(2\pi \times 100t + 135°)$ μA
(g) 0.040 μF

9. 0

11. **(a)** 20 Ω $\angle 0°$
(b) Resistance
(c) 20 Ω

13. **(a)** 0.2 Ω $\angle -90°$
(b) Inductance
(c) 0.53 mH

15. $R = X = 70.7\ \Omega$

17. **(a)** 0
(b) 10 mS
(c) Inductor

19. **(a)** Resistor
(b) Inductor
(c) Capacitor
(d) Inductor
(e) Capacitor

21. **(a)** Capacitor
(b) Resistor
(c) Inductor

23. **(a)** 0.5 A $\angle 210°$
(b) $0.5 \sin(2000t + 210°)$ A

25. 120 V_{dc}

27. 0

■ CHAPTER 5

3. 6.47 V $\angle -65.1°$

5. 0

7. **(a)** 31.7 Ω $\angle 39.4°$

9. **(a)** 267.18 Ω $\angle 20.7°$

11. **(a)** 34.82 V $\angle 11.3°$
(b) 43.8 V $\angle 101.3°$
(c) 34.63 V $\angle -78.7°$

13. **(a)** 65.12 mV $\angle 77.4°$
(b) 49.8 mV $\angle -12.6°$
(c) 0.16 mA $\angle 77.4°$
(d) 0.16 mA $\angle 77.4°$

15. **(a)** 0
(b) 0
(c) 0
(d) 60 V

17. **(a)** 0.78 mA $\angle 114°$
(b) 1.94 V $\angle -156°$
(c) 3.11 V $\angle 114°$
(d) 15.56 V $\angle 24°$
(e) 6.64 V $\angle 44.6°$
(f) 1.95 V $\angle -156°$
6.6 V $\angle 44.6°$

19. 6.22 mA $\angle 35.4°$

21. **(a)** 250 μA
(b) 100 μA
(c) 410 μA

23. **(a)** 37.59 Ω $\angle 39.9°$

25. **(a)** 129.17 mS $\angle 12.1°$
(b) 7.74 Ω $\angle -12.1°$

27. **(a)** 8.82 mH

29. 4.5 mH

31. **(a)** 5.05 mS $\angle 7.82°$
(b) 1.82 mA $\angle -52.2°$
(c) 1.80 mA $\angle -60°$
(d) 1.45 mA $\angle 30°$
(e) 1.2 mA $\angle -150°$

CHAPTER 6

1. 29.16 kΩ $\angle -42.9°$

3. 9.33 kΩ

5. 1.16 kΩ $\angle 51.9°$

7. **(a)** 13.33 mA
(b) 32 V
(c) 16 V

9. 250 Ω $\angle 53.1°$
0.2 sin(2000t − 8.1°) A

11. 18 kΩ
1.33 mA

CHAPTER 7

1. **(a)** $(15.62 \angle 50.2°)I_1 - (0)I_2 = 10\angle 0°$
$(0)I_1 + (25\angle 36.9°)I_2 = 10\angle 0°$
(b) 0.64 A $\angle -50.2°$
0.4 A $\angle -36.9°$

3. **(a)** $(2.84\text{K}\angle -10.1°)I_1 - (1.5\text{K}\angle 90°)I_2 = 0$
$(-1.5\text{K}\angle 90°)I_1 + (2.82\text{K}\angle 73.5°)I_2 = 40\angle 0°$
(b) 6.49 mA $\angle 39.2°$
12.29 mA $\angle -60.9°$
(c) 6.49 mA $\angle 39.2°$
(d) 12.29 mA $\angle 119.1°$
(e) 14.87 mA $\angle -86.3°$

5. **(a)** $(5.32\text{K}\angle -48.8°)I_1 - (4\text{K}\angle -90°)I_2 = 8\angle 0°$
$(-4\text{K}\angle -90°)I_1 + (2.55\text{K}\angle -11.3°)I_2 = -12\angle 0°$
(b) 1.89 mA $\angle 92.9°$
1.76 mA $\angle -173.6°$
(c) 2.66 mA $\angle -128.4°$

7. **(a)** $(144.2\angle -33.7°)I_1 - (120\angle 0°)I_2 - (40\angle 90°)I_3 = 200\text{ m}\angle 0°$
$(-120\angle 0°)I_1 + (360.6\angle -33.7°)I_2 - (200\angle -90°)I_3 = 240\text{ m}\angle 0°$
$(-40\angle 90°)I_1 - (200\angle -90°)I_2 + (170.9\angle -69.4°)I_3 = -180\text{ m}\angle 0°$
(b) 2.92 mA $\angle 46.4°$
0.73 mA $\angle 72.1°$
1.09 mA $\angle -106.4°$
(c) 274.11 mV $\angle -141.6°$

9. **(a)** $(0.55\angle 8.4°)V_1 - (0.5\angle 0°)V_2 = 14\angle 30°$
$(-0.5\angle 0°)V_1 + (0.625\angle 0°)V_2 = -20\angle 60°$
(b) $50\angle -88.1°$ V
$70\angle -102.4°$ V
(c) 4.47 A $\angle -24.7°$
(d) 8.75 A $\angle -102.4°$

11. **(a)** $(0.55 \times 10^{-3}\angle 61.6°)V_1 - (0.45 \times 10^{-3}\angle 63.4°)V_2 = 10 \times 10^{-3}\angle 0°$
$(-0.45 \times 10^{-3}\angle 63.4°)V_1 + (0.32 \times 10^{-3}\angle 27.9°)V_2 = 12 \times 10^{-3}\angle 0°$
(b) 66.30 V $\angle 43.9°$
89.52 V $\angle 55.9°$
(c) 28.26 V $\angle -94.9°$

13. **(a)** $(0.11 \times 10^{-3}\angle 7.8°)V_1 - (0.083 \times 10^{-3}\angle 0°)V_2 - (0)V_3 = 10 \times 10^{-3}\angle 0°$
$(-0.083 \times 10^{-3}\angle 0°)V_1 + (0.14 \times 10^{-3}\angle -14.8°)V_2 - (0.066 \times 10^{-3}\angle 66.8°)V_3 = 20 \times 10^{-3}\angle 40°$
$(-0)V_1 - (0.066 \times 10^{-3}\angle 66.8°)V_2 + (0.088 \times 10^{-3}\angle 7.6°)V_3 = 5 \times 10^{-3}\angle 50°$
(b) 212.6 V $\angle 37.6°$
215.1 V $\angle 68.7°$
175.6 V $\angle 109.2°$
(c) 6.08 mA $\angle 68.6°$

15. **(a)** $(136 \times 10^{-3})V_1 - (27.7 \times 10^{-3})V_2 - (0)V_3 = 2$
$(-27.7 \times 10^{-3})V_1 + (98.6 \times 10^{-3})V_2 - (50 \times 10^{-3})V_3 = -2$
$(0)V_1 - (50 \times 10^{-3})V_2 + (83.3 \times 10^{-3})V_3 = -1.6$
(b) 6.45 V
−40.7 V
−43.5 V
(c) 1.57 A

17. 2.62 mA $\angle -128.4°$

19. 4.73 V $\angle -37.2°$

21. 0.09 mA

23. 11.4 mA $\angle 57.1°$

25. 32 + 49.6 sin(ωt − 180°) μA

27. 13.33 V
5 kΩ

29. **(a)** 113.1 V $\angle -8.1°$
2.83 kΩ $\angle -61.3°$
(b) 10.4 mA $\angle -38.6°$

31. (0.44 + 0.42 sin(ωt − 5.4°) mA

33. −8.7 V

35. $35.8 \text{ V} \angle -116.6^\circ$

37. DC: 0.429 A, 5.1 Ω
AC: $0.101 \text{ A} \angle 126^\circ$, $8\ \Omega \angle 23^\circ$

39. $1.52 \text{ A} \angle 133.3^\circ$

41. $(36.1\angle 33.7^\circ)I_1 - (63.2\angle 71.6^\circ)I_2 = 200 \text{ m} \angle 0^\circ$
$(-63.2\angle 71.6^\circ)I_1 + (64\angle 38.7^\circ)I_2 = -100 \text{ m} \angle 30^\circ$

43. $72.1 \text{ mV} \angle -163.3^\circ$
$24\ \Omega \angle -10.3^\circ$

■ CHAPTER 8

		Characteristic +	*Mantissa*	=	*Logarithm*
1.	**(a)**	0	0.575	=	0.575
	(b)	1	0.366	=	1.366
	(c)	3	0.129	=	3.129
	(d)	−2	0.0899	=	−1.910
	(e)	−1	0.9943	=	−0.0057

3. **(a)** 7.56
(b) 260
(c) 155
(d) 6440
(e) 4.66×10^{-4}

7. 32 dB
35.6 dB
41.6 dB
109.2 dB
109.2 dBm
223,000 V

9. −97.4 dBm

13. **(a)** 6370 pF
(b) 1443 Hz
(c) 3676 Hz
(d) −55.8°

15. **(b)** 637 mH

16. **(a)** 0.0113 μF
(b) 5197 Hz
(c) 1302 Hz
(d) 66.5°

19. **(a)** 34.3 kHz
(b) 17 kHz
(c) 11.2 kHz

■ CHAPTER 9

1. **(a)** 1.0×10^4 Hz, 6.32×10^4 rad/s
(b) 2.01×10^3 Hz, 1.26×10^4 rad/s
(c) 2.91×10^5 Hz, 1.83×10^6 rad/s
(d) 2.37×10^6 Hz, 1.49×10^7 rad/s

3. **(a)** 63.4 mH
(b) 398 Ω
(c) 1.01 V
(d) 0.0101 V

5. **(a)** 3979 pF, 1.59 μH
(b) 0.999 Ω
(c) 1.95 MHz, 2.05 MHz

7. **(a)** 20
(b) 15.9 mH
(c) .398 μF
(d) 1950 Hz, 2050 Hz

9. **(a)** 25
(b) 3.183 mH
(c) 0.0796 μF
(d) 9800 Hz, 10,200 Hz

11. **(a)** 253 pF
(b) 1.9 MHz, 2.1 MHz
(c) 200 kHz
(d) 31.42 V, 0.1 A

13. **(a)** 25
(b) 1.59 mH
(c) 0.159 μF
(d) 9800 Hz, 10,200 Hz

■ CHAPTER 10

3. Step up—voltage increased
Step down—voltage decreased

5. **(a)** Step down
(b) 125 T
(c) 400 V
(d) 1.25 A
(e) 0.16 A

7. **(a)** 240 V
(b) 600 V
(c) 300 V
(d) 3 A
(e) 3.75 A
(f) 3 A
(g) 1.6125 A
(h) 9.76 kΩ

9. 9.5

11. **(a)** 0.166 A
(b) 0.666 A
(c) 383.5 V

13. **(a)** 0.08 Ω
(b) 0.18 Ω
(c) 95.0%
(d) 800 W
(e) 300 W
(f) 84.5%

■ CHAPTER 11

5. **(a)** 850 mA
(b) 8.67 W
(c) 11.56 VAR
(d) 14.45 VA
(e) 0.6 lag

7. **(a)** $3.12 \text{ k}\Omega \angle -14.5^\circ$
(b) 1.46 W

9. **(a)** 20 kΩ
(b) 8.9 mW

13. **(a)** 124.42 A

(b) 114.55 A
(c) 12.38 kW, 10.50 kW
(d) 1.88 kW

15. (a) 0.89 lag
(b) 34.83 μF

17. (a) 0.8 lag
(b) 1.66 mF
(c) 1.66 mF

■ CHAPTER 12

1. (a) $\dfrac{10s^2 + 20s + 4000}{s^3 + 400s}$
(b) $\dfrac{20}{s + 5} + \dfrac{10}{s + 10}$
(c) $\dfrac{3s^2}{s^4 + 40{,}000} + \dfrac{4}{s^3}$
(d) $\dfrac{2.55s - 443}{s^2 + 900}$

3. (a) $10\,U(t) = i(t)R + V_c(t)$
(b) $\dfrac{10}{s} = \boldsymbol{I}(s)R + \dfrac{\boldsymbol{I}(s)}{sC} + \dfrac{\mathbf{v}(0^+)}{s}$
(c) $0.00682e^{-90.9t}$
(d) $V_c = 10 - 15e^{-90.9t}$

5. $15e^{-90.9t}$
17.7 ms

7. (a) $10e^{-2t} = i(t)R + V_c(t)$
(b) $\dfrac{10}{s + 2} = \boldsymbol{I}(s)R + \dfrac{\boldsymbol{I}(s)}{sC} + \dfrac{\mathbf{v}(0^+)}{s}$
(c) $(0.0069e^{-90.9t} - 0.0001e^{-2t})$ A

9. (a) $50e^{-11t}$
(b) $45.5(1 - e^{-11t})$
(c) $V_c = 45.5$ V, $V_R = 0$

11. (a) 2.27×10^{-6}s
(b) $9e^{-440{,}000t}$
(c) 0.57 μs

13. (a) $-2.39 \times 10^{-3} \cos(100t) + 10.49 \sin(100t) + 2.39 \times 10^{-3}e^{-440{,}000t}$
(b) $-0.49 \sin(100t) + 2.39 \times 10^{-3} \cos(100t) - 2.39 \times 10^{-3}e^{-440{,}000t}$

■ CHAPTER 13

1. (a) 1.48 cm/s
(b) 294 m
(c) 35.4 W/m

3. 0.45 cm

5. 3.72 cm

7. 3.06%

9. 1 W/m

11. 141 cm^2 per plate

INDEX